Cornerstones in Quantitative Empirical Methods

Cornerstones in Quantitative Empirical Methods - Volume I: Foundations provides a complete, self-contained path from first principles to modern statistical inference, giving a comprehensive technical foundation for understanding and analyzing data problems using quantitative methods. The first part of the book provides a thorough introduction to probability theory and statistics with a focus on techniques and concepts, which are particularly useful for empirical analyses. Building directly on the first part, the second part of the book provides an introduction to estimation theory, hypothesis testing, likelihood theory, and Bayesian methods.

The nineteen concise chapters follow a transparent pattern: motivation and intuition, carefully chosen examples, main results stated as formal theorems, and a curated set of exercises that reinforce both computation and reasoning. Short illustrative datasets keep the exposition concrete while maintaining the book's primary focus on theory. Three appendices consolidate the required background in set theory, real analysis, and simulation techniques, making the volume fully self-contained.

Designed for graduate and advanced-undergraduate courses in e.g. economics, political science, business analytics, or data science—or for motivated self-study—**Volume I** equips readers with the concepts and techniques needed for quantitative empirical analysis. It also prepares readers for **Volume II**, which builds the foundation for modelling, prediction, and causal inference in empirical studies.

Key Features:

- Comprehensive coverage across the technical fields of probability theory, statistics, and mathematics, needed to understand and perform in-depth empirical modelling and analysis.
- Extensive treatment of estimation theory and various estimation methods, such as analog estimators, maximum likelihood estimators, and Bayesian methods.
- Elaborate discussion of hypothesis testing and recommendations to avoid pitfalls.
- Inclusion of technical details and proofs structured such that they can be skipped by readers who prefer a less technical, though still cohesive, approach.

Cornerstones in Quantitative Empirical Methods

Volume 1: Foundations

Mikkel Bennedsen and Allan H. Würtz

CRC Press
Taylor & Francis Group
Boca Raton London New York

CRC Press is an imprint of the
Taylor & Francis Group, an **informa** business

A CHAPMAN & HALL BOOK

Designed cover image: Mikkel Bennedsen and Allan H. Würtz

First edition published 2026
by CRC Press
2385 NW Executive Center Drive, Suite 320, Boca Raton, FL 33431

and by CRC Press
4 Park Square, Milton Park, Abingdon, Oxon, OX14 4RN

CRC Press is an imprint of Taylor & Francis Group, LLC

ISBN: 978-1-032-96772-1 (hbk)
ISBN: 978-1-032-96909-1 (pbk)
ISBN: 978-1-003-59119-1 (ebk)

DOI: 10.1201/9781003591191

Typeset in CMR10 font
by KnowledgeWorks Global Ltd.

Publisher's note: This book has been prepared from camera-ready copy provided by the authors.

To Fie, Frede, & Sylvester

To Anna & Roald

Contents

Preface   xv

Authors   xvii

1   Basic elements in probability theory   1
    1.1   Introduction   1
    1.2   The statistical experiment   2
    1.3   Interpretation of probabilities   4
    1.4   The probability model   4
      1.4.1   The sample space   4
      1.4.2   The event space   6
      1.4.3   The probability measure   8
      1.4.4   The probability model   9
    1.5   Random variables   9
      1.5.1   Defining outcomes of statistical experiment using numbers   9
      1.5.2   Linking a random variable to events   12
      1.5.3   Probability of outcomes of a random variable based on the probability model   13
    1.6   Proofs   16
    1.7   Exercises   16

2   Distribution of a random variable   17
    2.1   Introduction   17
    2.2   The cumulative distribution function   17
    2.3   Discrete random variables   21
      2.3.1   The probability mass function   22
      2.3.2   The distribution of a function of a discrete random variable   24
    2.4   Continuous random variables   25
      2.4.1   The probability density function   26
      2.4.2   The distribution of a function of a continuous random variable   30
    2.5   Mixed random variables   31
    2.6   Proofs   32
    2.7   Exercises   36

3   Joint distribution of random variables   39
    3.1   Introduction   39
    3.2   The joint event probability   39
    3.3   The joint cumulative distribution function   41
    3.4   The joint distribution of discrete random variables   42
      3.4.1   The joint probability mass function   42
      3.4.2   The marginal probability mass function   43

3.4.3   The relationship between the joint CDF and the joint PMF of two discrete random variables 45

3.4.4   The distribution of a function of two discrete random variables . . . 46

3.4.5   Multiple discrete random variables 48

3.5   The joint distribution of continuous random variables 49

3.5.1   The joint probability density function of two continuous random variables 49

3.5.2   The marginal probability density function 51

3.5.3   The relationship between the joint CDF and the joint PDF of two continuous random variables 52

3.5.4   The distribution of a function of two continuous random variables . 53

3.5.5   Multiple continuous random variables 55

3.6   The joint distribution of both continuous and discrete random variables . . 56

3.7   Proofs 56

3.8   Exercises 58

4   Conditional distribution of a random variable   60

4.1   Introduction 60

4.2   Conditional probability of events 60

4.3   Conditional distributions with two random variables 61

4.3.1   Two discrete random variables 62

4.3.2   Two continuous random variables 64

4.3.3   Two random variables of mixed type 65

4.4   Conditional distributions with three or more random variables 66

4.5   Relation between long and short conditioning sets 68

4.6   Bayes' theorem 70

4.6.1   Bayes' theorem for events 70

4.6.2   Bayes' theorem for random variables 71

4.7   Proofs 71

4.8   Exercises 73

5   Aspects of a univariate distribution   76

5.1   Introduction 76

5.2   The expected value and mean of a random variable 76

5.3   Rules of calculation for expected value 79

5.4   Variance 81

5.5   Moments 83

5.6   Quantiles 85

5.7   Other descriptive measures of a distribution 88

5.8   Proofs 89

5.9   Exercises 91

6   Aspects of a multivariate distribution   96

6.1   Introduction 96

6.2   Covariance 96

6.2.1   Definition of covariance 96

6.2.2   Expectation of a function of random variables 97

6.2.3   Covariance and the variance of a sum 99

6.3   The correlation coefficient 100

6.4   Conditional expectation 103

6.5   Conditional expectation on different sets of random variables 106

		6.5.1	The Law of Iterated Expectations	106
		6.5.2	Relationship between conditional expectation and covariance	108
		6.5.3	Long and short conditioning sets	108
	6.6	Conditional variance		109
	6.7	Conditional quantiles		112
	6.8	Proofs		113
	6.9	Exercises		117

7 Independence between random variables — **121**
	7.1	Introduction		121
	7.2	Statistical independence		122
		7.2.1	Independence between events	122
		7.2.2	Statistical independence between two random variables	123
		7.2.3	Statistical independence between many random variables	126
	7.3	Conditional statistical independence		128
	7.4	Mean independence and conditional mean independence		131
	7.5	Proofs		132
	7.6	Exercises		134

8 Commonly used univariate distributions — **137**
	8.1	Introduction		137
	8.2	Bernoulli and Binomial distributions		137
		8.2.1	Bernoulli distributions	137
		8.2.2	Binomial distributions	139
	8.3	Exponential, Erlang, and Poisson distributions		140
		8.3.1	Exponential distributions	140
		8.3.2	Erlang distributions	141
		8.3.3	Poisson distributions	143
	8.4	Uniform and beta distributions		145
		8.4.1	Uniform distributions	145
		8.4.2	Beta distributions	146
	8.5	Normal distributions		147
		8.5.1	Sums of independent normally distributed random variables	150
	8.6	Distributions derived from the normal distribution		151
		8.6.1	χ^2 distributions	151
		8.6.2	t-distributions	153
		8.6.3	F distributions	154
	8.7	Proofs		155
	8.8	Exercises		156

9 Bivariate normal distributions — **161**
	9.1	Introduction		161
	9.2	Density of a bivariate normal distribution		161
	9.3	Marginal distributions and linear combination of bivariate normal random variables		163
	9.4	Conditional distributions of bivariate normal distribution		165
	9.5	Constructing bivariate normal distribution from univariate normal distributions		166
	9.6	Proofs		167
	9.7	Exercises		171

10 Distribution of a sample   174

    10.1 Introduction 174
    10.2 Random samples 174
    10.3 Simple random sample 177
    10.4 Time series 178
        10.4.1 Time series population modeled by conditional distributions 180
        10.4.2 Time series population modeled as a process 181
        10.4.3 Time series sample 182
    10.5 Sampling methods used in practice 183
        10.5.1 Sampling without replacement 183
        10.5.2 Stratified sampling 185
    10.6 Proofs 186
    10.7 Exercises 187

11 Estimation theory   189

    11.1 Introduction 189
    11.2 Estimand, estimator, and estimate 189
    11.3 Distribution of an estimator 191
    11.4 Precision of an estimator 193
        11.4.1 Loss and risk functions 193
        11.4.2 Mean squared error, variance, and bias 194
        11.4.3 Other measures of precision 195
    11.5 Consistency of an estimator 197
        11.5.1 Relationship between MSE and consistency 198
        11.5.2 The Law of Large Numbers 199
    11.6 Comparison of estimators 201
    11.7 Monte Carlo simulation 203
        11.7.1 Elements in Monte Carlo simulation 204
        11.7.2 Random draws from a distribution 208
    11.8 Proofs 210
    11.9 Exercises 214

12 Estimating aspects of a univariate distribution   219

    12.1 Introduction 219
    12.2 Empirical distribution function 219
    12.3 The analogy principle 222
    12.4 Analog estimators of PDF, mean, p-quantiles, and higher-order moments . 223
        12.4.1 Analog estimator of a PDF 223
        12.4.2 Analog estimator of the mean 224
        12.4.3 Analog estimator of a p-quantile 225
        12.4.4 Analog estimators of the variance and higher order moments 226
    12.5 Analog estimators of parameters in distributions 227
    12.6 Adjustments to analog estimators 229
    12.7 Choice of variance estimator 230
    12.8 Proofs 231
    12.9 Exercises 233

13 Estimation of a sampling distribution — 237

13.1 Introduction — 237
13.2 Bootstrap distribution — 237
    13.2.1 The non-parametric bootstrap — 238
    13.2.2 The parametric bootstrap — 239
    13.2.3 Implementation of the bootstrap using Monte Carlo simulation — 240
    13.2.4 Other types of bootstrapping — 242
13.3 Asymptotic distribution — 243
    13.3.1 Convergence in distribution — 243
    13.3.2 The Central Limit Theorem — 244
    13.3.3 Extending asymptotic results beyond a sample average — 247
13.4 The standard error of an estimator — 250
13.5 Comparisons of estimators of a sampling distribution — 251
13.6 Proofs — 252
13.7 Exercises — 253

14 Confidence intervals — 256

14.1 Introduction — 256
14.2 The concept of a confidence interval — 256
14.3 Construction of a confidence interval — 257
14.4 Confidence intervals for mean and variance in the case of a normally distributed population — 258
    14.4.1 Confidence intervals for the mean — 258
    14.4.2 Confidence interval for the variance — 262
14.5 Bootstrap confidence intervals — 263
    14.5.1 Bootstrap confidence interval based on point estimator — 263
    14.5.2 Bootstrap confidence interval based on studentized point estimator — 265
14.6 Asymptotic confidence intervals — 266
    14.6.1 Constructing asymptotic confidence interval — 266
14.7 Comparisons of different confidence intervals — 268
14.8 Proofs — 269
14.9 Exercises — 271

15 Analysis of hypotheses — 274

15.1 Introduction — 274
15.2 Hypotheses — 274
15.3 Decision rules and test rules — 276
15.4 Properties of a test rule — 277
    15.4.1 Power function — 277
    15.4.2 Type I error and Type II error — 279
    15.4.3 Size of a test rule — 281
    15.4.4 Consistency of a test rule — 282
15.5 Loss of a test rule — 283
    15.5.1 Two approaches to measure loss — 283
    15.5.2 Loss function — 284
    15.5.3 Risk of a decision rule — 286
15.6 Selecting a test rule when the risk function is known — 288
    15.6.1 Admissibility — 288
    15.6.2 Minimax test rule — 291
15.7 Selecting a test rule when the risk function is unknown — 294
    15.7.1 Type I error focus — 295

15.7.2 Type II error balance 296
15.8 Significance of hypotheses and parameters 299
15.9 Proofs 299
15.10 Exercises 300

16 Test of hypotheses   **302**
16.1 Introduction 302
16.2 Generic approach to hypothesis testing 302
16.3 t-test of simple hypotheses 303
16.3.1 Known loss function 306
16.3.2 Unknown loss function 307
16.4 t-test and nuisance parameters 311
16.5 Hypotheses with multiple values of the parameter of interest 314
16.5.1 Multiple values of the parameter of interest under the alternative
hypothesis 314
16.5.2 Multiple values of the parameter of interest under both hypotheses 316
16.6 The p-value of a hypothesis test 319
16.6.1 p-value pitfalls 321
16.7 Estimation of the t-test statistics distribution 321
16.7.1 Bootstrap estimation of the t-test statistics distribution 322
16.7.2 Asymptotic estimation of the t-test statistics distribution 324
16.8 Likelihood ratio test of simple hypotheses 326
16.9 Proofs 330
16.10 Exercises 333

17 Maximum likelihood estimation   **337**
17.1 Introduction 337
17.2 Maximum likelihood estimation of one parameter 337
17.2.1 Likelihood function of one parameter 338
17.2.2 Maximum likelihood estimator of one parameter 341
17.2.3 Maximum likelihood estimator with differentiable likelihood function 342
17.2.4 Maximum likelihood estimator with non-differentiable likelihood
function 344
17.3 Maximum likelihood estimation of multiple parameters 345
17.3.1 Likelihood function of multiple parameters 345
17.3.2 Maximum likelihood estimator of multiple parameters 347
17.3.3 ML estimator of multiple parameters with differentiable likelihood
function 347
17.3.4 ML estimation with several population characteristics 349
17.4 ML estimation with time series 352
17.5 Properties of maximum likelihood estimators 354
17.5.1 Bias and variance of the ML estimator 354
17.5.2 Consistency of the ML estimator 357
17.6 Maximum-likelihood estimation viewed in an information context 359
17.7 Proofs 360
17.8 Exercises 361

18 Inference with maximum likelihood estimation **364**
  18.1 Introduction 364
  18.2 Bootstrap distribution of the ML estimator 364
  18.3 Asymptotic distribution of the ML estimator of one parameter 365
      18.3.1 Asymptotic distribution with one parameter 365
      18.3.2 Estimation of the variance of the asymptotic distribution 368
      18.3.3 ML estimation in case of misspecification 371
  18.4 Efficiency of the ML estimator 371
      18.4.1 The Cramér-Rao lower bound 371
      18.4.2 Asymptotic efficiency 373
  18.5 Confidence intervals 374
      18.5.1 Bootstrap confidence intervals 374
      18.5.2 Asymptotic confidence intervals 375
  18.6 Hypothesis testing 375
      18.6.1 The generalized likelihood ratio test 375
      18.6.2 The Wald test 377
      18.6.3 The Lagrange multiplier test 377
      18.6.4 The trinity of tests 379
  18.7 Proofs 379
  18.8 Exercises 386

19 Bayesian methods **389**
  19.1 Introduction 389
  19.2 The prior and posterior distributions 389
  19.3 Examples of prior distributions 391
      19.3.1 The degenerate prior 392
      19.3.2 The uniform prior 392
      19.3.3 The beta prior 393
      19.3.4 The normal prior 394
      19.3.5 The Laplace prior 394
      19.3.6 Improper priors 395
      19.3.7 Hyperparameters of a prior distribution 396
  19.4 Conjugate priors 396
      19.4.1 Normal prior and normal population 396
      19.4.2 Beta prior and Bernoulli population 397
  19.5 Reparametrization and the Jeffreys prior 399
      19.5.1 Reparametrization 399
      19.5.2 The Jeffreys prior 401
  19.6 Simulation-based estimation of the posterior distribution 402
  19.7 Estimators based on the posterior distribution 404
      19.7.1 Point estimators derived from the posterior distribution 404
      19.7.2 Interval estimators derived from the posterior distribution 405
      19.7.3 Bayesian decision theory and loss functions 405
  19.8 Bayesian analysis of hypotheses 407
  19.9 Proofs 408
  19.10 Exercises 411

Appendices   **413**

A Appendix A: Probability theory   **415**

  A.1   Sets and set operators   415
        A.1.1   Representation of a set   415
        A.1.2   Subsets   416
        A.1.3   Set operators   418
        A.1.4   Finite and infinite unions and intersections   420
  A.2   Measurability, random variables, and induced probability measures   422
        A.2.1   Sigma-algebras   423
        A.2.2   Measurability   425
        A.2.3   Formal definition of a random variable   426
        A.2.4   Induced probability measure   426
  A.3   The probability limit operator   427
  A.4   Proofs   428

B Appendix B: Real analysis   **429**

  B.1   Supremum and infimum of set   429
  B.2   Functions of one variable   430
        B.2.1   Domain, codomain, and range   430
        B.2.2   Supremum and infimum of a function   431
        B.2.3   Inverse image   432
        B.2.4   Inverse function   433
        B.2.5   Sequences and limits of sequences   434
        B.2.6   Limit of a function   435
        B.2.7   Continuity   438
        B.2.8   Derivative   439
        B.2.9   Inverse of differentiation   441
        B.2.10 Integration   442
        B.2.11 The Fundamental Theorem of Calculus   444
        B.2.12 The indicator function   447
  B.3   Functions of several variables   448
        B.3.1   Partial differentiation   448
        B.3.2   Integration over several variables   451
        B.3.3   Differentiation of an integral   452
  B.4   Proofs   453

C Appendix C: Simulation from a distribution   **454**

  C.1   Pseudo-random numbers   454
  C.2   Rejection sampling   455
  C.3   The Markov Chain Monte Carlo method   458
        C.3.1   Markov chains   458
        C.3.2   Making draws from a Markov chain   459
  C.4   The Gibbs sampler   462
  C.5   The Metropolis algorithm   464
  C.6   The Metropolis-Hastings algorithm   464
  C.7   Exercises   466

Index   **471**

Preface

This book originated in the graduate course *Advanced Quantitative Empirical Methods* that one of the authors taught at Aarhus University. The students were bright and highly motivated, yet most had had little formal exposure to probability or statistics. Existing texts forced them to choose between mathematically rigorous measure-theoretic treatments, on the one hand, and application-driven "cookbooks" that skipped foundations, on the other. Our goal is to offer a middle path: a book that is rigorous enough to prove the key results, but written in an intuitive, example-rich style that keeps the learner engaged.

The project has grown into two complementary volumes.

Volume I: Beginning with probability theory (Chapters 1–9), we progress through sampling, estimation, and frequentist inference (Chapters 10–16), culminating in likelihood-based and Bayesian methods (Chapters 17–19). This volume assumes no prior study of probability or statistics, although a semester of single-variable calculus will make life easier. Chapters are intentionally short (roughly twenty pages) so that each can be read in one or two sittings. The main text develops intuition and worked examples; formal proofs are collected at the end of every chapter.

Volume II: Building on the foundations, the second volume turns to statistical modeling, regression, prediction, causal inference, and further modern tools for quantitative social science.

How to use this book

A typical one-semester master's course can cover Chapters 1–16, taking students from the basics of probability through estimation and frequentist hypothesis testing. An ambitious graduate course may add Chapters 17 and 18 on maximum-likelihood estimation together with Chapter 19 on Bayesian methods.

Throughout, we present the main results as formally stated theorems. Readers interested primarily in application can focus on digesting the theorem statements and examples and consult the proofs only when deeper justification is desired.

Approach and philosophy

Our methodological workflow (Figure 1) starts with a causal model that gives rise to a well-defined population from which we observe a sample, which is used for estimation and inference of population parameters of interest. Whether the objective is description, prediction, or causal explanation, probability theory underpins the population model, while statistical theory links the sample to the quantities of interest. In Volume I, we abstract away from the causal model by taking the population as given, and develop the theory on how samples from the population may be used to conduct estimation and inference. In Volume II, causal analysis receives special attention: we emphasize the critical step of how a structural causal model underpins the population and how this structure can be used to specify estimands with causal interpretation that are identifiable from sample data.

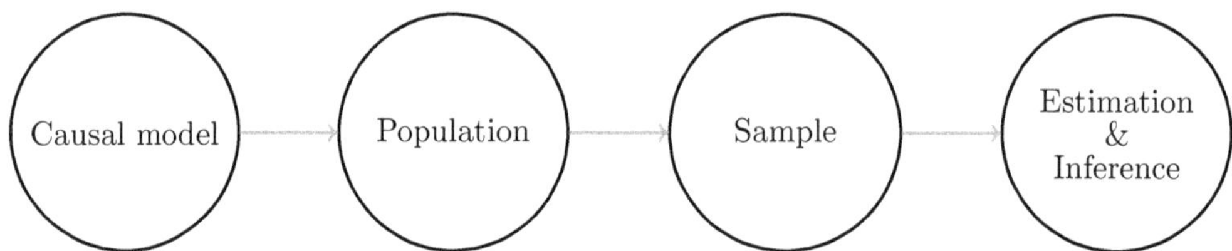

FIGURE 1

Our philosophy of statistical analysis: From a causal model to estimation and inference: the workflow linking modeling, populations, samples, and inference.

Organization of the chapters

The organization of Volume I is illustrated in the following table.

TABLE 1

Roadmap of Volume I.

Field	Topics	Chapters
Probability theory	Generic distribution theory	1–4
	Description of distributions	5–7
	Specific distributions	8–9
Statistical theory	Estimation of population quantities	10–12
	Inference and sampling uncertainty	13–14
	Hypothesis formulation and testing	15–16
	Likelihood-based and Bayesian estimation	17–19

Comprehensive appendices supply just-in-time mathematics: elementary set theory, advanced probability concepts, real analysis, and algorithms for Monte-Carlo simulation. These sections can be consulted as needed or as a focused basis for learning the material needed in the main text.

Acknowledgments

This text has benefited from the insightful comments of participants in the *Advanced Quantitative Empirical Methods* PhD course for students in Political Science from Aarhus University and University of Copenhagen. We also appreciate discussions with Ulrik Hvidman and Jesper Wulff.

Authors

Mikkel Bennedsen, PhD, is Associate Professor at the Department of Economics and Business Economics, Aarhus University. He has received several teaching honors, including the university's *Best Lecturer* award (2024). His research spans theoretical and applied econometrics, stochastic processes, and energy economics.

Allan H. Würtz, PhD, is Associate Professor and Head of the Econometrics and Business Analytics section at the Department of Economics and Business Economics, Aarhus University. An award-winning instructor, he is the author of the introductory text *An Insight into Statistics* and has published on econometric methodology.

1

Basic elements in probability theory

1.1 Introduction

Uncertainty is an inevitable condition of life. We are constantly faced with decisions where the outcome is uncertain. Is this egg still fresh? Is it safe to cross the street? Should I marry? Have kids? Should I invest in this stock? Read this book? In this chapter, we discuss how to model uncertainty using mathematics, more precisely *probability theory*. To quantify uncertainty mathematically, we define the *probability model* that serves as the mathematical foundation of probability. Since probabilities are not observable in nature as such, we discuss how probability can be interpreted. No matter the interpretation of probability, the assignment of probabilities has to satisfy the requirements of the probability model.

The probability model links the real-world context to the theoretical context, where mathematical modeling can be done; see Figure 1.1. The real-world context of our interest, to which the probability model will be applied, is framed as a statistical experiment. A statistical experiment describes from where, and how, data is obtained. The statistical experiment also serves as a point of departure when interpreting statistical results back into the real world.

The probability model allows us to define one of the most important concepts in probability theory, namely, a *random variable*. As we will see, a random variable represents uncertainty in a mathematically convenient manner such that we can use well-known techniques on numbers and functions when modeling uncertainty in complex situations. Random variables will be the main probabilistic tool used throughout this book. They provide a probabilistic description of the elements in the population of a statistical experiment.

The mathematical treatment of probability relies on the theory of *sets*. Chapter A in the appendix, and in particular Section A.1, provides brief background material on sets, which the interested reader can consult. With the exception of Theorem 1.2, which contains useful rules for calculations with probabilities, the technical details of this chapter will not be used much in the remainder of the book, however. The reason is that later chapters will generally specify the relevant random variables directly, instead of first specifying the probability model (Figure 1.1). As such, the reader not particularly interested in the underlying mathematical machinery behind probability theory can briefly skim this chapter without going into details. Readers wanting to know more about the technical details may consult Chapter A in the appendix which, after discussing the concept of sets, provides a more in-depth discussion of abstract probability theory.

DOI: 10.1201/9781003591191-1

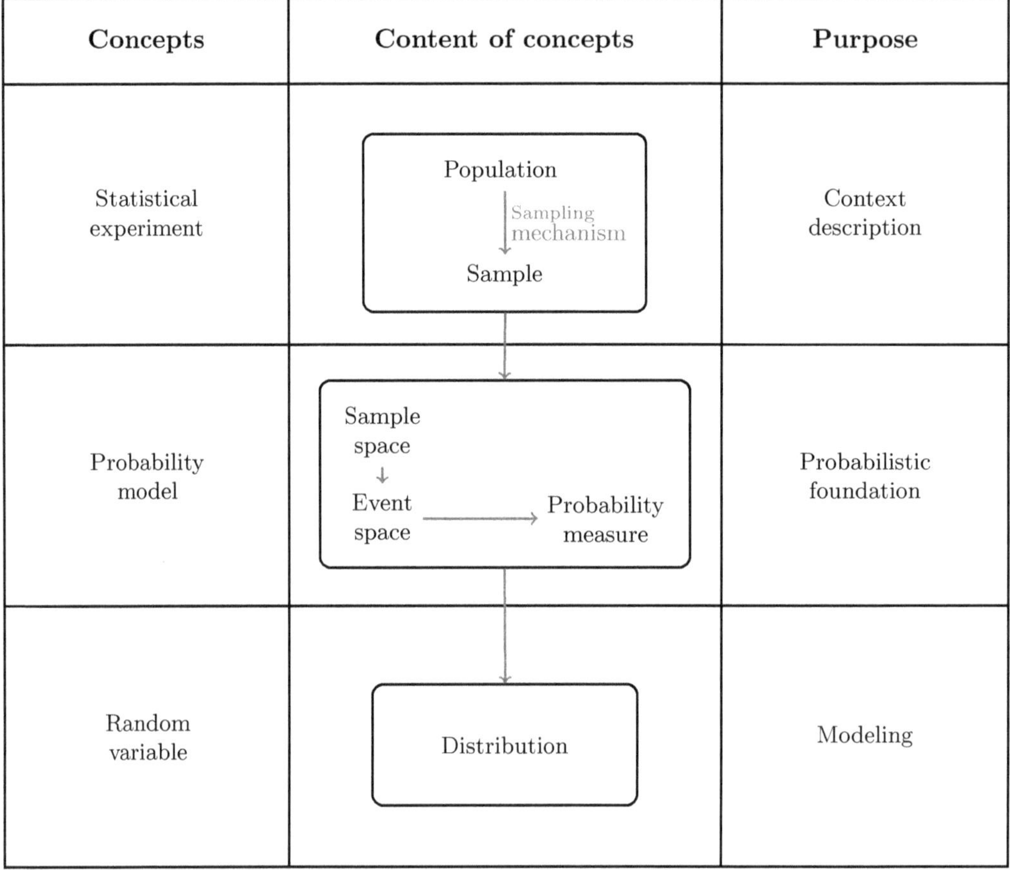

FIGURE 1.1

The basic elements in a mathematical treatment of a statistical experiment. The statistical experiment aims at describing relevant aspects of a population. This is done through a sample, which is a proper subset of the population. The sample is obtained using a sampling mechanism. The probability model provides the mathematical machinery for analyzing the outcomes, i.e. samples, from a statistical experiment. The main theoretical tool will be the random variable, representing possible outcomes of the statistical experiment in a manner particularly well-suited to mathematical and probabilistic analyses. Through its distribution (Chapter 2), the random variable provides a probabilistic description of the elements in the population.

1.2   The statistical experiment

A general framework for thinking about the real-world context and obtaining data is that of a *statistical experiment*. A statistical experiment consists of a *population* from which a *sampling mechanism* draws a *sample*. The population contains all the elements which we are interested in describing. In general, however, the population is not observed by the statistician. Instead, the statistician observes only part of the population, namely the sample. In this way, the sample offers a glimpse of the population, and our main goal is to use statistical tools to infer properties of the population by studying the sample.

A statistical experiment is illustrated in Figure 1.2, where the population consists of attitudes "Yes" and "No" of individuals, and a sampling mechanism draws a sample of attitudes of two individuals. Explicitly defining the statistical experiment in an empirical study is useful for choosing appropriate empirical methods and for the interpretation of their results. From a theoretical point of view, the concept of a statistical experiment provides a starting point for specifying the probability model presented below.

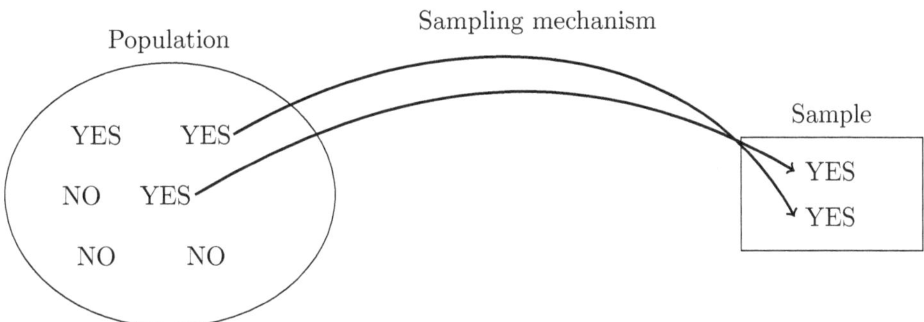

FIGURE 1.2
Illustration of the elements of a statistical experiment. The population consists of the attitudes of six individuals. A sampling mechanism samples attitudes from two different individuals. The sample is the outcome of the sample mechanism, i.e. it is the subset of the population chosen by the sampling mechanism.

Example 1.1 *Consider a population consisting of the heights of all citizens in the city of Paris. Suppose each citizen has a unique identification number. The sampling mechanism could consist of balls, each numbered with an identification number, put in a bowl, which is shaken, not stirred, and then three balls are drawn blindfolded. The sample is the three heights of the three citizens matching the three identification numbers drawn by the sampling mechanism.*

A population may refer to a real group like individuals, animals, firms, organizations, or countries. In this case, we refer to the population as a *real population*.

To extend the applicability of a statistical experiment, we allow for populations which are not real, that is, populations that do not exist. Such populations are referred to as *superpopulations*. For example, if we consider the Gross Domestic Product (GDP) of Denmark over a ten-year period, then the GDP is whatever it is. That is, from the perspective of a statistical experiment, there is only one Denmark, and the sample is the same as the population. If we think there is an element of randomness as to why Denmark's GDP is as it is, then we can accommodate such thinking in a statistical experiment by imagining that the Denmark observed is drawn from a superpopulation of "potential Denmarks". Whether this is a useful way of thinking depends on the question we would like to answer.

Example 1.2 *Suppose we have access to a register of all taxpayers in a municipality with information about their income and age. When we use the information for all taxpayers in the municipality, the sample is the same as the population. If we think that it is coincidental that the municipality looks exactly as it does, then it can be appropriate to model the municipality as a draw from a superpopulation of "potential municipalities". Another way of thinking about the sample is to imagine that the population consists of all municipalities in a country and that we happened to draw the one in which we had access to the register.*

In assessing the quality of a quantitative empirical study, an important element is a careful analysis of the population, sampling mechanism, and sample, as framed by a statistical experiment.

1.3 Interpretation of probabilities

We are going to use probabilities to convey knowledge about uncertainty. There are different interpretations of probabilities, and these different interpretations can be linked to different thought experiments.

The so-called *frequentist interpretation* of probability is based on the thought experiment about what would happen if the statistical experiment were repeated infinitely many times. Then the probability of a particular event occurring is defined as the fraction of times in which the event occurs among the samples from the repeated statistical experiment. For example, if the population consists of tails and heads of a coin, then the probability of tails is defined as the fraction of times tails occur if the coin is thrown infinitely many times. This interpretation is also intimately related to the Law of Large Numbers, which is discussed later in the book.

The so-called *subjectivist interpretation* of probability is based on the perception of an individual on how likely an event is to happen. Thus, the probabilities can be thought of as assigned based on the experience of the individual. Therefore, in the subjectivist interpretation, two individuals with the same information may differ in their perception of the probabilities of various events. Even if probabilistic judgments between individuals differ in this way, then, from a theoretical point of view, the probabilities of the two individuals are well-defined as long as they obey the formal mathematical requirements of the probability model, presented in the next section.

Although it will not feature prominently, we will, in the remainder of this book, mainly be working under a frequentist interpretation of probability. The exception will be Chapter 19, where we discuss Bayesian methods, which are more appropriately thought of as arising from a subjectivist interpretation of probability.

1.4 The probability model

The probability model provides the basis for modeling uncertainty using probabilities. The probability model is an elaborate mathematical structure. It ensures that calculations with probabilities can be done in a consistent manner.

In this section, we define the probability model. The probability model consists of three entities: A *sample space*, an *event space*, and a *probability measure*. They are illustrated in Figure 1.3. We discuss each of these entities in the next subsections, before combining them in Section 1.4.4, yielding the formal definition of the probability model.

1.4.1 The sample space

The sample space consists of all possible outcomes of the statistical experiment. As such, the sample space is completely determined by the statistical experiment. Formally, the sample space is modeled as a set, where each element in the set corresponds to a possible outcome

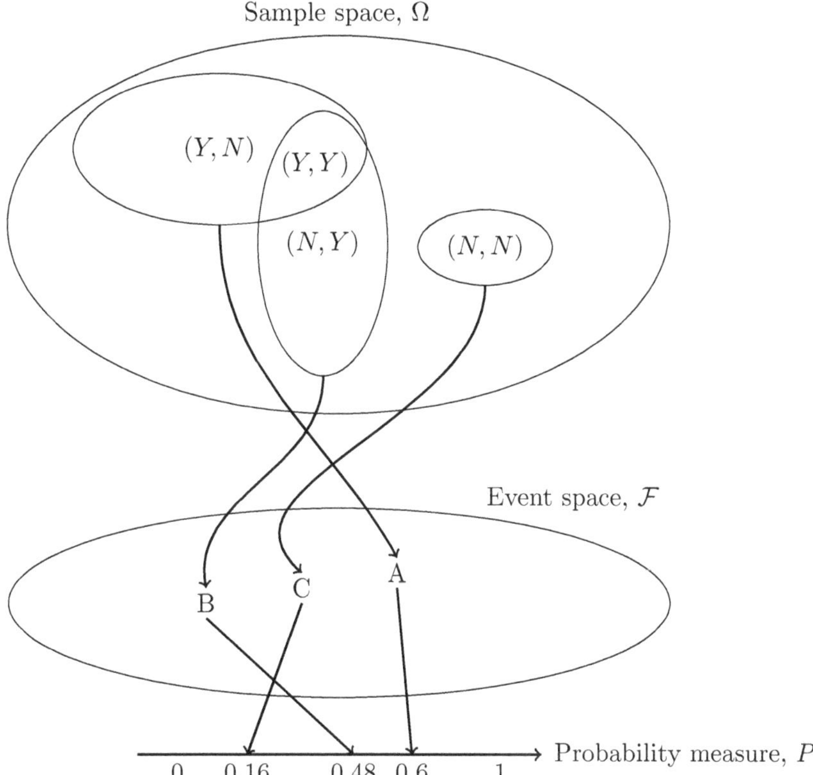

FIGURE 1.3
Elements in the probability model, see Example 1.3. The sample space Ω consists of all possible outcomes (samples) of the statistical experiment. The event space $\mathcal{F}$ consists of the events (sets of samples) which are of interest to the statistical analysis. The probability measure P assigns probabilities to the events in the event space.

of the statistical experiment. We will typically denote the sample space by Ω and a generic element in the sample space by ω.

Definition 1.1 (Sample space) *The **sample space** is a set where each element represents a possible sample (outcome) of the statistical experiment. The sample space contains all the possible samples of the statistical experiment.*

Example 1.3 *The statistical experiment illustrated in Figure 1.2 describes that two elements are drawn from the population of Yes and No elements. When two elements are drawn, there are four possible samples. They are shown within the sample space in Figure 1.3. The sample space can be written as*

$$\Omega = \{(Y, N), (Y, Y), (N, Y), (N, N)\}.$$

Example 1.4 *Consider a population of ages and incomes of individuals. Suppose the population consists of three individuals a, b, and c, with characteristics organized as* $\begin{pmatrix} age \\ income \end{pmatrix}$:

$$a = \begin{pmatrix} 27 \\ 304 \end{pmatrix}, \quad b = \begin{pmatrix} 54 \\ 406 \end{pmatrix}, \quad c = \begin{pmatrix} 35 \\ 708 \end{pmatrix}.$$

Suppose the sample mechanism draws two individuals, and that an individual can only be drawn one time. Sometimes this type of sampling mechanism is called sampling without replacement. *It leads to the following sample space*

$$\Omega = \{(a,b),(b,a),(a,c),(c,a),(b,c),(c,b)\}.$$

For example $(a,b) = \left(\begin{pmatrix} 27 \\ 304 \end{pmatrix}, \begin{pmatrix} 54 \\ 406 \end{pmatrix} \right)$ *is the element of the sample space corresponding to the situation where* $a = \begin{pmatrix} 27 \\ 304 \end{pmatrix}$ *is drawn as the first individual from the population, and* $b = \begin{pmatrix} 54 \\ 406 \end{pmatrix}$ *is drawn as the second individual from the population.*

Example 1.5 *Consider tossing a coin three times. This can be modeled as a statistical experiment, where one head* (H) *and one tail* (T) *constitute the population, and the sampling mechanism draws from the population three times with replacement. The sample space is*

$$\Omega = \{HHH, HHT, HTH, THH, HTT, THT, TTH, TTT\},$$

where, e.g. the element "HHT" represents the outcome of the statistical experiment where heads comes up in the first two tosses and tails in the third toss.

A sample space may be infinitely large, that is, it contains infinitely many elements. For example, if the outcome of a statistical experiment can be any non-negative number, then the sample space is infinitely large.

Example 1.6 *Consider rainfall measured in millimeters over three days. This can be modeled as a statistical experiment where the population is rainfall on a day and the sampling mechanism draws three times from this population. Thus, the population is a superpopulation. Since, in principle, no amount of rain can be ruled out on a given day, we allow the amount of rain on a day to be any non-negative number. Then the sample space can be written as any combination of three non-negative numbers* r_1 *(rain on day 1),* r_2 *(rain on day 2), and* r_3 *(rain on day 3):*

$$\Omega = \{(r_1, r_2, r_3),\ r_1, r_2, r_3 \in [0, \infty)\},$$

where $[0, \infty)$ *denotes the set of all the non-negative real numbers.*

1.4.2 The event space

Most often we are interested in certain aspects of the outcomes of a statistical experiment and the probability of those aspects. For example, if a sample consists of 1.000 replies on attitude ("positive", "indifferent", and "negative") toward a new road bridge, our interest could be the fraction of "positive" replies, and the probability that this fraction is above 50% in the population. It turns out that such aspects can be modeled using sets of samples. This is advantageous because the well-developed mathematical theory of sets may be applied to derive results. There is also a technical reason for using sets of samples to model aspects of the outcomes of a statistical experiment. It turns out that we cannot guarantee a consistent assignment of probabilities without imposing additional mathematical structure. That mathematical structure can be built on sets of samples.

We shall call a set of samples an *event*. The formal definition of an event is next.

Definition 1.2 (Event) *An* **event** A *is a subset of elements (samples) in the sample space* Ω:

$$A \subseteq \Omega$$

Sometimes an event is given a name that describes a particular aspect of the statistical experiment.

Example 1.7 *Figure 1.3 illustrates three different events. One event is $A = \{(Y,N),(Y,Y)\}$. It can be called "Yes in first draw". The two other events are $B = \{(Y,N),(N,Y)\}$, which can be called "one Yes and one No", and $C = \{(N,N)\}$, which can be called "all No".*

When the statistical experiment is performed and the outcome is realized, an event either happens or it does not happen. Mathematically, we say that an event *happens* (or that the event *obtains*) if the sample that is realized in the statistical experiment is an element of that event. Otherwise, the event does not happen.

Example 1.8 (Example 1.7, continued) *Suppose we perform the statistical experiment and (Y,Y) is drawn, i.e. both the first and the second draws result in "Yes", as illustrated in Figure 1.2. In this case, because the outcome (Y,Y) is an element of the set A, but not of the sets B and C, we say that the event A happened, whereas the events B and C did not happen.*

There are two special events, which have their own names, namely, the *sure event* and the *impossible event*. Since, by definition, the sample space Ω contains all possible outcomes of the statistical experiment, the event Ω is certain to happen, and Ω is therefore called the sure event. Conversely, since we require that *something* happens in our statistical experiment, i.e. precisely one of the elements of Ω is realized in the statistical experiment, an event without elements is certain not to happen. Therefore, the empty set, $\emptyset = \{\ \}$, is called the impossible event.

All the events to which we wish to be able to assign probabilities are collected in a set $\mathcal{F}$, called the "event space". That is, $\mathcal{F}$ is a set of events, i.e. a set of subsets of Ω. For technical reasons, we require that the event space is a so-called σ-*algebra* (pronounced "sigma-algebra"). The definition of a σ-algebra is given next.

Definition 1.3 (σ-algebra) *$\mathcal{F}$ is a σ-**algebra** on Ω if and only if the following conditions hold*:

1. $\Omega \in \mathcal{F}$.

2. $A \in \mathcal{F} \Rightarrow A^c \in \mathcal{F}$, where "$A^c$" denotes the complement of A.

3. $A_1, A_2, \ldots \in \mathcal{F} \Rightarrow \bigcup_{i=1}^{\infty} A_i \in \mathcal{F}$.

The requirement that the event space $\mathcal{F}$ satisfies Definition 1.3, i.e. that it is a σ-algebra, ensures that the event space $\mathcal{F}$ is "large enough", in the sense that, later on, it will allow us to make meaningful calculations with the probabilities of events in $\mathcal{F}$.

Conditions 2 and 3 of Definition 1.3 imply that the intersections of events in $\mathcal{F}$ belong to $\mathcal{F}$. This result is stated next.

Theorem 1.1 *Let $\mathcal{F}$ be a σ-algebra. Then, it holds that,*

$$A_1, A_2, \ldots \in \mathcal{F} \Rightarrow \bigcap_{i=1}^{\infty} A_i \in \mathcal{F}.$$

An important consequence of the event space $\mathcal{F}$ being a σ-algebra is, as we discuss in the next section, that probabilities can be assigned to events constructed using common set operations.

In the remainder of the book, we will implicitly assume that a relevant σ-algebra exists for the context studied.

1.4.3   The probability measure

In the probability model, probabilities are assigned to events in the event space $\mathcal{F}$ by a function $P()$. Since an event is a set, we require that the function $P()$ is a *set function*, meaning that it takes as input a set (event) and outputs a real number. Since the outputs of $P()$ will be interpreted as probabilities, we further require that the outputs of $P()$ are numbers from 0 to 1. We call functions which can assign probabilities in this manner for *probability measures*. The formal definition is given next.

Definition 1.4 (Probability measure) *Let Ω be a sample space and $\mathcal{F}$ a σ-algebra on Ω. The function*

$$P : \mathcal{F} \to [0, 1]$$

*is a **probability measure** on $(\Omega, \mathcal{F})$ if and only if the following conditions hold.*

1. *$P(\Omega) = 1$.*

2. *If $A \in \mathcal{F}$, then $P(A^c) = 1 - P(A)$.*

3. *If $A_1, A_2, \ldots \in \mathcal{F}$ are disjoint (i.e. $A_i \cap A_j = \emptyset$ for all $i \neq j$), then $P\left(\bigcup_{i=1}^{\infty} A_i \right) = \sum_{i=1}^{\infty} P(A_i)$.*

It may be noted how each requirement 1 to 3 for a probability measure (Definition 1.4) matches each requirement 1 to 3 for events to be in the σ-algebra (Definition 1.3). Since the probability measure $P()$ acts on sets in the σ-algebra $\mathcal{F}$, this ensures that the conditions of Definition 1.4 make sense. For instance, the second requirement of Definition 1.4 would be nonsensical if it was possible to attach a probability to an event A but not to its complement A^c. The second requirement of Definition 1.3 ensures that this is not the case, since it implies that if $A \in \mathcal{F}$, then also $A^c \in \mathcal{F}$.

The definition of the probability measure does not provide much in relation to interpretation of probabilities. In particular, the probability model is not informative about whether it is most useful to apply a frequentist or subjective approach to the interpretation of probabilities for solving a specific problem. The definition only ensures that we can do calculations with probabilities in a consistent manner which, after all, is not a small thing.

The requirements in Definition 1.4 have a range of implications for the probability measure. Some of these are collected in the next theorem.

Theorem 1.2 (Rules of calculations for probability measures) *Let Ω be a sample space, $\mathcal{F}$ a σ-algebra on Ω, and P a probability measure on $(\Omega, \mathcal{F})$. Then the following holds.*

1. *The probability of the impossible event $\emptyset$ is 0, that is $P(\emptyset) = 0$.*

2. *All events $A \in \mathcal{F}$ have a probability of no less than 0 and no more than 1, that is $0 \leq P(A) \leq 1$.*

3. *(Addition rule of probabilities) For any two events $A, B \in \mathcal{F}$,*

$$P(A \cup B) = P(A) + P(B) - P(A \cap B).$$

4. *(The Law of Total Probability) For any two events $A, B \in \mathcal{F}$,*

$$P(A) = P(A \cap B) + P(A \cap B^c),$$

where B^c is the complement of B.

Example 1.9 *A person is selected randomly from some population. Consider the following four events, pertaining to the person:*

$$A \quad : \quad High\ income; \quad A^c : Low\ income;$$
$$B \quad : \quad Long\ education; \quad B^c : Short\ education.$$

The following table displays the probabilities of the intersections between the four sets.

	B	B^c
A	$P(A \cap B) = 0.1$	$P(A \cap B^c) = 0.05$
A^c	$P(A^c \cap B) = 0.15$	$P(A^c \cap B^c) = 0.7$

Using Theorem 1.2, the probability that high income A occurs can be calculated as

$$P(A) = P(A \cap B) + P(A \cap B^c) = 0.1 + 0.05 = 0.15.$$

Similarly, the probability that long education B occurs can be calculated as

$$P(B) = P(A \cap B) + P(A^c \cap B) = 0.1 + 0.15 = 0.25.$$

The probability that either A or B, or both occur is

$$P(A \cup B) = P(A) + P(B) - P(A \cap B) = 0.15 + 0.25 - 0.1 = 0.3.$$

1.4.4 The probability model

We can now use the concepts defined above to state the formal definition of a probability model.

Definition 1.5 (Probability model) *Let Ω be a sample space (Definition 1.1), $\mathcal{F}$ a σ-algebra on Ω (Definition 1.3), and P a probability measure on $(\Omega, \mathcal{F})$ (Definition 1.4). The triple $(\Omega, \mathcal{F}, P)$ is called the **probability model**.*

The probability model is sometimes also referred to as a *probability space*.

The probability model is the foundation for probabilities as it will ensure that we can make useful calculations with probabilities much the same way as axioms for the number system ensure that we can make useful calculations with numbers.

1.5 Random variables

1.5.1 Defining outcomes of statistical experiment using numbers

A random variable with associated probabilities is undoubtedly the most useful representation of the outcomes of a statistical experiment. A random variable X is a function, which takes the outcome of a statistical experiment ω as input and returns a real number. The random variable is a practical way of coding an interesting event of a statistical experiment. It changes the event from a set to a real number.

The definition of a random variable consists of two parts. One part is about how a random variable represents outcomes as numbers, and the other part is a rather technical requirement about "measurability" to ensure that probabilities can be assigned to events related to the random variable. The definition is here.

Definition 1.6 (Random variable) *Let Ω be a sample space and $\mathcal{F}$ a σ-algebra on Ω. A function X is a **random variable** on $(\Omega, \mathcal{F})$ if and only if the following two conditions hold.*

1. *X has as domain the sample space Ω and as codomain the real numbers, i.e.*

$$X : \Omega \to \mathbb{R}.$$

2. *X is a measurable function on $(\Omega, \mathcal{F})$, where $\mathcal{F}$ is a σ-algebra on Ω.*

In the following, condition 1 is the focus since this condition encapsulates the big advantage of random variables compared to modeling outcomes as single events like A and B. Condition 2 is a technical requirement which ensures that the σ-algebra $\mathcal{F}$ contains subsets of Ω associated with the random variable X such that probabilities may be assigned. For further details on this technical condition, we refer the reader to Chapter A in the appendix.

Example 1.10 (Example 1.5, continued) *We can define a random variable $X(\omega)$ to be the total number of tails in the three tosses. The input ω to the random variable is an outcome of the statistical experiment, e.g. $\omega = THT$. Then the random variable can be specified as*

$$X(\omega) = \begin{cases} 0 & if & \omega \in \{HHH\}, \\ 1 & if & \omega \in \{THH, HTH, HHT\}, \\ 2 & if & \omega \in \{TTH, THT, HTT\}, \\ 3 & if & \omega \in \{TTT\}. \end{cases}$$

The function $X(\omega)$ is specified as a piecewise function, where the value of the function is given in the line where the condition is satisfied. For example, if the sample is $\omega = THT$, then the third line is satisfied because $THT \in \{TTH, THT, HTT\}$ and, thus, $X(THT) = 2$. Figure 1.4 illustrates how each possible sample of this statistical experiment is mapped into real numbers by X.

Example 1.10 illustrates how a random variable can be used to represent the relevant outcomes of a statistical experiment (here, the number of tails) as real numbers, instead of the more abstract elements in Ω (here, outcomes of coin tosses). Indeed, the real strength of using random variables is that they are real-valued functions and as such can be manipulated as any other function. For instance, we can apply the usual arithmetic operations for real numbers, i.e. addition, subtraction, multiplication, and division, to random variables. The next example illustrates this point.

Example 1.11 (Example 1.10, continued) *Let a random variable Y_1 indicate, by 0 and 1, whether the first toss is a tail. That is, define Y_1 as*

$$Y_1(\omega) = \begin{cases} 0 & if & \omega \in \{HHH, HTH, HHT, HTH\}, \\ 1 & if & \omega \in \{THH, TTH, THT, TTT\}. \end{cases}$$

Similarly, define the random variables $Y_2(\omega)$ and $Y_3(\omega)$ to indicate whether the second and third tosses, respectively, are tails:

$$Y_2(\omega) = \begin{cases} 0 & if & \omega \in \{HHH, THT, THH, HHT\}, \\ 1 & if & \omega \in \{TTH, HTH, HTT, TTT\}, \end{cases}$$

$$Y_3(\omega) = \begin{cases} 0 & if & \omega \in \{THH, TTH, HHH, HTH\}, \\ 1 & if & \omega \in \{THT, TTT, HHT, HHT\}. \end{cases}$$

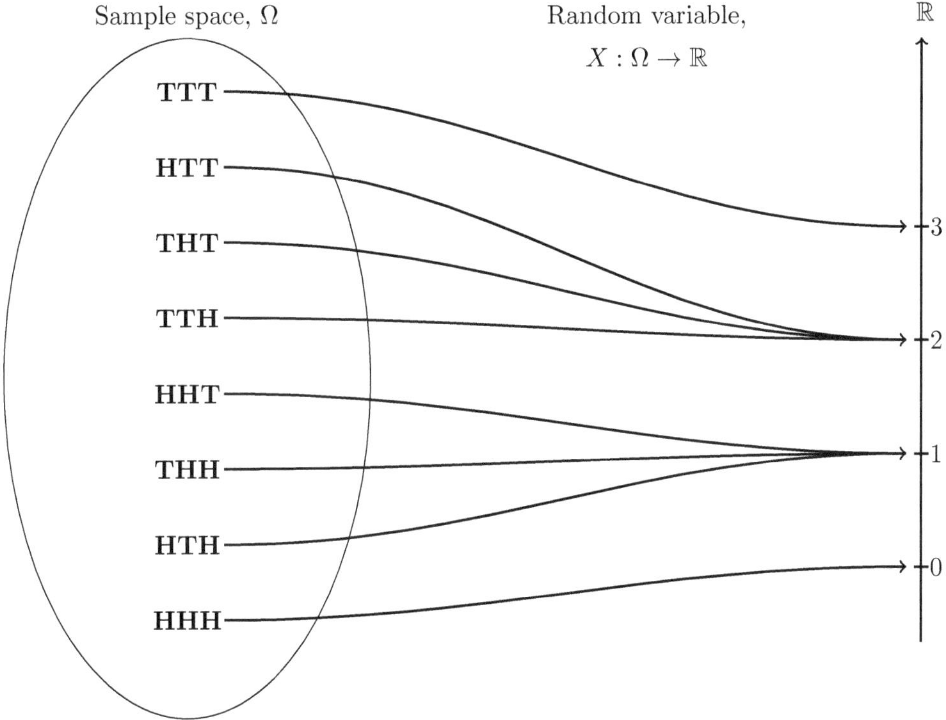

FIGURE 1.4
A random variable maps the sample space (Ω) into the real numbers ($\mathbb{R}$), see Example 1.11.

Now it can be utilized that the random variables are functions such that the random variable X, denoting the total number of tails, can be written as a sum of the random variables Y_1, Y_2, and Y_3:

$$X(\omega) = Y_1(\omega) + Y_2(\omega) + Y_3(\omega).$$

We may also be interested in the fraction of tails in the sample. This can be specified as the new random variable

$$Z(\omega) = \frac{Y_1(\omega) + Y_2(\omega) + Y_3(\omega)}{3} = \frac{X(\omega)}{3}.$$

For example, if $\omega = THT$, then

$$Z(THT) = \frac{1 + 0 + 1}{3} = \frac{2}{3} = 0.6667.$$

Example 1.12 (Example 1.4, continued) *Consider drawing two individuals from the population in Example 1.4. Let X_1 be the age of the first person drawn and let Y_1 be the income of the first person drawn. Similarly let X_2 be the age of the second person drawn and let Y_2 be the income of the second person drawn. Notice, that each of these random variables picks a particular aspect of the sample.*

The average income of the individuals in the sample can be represented by the random variable

$$Z = \frac{Y_1 + Y_2}{2}.$$

For example, if the outcome of the statistical experiment results in the sample

$$\omega = \left(\begin{pmatrix} 54 \\ 406 \end{pmatrix}, \begin{pmatrix} 35 \\ 708 \end{pmatrix} \right),$$

then

$$\begin{aligned} X_1(\omega) &= 54, \quad Y_1(\omega) = 406, \\ X_2(\omega) &= 35, \quad Y_2(\omega) = 708, \end{aligned}$$

and, thus,

$$Z(\omega) = \frac{Y_1(\omega) + Y_2(\omega)}{2} = \frac{406 + 708}{2} = 557.$$

1.5.2 Linking a random variable to events

In this subsection, we show how the value of a random variable is directly linked to a subset of the sample space and, thus, defines an event. This link is important for two reasons. Firstly, the link may inspire a substantive interpretation of the random variable. Secondly, the link is used to assign probabilities to the outcomes of the random variable based on the probability measure defined in the probability model.

The event corresponding to the outcome of a random variable can be derived by finding all the outcomes ω in the sample space Ω that maps to the same outcome of $X(\omega)$. To do so, we need to find the *inverse* of the random variable X, that is, the inverse of $X(\omega)$. Since a random variable may map different outcomes ω from the sample space Ω into the same number, the inverse function of a random variable may not exist. The resolution is to extend the concept of an inverse function to allow each value in the codomain of the function to be related to a set in the domain of the function. Figure 1.5 illustrates how to start at a point in the codomain of the function X (on the right) and trace all values in the domain (to the left) of X that leads to that particular point in the codomain.

To ease the mathematical exposition and to focus on the intuition behind the formalism, we will consider only the inverse image of a single point b, i.e. for a value $b \in \mathbb{R}$. That is, we find all elements $\omega \in \Omega$ resulting in $X(\omega) = b$. More generally, to also cover cases where the outcomes of a random variable are allowed to be a continuum of values, the theory requires that we are able to define the inverse image of a set B, i.e. for a set $B \subseteq \mathbb{R}$, we ask what elements $\omega \in \Omega$ results in $X(\omega) \in B$. This more involved theory is relegated to Chapter A in the appendix, see in particular Section A.2.2, where the interested reader may find additional technical details. We next give the definition of an inverse image of a point.

Definition 1.7 (Inverse image of a point) *Let $X : \Omega \to \mathbb{R}$ be a function with domain Ω and codomain $\mathbb{R}$. Let b be a point in $\mathbb{R}$ i.e. $b \in \mathbb{R}$. Then the **inverse image of a point** b is the set*

$$A = \{\omega \in \Omega : X(\omega) = b\}.$$

This may also be written as

$$X^{-1}(b) = \{\omega \in \Omega : X(\omega) = b\}.$$

Example 1.13 (Example 1.10, continued) *Based on the definition of X, the subset of Ω with the property that $X(\omega) = 2$ is*

$$X^{-1}(2) = \{\omega \in \Omega : X(\omega) = 2\} = \{TTH, THT, HTT\}.$$

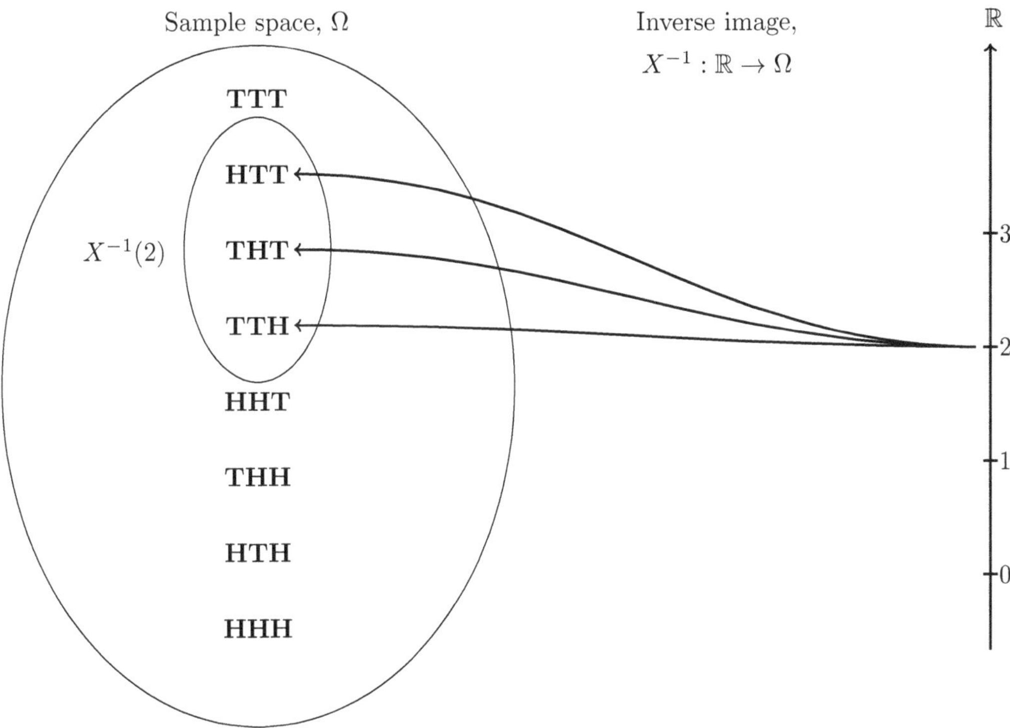

FIGURE 1.5
Inverse image of a random variable X, see Example 1.14. The ellipse represents the inverse image of $X(\omega) = 2$, $X^{-1}(2)$, i.e. the elements in ω such that $X(\omega) = 2$.

This case is shown in Figure 1.5. At the point 2 in the codomain of X (the real numbers $\mathbb{R}$, shown on the right), arrows trace backwards to all the values in the domain of X (the sample space Ω, shown on the left), enclosed by an ellipse, that leads to X being equal to 2.

Similarly, the subset of Ω with the property that $X(\omega) = 0$ is

$$X^{-1}(0) = \{\omega \in \Omega : X(\omega) = 0\} = \{HHH\}.$$

The subset of Ω with the property that $X(\omega) = 5$ is the empty set:

$$X^{-1}(5) = \{\omega \in \Omega : X(\omega) = 5\} = \emptyset.$$

The reason is that no sample point ω results in $X(\omega)$ being equal to 5.

1.5.3 Probability of outcomes of a random variable based on the probability model

By the definition of a random variable and by the definition of an inverse image, it is now possible to assign probabilities to the outcomes of the random variable solely based on the probability measure defined in the probability model. Indeed, it can be shown that a consequence of requirement 2 in the definition of a random variable (Definition 1.6) is that, under mild conditions, sets of the form $X^{-1}(A)$ are in $\mathcal{F}$. That is, we can assign probabilities to these sets using the probability model $(\Omega, \mathcal{F}, P)$.

Suppose we are interested in the probability that X equals a value $a \in \mathbb{R}$. This is denoted $\Pr(X = a)$. Since $X^{-1}(a) \in \mathcal{F}$, we can apply P to $X^{-1}(a)$ to arrive at this probability. That is,

$$\Pr(X = a) = P(A),$$

where

$$A = X^{-1}(a) = \{\omega \in \Omega : X(\omega) = a\}.$$

Notice, we use $\Pr()$ to denote the probability of outcomes of a random variable, whereas we write $P()$ to refer to the probability measure of the underlying probability model $(\Omega, \mathcal{F}, P)$.

Example 1.14 (Example 1.10, continued) *Assume tails and heads have the same chance of occurring. Then all elements in the sample space have the same probability, namely, $\frac{1}{8}$.*

Together with the rules of calculation for probability measures (Theorem 1.2), we may use this to find the probabilities for each of the outcomes of X. They are

$$
\begin{aligned}
\Pr(X = 0) = P(X^{-1}(0)) &= P(\{HHH\}) = \frac{1}{8}, \\
\Pr(X = 1) = P(X^{-1}(1)) &= P(\{THH, HTH, HHT\}) \\
&= P(\{THH\}) + P(\{HTH\}) + P(\{HHT\}) \\
&= \frac{1}{8} + \frac{1}{8} + \frac{1}{8} \\
&= \frac{3}{8}, \\
\Pr(X = 2) = P(X^{-1}(2)) &= P(\{TTH, THT, HTT\}) \\
&= P(\{TTH\}) + P(\{THT\}) + P(\{HTT\}) \\
&= \frac{1}{8} + \frac{1}{8} + \frac{1}{8} \\
&= \frac{3}{8}, \\
\Pr(X = 3) = P(X^{-1}(3)) &= P(\{TTT\}) = \frac{1}{8}, \\
\Pr(X = 5) = P(X^{-1}(5)) &= P(\emptyset) = 0.
\end{aligned}
$$

Example 1.14 illustrates how probabilities related to the object of interest, that is, the outcomes of the random variable can be calculated by using the probability measure $P()$, evaluated at the relevant element in $\mathcal{F}$, e.g. $X^{-1}(2) = \{\omega \in \Omega : X(\omega) = 2\}$. Figure 1.6 illustrates the case of assigning probability to $X = 2$. In the figure, start with the outcome $X = 2$ in the codomain of X. Then trace backwards to find the inverse image of $X = 2$. Such an inverse image is a subset of the sample space Ω and, using that X is a random variable (condition 2 in Definition 1.6), it is also an element A in the σ-algebra $\mathcal{F}$. To every element in $\mathcal{F}$, the probability measure P provides a probability, $P(A)$. Finally, assign $P(A)$ to the probability $\Pr(X = 2)$.

It can be shown that the set function $\Pr()$ defined in this way is indeed a proper probability measure in the sense of Definition 1.4. The interested reader may consult Section A.2 in the appendix for further details. The upshot of this is that the rules for calculations with probability measures, such as those given in Theorem 1.2, holds also for $\Pr()$.

Strictly speaking, the probability measure $P()$ is the more basic object since $\Pr()$ is derived from $P()$. However, in the remainder of this book, we almost always specify the probability $\Pr()$ directly, instead of first specifying the underlying probability model $(\Omega, \mathcal{F}, P)$. Specifying $\Pr()$ is typically considerable easier because outcomes of X are numbers whereas

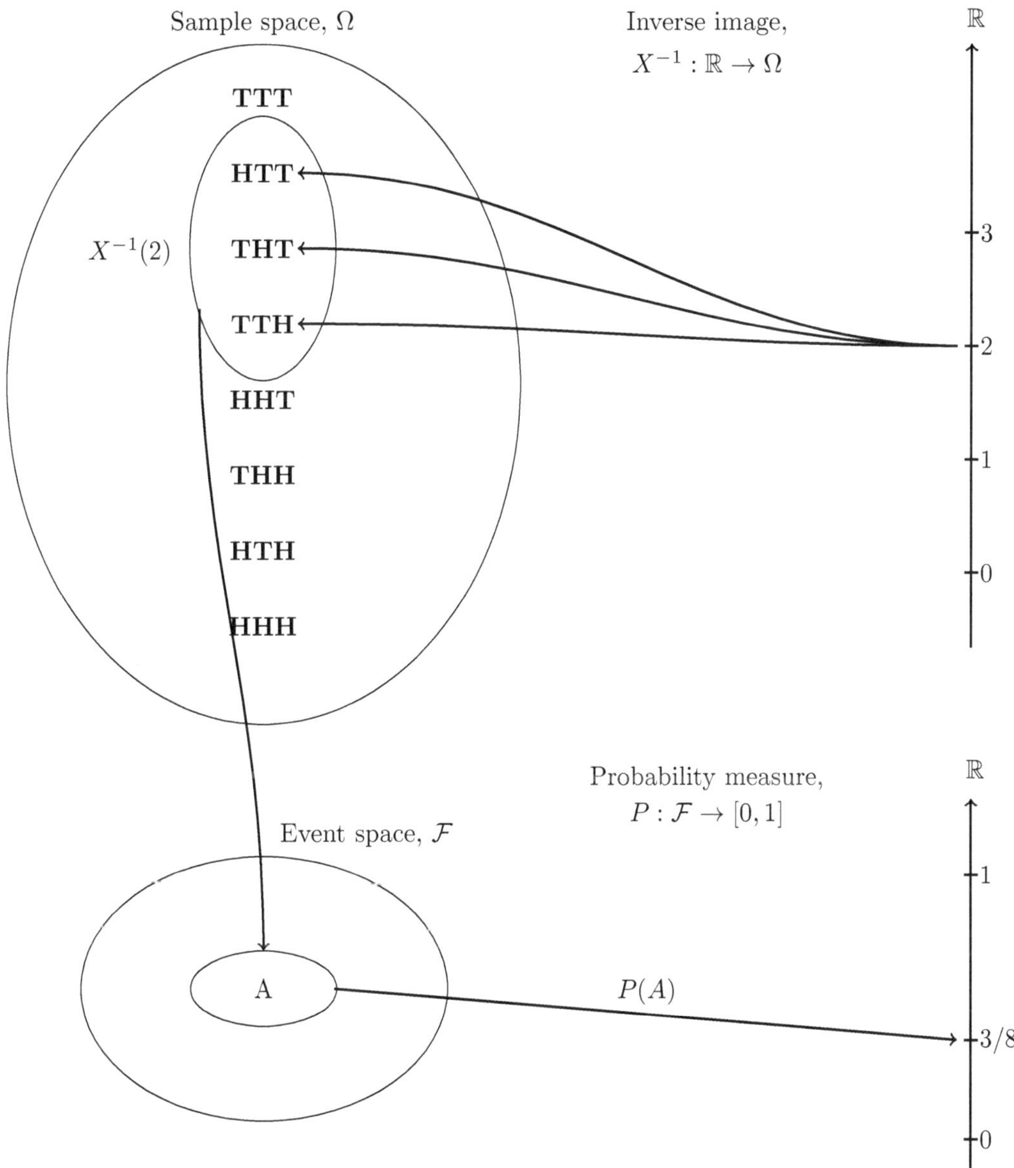

FIGURE 1.6
Assigning probability to outcomes of random variable X, see Example 1.14. Assigning probability to "$X = 2$", $\Pr(X = 2)$, is done by first finding the elements ω in Ω such that $X(\omega) = 2$, i.e. $X^{-1}(2)$. By the second condition in the definition of a random variable (Definition 1.6), the set of these ω's, A, is in the event space $\mathcal{F}$. The probability measure P assigns probabilities to the sets in $\mathcal{F}$, e.g. $P(A)$ to A, which allows us to equate $\Pr(X = 2)$ to $P(A)$.

sets are needed as input to the probability $P()$. Probabilities associated with $\Pr()$ are represented as a *distribution* of the random variable X, which we discuss in the next chapter.

1.6   Proofs

Proof of Theorem 1.1   *The proof of Theorem 1.1 is left as an exercise to the reader (Problem 1.7.1).*

Proof of Theorem 1.2   *The proof of Theorem 1.2 is left as an exercise to the reader (Problem 1.7.4).*

1.7   Exercises

Problem 1.7.1 *Prove Theorem 1.1. (Hint: Apply De Morgan's laws, see Theorem A.2 in Chapter A of the appendix.)*

Problem 1.7.2 *Consider the sample space $\Omega = \{0, 1, 2, 3\}$. Prove that*

$$\mathcal{F} = \{\{0\}, \{1\}, \{2, 3\}, \{0, 2, 3\}, \{1, 2, 3\}, \{0, 1\}, \Omega, \emptyset\},$$

is a σ-algebra on Ω.

Problem 1.7.3 *Let Ω be a set and $\mathcal{P}(\Omega)$ be the power set of Ω (i.e. $\mathcal{P}(\Omega)$ is the set of all subsets of Ω). Show that $\mathcal{P}(\Omega)$ is a σ-algebra on Ω.*

Problem 1.7.4 *Prove Theorem 1.2.*

2

Distribution of a random variable

2.1 Introduction

Random variables, as introduced in the previous chapter, are the core building blocks in the modeling of uncertainty. In this chapter, we discuss how to model the probability that a certain random variable will have a given outcome. The collection of these probabilities are called the *distribution* of the random variable. Modeling random variables through their distribution is an extremely powerful tool to specify and model uncertainty of statistical experiments. In particular, as discussed in the previous chapter, directly modeling the distribution of the random variable bypasses the need for specifying an underlying probability model.

The probabilities of the outcomes of a random variable can always be represented by a so-called *cumulative distribution function*. Therefore, we start by defining this function and illustrate its use. To obtain more insight into the distribution of a random variable, we divide random variables into three types: discrete, continuous, and mixed. We then consider aspects of their probabilities with techniques especially suited for each type.

Section B.2 in Chapter B of the appendix contains mathematical background about functions of one variable. This section reviews the concepts of inverse, limit, continuity, differentiation, and integration of a function. These concepts will be used throughout this chapter.

2.2 The cumulative distribution function

The cumulative distribution function of a random variable contains all the information needed to calculate the probabilities that the random variable will have certain outcomes, i.e. about the distribution of the random variable. Specifically, the cumulative distribution function gives the probability that a random variable X has an outcome less than or equal to any real number, say, x. Using the framework introduced in the previous chapter, this probability is denoted

$$\Pr(X \leq x).$$

The word "cumulative" refers to the fact that the probability in question is of the event $\{X \leq x\}$, that is, accumulating the probability over all outcomes up to x. Though focusing on $\Pr(X = x)$ might seem more intuitive than on $\Pr(X \leq x)$, it turns out that $\Pr(X = x)$ can only describe the distribution of one type of random variables, whereas $\Pr(X \leq x)$ works for all types of random variables.

Considering the probability $\Pr(X \leq x)$ as a function of x leads directly to the cumulative distribution function, defined next.

Definition 2.1 (Cumulative distribution function, CDF) *A function F is a **cumulative distribution function** (**CDF**) for the random variable X if and only if*

$$F(x) = \Pr(X \leq x) \text{ for all } x \in \mathbb{R}.$$

Often we will denote the CDF of a random variable by a capital letter, e.g. F. If we need to be specific about which random variable a CDF refers, we denote this by a subscript, e.g. F_X is the CDF of the random variable X and F_Y is the CDF of the random variable Y.

Example 2.1 (Example 1.5, continued) *Let X be the number of tails in three coin tosses. Assuming the coin is fair, i.e. the probability of getting heads and tails in one toss are both 50%, the CDF $F_X(x)$ for this random variable X is*

$$F_X(x) = \begin{cases} 0 & \text{if} & x < 0, \\ \frac{1}{8} & \text{if} & 0 \leq x < 1, \\ \frac{4}{8} & \text{if} & 1 \leq x < 2, \\ \frac{7}{8} & \text{if} & 2 \leq x < 3, \\ 1 & \text{if} & 3 \leq x. \end{cases}$$

Figure 2.1 illustrates the CDF F_X. When there is a jump in the graph, it is the solid circle that indicates the value of the function at the jump.

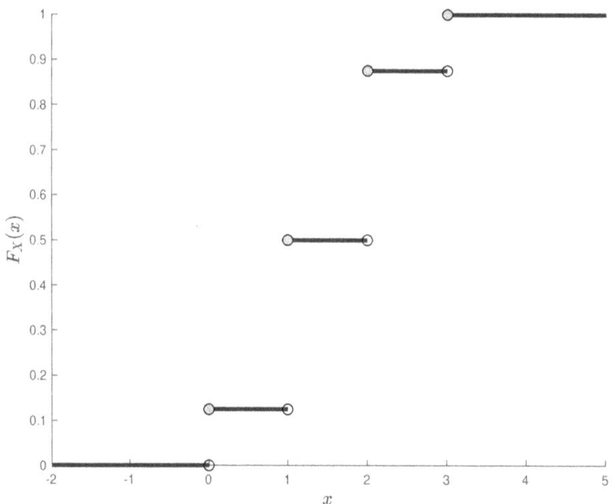

FIGURE 2.1
CDF $F_X(x)$ for the number of tails in three coin tosses, see Example 2.1.

For instance,

$$\Pr(X \leq 1) = F_X(1) = \frac{4}{8} = \frac{1}{2},$$

because the probability for an outcome of $X \leq 1$ is the same as the probability of $X = 0$ added to the probability of $X = 1$. Notice, e.g. $F_X(1.7) = \frac{4}{8}$ is the same as $F_X(1)$ because there are no outcomes possible of X larger than 1 and less than or equal to 1.7.

The CDF has a number of properties, which are useful for doing calculations and for identifying when a function can be a CDF. These are given next.

Theorem 2.1 (Properties of CDF) *Let $a, b \in \mathbb{R}$ such that $a < b$. A cumulative distribution function $F(x)$ has the following properties.*

1. *$F : \mathbb{R} \to [0, 1]$.*

2. *$F(x)$ is a non-decreasing function.*

3. *$\lim_{x \to -\infty} F(x) = 0, \ \lim_{x \to \infty} F(x) = 1$.*

4. *$F(x)$ is continuous from the right.*

5. *$\Pr(X > a) = 1 - F_X(a)$.*

6. *$\Pr(a < X \leq b) = \Pr(X \leq b) - \Pr(X \leq a) = F_X(b) - F_X(a)$.*

7. *$F_X(a) = \Pr(X \leq a) = \Pr(X < a) + \Pr(X = a)$.*

8. *$\Pr(X < a) = \lim_{x \to a^-} F_X(x)$.*

Sometimes, the notation in property 3 of Theorem 2.1 is stated differently, where $\lim_{x \to -\infty} F_X(x)$ is written as $F_X(-\infty)$ and $\lim_{x \to \infty} F_X(x)$ as $F_X(\infty)$. Thus, $F(-\infty) = 0$ means that the probability of getting an outcome less than a number x goes to 0 as x decreases toward $-\infty$. Similarly, $F(\infty) = 1$ means that the probability of getting an outcome less than x goes to 1 as x increases toward ∞.

In property 4, the mathematical formalization of "continuous from the right" is

$$\lim_{x \to a^+} F_X(x) = F_X(a)$$

for all $a \in (-\infty, \infty)$. For example, in Figure 2.1 at $x = 1$, $F(1) = 0.5$, and the values of $F(x)$ for x's "just right" of $x = 1$ are also close to 0.5 (in the example equal to 0.5) and, therefore, $F(x)$ is continuous from the right at $x = 1$. Notice, the values of $F(x)$ for x's "just left" of $x = 1$ are close to $1/8$ (in the example equal to $1/8$) and, therefore, $F_X(x)$ is not continuous from the left at $x = 1$.

The next example shows how Theorem 2.1 can be used to calculate various probabilities

Example 2.2 *Consider the CDF from Example 2.1. Using Theorem 2.1, here are examples of probability calculations:*

$$\Pr(X \leq 2) = F_X(2) = \frac{7}{8},$$

$$\Pr(X < 2) = \lim_{x \to 2^-} F_X(x) = \frac{4}{8},$$

$$\Pr(X > 2) = 1 - \Pr(X \leq 2) = 1 - F_X(2) = \frac{1}{8},$$

$$\Pr(X \leq 2.8) = F_X(2.8) = \frac{7}{8},$$

$$\Pr(X \leq -6) = F_X(-6) = 0,$$

$$\Pr(1.8 < X \leq 2.8) = F_X(2.8) - F_X(1.8) = \frac{7}{8} - \frac{4}{8} = \frac{3}{8}.$$

It turns out that the first four properties in Theorem 2.1 are necessary and sufficient conditions for a function $F(x)$ to be a valid CDF, where we by "valid" mean that it is possible for a random variable to have $F(x)$ as its CDF. This is stated in the next theorem.

Theorem 2.2 (Validity of a CDF) *Let $F(x)$ be some function. If $F(x)$ satisfies the properties 1 to 4 stated in Theorem 2.1, then $F(x)$ is valid as a CDF.*

The next example shows how to use Theorem 2.2 to verify that a function is valid as a *CDF*.

Example 2.3 *Consider the function*

$$F_X(x) = \begin{cases} 0 & \text{if} \quad x < 0, \\ 0.5 \cdot x & \text{if} \quad 0 \le x < 2, \\ 1 & \text{if} \quad 2 \le x. \end{cases}$$

The left panel of Figure 2.2 illustrates $F_X(x)$.

We check if this function is valid as a CDF using Theorem 2.2 and, thus, checking properties 1 to 4 of Theorem 2.1. Regarding 1, $F_X(x)$ is well defined for any number $x \in \mathbb{R}$. That is, for any input value x, there is an output value $F_X(x)$. Further, for all $x \in \mathbb{R}$, it holds that $F_X(x) \in [0,1]$. Regarding 2, $F_X(x)$ is constant for $x < 0$ and $x > 2$ and, therefore, non-decreasing in each of those segments. $F_X(x)$ is increasing for $0 \le x \le 2$, starting at 0 and ending at 1. In total, $F_X(x)$ is a non-decreasing function. Regarding 3, since $F_X(x) = 0$ for any $x \le 0$, then $\lim_{x \to -\infty} F(x) = 0$. Similarly, since $F_X(x) = 1$ for any $x \ge 2$, then $\lim_{x \to \infty} F(x) = 1$. Finally, regarding 4, $F_X(x)$ is a continuous function and, thus, also continuous from the right. Thus, $F_X(x)$ is a valid CDF.

The left panel of Figure 2.2 illustrates the CDF $F_X(x)$ used in this example.

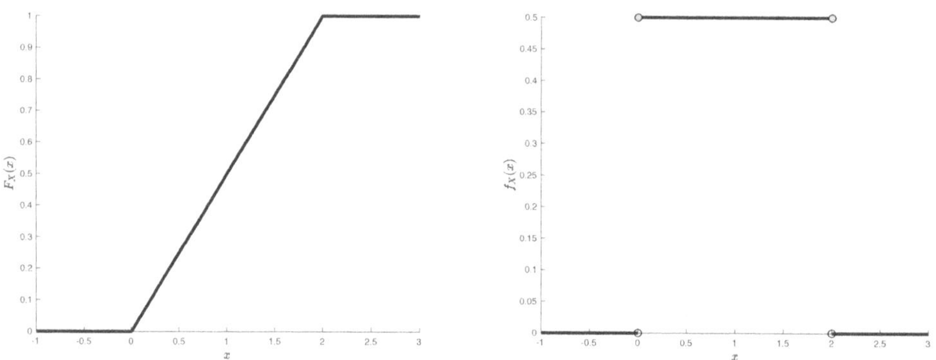

FIGURE 2.2
Left: Graph of CDF $F_X(x) = 0.5x$ for $x \in [0, 2]$, see Example 2.3. Right: PDF $f_X(x) = 0.5$ for $x \in [0, 2]$, see Example 2.9.

Despite the fact that the CDF contains all the information needed to calculate probabilities for outcomes of a random variable, the following sections will show how particular functions derived from the CDF are considerably easier to interpret and use for calculation of probabilities. It turns out that the appropriate function to consider depends on what outcomes the random variable can have. This leads to the definition of three types of random variables, namely, discrete, continuous, and mixed random variables. These three types of random variables will be discussed in turn in the following sections.

2.3 Discrete random variables

A random variable is discrete if the set of possible outcomes of the random variable is finite or countably infinite. A finite set means that the number of elements in the set is finite, e.g. the set $\{2, 4, 6\}$ has a finite number of elements, namely three. In contrast, an infinite set means that there are infinitely many elements in the set. An infinite set which is countable, means that each element of the set can be put in a one-to-one correspondence with the natural numbers, $1, 2, 3, \ldots$ For example, the set $\{2, 4, 6, 8, \ldots\}$ is infinite because the sequence $2, 4, 6, 8, \ldots$ does not have an end, and it is countable because each element can be put in a one-to-one correspondence with the natural numbers, as shown here:

$$
\begin{array}{ccccc}
\{ \quad 2, & 4, & 6, & 8, & \ldots \quad \} \\
\uparrow & \uparrow & \uparrow & \uparrow & \uparrow \\
1, & 2, & 3, & 4, & \ldots
\end{array}
$$

An example of a discrete random variable is given in Example 2.1. In contrast to a countably infinite set, an uncountably infinite set cannot be put in one-to-one correspondence with the natural numbers. The set of real numbers between 0 and 2, i.e. $[0, 2]$, and the set of non-negative real numbers, i.e. $[0, \infty)$, are examples of uncountably infinite sets. We shall discuss such sets of outcomes in Section 2.4 when discussing continuous random variables.

The CDFs of random variables with countable many outcomes are all characterized by their functional forms being stair functions. A stair function is a function consisting only of jumps and flat parts in-between those jumps. We shall use this characteristic to formally define a discrete random variable such that the definition of a discrete random variable is directly linked to its distribution as represented by its CDF. The functional form of a stair function can be written using an indicator function $I(Q)$, where Q is a logical statement which is either true or false. If the logical statement Q is true, then the indicator function equals 1 and if it is false, then the indicator function equals 0:

$$
I(Q) = \begin{cases} 1 & \text{if} \quad Q \text{ is true,} \\ 0 & \text{if} \quad Q \text{ is false.} \end{cases}
$$

For instance, $I(42 > 5) = 1$ and $I(5 > 42) = 0$. The definition of a discrete random variable can be stated next.

Definition 2.2 (Discrete random variable) *A random variable X is a **discrete random variable** if and only if the CDF $F_X(x)$ of X can be written in the following form*

$$
F_X(x) = \sum_{j=1}^{J} b_j I(a_j \leq x),
$$

where $a_j \in \mathbb{R}$, $b_j \geq 0$ for all j, $\sum_{j=1}^{J} b_j = 1$, and J is a natural number or $+\infty$.

The a_j's in the definition are the location of the jump points, and the size of the jump at a_j is b_j. This is illustrated in the next example.

Example 2.4 *Consider the CDF in Example 2.1, where X is the number of tails in three coin tosses. The CDF for X is*

$$
F_X(x) = \begin{cases} 0 & \textit{if} & x < 0, \\ \frac{1}{8} & \textit{if} & 0 \leq x < 1, \\ \frac{4}{8} & \textit{if} & 1 \leq x < 2, \\ \frac{7}{8} & \textit{if} & 2 \leq x < 3, \\ 1 & \textit{if} & x \geq 3. \end{cases}
$$

The graph of the CDF is illustrated in Figure 2.1, where it can be seen that the graph is a "stair".

The CDF can be written in the form of the Definition 2.2 for a discrete random variable. It is

$$F_X(x) = \frac{1}{8} \cdot I(0 \leq x) + \frac{3}{8} \cdot I(1 \leq x) + \frac{3}{8} \cdot I(2 \leq x) + \frac{1}{8} \cdot I(3 \leq x).$$

Thus, the values of the J, $a's$ and $b's$ in Definition 2.2 are $J = 4$, $a_1 = 0, a_2 = 1, a_3 = 2, a_4 = 3$, and $b_1 = \frac{1}{8}, b_2 = \frac{3}{8}, b_3 = \frac{3}{8}, b_4 = \frac{1}{8}$.

2.3.1   The probability mass function

In this section, we show how a *probability mass function*, often just called a *probability function*, is directly informative about the probability of obtaining a particular outcome of a discrete random variable.

The probability mass function is only defined for a discrete random variable. We state the definition of a probability mass function based on a CDF. While this is a technical definition, we show right after how to interpret the probability mass function as probabilities.

Definition 2.3 (Probability mass function, PMF) *Assume X is a discrete random variable with CDF $F_X(x)$. Then $f_X(x)$ is the* **probability mass function (PMF)** *of X if and only if*

$$i) \; f_X(a) \; \geq \; 0,$$
$$ii) \; F_X(a) \; = \; \sum_{x \leq a} f_X(x),$$

for all $a \in \mathbb{R}$. In ii), the sum is understood to be taken over all the outcomes x of X, which satisfy $x \leq a$.

The PMF gives the probability of each outcome of a discrete random variable. This is stated in the next theorem.

Theorem 2.3 (PMF and probability) *Let X be a discrete random variable with PMF $f_X(x)$. Then*

$$f_X(x) = \Pr(X = x).$$

When we work with discrete random variables, we are often interested in the probabilities of specific outcomes of the discrete random variable. According to Theorem 2.3, this is exactly what the PMF provides, and this is the reason that PMFs are more used than CDFs when working with discrete random variables.

Example 2.5 *Consider a random variable X with two outcomes 0 and 1. Suppose $\Pr(X = 0) = 0.4$ and $\Pr(X = 1) = 0.6$. Then*

$$f_X(x) = \begin{cases} 0.4 & \text{if } x = 0, \\ 0.6 & \text{if } x = 1, \\ 0 & \text{otherwise.} \end{cases}$$

In Example 2.5, the last statement in the function definition "0 otherwise" is included because $f_X(x)$ must have, by definition, domain $\mathbb{R}$. Often, this qualification is understood, and therefore omitted, when saying $f(x)$ is a PMF. For instance, we may write the PMF from Example 2.5, as

$$f_X(x) = \begin{cases} 0.4 & \text{if } x = 0, \\ 0.6 & \text{if } x = 1. \end{cases}$$

Indeed, in the remainder of the book we may not explicitly mention the "0 otherwise" part but it is implicitly understood to be present when dealing with PMFs.

A PMF has the following properties.

Theorem 2.4 (Properties of a PMF) *A probability mass function $f_X(x)$ has the following properties.*

1. *$f_X(x) \geq 0$ for all x.*

2. *$\sum_x f_X(x) = 1$, where the sum is taken over all x's where $f_X(x) > 0$.*

3. *$\sum_{a < x \leq b} f_X(x) = F_X(b) - F_X(a)$ for all a, b such that $a < b$.*

4. *$\sum_{x > a} f_X(x) = 1 - F_X(a)$ for all a.*

It turns out that properties 1 and 2 in Theorem 2.4 are necessary and sufficient for a function $f_X(x)$ to be a valid PMF, where we by "valid" mean that it is possible for a discrete random variable to have $f(x)$ as its PMF.

Theorem 2.5 (Validity of a PMF) *Let $f(x)$ be some function such that $f_X(x) > 0$ for only countably many x's. If $f(x)$ satisfies properties 1 and 2 in Theorem 2.4, then $f(x)$ is valid as a PMF.*

For a discrete random variable, the PMF contains exactly the same information as the CDF. The difference between the two is how the information is conveyed. Definition 2.3 shows the link from the PMF to the CDF. The link from the CDF to the PMF is given by the next theorem.

Theorem 2.6 (From CDF to PMF) *Assume X is a discrete random variable with CDF $F_X(x)$. Then*

$$f_X(x) = F_X(x) - \lim_{a \to x^-} F_X(a)$$

is the PMF of X.

As Theorem 2.6 shows, the value of $f_X(x)$ equals the size of the jump of $F_X(x)$. In other words, the height of a step of $F_X(x)$ equals $f_X(x)$ in case there is a step at x. Figure 2.3 and the next example illustrate the links between a PMF and a CDF.

Example 2.6 *Consider the Example 2.1 of tossing a coin three times, where X is the number of tails with CDF*

$$F_X(x) = \begin{cases} 0 & \text{if} & x < 0, \\ \frac{1}{8} & \text{if} & 0 \leq x < 1, \\ \frac{4}{8} & \text{if} & 1 \leq x < 2, \\ \frac{7}{8} & \text{if} & 2 \leq x < 3, \\ 1 & \text{if} & 3 \leq x. \end{cases}$$

The PMF is

$$f_X(x) = \begin{cases} \frac{1}{8} & \text{if} & x = 0, \\ \frac{3}{8} & \text{if} & x = 1, \\ \frac{3}{8} & \text{if} & x = 2, \\ \frac{1}{8} & \text{if} & x = 3, \\ 0 & & \text{otherwise.} \end{cases}$$

The PMF $f_X(x)$ and CDF $F_X(x)$ are depicted in Figure 2.3. The CDF (right panel) may be found by summing up the probabilities, as given by the PMF (left panel), i.e. $F_X(a) = \sum_{x \leq a} f_X(x)$, see Definition 2.3. Conversely, the PMF (left panel) may be found by the heights of the jumps of the CDF (right panel), i.e. $f_X(x) = F_X(x) - \lim_{a \to x^-} F_X(a)$, see Theorem 2.6.

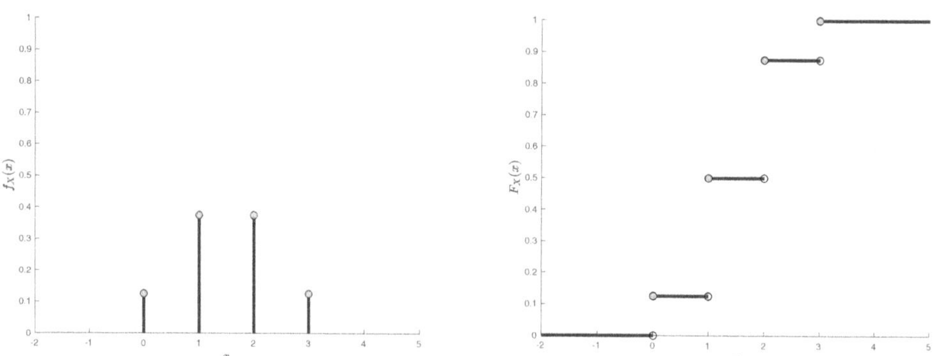

FIGURE 2.3
Relationship between PMF $f_X(x)$ (left panel) and CDF $F_X(x)$ (right panel), see Example 2.6. The CDF (right) may be recovered from the PMF (left) by summing up the values of the PMF, see Definition 2.3. Conversely, the PMF (left) may be recovered from the CDF (right) by reading off the heights of the jumps of the CDF, see Theorem 2.6.

2.3.2   The distribution of a function of a discrete random variable

It can be useful to be able to derive the distribution of a new random variable, which is specified as a function of another random variable with known distribution. In this section, we consider how to do so in the case where X is a discrete random variable. We begin by stating the result in terms of the CDF of the new random variable.

Theorem 2.7 (CDF of a function of a discrete random variable) *Assume X is a discrete random variable with CDF $F_X(x)$ and PMF $f_X(x)$. Let $g()$ be a function and consider the random variable $Y = g(X)$. Then the CDF $F_Y(y)$ of Y is given by*

$$F_Y(y) = \sum_x I\left(g(x) \leq y\right) f_X(x),$$

where $I()$ is the indicator function and "$\sum_x$" denotes the sum over all the x-values for which $f_X(x) > 0$.

Assume, in addition, that $g(x)$ is a continuous one-to-one function with the inverse function $g^{-1}(y)$. If $g(x)$ is increasing, then

$$F_Y(y) = \sum_{x \leq g^{-1}(y)} f_X(x) = F_X\left(g^{-1}(y)\right),$$

whereas if $g(x)$ is decreasing, then

$$F_Y(y) = 1 - \sum_{x \leq g^{-1}(y)} f_X(x) = 1 - F_X\left(g^{-1}(y)\right).$$

The PMF of a new random variable specified as a function of X is given in the next theorem.

Theorem 2.8 (PMF of function of a discrete random variable) *Assume $g(x)$ is a one-to-one function and that the discrete random variable X has the PMF $f_X(x)$. Then the PMF $f_Y(y)$ of the random variable Y given by $Y = g(X)$ is*

$$f_Y(y) = f_X\left(g^{-1}(y)\right),$$

where $g^{-1}(y)$ is the inverse function of $g(x)$.

Example 2.7 *Let X be a discrete random variable and define $Y = 2X + 10$. The function $g(x) = 2x + 10$ is a strictly increasing function, and, thus, also one-to-one. The inverse function can be found by solving*

$$g(x) = 2x + 10 = y,$$

yielding

$$g^{-1}(y) = 0.5y - 5.$$

If $F_X(x)$ and $f_X(x)$ are the CDF and PMF of X, respectively, then

$$F_Y(y) = \sum_{x \leq 0.5y - 5} f_X(x) = F_X(0.5y - 5),$$

and

$$f_Y(y) = f_X(0.5y - 5).$$

Suppose $f_X(x)$ is given as in Example 2.6. Then, in the expression for $f_X(x)$, we replace all occurrences of x with $0.5y - 5$ to obtain $f_Y(y)$. The result is

$$f_Y(y) = f_X(0.5y - 5) = \begin{cases} \frac{1}{8} & \text{if} \quad 0.5y - 5 = 0 \\ \frac{3}{8} & \text{if} \quad 0.5y - 5 = 1 \\ \frac{3}{8} & \text{if} \quad 0.5y - 5 = 2 \\ \frac{1}{8} & \text{if} \quad 0.5y - 5 = 3 \\ 0 & \text{otherwise} \end{cases} = \begin{cases} \frac{1}{8} & \text{if} \quad y = 10, \\ \frac{3}{8} & \text{if} \quad y = 12, \\ \frac{3}{8} & \text{if} \quad y = 14, \\ \frac{1}{8} & \text{if} \quad y = 16, \\ 0 & \text{otherwise.} \end{cases}$$

2.4 Continuous random variables

A continuous random variable handles situations where there is a continuum of possible outcomes. The CDF of a continuous random variable is continuous. This should be contrasted with the CDF of a discrete random variable, where the CDF is a stair function (Definition 2.2). The formal definition of a continuous random variable is given next.

Definition 2.4 (Continuous random variable) *A random variable X is a **continuous random variable** if and only if the CDF $F_X(x)$ is a continuous function.*

Example 2.8 *Consider the CDF from Example 2.3. Since $F_X(x)$ is a CDF and it is a continuous function, then, by Definition 2.4, the random variable X is continuous.*

One particular characteristics of a continuous random variable is that the probability of any outcome $a \in \mathbb{R}$ is zero. This is stated in the next theorem.

Theorem 2.9 (Zero probability of point outcome for continuous random variable) *Assume X is a continuous random variable. Then for any value $x \in \mathbb{R}$,*

$$\Pr(X = x) = 0.$$

At first it may seem strange that any outcome of a continuous random variable X has zero probability, including the one that is realized by the statistical experiment. One way to perceive the result is by thinking about probability in the context of the statistical experiment. The result says that getting the same outcome by re-running the statistical experiment, to the precision of infinitely many digits, is zero.

On a technical note, since $\Pr(X = x) = 0$ for a continuous random variable, we cannot build a general theory of distributions of random variables based on $\Pr(X = x)$. This is the main reason that we use the cumulative probability $\Pr(X \leq x)$ as our starting point, and thus, the CDF as the foundation for discussing the distributions of random variables. This also shows that the probability mass function (Definition 2.3), which is a central concept for a discrete random variable, is not a useful concept for continuous random variables. The next section introduces the analogous concept to a PMF for continuous random variables called a "probability density function".

2.4.1   The probability density function

For a continuous random variable, the CDF $F_X(x)$ contains all the information about the distribution of the random variable X. As previously mentioned, there are often more informative ways to convey certain aspects of the distribution of a random variable. In this section, we show how the *probability density function*, often just called a *density function*, is directly informative about how the probability of outcomes changes around a value x of X.

We start by defining a probability density function.

Definition 2.5 (Probability density function, PDF) *Assume X is a continuous random variable with CDF $F_X(x)$. Then $f_X(x)$ is a **probability density function** (**PDF**) of X if and only if*

$$i)\ f_X(a)\ \geq\ 0,$$
$$ii)\ F_X(a)\ =\ \int_{-\infty}^{a} f_X(x)dx,$$

for all $a \in \mathbb{R}$.

By ii) in Definition 2.5, for a function $f(x)$ to be a PDF, it must integrate to the CDF $F(x)$. Since the PDF is non-negative by i) in Definition 2.5, this means that the CDF $F_X(x)$ equals the area under the PDF $f_X()$, calculated over the interval $(-\infty, x]$. This is illustrated in the left panel of Figure 2.4 for a value of $x = -1$. The area under the density function left of x equals the value of $F(x)$.

A PDF has the following properties.

Theorem 2.10 (Properties of PDF) *Let $f_X(x)$ be a PDF. Then*

1. $f_X(x) \geq 0$ for all $x \in \mathbb{R}$.

2. $\int_{-\infty}^{\infty} f_X(x)dx = 1$.

3. $\int_{a}^{b} f_X(x)dx = F_X(b) - F_X(a)$.

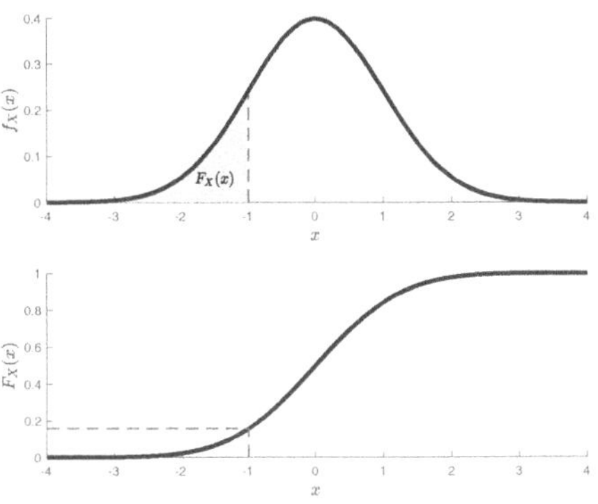
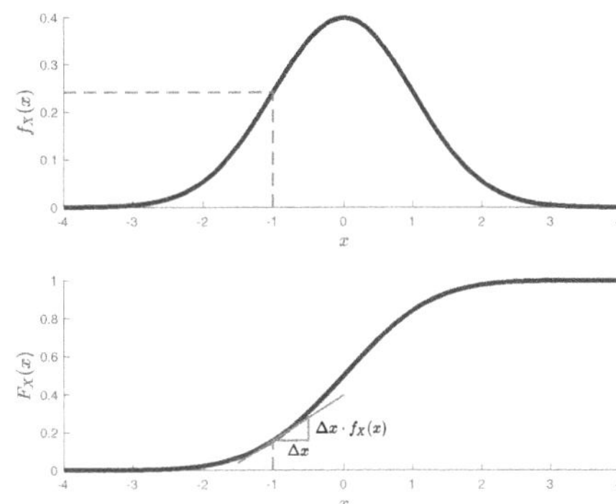

FIGURE 2.4
Relationship between CDF $F_X(x)$ and PDF $f_X(x)$. Left: The value of the CDF $F_X(x)$ at a point $x = -1$ (bottom) equals the area under the PDF $f_X(x)$ up to that point $x = -1$ (top). Right: The value of the PDF $f_X(x)$ at a point $x = -1$ (top) is equal to the slope (i.e. derivative) of the CDF $F_X(x)$ at that point (bottom).

4. $\int_a^\infty f_X(x)dx = 1 - F_X(a)$.

Theorem 2.10 shows how the area under a PDF over an interval gives the probability that the random variable will have an outcome in that interval. For instance, combining the definition of the CDF (Definition 2.1) with Theorem 2.10, we have for any $a \in \mathbb{R}$ that

$$\Pr(X \le a) = \int_{-\infty}^{a} f_X(x)dx$$

and, for $a, b \in \mathbb{R}$ such that $a \le b$,

$$\Pr(a < X \le b) = \int_a^b f_X(x)dx.$$

Notice, for a continuous random variable, $\Pr(a \le X \le b) = \Pr(a < X \le b)$ since $\Pr(X = a) = 0$.

It turns out that properties 1 and 2 in Theorem 2.10 are necessary and sufficient conditions for a function $f(x)$ to be a valid PDF, where we by "valid" mean that it is possible for a continuous random variable to have $f(x)$ as its PDF.

Theorem 2.11 (Constructing a PDF) *Let $f(x)$ be some function. If $f(x)$ satisfies properties 1 and 2 in Theorem 2.10, then $f(x)$ is valid as a PDF.*

Example 2.9 (Example 2.3, continued) *Consider the random variable X in Example 2.3 and define the function*

$$f_X(x) = \begin{cases} 0 & \text{if} & x < 0, \\ 0.5 & \text{if} & 0 \le x < 2, \\ 0 & \text{if} & 2 \le x. \end{cases}$$

We check that $f_X(x)$ is a PDF for X by verifying the two requirements of Definition 2.5. Requirement 1) is satisfied since the function only admits the values 0 and 0.5. To check requirement 2), notice that

$$\int_{-\infty}^{\infty} f_X(x)dx = \int_{-\infty}^{0} f_X(x)dx + \int_0^2 f_X(x)dx + \int_2^\infty f_X(x)dx.$$

Considering each term,

$$\int_{-\infty}^{0} f_X(x)dx = \int_{-\infty}^{0} 0\,dx = 0,$$

$$\int_{0}^{2} f_X(x)dx = \int_{0}^{2} 0.5\,dx = 0.5(2-0) = 1,$$

$$\int_{2}^{\infty} f_X(x)dx = \int_{2}^{\infty} 0\,dx = 0.$$

Therefore,

$$\int_{-\infty}^{\infty} f_X(x)dx = 0 + 1 + 0 = 1,$$

showing that requirement 2) of Definition 2.5 is satisfied. We conclude that $f_X(x)$ is a PDF for X. The PDF and CDF of X are illustrated in the right and left panels of Figure 2.2, respectively.

Example 2.10 *Consider the CDF for a continuous random variable X, given by*

$$F_X(x) = \begin{cases} 0 & \text{if } x < 0, \\ 1 - e^{-2x} & \text{if } x \geq 0. \end{cases}$$

The function

$$f_X(x) = \begin{cases} 0 & \text{if } x < 0, \\ 2e^{-2x} & \text{if } x \geq 0, \end{cases} \tag{2.1}$$

is a PDF for X because, for $a \geq 0$,

$$\int_{-\infty}^{a} f_X(x)dx = \int_{-\infty}^{0} 0\,dx + \int_{0}^{a} 2e^{-2x}dx = 0 + \left(1 - e^{-2a}\right) = F_X(a),$$

and $f_X(x) \geq 0$ for all $x \in \mathbb{R}$. Figure 2.5 illustrates the CDF and PDF of X.

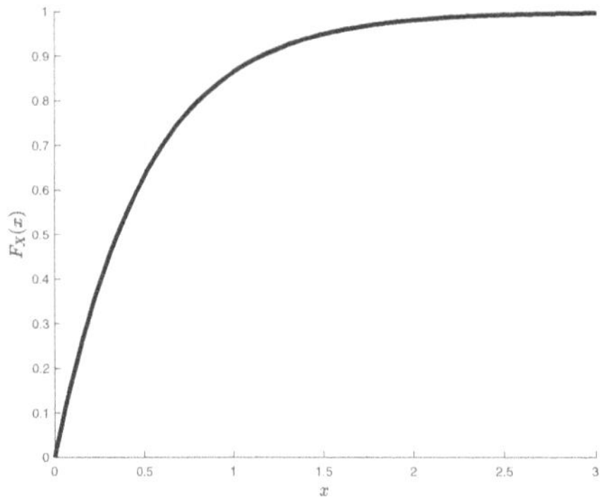
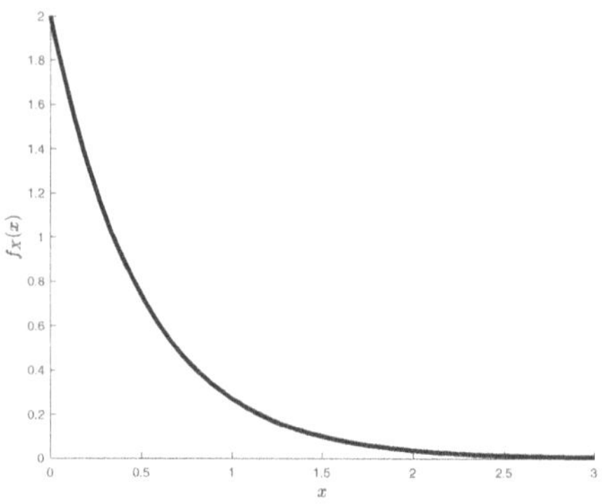

FIGURE 2.5
CDF and PDF from Example 2.10. Left: CDF $F_X(x) = 1 - e^{-2x}$ for $x \geq 0$. Right: PDF $f_X(x) = 2e^{-2x}$ for $x \geq 0$.

There exists many different PDFs for the same random variable. These PDFs do not, however, differ in any important ways because they all lead to the same probability results. The next example illustrates this point.

Example 2.11 (Example 2.10, continued) *The CDF from Example 2.10 also has the function*

$$g(x) = \begin{cases} 0 & if \quad x < 0, \\ 50 & if \quad x = 0, \\ 2e^{-2x} & if \quad x > 0, \end{cases}$$

as a PDF for X. The only difference between the PDF $f(x)$ in Example 2.10 and $g(x)$ is at $x = 0$. Since this difference is only at a point, and the integral of a function at a point is 0, we have

$$\int_{-\infty}^{a} g(x)dx = \int_{-\infty}^{0} 0dx + \int_{0}^{0} 50dx + \int_{0}^{a} 2e^{-2x}dx = 0 + 0 + \left(1 - e^{-2x}\right) = F_X(a),$$

showing that the particular value of the PDF at that point does not matter for the probabilities.

In many cases, we can derive a PDF for a continuous random variable by taking the derivative of the CDF. This is stated in the next theorem.

Theorem 2.12 (From CDF to PDF) *Let $F_X(x)$ be the CDF of a continuous random variable X. Assume $F_X(x)$ is a differentiable function at all but a finite number of x's. Set*

$$f_X(x) = \frac{dF_X(x)}{dx}$$

at all points where $F_X(x)$ is differentiable. At the non-differentiable points, any non-negative value assigned to $f_X(x)$ at those points makes $f_X(x)$ a density function for X.

The theorem provides us with a way of deriving the PDF if we know the CDF. This relationship from CDF to PDF is illustrated in the right panel of Figure 2.4.

Example 2.12 (Example 2.10, continued) *The CDF*

$$F_X(x) = \begin{cases} 0 & if \quad x < 0, \\ 1 - e^{-2x} & if \quad x \geq 0, \end{cases}$$

is differentiable at all points except at $x = 0$. For $x > 0$, define the function $h(x)$ as

$$h(x) = \frac{dF_X(x)}{dx} = \frac{d\left(1 - e^{-2x}\right)}{dx} = 2e^{-2x},$$

while, for $x < 0$,

$$h(x) = \frac{dF_X(x)}{dx} = \frac{d0}{dx} = 0.$$

According to Theorem 2.12, we can set $h(0)$ equal to any non-negative real number and the resulting function $h(x)$ will be a PDF for the random variable with CDF $F_X(x)$. In particular, if we choose $h(0) = 2$, then $h(x)$ equals the function $f(x)$ given in Equation (2.1).

On a technical note, there are continuous random variables which do not have a density function. That is, there are continuous CDFs $F_X(x)$ for which no function $f(x)$ satisfies Definition 2.5. In general, a function $F(x)$ that can be written as an integral of a function $f(x)$ is called *absolutely continuous* (see Definition B.20 in Chapter B of the appendix).

Therefore, if X is a continuous random variable with an absolutely continuous CDF $F_X(x)$, it has a PDF $f_X(x)$, and the relationships between the two are

$$f_X(x) = \frac{dF_X(x)}{dx},$$

for all x where $F_X(x)$ is differentiable, and

$$F_X(a) = \int_{-\infty}^{a} f_X(x)dx.$$

A PDF $f(x)$ can be interpreted as the intensity of the probability of an outcome in a small neighborhood of x. For a continuous random variable, the probability of a particular outcome is 0, that is, $\Pr(X = x) = 0$ for any $x \in \mathbb{R}$. The PDF $f(x)$ at that point may not be equal to 0, however. Hence, the PDF cannot be interpreted as a probability. Instead it can be interpreted as the probability of an outcome in a small neighborhood around x, that is

$$\Pr\left(X \in \left[x - \frac{\Delta x}{2}, x + \frac{\Delta x}{2}\right]\right) = \underbrace{\Pr\left(x - \frac{\Delta x}{2} < X < x + \frac{\Delta x}{2}\right)}_{=F_X\left(x + \frac{\Delta x}{2}\right) - F_X\left(x - \frac{\Delta x}{2}\right)} \approx f_X(x) \cdot \Delta x,$$

where Δx is a small number and "$\approx$" means approximately. This statement follows from Theorem 2.12 and an approximation of a derivative $\frac{dF(x)}{dx}$, see, e.g. Definition B.15 in Chapter B of the appendix.

At first sight, it may appear confusing that the same notation, $f(x)$, is used for the PMF of a discrete random variable and the PDF of a continuous random variable. However, it will usually be immediately clear from the context whether a given function is a PMF or PDF. Indeed, the former is always associated with discrete random variables and the latter is always associated with continuous random variables.

2.4.2   The distribution of a function of a continuous random variable

As discussed with discrete random variables, it may be useful to be able to derive the distribution of a random variable, which is defined as a function of another random variable. In this section, we consider how to do so in the case where X is a continuous random variable. We begin by stating the result in terms of the CDF.

Theorem 2.13 (CDF of a function of a continuous random variable) *Assume X is a continuous random variable with CDF $F_X(x)$ and PDF $f_X(x)$. Let $g()$ be a function and consider the random variable $Y = g(X)$. Then the CDF $F_Y(y)$ of Y is given by*

$$F_Y(y) = \int_{-\infty}^{\infty} I\left(g(x) \leq y\right) f_X(x)dx,$$

where $I()$ is the indicator function.

Assume, in addition, that $g(x)$ is a continuous one-to-one function with the inverse function $g^{-1}(y)$. If $g(x)$ is increasing, then

$$F_Y(y) = \int_{-\infty}^{g^{-1}(y)} f_X(x)dx = F_X\left(g^{-1}(y)\right),$$

whereas if $g(x)$ is decreasing, then

$$F_Y(y) = 1 - \int_{-\infty}^{g^{-1}(y)} f_X(x)dx = 1 - F_X\left(g^{-1}(y)\right).$$

The PDF of a function of a continuous random variable X is given in the next theorem, provided the function of X is differentiable.

Theorem 2.14 (PDF of function of a continuous random variable) *Assume $g(x)$ is a differentiable one-to-one function and that the continuous random variable X has the PDF $f_X(x)$. Then the PDF $f_Y(y)$ of the random variable Y given by $Y = g(X)$ is*

$$f_Y(y) = f_X\left(g^{-1}(y)\right) \cdot \left| \frac{dg^{-1}(y)}{dy} \right|,$$

where $g^{-1}(y)$ is the inverse function of $g(x)$ and $|\cdot|$ is the absolute value.

Example 2.13 *Let X be a continuous random variable and define $Y = 2X + 10$. The function $g(x) = 2x + 10$ is a strictly increasing function, and, thus, also one-to-one. The inverse function is found by solving*

$$g(x) = 2x + 10 = y,$$

yielding

$$g^{-1}(y) = 0.5y - 5.$$

If $F_X(x)$ and $f_X(x)$ are the CDF and PDF of X, respectively, then

$$F_Y(y) = \int_{-\infty}^{0.5y-5} f_X(x)dx = F_X(0.5y - 5),$$

and

$$f_Y(y) = f_X(0.5y - 5) \cdot \frac{dg^{-1}(y)}{dy} = f_X(0.5y - 5) \cdot \frac{d(0.5y - 5)}{dy} = f_X(0.5y - 5) \cdot 0.5.$$

Suppose $f_X(x)$ and $F_X(x)$ are given as in Example 2.10. Then in the expression for $f_X(x)$, we replace all occurrences of x with $0.5y - 5$ to obtain $f_Y(y)$ and, similarly, for $F_Y(y)$. The results are

$$F_Y(y) = F_X(0.5y-5) = \begin{cases} 0 & \text{if } 0.5y - 5 < 0 \\ 1 - e^{-2(0.5y-5)} & \text{if } 0.5y - 5 \geq 0 \end{cases} = \begin{cases} 0 & \text{if } y < 10, \\ 1 - e^{-y+10} & \text{if } y \geq 10. \end{cases}$$

The function

$$f_Y(y) = f_X(0.5y-5) \cdot 0.5 = 0.5 \cdot \begin{cases} 0 & \text{if } 0.5y - 5 < 0 \\ 2e^{-2(0.5y-5)} & \text{if } 0.5y - 5 \geq 0 \end{cases} = \begin{cases} 0 & \text{if } y < 10, \\ e^{-y+10} & \text{if } y \geq 10. \end{cases}$$

2.5 Mixed random variables

There are empirically relevant situations where the possible outcomes of a statistical experiment both has a discrete and a continuous nature. The CDF for such a situation does not satisfy the requirement for X to be neither discrete (Definition 2.2) nor continuous (Definition 2.4). For such a CDF, the corresponding random variable is said to be of *mixed type*. It turns out that the CDF for a mixed type random variable X can be written as a weighted sum of two CDFs, where one is a CDF for a continuous random variable and the other is a CDF for a discrete random variable. This is stated in the next theorem.

Theorem 2.15 (Decomposing a CDF) *For any CDF $F_X(x)$, there exists a unique CDF $F_X^d(x)$ for a discrete random variable, a unique CDF $F_X^c(x)$ for a continuous random variable, and an $\alpha \in [0,1]$, such that, for all $x \in \mathbb{R}$,*

$$F_X(x) = \alpha \cdot F_X^d(x) + (1 - \alpha) \cdot F_X^c(x).$$

Note that Theorem 2.15 applies to random variables of any type. In case X is discrete, then $\alpha = 1$ and $F_X(x) = F_X^d(x)$ and, similarly, if X is continuous, then $\alpha = 0$ and $F_X(x) = F_X^c(x)$. The values of α in between 0 and 1 leads to the definition of a mixed type random variable, given next.

Definition 2.6 (Mixed type random variable) *A random variable X is called a **mixed type random variable** if and only if the CDF has a decomposition given in Theorem 2.15 with an α satisfying $0 < \alpha < 1$.*

Example 2.14 *Suppose 2 liters of water is offered to each individual after a run. Each individual may drink any amount corresponding to nothing, i.e. exactly 0, in between 0 and 2, or the full amount, i.e. exactly 2. Assume 10% of the runners drink nothing and 30% of them drink the whole 2 liters. These outcomes can be modeled as a discrete random variable. The remaining 60% of runners drink some amount between 0 and 2 liters. Since we cannot rule out any specific amount between 0 and 2 liters, we choose to model the amount drunk by these runners as a continuous random variable. For example, suppose*

$$F_X^c(x) = \begin{cases} 0 & if & x < 0, \\ 0.5 \cdot x & if & 0 \le x < 2, \\ 1 & if & 2 \le x, \end{cases}$$

and

$$F_X^d(x) = \begin{cases} 0 & if & x < 0, \\ 0.25 & if & 0 \le x < 2, \\ 1 & if & 2 \le x. \end{cases}$$

Let X be the random variable denoting the amount of water drunk by a randomly chosen runner. Then

$$F_X(x) = 0.4 \cdot F_X^d(x) + 0.6 \cdot F_X^c(x).$$

For example,

$$\begin{aligned} \Pr(X \le 2) &= F_X(2) = 0.4 F_X^d(2) + 0.6 F_X^c(2) = 0.4 \cdot 1 + 0.6 \cdot 1 = 1, \\ \Pr(X \le 1) &= F_X(1) = 0.4 F_X^d(1) + 0.6 F_X^c(1) = 0.4 \cdot 0.25 + 0.6 \cdot 0.5 = 0.4, \\ \Pr(X \le 0) &= F_X(0) = 0.4 F_X^d(0) + 0.6 F_X^c(0) = 0.4 \cdot 0.25 + 0.6 \cdot 0 = 0.1. \end{aligned}$$

Hence, there is a 0.4 probability that an individual drawn from this population has drunk no more than 1 liter of water and a 0.1 probability that the individual has drunk nothing (since one cannot drink a negative amount).

2.6   Proofs

Proof of Theorem 2.1   *The proof of Theorem 2.1 is left as an exercise to the reader (Problem 2.7.8).*

Proof of Theorem 2.2 *Let $F(x)$ be a function satisfying conditions 1. to 4. of Theorem 2.1. We need to show that there exists a random variable X with F as its CDF, i.e. such that (Definition 2.1)*

$$\Pr(X \leq x) = F(x)$$

for all $x \in \mathbb{R}$. Our proof will be a so-called constructive *proof, meaning that we will construct a random variable X such that the function F is its CDF. To this end, consider the generalized inverse of F, $F^{-1} : (0,1) \to \mathbb{R}$, given by*

$$F^{-1}(u) = \inf\{x \in \mathbb{R} : F(x) \geq u\}.$$

Note that this function exists due to conditions 1, 2, and 4 of Theorem 2.1. Now, let U be a continuous random variable with CDF $F_U(x) = x$ for $x \in [0,1]$ and $F_U(x) = 0$ otherwise. Such a random variable U is known as a uniform random variable (see Definition 8.6 in Chapter 8). Define the function X as

$$X = F^{-1}(U).$$

Conditions 1 to 4 ensure that X is a random variable (Definition 1.6). Let $x \in \mathbb{R}$ and $u \in (0,1)$. Conditions 2. and 4. ensure that

$$F^{-1}(u) \leq x \iff u \leq F(x).$$

Therefore,

$$\Pr(X \leq x) = \Pr(F^{-1}(U) \leq x) = \Pr(U \leq F(x)) = F(x),$$

which is what we wanted to show.

Proof of Theorem 2.4 *The proof of Theorem 2.4 is left as an exercise to the reader (Problem 2.7.9).*

Proof of Theorem 2.3 *Property 7 and 8 of Theorem 2.1 give that*

$$\Pr(X = a) = F_X(a) - \Pr(X < a) = F_X(a) - \lim_{x \to a^-} F_X(x).$$

By Definition 2.3,

$$F_X(a) = \sum_{x \leq a} f_X(x) = \sum_{x < a} f_X(x) + f_X(a).$$

Since

$$\lim_{x \to a^-} F_X(x) = F_X(a^-) = \sum_{x < a} f_X(x),$$

then

$$\Pr(X = a) = \sum_{x < a} f_X(x) + f_X(a) - \sum_{x < a} f_X(x) = f_X(a).$$

Proof of Theorem 2.5 *Define the function F_X via*

$$F_X(a) = \sum_{x \leq a} f_X(x) = \sum_{x} f_X(x) I(x \leq a).$$

The conditions of the theorem imply that conditions 1 to 4 of Theorem 2.1 hold. By Theorem 2.2 there exists a random variable X with $F_X()$ as its CDF. Further, the conditions imply that $F_X()$ may be written on the form given in Definition 2.2, which shows that X is a discrete random variable.

Proof of Theorem 2.6   *This follows readily from Definitions 2.2 and 2.3.*

Proof of Theorem 2.7   *Note that since X is a discrete random variable, then Y will be a discrete random variable. Further, Y has the PMF*

$$f_Y(a) = \Pr(Y = a) = \Pr(g(X) = a) = \sum_x f_X(x) I(g(x) = a).$$

Using this, we deduce that the CDF of Y is

$$F_Y(a) = \Pr(Y \le a) = \sum_{y \le a} f_Y(y) = \sum_{y \le a} \sum_x f_X(x) I(g(x) = y) = \sum_x f_X(x) \sum_{y \le a} I(g(x) = y).$$

Since $g()$ is a function then, for a fixed x, we will at most have $I(g(x) = y) = 1$ for one value of y such that $y \le a$. Hence $\sum_{y \le a} I(g(x) = y) = I(g(x) \le a)$ and

$$F_Y(a) = \Pr(Y \le a) = I(g(x) \le a),$$

which is what we wanted to show.

    Suppose now that $g(x)$ is continuous and one-to-one. Assume that $g(x)$ is increasing (the decreasing case follows similarly). Then the inverse $g^{-1}(y)$ of $g()$ exists and we may write

$$F_Y(a) = \Pr(Y \le a) = \Pr(g(X) \le a) = \Pr(X \le g^{-1}(a)) = \sum_{x \le g^{-1}(a)} f_X(x)$$

by definition of the PMF of X (Definition 2.3). This concludes the proof.

Proof of Theorem 2.8   *By the properties of the PMF,*

$$f_Y(y) = \Pr(Y = y) = \Pr(g(X) = y) = \Pr(X = g^{-1}(y)) = \sum_{x = g^{-1}(y)} f_X(x),$$

as we wanted to show.

Proof of Theorem 2.9   *Let $a \in \mathbb{R}$. It follows from property 7 in Theorem 2.1 that*

$$\Pr(X = a) = F_X(a) - \Pr(X < a), \tag{2.2}$$

and, from property 8 in Theorem 2.1,

$$\Pr(X < a) = \lim_{x \to a^-} F_X(x).$$

Since a continuous random variable is defined as having a continuous CDF $F_X(x)$, then

$$\lim_{x \to a^-} F_X(x) = F_X(a).$$

Insert this into (2.2) for $\Pr(X < a)$ *to get*

$$\Pr(X = a) = F_X(a) - \lim_{x \to a^-} F_X(x) = F_X(a) - F_X(a) = 0.$$

Proof of Theorem 2.10 *The proof of Theorem 2.10 is left as an exercise to the reader (Problem 2.7.10).*

Proof of Theorem 2.11 *The proof follows similar arguments as those in the proof of Theorem 2.5. In particular, define the function F_X via*

$$F_X(a) = \int_{-\infty}^{a} f_X(x)dx.$$

for $a \in \mathbb{R}$. Note that this function is continuous by construction. The conditions of the theorem imply that conditions 1 to 4 of Theorem 2.1 hold. By Theorem 2.2 there exists a random variable X with $F_X()$ as its CDF. Further, since $F_X()$ is continuous, the random variable X is continuous by Definition 2.4.

Proof of Theorem 2.12 *This follows from the so-called Fundamental Theorem of Calculus (see Theorem B.8 in Chapter B of the appendix).*

Proof of Theorem 2.13 *This follows similar arguments as those in the proof of Theorem 2.7. We omit the details.*

Proof of Theorem 2.14 *Since the function $g(x)$ is differentiable and one-to-one, it is monotone (i.e. either decreasing or increasing) and thus the same goes for its inverse $g^{-1}(y)$. Suppose first that $g^{-1}(y)$ is increasing. By the properties of the CDF, we have that*

$$F_Y(y) = \Pr(Y \leq y) = \Pr(g(X) \leq y) = \Pr(X \leq g^{-1}(y)) = F_X(g^{-1}(y)).$$

From Theorem 2.12 we know that $F_Y'(y) = f_Y(y)$, so

$$f_Y(y) = F_Y'(y) = \frac{d}{dy}F_X(g^{-1}(y)) = F_X'^{-1}(y))\frac{d}{dy}g^{-1}(y) = f_X(g^{-1}(y))\frac{d}{dy}g^{-1}(y),$$

where we used the chain rule of differentiation and that $F_X'(x) = f_X(x)$. Suppose now that $g^{-1}(y)$ is decreasing. This time, we have

$$F_Y(y) = \Pr(Y \leq y) = \Pr(g(X) \leq y) = \Pr(X \geq g^{-1}(y)) = 1 - F_X(g^{-1}(y)).$$

And

$$f_Y(y) = F_Y'(y) = \frac{d}{dy}\left(1 - F_X(g^{-1}(y))\right) = -F_X'^{-1}(y))\frac{d}{dy}g^{-1}(y) = -f_X(g^{-1}(y))\frac{d}{dy}g^{-1}(y),$$

Combining these two results, whether or not $g^{-1}(y)$ is increasing or decreasing, we may write

$$f_Y(y) = f_X(g^{-1}(y))\left|\frac{d}{dy}g^{-1}(y)\right|,$$

which is what we wanted to show.

Proof of Theorem 2.15 *Let $F_X(x)$ be the CDF of the random variable X. If X is discrete, the results follows with $\alpha = 1$ and $F_X^d(x) = F_X(x)$. If X is continuous, the results follows with $\alpha = 0$ and $F_X^c(x) = F_X(x)$. Suppose therefore that X is of mixed type, i.e. the CDF $F_X(x)$ has continuous parts and discontinuities. Since X is a random variable (Definition 1.6), the function $F_X(x)$ may only have countably many points of discontinuity. Let the function $g_X^d(x)$ denote the size of these discontinuities, which by the right-continuity of $F_X(x)$ (Theorem 2.1) is given by,*

$$g_X^d(x) = F_X(x) - \lim_{a \to x-} F_X(x).$$

Define

$$\tilde{F}_X^d(x) = \sum_{a \le x : g_X^d(a) > 0} g_X^d(a).$$

Further, define

$$\tilde{F}_X^c(x) = F_X(x) - \tilde{F}_X^d(x).$$

Here, $\tilde{F}_X^d(x)$ represents the "discrete part" of $F_X(x)$, i.e. the sum of the jumps of $F_X()$ from $-\infty$ to x, while $\tilde{F}_X^c(x)$ represents the "continuous part" of $F_X()$, i.e. the part that is left of $F_X(x)$ after the jumps have been removed. By the above, we can write

$$F_X(x) = \tilde{F}_X^d(x) + \tilde{F}_X^c(x). \tag{2.3}$$

The functions $\tilde{F}_X^c(x)$ and $\tilde{F}_X^d(x)$ are not proper CDFs because their limit as x tends to infinity is not 1 (see Theorem 2.1). Define the constant $\alpha \in (0,1)$ by

$$\alpha = \sum_{x : g_X^d(x) > 0} g_X^d(x) = \lim_{x \to \infty} F_X^d(x).$$

Note that, since $F_X(x)$ is a CDF, then (Theorem 2.1)

$$\lim_{x \to \infty} \tilde{F}_X^c(x) = \lim_{x \to \infty} \left(F_X(x) - \tilde{F}_X^d(x) \right) = 1 - \alpha.$$

The functions

$$\begin{aligned}
F_X^d(x) &= \alpha^{-1} \tilde{F}_X^d(x), \\
F_X^c(x) &= (1-\alpha)^{-1} \tilde{F}_X^c(x),
\end{aligned}$$

are proper CDFs. Using these and (2.3), we may write

$$F_X(x) = \alpha F_X^d(x) + (1-\alpha) F_X^c(x).$$

Further, we see from the constructions of $F_X^c(x)$ and $F_X^d(x)$ that they are unique. This completes the proof.

2.7 Exercises

Problem 2.7.1 *Let*

$$F_X(x) = \begin{cases} 0 & \text{if} & x < 0, \\ 0.7 & \text{if} & 0 \le x < 5, \\ 0.9 & \text{if} & 5 \le x < 10, \\ 1 & \text{if} & x \ge 10. \end{cases}$$

1. *Sketch the graph of $F_X(x)$.*

2. *Prove that $F_X(x)$ is a valid CDF.*

3. *Calculate*

$$\Pr(X \le 5), \ \Pr(X = 5), \ \Pr(X < 5), \ \Pr(X \ge 5), \ \text{and} \ \Pr(X > 5).$$

4. *Explain how a PMF $f_X(x)$ can be derived from $F_X(x)$, and derive $f_X(x)$. Sketch the PMF.*

Problem 2.7.2 *Consider the function $f(x)$ given by $f(x) = \frac{1}{2^x}$ for $x = 1, 2, 3, \ldots$.*

1. *Show that $f(x)$ is a valid PMF. (Hint: You might need to use the expression for the geometric series.)*

2. *Let X be a random variable with probability function $f(x)$. Is X a discrete or continuous random variable? Justify your answer.*

3. *Compute the CDF $F(x)$ for $x = 1, 2, 3, \ldots$.*

4. *Calculate $\Pr(X = 5)$, $\Pr(X \le 5)$, $\Pr(X \ge 2)$, $\Pr(5 < X \le 10)$, and $\Pr(5 \le X \le 10)$.*

Problem 2.7.3 *Consider the functions*

$$
f_X(x) \;=\;
\begin{cases}
\frac{1}{4}x & \text{if } \ 0 < x < 2, \\
-\frac{1}{4}x + 1 & \text{if } \ 2 \le x \le 4, \\
0 & \text{otherwise,}
\end{cases}
$$

$$
F_X(x) \;=\;
\begin{cases}
0 & \text{if} \quad x < 0, \\
\frac{1}{8}x^2 & \text{if} \ \ 0 \le x < 2, \\
-\frac{1}{8}x^2 + x - 1 & \text{if} \ \ 2 \le x < 4, \\
1 & \text{if} \quad x \ge 4.
\end{cases}
$$

1. *Sketch the graphs of $f_X(x)$ and $F_X(x)$.*

2. *Prove that $f_X(x)$ is a valid PDF.*

3. *Prove that $F_X(x)$ is a valid CDF.*

4. *Argue that $f_X(x)$ and $F_X(x)$ are the PDF/CDF of the same random variable X.*

5. *Let X be the annual return (in kr) on a share. The CDF of X is $F_X(x)$. What is the probability to obtain a return on a share of at least 1 kr? Illustrate your result in the sketches of the graphs of $f_X(x)$ and $F_X(x)$.*

6. *Let the annual administrative cost be 1.5 kr per share. If someone owns 4 shares, what is the probability that the net-return (return minus cost) is at least 1 kr?*

   Hint: Let the random variable Y be the net-return on the 4 shares. Then $Y = 4 \cdot X - 4 \cdot 1.5$. Calculate $\Pr(Y \ge 1)$ by rewriting it to a probability statement of X

Problem 2.7.4 *Let $F_X(x)$ be the CDF of a random variable X, given by $F_X(x) = \sum_{j=1}^{J} b_j I(a_j \le x)$.*

1. *Is X a discrete or continuous random variable? Justify your answer.*

2. *Show that the PMF of X is given by*

$$f_X(x) = \begin{cases} b_j & \text{if } x = a_j, \ j = 1, \ldots, J, \\ 0 & \text{otherwise.} \end{cases}$$

Problem 2.7.5 *Let*

$$F_X(x) = \begin{cases} 0 & \text{if} & x < 1, \\ -\frac{1}{8}x^2 + x - \frac{7}{8} & \text{if} & 1 \le x < 3, \\ 1 & \text{if} & x \ge 3. \end{cases}$$

1. *Sketch the graph of $F_X(x)$.*

2. *Prove that $F_X(x)$ is a valid CDF.*

3. *Calculate the following:*

$$\begin{aligned} \Pr(X &\le 2), \\ \Pr(X &= 2), \\ \Pr(X &< 2), \\ \Pr(X &\ge 2), \\ \Pr(1 &\le X \le 2), \\ \Pr(2 &\le X \le 3). \end{aligned}$$

4. *Explain why a PDF $f_X(x)$ can be derived from $F_X(x)$ using differentiation, and derive $f_X(x)$. Sketch the PDF.*

5. *Show how the PDF $f_X(x)$ you just derived can be used to derive the CDF $F_X(x)$ using integration.*

Problem 2.7.6 *Consider the function*

$$f(x) = \frac{3}{8}x^2, \qquad x \in (0, 2].$$

1. *Show that $f(x)$ is a valid PDF.*

Let the continuous random variable X have PDF $f(x)$.

2. *What is the probability that an outcome of X takes the value $\sqrt{2}$?*

3. *Calculate the CDF of X.*

4. *Calculate the probabilities $\Pr\left(\frac{1}{2} < X < \frac{3}{2}\right)$ and $\Pr\left(X > \frac{3}{2}\right)$.*

Problem 2.7.7 *Consider the function*

$$f(x) = \begin{cases} 1 & \text{if } x \in [0, 1], \\ 0 & \text{otherwise.} \end{cases}$$

1. *Show that $f(x)$ is a valid PDF.*

2. *Let X be a continuous random variable with PDF $f(x)$. Let $Y = -\ln X$. Derive the PDF of Y. (Hint: Use Theorem 2.14.)*

Problem 2.7.8 *Prove Theorem 2.1.*

Problem 2.7.9 *Prove Theorem 2.4.*

Problem 2.7.10 *Prove Theorem 2.10.*

3

Joint distribution of random variables

3.1 Introduction

Investigating relationships lies at the core of empirical studies. When uncertainty is involved, the relationships can be characterized using probability distributions and functions of those distributions. The basic building block when considering probabilistic relationships is the *joint distribution*. We first consider joint distributions of events. Studying events gives a basic interpretation of joint probabilities. In practice, relationships are more complicated than can be modeled by single events and therefore we use random variables for the subsequent analysis. Different techniques must be used, depending on the type of random variables. Therefore, we first consider discrete random variables followed by continuous random variables. We end the discussion with a situation where both discrete and continuous random variables are involved.

Modeling distributions of multiple random variables is done using functions of multiple variables. For readers not familiar with functions of several variables, Section B.3 in Chapter B of the appendix provides the necessary background detail. Furthermore, knowledge of partial differentiation, covered in Section B.3.1 in the appendix, will be needed in some of the chapters to come.

3.2 The joint event probability

To give a basic intuition, we first consider the joint probability in the context of two events. The joint probability of two events is the probability that both events occur.

Let Ω be the sample space and consider two events $A, B \subseteq \Omega$. Based on these two events, four mutually exclusive events can be constructed by taking the four intersections

$$A \cap B, \ A \cap B^c, \ A^c \cap B \text{ and, } A^c \cap B^c, \tag{3.1}$$

where A^c and B^c are the complements of A and B, respectively. These joint events are illustrated in the Venn diagram in Figure 3.1.

Example 3.1 *Let A be the event that the current coach of a sports team is replaced by a new coach, and let event B be the event that a team wins the next game. Then $A \cap B$ is the event that the coach is replaced and the team wins the next game, $A \cap B^c$ is the event coach is replaced and the team does not win the next game, $A^c \cap B$ is the event that a coach is not replaced and the team wins the next game, and $A^c \cap B^c$ is the event that a coach is not replaced and the team does not win the next game.*

DOI: 10.1201/9781003591191-3

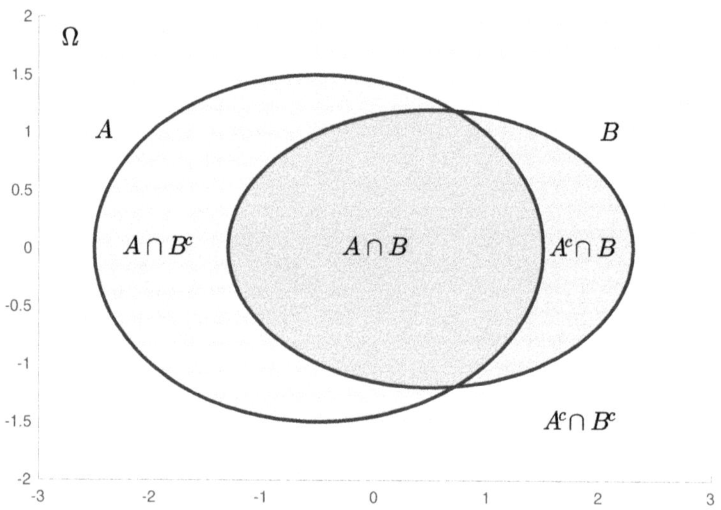

FIGURE 3.1
Venn diagram illustrating the sample space Ω (entire space), the events A (left ellipse) and
B (right ellipse), and the four joint events $A \cap B$, $A \cap B^c$, $A^c \cap B$, and $A^c \cap B^c$.

Since the four events in (3.1) are mutually exclusive, and that one, and only one, of
them happens (see Problem 3.8.1), their probabilities sum to 1 according to Definition 1.4:

$$P\left(A \cap B\right) + P\left(A \cap B^c\right) + P\left(A^c \cap B\right) + P\left(A^c \cap B^c\right) = 1.$$

Each of these probabilities are called *joint event probabilities* because they are the probabilities of the intersection of two events.

Suppose we know the joint event probabilities and would like to know the probability of
a particular event. For example, suppose we would like to know the probability of the event
A. Then we can add the two joint event probabilities in which A happens, namely $A \cap B$
and $A \cap B^c$, according to the law of total probability (Theorem 1.2). That is,

$$P\left(A\right) = P\left(A \cap B\right) + P\left(A \cap B^c\right).$$

Similarly, if we want the probability of B, then

$$P\left(B\right) = P\left(B \cap A\right) + P\left(B \cap A^c\right).$$

Example 3.2 (Example 3.1, continued) *Suppose the joint probabilities are*

$$\begin{aligned}
P\left(A \cap B\right) &= 0.1, \\
P\left(A \cap B^c\right) &= 0.2, \\
P\left(A^c \cap B\right) &= 0.4, \\
P\left(A^c \cap B^c\right) &= 0.3.
\end{aligned}$$

Then the probability the coach being replaced is

$$P\left(A\right) = P\left(A \cap B\right) + P\left(A \cap B^c\right) = 0.1 + 0.2 = 0.3.$$

This includes both when the team wins and when the team does not win.

The probability the team wins the next game is

$$P\left(B\right) = P\left(A \cap B\right) + P\left(A^c \cap B\right) = 0.1 + 0.4 = 0.5.$$

This includes both when the coach is replaced and when the coach is not replaced.

In the context of joint event probabilities, we sometimes refer to the probability of one event as a *marginal event probability*. For instance, $P\left(A \cap B\right)$ is a joint event probability and $P\left(A\right)$ is a marginal event probability.

3.3 The joint cumulative distribution function

We extend the concept of a joint event probability to multiple events using random variables. We start by extending the cumulative distribution function for one random variable to a cumulative distribution function for multiple random variables. As discussed in Chapter 2, considering the cumulative distribution function, instead of the probability mass function or probability density function, has the advantage that it is valid no matter the type of random variables involved.

Suppose we want to know the probability that both $X \leq 3$ and $Y \leq 2$. Formally, we want to know the probability of the joint event $\{X \leq 3\} \cap \{Y \leq 2\}$. It is denoted

$$\Pr(X \leq 3, Y \leq 2),$$

where the "," in the expression means "AND". For example, X could be the number of questions asked in parliament by party 1 to the Prime Minister and Y the number of questions asked by party 2. Hence, this is the probability that party 1 at most asks 3 questions and party 2 at most asks 2 questions. Notice the difference to, say, $\Pr(X \leq 3)$. This is the probability that party 1 asks at most 3 questions however many questions party 2 asks.

The above can be written in a more general form. Let x and y be two real numbers. Then

$$\Pr(X \leq x, Y \leq y)$$

is the probability of the joint event $\{X \leq x\} \cap \{Y \leq y\}$, i.e. that both X is at most x and Y is at most y. We can consider this probability as a function of x and y. To make this explicit, define a function $F_{X,Y}$ of two inputs x and y, by

$$F_{X,Y}(x,y) = \Pr(X \leq x, Y \leq y).$$

Hence, $F_{X,Y}(3,2)$ gives the probability that party 1 at most asks 3 questions and party 2 at most asks 2 questions. This function is called the *joint cumulative distribution function* of X and Y. Next, it is defined for the general case of k random variables.

Definition 3.1 (Joint CDF) *Let $X_1, \ldots, X_k$ be k random variables. Their **joint cumulative distribution function** (**joint CDF**) $F_{X_1,\ldots,X_k}(x_1,\ldots,x_k)$ is given by*

$$F_{X_1,\ldots,X_k}(x_1,\ldots,x_k) = \Pr(X_1 \leq x_1, X_2 \leq x_2, \ldots, X_k \leq x_k)$$

for all $x_1, x_2, \ldots, x_k \in \mathbb{R}$.

The subscripts on F will often be omitted when it is clear as to which random variables the CDF refers.

We denote a distribution of multiple random variables, e.g. represented by their joint CDF, as a *multivariate distribution*. In case there are two random variables, we may also say they have a *bivariate distribution*. If there is only one random variable, as in the previous chapter, we may also refer to such a distribution as a *univariate distribution*.

As already mentioned, an advantage of representing a distribution using a joint CDF is that it is valid no matter the type of random variables. For example, X_1 and X_3 are allowed to be discrete, X_2 to be continuous, and X_4 of the mixed type..

It is often useful and informative to represent a distribution differently, e.g. using probability mass functions when all random variables are discrete and using probability density functions when all random variables are continuous. These functions are discussed in the following sections, including how they relate to the joint CDF.

3.4   The joint distribution of discrete random variables

In this section, we consider the case where all the random variables are discrete. All the concepts can be introduced and discussed with two random variables only. In the last subsection, the extension to multiple random variables is described.

3.4.1   The joint probability mass function

When all random variables are discrete, it is meaningful to consider the probability of particular outcomes of the random variables. For example, in case X and Y are number of questions asked by party 1 and 2, respectively, both X and Y are discrete random variables. Then it is meaningful to calculate, say, the probability that party 1 asks precisely 3 questions and party 2 asks precisely 2 questions. This is denoted

$$\Pr(X = 3, Y = 2).$$

Such a probability can be considered a function $f(x, y)$ of two inputs x and y. This gives rise to the definition of a *joint probability mass function*.

Definition 3.2 (Joint PMF, two random variables) *Let X and Y be two discrete random variables. Then $f_{X,Y}(x, y)$ is the **joint probability mass function (joint PMF)** of X and Y if and only if*

$$f_{X,Y}(x, y) = \Pr(X = x, Y = y),$$

for all $x, y \in \mathbb{R}$.

Example 3.3 *Let X indicate a "low" or "high" grade point average (GPA) taken over all students in a randomly chosen school and let Y be the ownership form of the school. They are defined as*

$$X = \begin{cases} 0 & \text{if GPA low,} \\ 1 & \text{if GPA high,} \end{cases}$$

$$Y = \begin{cases} 0 & \text{if school private,} \\ 1 & \text{if school public.} \end{cases}$$

Let the joint PMF $f_{X,Y}(x, y)$ for X and Y be given by the numbers in the table below.

$f_{X,Y}(x, y)$

		X	
		0	*1*
Y	*0*	*0.10*	*0.25*
	1	*0.20*	*0.45*

For example, the probability of a high GPA and the school being private is $f_{X,Y}(1, 0) = 0.25$. Suppose the probabilities correspond to the fraction of the population with specific characteristics. Then we can also say that 25% of all the schools are private with a high GPA.

The next theorem provides necessary and sufficient conditions for a function $f_{X,Y}(x, y)$ to be a valid joint PMF, where we by "valid" mean that it is possible for two random variables X and Y to have $f_{X,Y}(x, y)$ as their joint PMF. This is stated in the next theorem.

Theorem 3.1 *Let $f(x, y)$ be a function of two arguments x and y such that $f(x, y) > 0$ for only countably many pairs (x, y). Then $f(x, y)$ is a valid joint PMF if and only if the following two conditions hold.*

1. $f(x, y) \geq 0$ for all x, y.

2. $\sum_x \sum_y f(x, y) = 1$, where "$\sum_x \sum_y$" denotes the sum over all the values x, y such that $f(x, y) > 0$.

Example 3.4 (Example 3.3, continued) *We can check whether the function $f_{X,Y}(x, y)$ represented by the table is a valid joint PMF. First, all probabilities are positive. Second, the sum of all the probabilities are*

$$
\begin{aligned}
\sum_{x=0}^{1} \sum_{y=0}^{1} f_{X,Y}(x, y) &= \sum_{x=0}^{1} \left(\sum_{y=0}^{1} f_{X,Y}(x, y) \right) \\
&= \sum_{x=0}^{1} \left(f_{X,Y}(x, 0) + f_{X,Y}(x, 1) \right) \\
&= f_{X,Y}(0, 0) + f_{X,Y}(0, 1) + f_{X,Y}(1, 0) + f_{X,Y}(1, 1). \\
&= 0.10 + 0.20 + 0.25 + 0.45 \\
&= 1.
\end{aligned}
$$

Hence, by Theorem 3.1, the function $f_{X,Y}(x, y)$ is a valid joint PMF.

3.4.2 The marginal probability mass function

It is possible to calculate the distribution of a random variable, say, X, if the joint distribution of X and some other random variable, say, Y is known. A distribution of a random variable is called a *marginal distribution* when it is discussed in the context of a joint distribution.

In the case of two discrete random variables X and Y, suppose we know the joint PMF $f_{X,Y}(x, y)$. The next theorem shows how the marginal PMFs of X and Y, respectively, can be calculated using the knowledge of $f_{X,Y}(x, y)$.

Theorem 3.2 (Joint to marginal PMF) *Let $f_{X,Y}(x,y)$ by a joint PMF for the two discrete random variables X and Y. Then the PMF $f_X(x)$ of X is*

$$f_X(x) = \sum_y f_{X,Y}(x,y),$$

for all x, where $\sum_y$ is the summation over all possible outcomes of Y. Similarly, the PMF $f_Y(y)$ of Y is

$$f_Y(y) = \sum_x f_{X,Y}(x,y),$$

for all y, where $\sum_x$ is the summation over all possible outcomes of X.

Example 3.5 (Example 3.3, continued) *The probability of a low GPA is*

$$f_X(0) = \sum_{y=0}^{1} f_{X,Y}(0,y) = f_{X,Y}(0,0) + f_{X,Y}(0,1) = 0.10 + 0.20 = 0.30,$$

and the probability of a high GPA is

$$f_X(1) = \sum_{y=0}^{1} f_{X,Y}(1,y) = f_{X,Y}(1,0) + f_{X,Y}(1,1) = 0.25 + 0.45 = 0.70.$$

We can also calculate the probabilities for a school being private and public:

$$f_Y(0) = f_{X,Y}(0,0) + f_{X,Y}(1,0) = 0.10 + 0.25 = 0.35,$$

$$f_Y(1) = f_{X,Y}(0,1) + f_{X,Y}(1,1) = 0.20 + 0.45 = 0.65.$$

Notice that $f_X(x)$ adds up to 1 for all possible values of x, and similarly for $f_Y(y)$, as they should in order to be proper PMFs (Theorem 2.5).

Example 3.6 *Let X be Public Service Motivation (motivation to work for the benefit of the public) and Y be Performance of a randomly chosen individual in the population. Suppose there are the following possible outcomes for the two random variables*

$$X = \begin{cases} 1 & \text{if} & \text{low,} \\ 2 & \text{if} & \text{middle,} \\ 3 & \text{if} & \text{high,} \end{cases} \qquad Y = \begin{cases} 0 & \text{if} & \text{poor,} \\ 1 & \text{if} & \text{good.} \end{cases}$$

Suppose further that the joint PMF of X and Y is $f_{X,Y}(x,y)$, given by the table below.

$f_{X,Y}(x,y)$		Y	
		0	1
X	1	0.40	0.10
	2	0.16	0.24
	3	0.04	0.06

The top panel of Figure 3.2 illustrates the joint PMF of X and Y.

The calculation of the marginal probability function for X can illustrated directly in the table. To obtain $f_X(x)$, sum horizontally, and, to obtain $f_Y(y)$, sum vertically. This is shown in the table below.

		Y			
		0	1		$f_X(x)$
X	1	0.40	0.10	$\rightarrow$	0.50
	2	0.16	0.24	$\rightarrow$	0.40
	3	0.04	0.06	$\rightarrow$	0.10
		$\downarrow$	$\downarrow$		
	$f_Y(y)$	0.60	0.40		

The marginal PMFs are shown in the bottom panel of Figure 3.2. The figure illustrates how $f_Y(0)$ is calculated by adding all the joint probabilities in the direction where $y = 0$, i.e. $f_Y(0) = \sum_x f(x, 0)$.

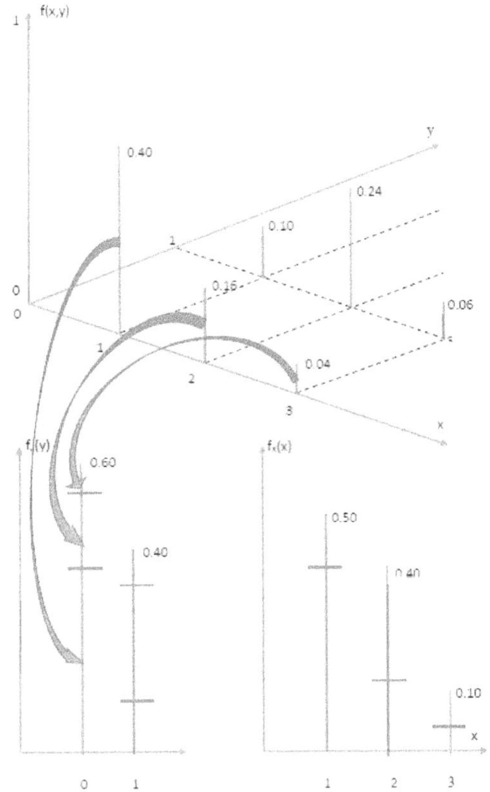

FIGURE 3.2
From joint to marginal PMF of discrete random variables, see Example 3.6. Top: Joint PMF $f(x, y)$ of two discrete random variables, X and Y. Bottom: Marginal PMF $f_Y(y)$ of Y (left) and marginal PMF $f_X(x)$ of X (right). The figure illustrates how the marginal probability of $Y = 0$, $f_Y(0)$, is found by summing up the joint probabilities $f(x, 0)$ over all values of x.

3.4.3　The relationship between the joint CDF and the joint PMF of two discrete random variables

The joint PMF contains the same information about the distribution of two discrete random variables X and Y as their joint CDF. The difference is how the information is presented. The next theorem shows the relationship between the joint PMF and the joint CDF.

Theorem 3.3 (Relationship between joint PMF and joint CDF) *Let X and Y be two discrete random variables. The joint CDF $F_{X,Y}(x,y)$ can be calculated from the joint PMF $f_{X,Y}(x,y)$ by*

$$F_{X,Y}(a,b) = \sum_{x \leq a} \sum_{y \leq b} f_{X,Y}(x,y),$$

for all values of a and b, where $\sum_{x \leq a}$ denotes summation over all possible outcomes of the random variable X less than or equal to a, and similarly for $\sum_{y \leq b}$.

Conversely, the joint PMF can be calculated from the joint CDF by

$$f_{X,Y}(a,b) = F_{X,Y}(a,b) - \lim_{x \to a^-} \lim_{y \to b^-} F_{X,Y}(x,y),$$

for all values of a and b, where $\lim_{x \to a^-}$ denotes the limit from the left at a of the function and similarly for $\lim_{y \to b^-}$.

On a technical note, we could have used Theorem 3.3 as the definition of a joint PMF. This definition would have tied the joint CDF to the joint PMF directly. Then Definition 3.2 of the joint PMF could instead have been stated as a theorem, since it would have been a consequence of that new definition. We chose Definition 3.2 because it immediately makes the joint PMF interpretable.

Example 3.7 (Example 3.6, continued) *We can use Theorem 3.3 to calculate the joint CDF $F_{X,Y}(x,y)$ of X and Y from our knowledge of the joint PMF $f_{X,Y}(x,y)$. For instance,*

$$F_{X,Y}(1,0) = \sum_{x \leq 1} \sum_{y \leq 0} f_{X,Y}(x,y) = f_{X,Y}(1,0) = 0.40,$$

and

$$\begin{aligned}
F_{X,Y}(2,1) = \sum_{x \leq 2} \sum_{y \leq 1} f_{X,Y}(x,y) &= f_{X,Y}(1,0) + f_{X,Y}(1,1) + f_{X,Y}(2,0) + f_{X,Y}(2,1) \\
&= 0.40 + 0.10 + 0.16 + 0.24 \\
&= 0.90.
\end{aligned}$$

The table below shows $F_{X,Y}(x,y)$ for all relevant x and y.

$F_{X,Y}(x,y)$

		Y	
		0	1
X	1	0.40	0.50
	2	0.56	0.90
	3	0.60	1

3.4.4   The distribution of a function of two discrete random variables

It may be useful to be able to define a new random variable as a function of other random variables. The distribution of the new random variable is derived from the joint distribution of the random variables, which are transformed into the new random variable. The next example illustrates.

Example 3.8 *Let X and Y be the number of questions asked by two different political parties in parliament. Let the joint PMF $f_{X,Y}(x,y)$ be given by the numbers in the following table.*

$$f_{X,Y}(x,y)$$

		Y		
	0	1	2	3
0	**0.00**	**0.08**	**0.02**	0
1	**0.03**	**0.13**	0.14	0.09
2	**0.01**	0.10	0.12	0.11
3	0	0.04	0.07	0.06

(with X labelling the rows 0–3)

For example, the probability of party 1 asking 3 questions and party 2 asking 2 questions is $f_{X,Y}(3,2) = 0.07 = 7\%$.

Suppose our interest is the total number of questions Z asked by the two parties. Then Z is a discrete random variable that can be defined by the following function of X and Y:

$$Z = X + Y.$$

The probability that no more than 2 questions asked in total is

$$\Pr(Z \leq 2) = \Pr(X + Y \leq 2).$$

The possible combinations of X and Y for which their sum is at most 2 are marked in bold in the table above. By the properties of probability measures, the probability $\Pr(Z \leq 2)$ is the sum of those bold faced probabilities. The sum of those numbers is 0.27.

A general approach to calculate probabilities for a random variable Z constructed from two other random variables X and Y is given in the next theorem.

Theorem 3.4 (Distribution of a function of two discrete random variables)
Let X and Y be two discrete random variables and define the new discrete random variable Z by

$$Z = h(X,Y),$$

where $h(x,y)$ is a function. Then the CDF of Z can be calculated as

$$
\begin{aligned}
F_Z(z) &= \Pr(Z \leq z) \\
&= \Pr(h(X,Y) \leq z) \\
&= \sum_x \sum_y I(h(x,y) \leq z) f_{X,Y}(x,y),
\end{aligned}
$$

where $I()$ is the indicator function and $f_{X,Y}(x,y)$ is the joint PMF of X and Y.
The PMF of Z can be calculated as

$$
\begin{aligned}
f_Z(z) &= \Pr(Z = z) \\
&= \Pr(h(X,Y) = z) \\
&= \sum_x \sum_y I(h(x,y) = z) f_{X,Y}(x,y).
\end{aligned}
$$

Example 3.9 (Example 3.8, continued) *The probability of no more than 2 questions* *asked in total can be calculated as*

$$\Pr(X + Y \leq 2) = \sum_{x=0}^{3}\sum_{y=0}^{3} I(x + y \leq 2) f_{X,Y}(x, y) =$$

$$\begin{aligned}
&= 1 \cdot f_{X,Y}(0,0) + 1 \cdot f_{X,Y}(0,1) + 1 \cdot f_{X,Y}(0,2) + 0 \cdot f_{X,Y}(0,3) \\
&\quad + 1 \cdot f_{X,Y}(1,0) + 1 \cdot f_{X,Y}(1,1) + 0 \cdot f_{X,Y}(1,2) + 0 \cdot f_{X,Y}(1,3) \\
&\quad + 1 \cdot f_{X,Y}(2,0) + 0 \cdot f_{X,Y}(2,1) + 0 \cdot f_{X,Y}(2,2) + 0 \cdot f_{X,Y}(2,3) \\
&\quad + 0 \cdot f_{X,Y}(3,0) + 0 \cdot f_{X,Y}(3,1) + 0 \cdot f_{X,Y}(3,2) + 0 \cdot f_{X,Y}(3,3) \\
&= 0 + 0.08 + 0.02 + 0 + 0.03 + 0.13 + 0 + 0 + 0.01 + 0 + 0 + 0 + 0 + 0 + 0 \\
&= 0.27.
\end{aligned}$$

The probability of exactly two questions asked in total is

$$\begin{aligned}
\Pr(X + Y = 2) &= \sum_{x}\sum_{y} I(x + y = 2) f_{X,Y}(X, y) \\
&= 1 \cdot f_{X,Y}(0,2) + 1 \cdot f_{X,Y}(1,1) + 1 \cdot f_{X,Y}(2,0) \\
&= 0.02 + 0.13 + 0.01 = 0.16.
\end{aligned}$$

3.4.5   Multiple discrete random variables

We now extend the results above to the general case of two or more discrete random variables. The interpretations of the concepts are the same as above.

The definition of a joint PMF for $k \geq 2$ discrete random variables is given first.

Definition 3.3 (Joint PMF of multiple discrete random variables) *Let $X_1, \ldots, X_k$ be k discrete random variables. Then the joint probability mass function (joint PMF) for $X_1, \ldots, X_k$ is*

$$f_{X_1,\ldots,X_k}(x_1, \ldots, x_k) = \Pr(X_1 = x_1, \ldots, X_k = x_k),$$

for all $x_1, \ldots, x_k$.

Any function can be a joint PMF if it satisfies the two conditions given in the next theorem.

Theorem 3.5 *Let $f(x_1, \ldots, x_k)$ be a function of k arguments $x_1, \ldots, x_k$ such that $f(x_1, \ldots, x_k) > 0$ for only countably many values $(x_1, \ldots, x_k)$. Then $f(x_1, \ldots, x_k)$ is a valid joint PMF if and only if the following two conditions hold.*

1. $f(x_1, \ldots, x_k) \geq 0$ for all $x_1, \ldots, x_k \in \mathbb{R}$.

2. $\sum_{x_1} \cdots \sum_{x_k} f(x_1, \ldots, x_k) = 1$.

When we know the distribution of a set of random variables $X_1, \ldots, X_k$, we can calculate the distribution for any combination of them. We refer to such a distribution of a subset of the random variables as a marginal distribution. For example, we may calculate $f_{X_1}(x_1)$ from $f_{X_1,X_2,X_3}(x_1, x_2, x_3)$. We may also calculate the PMF $f_{X_1,X_2}(x_1, x_2)$ for X_1 and X_2 from $f_{X_1,X_2,X_3}(x_1, x_2, x_3)$. The next theorem contains such calculations.

Theorem 3.6 (Marginal PMF from joint PMF) *Let $X_1, \ldots, X_k$ be k discrete random variables with joint PMF $f_{X_1,\ldots,X_k}(x_1, \ldots, x_k)$. For all a, the marginal PMF for X_j is given by*

$$f_{X_j}(a) = \sum_{x_1} \cdots \sum_{x_{j-1}} \sum_{x_{j+1}} \cdots \sum_{x_k} f_{X_1,\ldots X_j,\ldots,X_k}(x_1, \ldots, x_{j-1}, a, x_{j+1}, \ldots, x_k).$$

The marginal probability function for a subset of the random variables, say, $X_1, \ldots, X_m$, $m < k$, is given by

$$f_{X_1,\ldots,X_m}(a_1,\ldots,a_m) = \sum_{x_{m+1}} \cdots \sum_{x_k} f_{X_1,\ldots,X_m,X_{m+1},\ldots,X_k}(a_1,\ldots,a_m,x_{m+1},\ldots,x_k),$$

for all $a_1, \ldots, a_m$.

The relationship between the joint CDF and the joint PMF is given in the next theorem.

Theorem 3.7 (Relationship between joint PMF and joint CDF) *Let $X_1, \ldots, X_k$ be k discrete random variables. Then*

$$F_{X_1,\ldots,X_k}(a_1,\ldots,a_k) = \sum_{x_1 \leq a_1} \cdots \sum_{x_k \leq a_k} f_{X_1,\ldots,X_k}(x_1,\ldots,x_k),$$

and

$$f_{X_1,\ldots,X_k}(a_1,\ldots,a_k) = F_{X_1,\ldots,X_k}(a_1,\ldots,a_k) - \lim_{x_1 \to a_1^-} \cdots \lim_{x_k \to a_k^-} F_{X_1,\ldots,X_k}(x_1,\ldots,x_k),$$

where $F_{X_1,\ldots,X_k}(x_1,\ldots,x_k)$ is the joint CDF and $f_{X_1,\ldots,X_k}(x_1,\ldots,x_k)$ is the joint PMF of the random variables $X_1, \ldots, X_k$.

3.5 The joint distribution of continuous random variables

In this section, we consider the case where all the random variables are continuous. Similarly to above, all the concepts can be introduced and discussed with two random variables only. In the last subsection, the extension to multiple random variables is described.

3.5.1 The joint probability density function of two continuous random variables

Similarly to the case of one continuous random variable, it turns out to be practical to define a probability density function when working with a group of continuous random variables. We start by defining a *joint probability density function* in the case of two continuous random variables. Immediately after the definition, we state a theorem, which is useful in modeling multiple continuous random variables. We start with the definition.

Definition 3.4 (Joint PDF, two random variables) *Let X and Y be two continuous random variables. Then $f_{X,Y}(x,y)$ is a **joint probability density function (joint PDF)** for X and Y if and only if*

$$f_{X,Y}(a,b) \geq 0,$$

$$F_{X,Y}(a,b) = \int_{-\infty}^{a} \int_{-\infty}^{b} f_{X,Y}(x,y)\,dx\,dy,$$

for all values of a and b, where $F_{X,Y}(x,y)$ is the joint CDF of X and Y.

When working with the distribution of continuous random variables, we often directly specify a function and claim that it is a joint PDF. The next theorem gives the conditions for such a claim to be valid, that is, the conditions a function $f(x, y)$ must satisfy to be valid as a joint PDF. As above, by "valid" we mean that it is possible for two continuous random variables X and Y to have $f(x, y)$ as their joint PDF.

Theorem 3.8 *A function $f(x, y)$ of two arguments x and y is a valid joint PDF if and only if the following two conditions are satisfied.*

1. *$f(x, y) \geq 0$ for all x and y.*

2. *$\int_{-\infty}^{\infty} \int_{-\infty}^{\infty} f(x, y)\, dx dy = 1$.*

Graphically, the second condition says that the area under the graph of the function $f(x, y)$ is 1. In the case of two variables, the area corresponds to the volume under the graph since the graph is a three-dimensional object.

Example 3.10 *Let*

$$f(x, y) = \begin{cases} 0.5 & if \ \ 1 \leq x \leq 2, \ 4 \leq y \leq 6, \\ 0 & otherwise. \end{cases}$$

Graphically, this is a box with length (x-coordinate) being 1 ($= 2 - 1$), width (y-coordinate) being 2 ($= 6 - 4$) and height (function value, $f(x, y)$) being 0.5, see the top left panel of Figure 3.3. The volume of this box is $1 \cdot 2 \cdot 0.5 = 1$.

When the joint PDF is more complicated, it is necessary to do the formal integration when verifying condition 2 from Theorem 3.8. This is illustrated in the next example.

Example 3.11 *Consider the function*

$$f(x, y) = \begin{cases} 3xe^{-xy}e^{-3x} & if \ \ x > 0, \ y > 0, \\ 0 & otherwise. \end{cases}$$

The graph of the function is shown in the top right panel of Figure 3.3.

Using Theorem 3.8, we can verify that this function is a valid joint PDF for two continuous random variables.

By the properties of the exponential function, we have $f(x, y) > 0$ for $x, y > 0$ and $f(x, y) = 0$ otherwise. Hence, the first condition of Theorem 3.8 is satisfied.

To calculate the double integral, first integrate with respect to one of the variables, here y, keeping the other variable as a fixed number, and then integrate the result of the first integral with respect to the second variable, here x. The calculation is

$$\begin{aligned} \int_0^\infty \int_0^\infty 3xe^{-xy}e^{-3x}\, dx dy &= \int_0^\infty \left[\int_0^\infty xe^{-xy} dy \right] 3e^{-3x} dx \\ &= \int_0^\infty \left[-e^{-xy} \right]_0^\infty 3e^{-3x} dx \\ &= \int_0^\infty \left[0 - (-1) \right] 3e^{-3x} dx \\ &= \int_0^\infty 3e^{-3x} dx \\ &= \left[-e^{-3x} \right]_0^\infty \\ &= 0 - (-1) \\ &= 1. \end{aligned}$$

Hence, the second condition of Theorem 3.8 is satisfied and we conclude that $f(x, y)$ is a valid joint PDF.

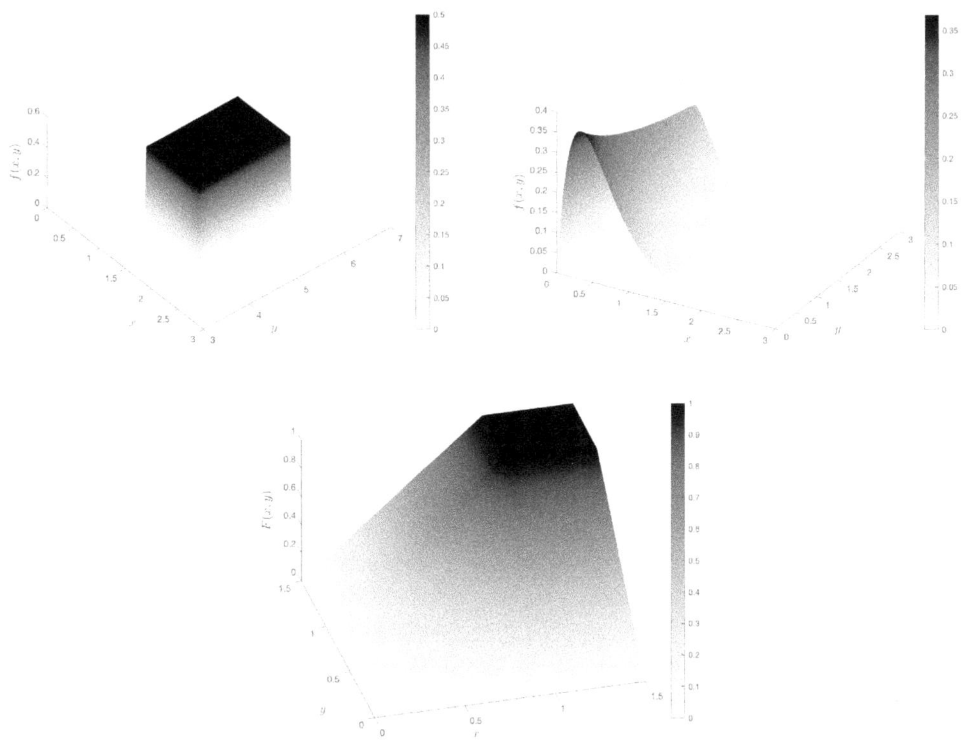

FIGURE 3.3
Joint distributions. Top left: Joint PDF $f(x, y)$ from Example 3.10. Top right: Joint PDF $f(x, y)$ from Example 3.11. Bottom: Joint CDF $F(x, y)$ from Example 3.13.

3.5.2   The marginal probability density function

From a joint PDF, we can calculate the PDF for each of the random variables, called the *marginal PDFs*. The next theorem contains that result.

Theorem 3.9 (Joint to marginal PDF) *Let the joint PDF of X and Y be $f_{X,Y}(x, y)$. Then the PDF $f_X(x)$ of X is*

$$f_X(x) = \int_{-\infty}^{\infty} f_{X,Y}(x, y) dy,$$

for all x. Similarly, the PDF $f_Y(y)$ of Y is

$$f_Y(y) = \int_{-\infty}^{\infty} f_{X,Y}(x, y) dx.$$

for all y.

As Theorem 3.9 shows, the marginal PDF of a random variable is obtained by "integrating out" the other variable from the joint PDF.

Example 3.12 (Example 3.11, continued) *Let $x > 0$. To obtain the marginal PDF $f_X(x)$ of X, we use Theorem 3.9 and integrate Y out of the joint PDF:*

$$
\begin{aligned}
f_X(x) = \int_{-\infty}^{\infty} f_{X,Y}(x,y)dy = \int_0^{\infty} 3xe^{-xy}e^{-3x}dy &= 3e^{-3x}\int_0^{\infty} xe^{-xy}dy \\
&= 3e^{-3x}\left[-e^{-xy}\right]_0^{\infty} \\
&= 3e^{-3x}.
\end{aligned}
$$

3.5.3   The relationship between the joint CDF and the joint PDF of two continuous random variables

While Definition 3.4 provides the joint CDF from the joint PDF, we next show how the joint CDF can provide the joint PDF. This can be done using partial differentiation if the joint CDF is differentiable. The next theorem states the result.

Theorem 3.10 (From joint CDF to joint PDF) *Let $F_{X,Y}(x,y)$ be the joint CDF of two continuous random variables X and Y. If $F_{X,Y}(x,y)$ is differentiable with respect to x and y, then the joint PDF $f_{X,Y}(x,y)$ of X and Y is given by*

$$
f_{X,Y}(x,y) = \frac{\partial^2 F_{X,Y}(x,y)}{\partial x \partial y},
$$

where $\frac{\partial^2 F_{X,Y}(x,y)}{\partial x \partial y}$ is the partial derivative of $F_{X,Y}(x,y)$ with respect to x and y.

In cases where the joint CDF is not differentiable at all points of its domain, we can often still derive a joint PDF by differentiating at all points, where the joint CDF is differentiable. The next example illustrates.

Example 3.13 *Let the joint CDF $F_{X,Y}(x,y)$ of two random variables X and Y be given by*

$$
F_{X,Y}(x,y) = \begin{cases} 0 & if & x < 0 \ or \ y < 0, \\ x \cdot y & if & 0 \le x \le 1, \ 0 \le y \le 1, \\ x & if & 0 \le x \le 1 \ , \ y > 1, \\ y & if & 0 \le y \le 1 \ , \ x > 1, \\ 1 & if & x > 1, \ y > 1. \end{cases}
$$

The graph of the CDF is shown in the bottom panel of Figure 3.3.

Since this joint CDF is continuous, then X and Y are continuous random variables. This joint CDF is not differentiable at all points (x,y). In particular, it is non-differentiable in the points (x,y) where we go from one piece of the definition of $F(x,y)$ to another piece, since the CDF has kinks at these points. For example, if $y = 1.5$, then $F(x,y)$ is not differentiable at $x = 0$ because if x is slightly below 0, then $F(x,y)$ equals 0 whereas $F(x,y)$ equals x when x is slightly above 0. Thus, the slope at $(x,y) = (0, 1.5)$ is 0 when x comes from the left, and 1 when x comes from the right.

The partial derivative at all differentiable points is

$$
f(x,y) = \frac{\partial^2 F_{X,Y}(x,y)}{\partial x \partial y} = \begin{cases} 0 & if & x < 0 \ or \ y < 0, \\ 1 & if & 0 < x < 1, \ 0 < y < 1, \\ 0 & if & 0 < x < 1 \ , \ y > 1, \\ 0 & if & 0 < y < 1 \ , \ x > 1, \\ 0 & if & x > 1, \ y > 1. \end{cases}
$$

Notice, that since each of the pieces in $f(x,y)$ are defined by sharp inequalities compared to the pieces in $F_{X,Y}(x,y)$, the non-differentiable points are avoided.

To complete a candidate for joint density function, we must assign values at the remaining points (the non-differentiability points). Since point probabilities for continuous random variables are zero (Theorem 2.9), it does not matter what values we assign to f at these points. For instance, we could let $f(x,y) = 0$ at all non-differentiability points. In this case, the function $f(x,y)$ becomes

$$f(x,y) = \begin{cases} 1 & \text{if } 0 < x < 1,\ 0 < y < 1, \\ 0 & \text{otherwise.} \end{cases}$$

We can use Theorem 3.8 to show that this is indeed a valid joint PDF (Problem 3.8.3).

3.5.4 The distribution of a function of two continuous random variables

Similarly to the case with discrete random variables, it is useful to be able to define a new random variable as a function of other continuous random variables. To find the distribution of the new random variable, we need the joint distribution of the random variables which are transformed into the new random variable.

The next theorem provides a general method to calculate the joint CDF of a random variable created from two continuous random variables.

Theorem 3.11 (CDF of a function of two continuous random variables) *Let X and Y be two continuous random variables and let $f_{X,Y}(x,y)$ be the joint PDF of X and Y. Assume the random variable Z is given by*

$$Z = h(X,Y),$$

where $h(x,y)$ is a function. Then the CDF of Z can be calculated as

$$F_Z(z) = \Pr(h(X,Y) \le z) = \int_{-\infty}^{\infty} \int_{-\infty}^{\infty} I(h(x,y) \le z) f_{X,Y}(x,y)\, dx\, dy,$$

where $I()$ is the indicator function.

In Theorem 3.11, the random variable Z may be continuous, discrete, or of mixed type. The type depends on the function $h(x,y)$.

Example 3.14 *Let X and Y be two random variables and define Z as*

$$Z = \begin{cases} 1 & \text{if } X > 0 \text{ and } Y > 0, \\ -1 & \text{otherwise.} \end{cases}$$

Then Z is a discrete random variable no matter if X and Y are continuous or discrete.

The next example illustrates a case where the new random variable Z is continuous.

Example 3.15 *Consider the two continuous random variables X and Y, and a new random variable Z, defined as*

$$Z = \frac{Y}{X}.$$

Suppose the joint PDF $f_{X,Y}(x,y)$ of X and Y is given by

$$f_{X,Y}(x,y) = \begin{cases} 1 & \text{if } x \in (0,1), y \in (0,1), \\ 0 & \text{otherwise.} \end{cases}$$

Using the indicator function, this can be rewritten as

$$f_{X,Y}(x,y) = 1 \cdot I\left(x \in (0,1)\right) \cdot I\left(y \in (0,1)\right).$$

Clearly, Z is non-negative. Let $z > 0$. We wish to derive the CDF for Z, i.e. $\Pr(Z \leq z)$. Using Theorem 3.11, this probability can be expressed as

$$\begin{aligned}
\Pr(Z \leq z) &= \Pr\left(\frac{Y}{X} \leq z\right) \\
&= \int_{-\infty}^{\infty} \int_{-\infty}^{\infty} I\left(\frac{y}{x} \leq z\right) \cdot f_{X,Y}(x,y)\,dy\,dx \\
&= \int_{-\infty}^{\infty} \int_{-\infty}^{\infty} I\left(\frac{y}{x} \leq z\right) \cdot 1 \cdot I\left(x \in (0,1)\right) \cdot I\left(y \in (0,1)\right)\,dy\,dx \\
&= \int_{-\infty}^{\infty} I\left(y \in (0,1)\right) \underbrace{\left[\int_{-\infty}^{\infty} I\left(\frac{y}{x} \leq z\right) \cdot I\left(x \in (0,1)\right)\,dx\right]}_{=A}\,dy.
\end{aligned}$$

Consider the term A. Since $x, z > 0$, it can be written as

$$\begin{aligned}
A = \int_{-\infty}^{\infty} I\left(\frac{y}{x} \leq z\right) \cdot I\left(x \in (0,1)\right)\,dx &= \int_{-\infty}^{\infty} I\left(\frac{y}{z} \leq x\right) \cdot I\left(x \in (0,1)\right)\,dx \\
&= \int_{\frac{y}{z}}^{\infty} I\left(x \in (0,1)\right)\,dx.
\end{aligned}$$

Two cases arise. If $y/z < 1$, then the integral becomes

$$\int_{\frac{y}{z}}^{1} dx = 1 - \frac{y}{z}.$$

Conversely, if $y/z \geq 1$, then the integral evaluates to 0 because it implies $x \geq 1$ and, thus, $I\left(x \in (0,1)\right) = 0$.

Now, collect the above terms to get

$$\begin{aligned}
\Pr(Z \leq z) &= \int_{-\infty}^{\infty} I\left(y \in (0,1)\right) \cdot I\left(\frac{y}{z} \leq 1\right) \cdot \left(1 - \frac{y}{z}\right)\,dy \\
&= \begin{cases} \int_0^z \left(1 - \frac{y}{z}\right) dy = z - \frac{1}{2z}z^2 = \frac{1}{2}z & \text{if } 0 < z \leq 1, \\ \int_0^1 \left(1 - \frac{y}{z}\right) dy = 1 - \frac{1}{2z} & \text{if } z > 1. \end{cases}
\end{aligned}$$

Thus, the CDF of Z is

$$F_Z(z) = \begin{cases} 0 & \text{if } z \leq 0, \\ \frac{z}{2} & \text{if } 0 < z \leq 1, \\ 1 - \frac{1}{2z} & \text{if } z > 1. \end{cases}$$

Notice that since $F_Z(z)$ is a continuous function, we may deduce that Z is a continuous random variable (Definition 2.4). Further, by Theorem 2.12, since $F_Z(z)$ is differentiable, also at $z = 0$ and $z = 1$, a valid density function $f_Z(z)$ of Z is

$$f_Z(z) = \begin{cases} \frac{1}{2} & \text{if } 0 < z \leq 1, \\ \frac{1}{2z^2} & \text{if } z > 1. \end{cases}$$

In Problem 3.8.5 at the end of this chapter, you are asked to verify that this function $f_Z(z)$ is indeed a proper PDF.

3.5.5 Multiple continuous random variables

We now extend the results above to the general case of two or more continuous random variables. The interpretations of the concepts are the same as the cases with two continuous random variables. The only difference is the number of integrals or partial derivatives involved.

The definition of a joint PDF for $k \geq 2$ continuous random variables is given first.

Definition 3.5 (Joint PDF of multiple continuous random variables) *Let $X_1, \ldots, X_k$ be k continuous random variables. Then $f_{X_1,\ldots,X_k}(x_1,\ldots,x_k)$ is a joint probability density function (joint PDF) for $X_1, \ldots, X_k$ if and only if*

$$
\begin{aligned}
f_{X_1,\ldots,X_k}(a_1,\ldots,a_k) &\geq 0, \\
F_{X_1,\ldots,X_k}(a_1,\ldots,a_k) &= \int_{-\infty}^{a_1} \cdots \int_{-\infty}^{a_k} f_{X_1,\ldots,X_k}(x_1,\ldots,x_k)dx_1\ldots dx_k,
\end{aligned}
$$

for all of $a_1, \ldots, a_k \in \mathbb{R}$.

A function can be a joint PDF if it satisfies the two conditions given in the next theorem.

Theorem 3.12 *A function $f(x_1,\ldots,x_k)$ is a valid joint PDF if and only if*

1. $f(x_1,\ldots,x_k) \geq 0$ for all $x_1,\ldots,x_k$.

2. $\int_{-\infty}^{\infty} \cdots \int_{-\infty}^{\infty} f(x_1,\ldots,x_k)\, dx_1 \ldots dx_k = 1$.

Definition 3.5 describes how to obtain the joint CDF from a joint PDF. The opposite direction of obtaining the joint PDF from a joint CDF is given in the next theorem, provided that the joint CDF is differentiable.

Theorem 3.13 (From joint CDF to joint PDF) *Let $F_{X_1,\ldots,X_k}(x_1,\ldots,x_k)$ be the joint CDF of $X_1, \ldots, X_k$. Assume $F_{X_1,\ldots,X_k}(x_1,\ldots,x_k)$ is differentiable. Then a joint PDF of $X_1, \ldots, X_k$ is*

$$
f_{X_1,\ldots,X_k}(x_1,\ldots,x_k) = \frac{\partial^k F_{X_1,\ldots,X_k}(x_1,\ldots,x_k)}{\partial x_1 \partial x_2 \ldots \partial x_k},
$$

where $\frac{\partial^k F_{X_1,\ldots,X_k}(x_1,\ldots,x_k)}{\partial x_1 \partial x_2 \ldots \partial x_k}$ is the partial derivative of F with respect to each of the variables $x_1, \ldots, x_k$.

The joint PDF of a subset of the random variables, say, $X_1, \ldots, X_m$ can be calculated from the joint PDF of all the k random variables by integrating out the variables $x_{m+1}, \ldots, x_k$. The next theorem states this result.

Theorem 3.14 (Marginal PDF from joint PDF) *Let $f_{X_1,\ldots,X_k}(x_1,\ldots,x_k)$ be the joint PDF of $X_1, \ldots, X_k$. The joint PDF of $X_1, \ldots, X_m$, $m < k$, is*

$$
f_{X_1,\ldots,X_m}(x_1,\ldots,x_m) = \int_{-\infty}^{\infty} \cdots \int_{-\infty}^{\infty} f_{X_1,\ldots,X_k}(x_1,\ldots,x_m,x_{m+1},\ldots,x_k)\, dx_{m+1}\ldots dx_k
$$

for all $x_1, \ldots, x_m \in \mathbb{R}$.

3.6 The joint distribution of both continuous and discrete random variables

To consider a case of a joint distribution with different types of random variables, let X be a discrete random variable and Y be a continuous random variable. The joint CDF $F_{X,Y}(x,y)$ is well defined for all types of random variables, so, for all $a,b \in \mathbb{R}$,

$$F_{X,Y}(a,b) = \Pr(X \le a, Y \le b).$$

The CDF of X can be calculated as

$$F_X(x) = \lim_{y \to \infty} F_{X,Y}(x,y) = F_{X,Y}(x,\infty), \quad x \in \mathbb{R},$$

and, similarly, the CDF of Y can be calculated as

$$F_Y(y) = \lim_{x \to \infty} F_{X,Y}(x,y) = F_{X,Y}(\infty,y), \quad y \in \mathbb{R}.$$

The PMF for the discrete random variable X can be derived from $F_X(x)$, using

$$f_X(a) = F_{X,Y}(a,\infty) - \lim_{x \to a^-} F_{X,Y}(x,\infty),$$

for all a. Similarly, if it exists, the PDF for the continuous random variable Y can be derived from $F_Y(y)$, using

$$f_Y(y) = \frac{\partial F_{X,Y}(\infty,y)}{\partial y},$$

for all y.

3.7 Proofs

Proof of Theorem 3.1 *Formally, we can consider the two discrete random variables X and Y as a single "two-dimensional" discrete random variable*

$$(X,Y) : \Omega \to \mathbb{R}^2.$$

Using this object, the proof is similar to the proof of Theorem 2.5 in the previous chapter. We omit the details.

Proof of Theorem 3.2 *Using that $f_X(x) = \Pr(X = x)$ and $f_{X,Y}(x,y) = \Pr(X = x \cap Y = y)$ and the fact that $\Pr()$ is a probability measure (see Theorem A.4 in Chapter A of the appendix), this is a consequence of the law of total probability (Theorem 1.2).*

Proof of Theorem 3.3 *Similarly to the proof of Theorem 3.2, using that $F_{X,Y}(x,y) = \Pr(X \le x \cap Y \le y)$, $f_{X,Y}(x,y) = \Pr(X = x \cap Y = y)$, and the fact that $\Pr()$ is a probability measure (see Theorem A.4 in Chapter A of the appendix), this is a consequence of the law of total probability (Theorem 1.2).*

Proof of Theorem 3.4 *Similarly to the proofs of the previous two theorems, this is a consequence of the fact that* $\Pr()$ *is a probability measure (see Theorem A.4 in Chapter A of the appendix). For instance, by the third property in the definition of a probability measure (Definition 1.4), we have*

$$
\begin{aligned}
F_Z(z) = \Pr(Z \le z) = \Pr(h(X,Y) \le z) \;&=\; \Pr\left(\cup_{k=-\infty}^{z}\{h(X,Y)=k\}\right)\\
&=\; \sum_{k=-\infty}^{z} \Pr\left(h(X,Y)=k\right)\\
&=\; \sum_x \sum_y I(h(x,y) \le z) f_{X,Y}(x,y),
\end{aligned}
$$

as we wanted to show. A similar calculation can be made for $f_Z(z)$.

Proof of Theorem 3.5 *We omit the details of this proof, but see the comments to the proof of Theorem 3.1.*

Proof of Theorem 3.6 *The proof follows along similar lines as those in the proof of Theorem 3.2.*

Proof of Theorem 3.7 *The proof follows along similar lines as those in the proof of Theorem 3.3.*

Proof of Theorem 3.8 *We omit the details of this proof, but see the comments to the proof of Theorem 3.1.*

Proof of Theorem 3.9 *Recall that we may write*

$$
F_X(x) = \lim_{y \to \infty} F_{X,Y}(x,y).
$$

From the definition of the joint PDF (Definition 3.4), we have

$$
F_X(x) = \lim_{y \to \infty} \int_{-\infty}^{x} \int_{-\infty}^{y} f_{X,Y}(t,s)\,ds\,dt = \int_{-\infty}^{x} \int_{-\infty}^{\infty} f_{X,Y}(t,y)\,dy\,dt.
$$

Using that $f_X(x) = F_X'(x)$ *(Theorem 2.12) and the so-called Leibniz integral rule for differentiating an integral (see Section B.3.3 in Chapter B of the appendix), we have that*

$$
f_X(x) = F_X'(x) = \frac{d}{dx} \int_{-\infty}^{x} \int_{-\infty}^{\infty} f_{X,Y}(t,y)\,dy\,dt = \int_{-\infty}^{\infty} f_{X,Y}(x,y)\,dy,
$$

as we wanted to show.

Proof of Theorem 3.10 *This follows from the so-called Fundamental Theorem of Calculus (see Theorem B.8 in Chapter B of the appendix).*

Proof of Theorem 3.11 *This can be proved using similar arguments as in the proof of Theorem 3.4, although the details are more involved because we are forced to use integrals instead of sums. We omit the details.*

Proof of Theorem 3.12 *We omit the details of this proof, but see the comments to the proof of Theorem 3.1.*

Proof of Theorem 3.13 *This follows from the so-called Fundamental Theorem of Calculus (see Theorem B.8 in Chapter B of the appendix).*

Proof of Theorem 3.14 *The proof follows along similar lines as those in the proof of Theorem 3.9.*

3.8 Exercises

Problem 3.8.1 *Let Ω be a sample space and $A, B \subseteq \Omega$ two events. Consider the events $A \cap B$, $A \cap B^c$, $A^c \cap B$ and $A^c \cap B^c$. Show each pair of these events are disjoint and show that*
$$(A \cap B) \cup (A \cap B^c) \cup (A^c \cap B) \cup (A^c \cap B^c) = \Omega.$$

Problem 3.8.2 *Consider the joint PDF $f(x, y)$ of two random variables X and Y from Example 3.11, i.e.*
$$f(x, y) = \begin{cases} 3xe^{-xy}e^{-3x} & if \quad x > 0, y > 0, \\ 0 & otherwise. \end{cases}$$

1. *Derive the marginal PDF $f_Y(y)$ of Y. (Hint: Use Theorem 3.9 and partial integration.)*

2. *Use Theorem 2.11 to verify that the function $f_Y(y)$ you just found is a valid PDF.*

Problem 3.8.3 *Consider the function $f(x, y)$ derived in Example 3.13. Show that this function is a valid PDF.*

Problem 3.8.4 *Consider the two continuous random variables X and Y with joint PDF $f_{X,Y}(x, y)$ given by*
$$f_{X,Y}(x, y) = \begin{cases} 1 & if \quad x \in [0, 1], \ y \in [0, 1], \\ 0 & otherwise. \end{cases}$$

Define the new random variable Z, via
$$Z = \begin{cases} Y & if \quad X > \frac{1}{2}, \\ 0 & otherwise. \end{cases}$$

1. *Derive the CDF of Z.*

2. *What type of random variable is Z?*

Problem 3.8.5 *Consider the PDF of the random variable Z from Example 3.15,*

$$f_Z(z) = \begin{cases} \frac{1}{2} & if \quad 0 < z \leq 1, \\ \frac{1}{2z^2} & if \quad z > 1. \end{cases}$$

Show that $f_Z(z)$ is a valid PDF.

Problem 3.8.6 *Consider the two continuous random variables X and Y with joint PDF $f_{X,Y}(x, y)$ given by*

$$f_{X,Y}(x, y) = \begin{cases} 1 & if \quad x \in [0, 1], y \in [0, 1], \\ 0 & otherwise. \end{cases}$$

Define the new random variable Z, via

$$Z = X \cdot Y.$$

1. *Derive the CDF $F_Z(z)$ of Z.*

2. *Use the CDF $F_Z(z)$ to show that Z is a continuous random variable.*

3. *Derive the PDF of Z.*

4

Conditional distribution of a random variable

4.1 Introduction

A particularly useful approach to investigating relationships among many random variables is to consider the distribution of one of them as a function of the others. This leads to the concept of a *conditional distribution*. A conditional distribution is a certain aspect of a joint distribution. Knowing or specifying a conditional distribution is often sufficient in order to do prediction or calculate causal effects, and this can be a big advantage in practice. The reason is that specifying a reasonable conditional distribution may be much easier than specifying a reasonable joint distribution. We will exploit this in Volume II of the book when modeling and estimating relationships between variables. In this chapter, we investigate how to interpret a conditional distribution and how it can be derived from the joint distribution.

We first study relationships between events to obtain a basic understanding of a conditional probability and probabilistic relationships. In practice, probabilistic relationships are typically more effectively described by random variables than by events, and we will therefore use random variables for the subsequent analysis. Depending on the types of random variables, discrete, continuous, or mixed, conditional distributions are expressed and derived using different mathematical techniques. Although the techniques are different, the interpretations are basically the same no matter the type of random variables.

4.2 Conditional probability of events

Whether the aim is to describe, predict, or to understand causal links, a key probabilistic concept is that of "conditional probability". The probability of an event A occurring given the knowledge that another event B has occurred, or will occur, is called the *conditional probability of A given B*. Such a conditional probability is denoted $P(A|B)$, where the vertical line "|" is pronounced "given" or "conditional on". Hence, the notation "$P(A|B)$" may be read "the probability that the event A occurs given that event B has occurred". The mathematical definition is presented next.

Definition 4.1 (Conditional probability of events) *Let A and B be two events, and assume $P(B) > 0$. Then the **conditional probability of A given B** is defined as*

$$P(A|B) = \frac{P(A \cap B)}{P(B)}.$$

DOI: 10.1201/9781003591191-4

The conditional probability can be interpreted as the probability of an event occurring, given that we have information, and use it, about whether another event has occurred.

Example 4.1 (Example 1.9, continued) *Let $A = \{high\ income\}$ and $B = \{long\ education\}$, with the following joint probabilities:*

	B	B^c
A	$P(A \cap B) = 0.1$	$P(A \cap B^c) = 0.05$
A^c	$P(A^c \cap B) = 0.15$	$P(A^c \cap B^c) = 0.7$

Using Definition 4.1, we can calculate the probability of a randomly selected individual having high income, given the information that the individual has a long education, as being

$$P(A|B) = \frac{P(A \cap B)}{P(B)} = \frac{0.1}{0.25} = 0.4.$$

If we compare this to the probability of having a high income, $P(A) = 0.15$, we see that the information that the person has a long education makes it more likely that the person also has a high income.

The probability of a high income, given the individual has a short education, is

$$P(A|B^c) = \frac{P(A \cap B^c)}{P(B^c)} = \frac{0.05}{1 - 0.25} = \frac{0.05}{0.75} = 0.067.$$

In this example, the probability of an individual having a high income is larger if we know the person has a long education compared to if we know the person has a short education.

Definition 4.1 of conditional probability of events can be extended to more events than two, by considering A or B as an event comprised of other events. For example, if

$$B = C \cap D,$$

where C and D are two events, then insert for B in Definition 4.1 to get

$$P(A \mid C \cap D) = \frac{P(A \cap C \cap D)}{P(C \cap D)}.$$

This can be interpreted as the probability that the event A occurs given that it is known that both event C and event D have occurred. Note that we may sometimes write $P(A|C, D)$ instead of $P(A|C \cap D)$.

4.3 Conditional distributions with two random variables

We next extend the concept of conditional probability with events to conditional distributions for two random variables. One question in applications may be if knowledge of the outcome of one random variable, say, X, changes the assessment of the likelihood of the outcomes of another random variable, say, Y. The answer to this question is found by considering the conditional distribution of Y given that X has a particular outcome, x, say.

4.3.1   Two discrete random variables

First we consider the conditional distribution of two discrete random variables. Next, we provide the definition of the *conditional probability mass function.*

Definition 4.2 (Conditional distribution, two discrete random variables) *Assume X and Y are discrete random variables with joint PMF $f_{X,Y}(x,y)$, and let $x \in \mathbb{R}$ be a real number such that $f_X(x) > 0$, where $f_X(x)$ is the marginal PMF of X. The* **conditional probability mass function** *(***conditional PMF***) of Y, conditional on $X = x$, is defined as*

$$f_{Y|X}(y|x) = \frac{f_{X,Y}(x,y)}{f_X(x)},$$

for all $y \in \mathbb{R}$.

In a similar manner, the conditional PMF $f_{X|Y}(x|y)$ of X conditional on $Y = y$ can be defined. It is worth stressing that, in general, $f_{Y|X}(y|x) \neq f_{X|Y}(x|y)$.

Example 4.2 (Example 3.3, continued) *Let the random variables Y and X denote, respectively, the GPA and ownership type of a randomly chosen school, such that*

$$Y = \begin{cases} 0 & \text{if GPA low,} \\ 1 & \text{if GPA high,} \end{cases}$$

$$X = \begin{cases} 0 & \text{if school private,} \\ 1 & \text{if school public.} \end{cases}$$

The joint distribution $f_{X,Y}(x,y)$ of X and Y, together with their marginal distributions, are given in the following table.

$f_{X,Y}(x,y)$		Y		$f_X(x)$
		0	*1*	
X	*0*	*0.10*	*0.25*	*0.35*
	1	*0.20*	*0.45*	*0.65*
$f_Y(y)$		*0.30*	*0.70*	

Based on these probabilities, the conditional PMF of Y given X can be calculated using Definition 4.2. For example, the probability of a low GPA given a public school is

$$f_{Y|X}(0|1) = \frac{f_{X,Y}(1,0)}{f_X(1)} = \frac{0.20}{0.65} \approx 0.31.$$

The next table contains the values of $f_{Y|X}(y|x)$ for all outcomes of x and y.

$f_{Y|X}$		Y *grade*	
		0	*1*
Ownership X	*0*	*0.29*	*0.71*
	1	*0.31*	*0.69*

It can be seen that 71% of the privately owned schools have high GPA whereas it is 69% among the publicly owned schools.

Example 4.3 (Example 3.6, continued) *Let the random variables X and Y denote, respectively, Public Service Motivation and Performance of a randomly chosen individual. The joint PMF $f_{X,Y}(x,y)$ of X and Y, and the marginal PMFs $f_X(x)$ of X and $f_Y(y)$ of Y, are given in the following table.*

$f_{X,Y}(x,y)$		Y		$f_X(x)$
		0	1	
X	1	0.40	0.10	0.5
	2	0.16	0.24	0.4
	3	0.04	0.06	0.1
$f_Y(y)$		0.60	0.40	1

Based on these probabilities, the three conditional distributions of Y given X can be calculated using Definition 4.2. They are given in the next table.

$f_{Y|X}(y|x)$		Y	
		0	1
X	1	0.80	0.20
	2	0.40	0.60
	3	0.40	0.60

Figure 4.1 illustrates the conditional distributions of Y given the three possible values of X. It is seen that the assessment of Performance is the same whether Public Service Motivation is medium ($X = 2$) or high ($X = 3$). Thus, the difference in Performance is associated with low, or not low, Public Service Motivation.

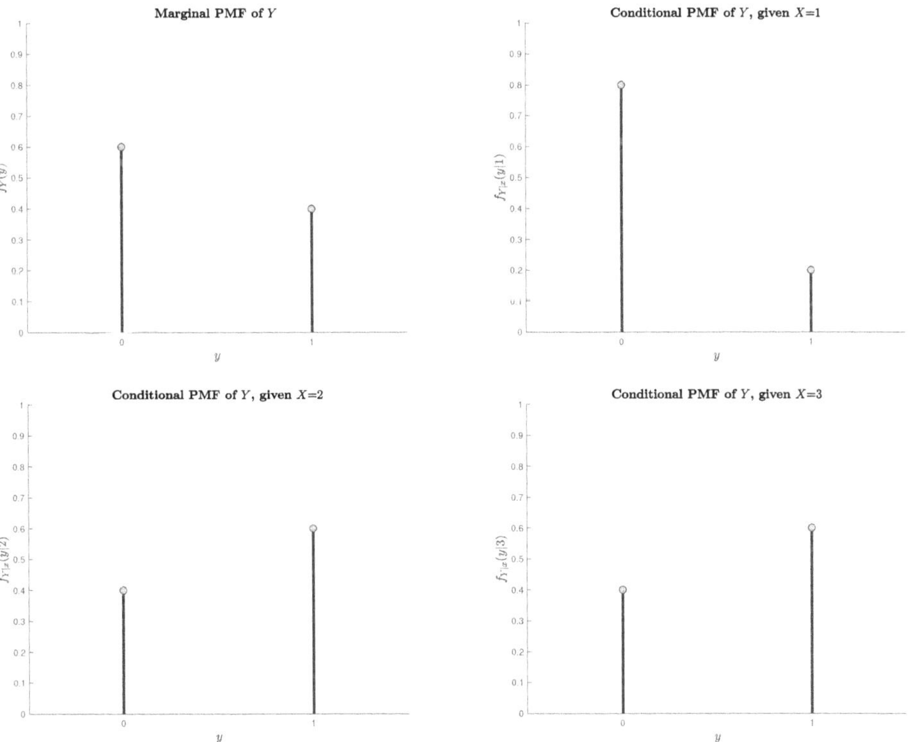

FIGURE 4.1

Marginal and conditional PMFs from Example 4.3. Top left: Marginal PMF $f_Y(y)$. Top right: Conditional PMF of Y given $X = 1$, $f_{Y|X}(y|1)$. Bottom left: Conditional PMF of Y given $X = 2$, $f_{Y|X}(y|2)$. Bottom right: Conditional PMF of Y given $X = 3$, $f_{Y|X}(y|3)$.

The joint distribution $f_{X,Y}$ contains more information about X and Y than the conditional distribution $f_{Y|X}$ because $f_{Y|X}$ can be derived from $f_{X,Y}$ whereas $f_{X,Y}$ cannot be derived from $f_{Y|X}$ unless we also know f_X. In practice, however, specifying the conditional distribution is often sufficient for many purposes, and specifying the conditional distribution may be easier than specifying the joint distribution.

4.3.2   Two continuous random variables

For two continuous random variables X and Y with joint PDF $f_{X,Y}(x,y)$, a *conditional probability density function* of Y given X can be defined.

Definition 4.3 (Conditional distribution, two continuous random variables)
Assume X and Y are continuous random variables with joint PDF $f_{X,Y}(x,y)$, and let $x \in \mathbb{R}$ be a real number such that $f_X(x) > 0$, where $f_X(x)$ is the marginal PDF of X. The **conditional probability density function** (**conditional PDF**) **of Y, conditional on** *$X = x$, is defined as*

$$f_{Y|X}(y|x) = \frac{f_{X,Y}(x,y)}{f_X(x)},$$

for all $y \in \mathbb{R}$.

The only difference in the definition of conditional distribution of two discrete random variables and conditional distribution of two continuous random variables is, that PMFs are used in the former case, whereas PDFs are used in the latter case.

Example 4.4 (Example 3.12, continued) *For $x, y > 0$, the joint PDF of the continuous random variables X and Y is*

$$f_{X,Y}(x,y) = 3xe^{-xy}e^{-3x}.$$

As shown in Example 3.12, the marginal PDF of X is

$$f_X(x) = 3e^{-3x}.$$

Let $x > 0$. Using Definition 4.3, the conditional PDF of Y given $X = x$ is

$$f_{Y|X}(y|x) = \frac{f_{X,Y}(x,y)}{f_X(x)} = \frac{3xe^{-xy}e^{-3x}}{3e^{-3x}} = xe^{-xy}.$$

In Problem 4.8.4 at the end of this chapter, you are asked to show that the marginal PDF of Y is

$$f_Y(y) = \frac{3}{(3+y)^2}, \quad y > 0.$$

Figure 4.2 illustrates the marginal PDF $f_Y(y)$ of Y and compares with two conditional PDFs of Y given X, $f_{Y|X}(y|1)$ and $f_{Y|X}(y|2)$, that is one for $x = 1$ and one for $x = 2$. As can be seen from the figure by comparing the solid line to the dashed lines, the knowledge of the value of X drastically changes the (conditional) distribution of Y, compared to the marginal (unconditional) distribution of Y.

From Figure 4.2, it can also be seen that for $y \leq 0.5$, the conditional density of Y given $X = 2$ is above the conditional density given $X = 1$. Recall that the area under the PDF is the cumulative probability, e.g. $P(Y \leq 0.5 \mid X = x)$ can be calculated as the area under the conditional PDF from $y = 0$ to $y = 0.5$. Hence, in this case, the assessment of the probability of an outcome $Y \leq 0.5$ is higher if $X = 2$ than if $X = 1$.

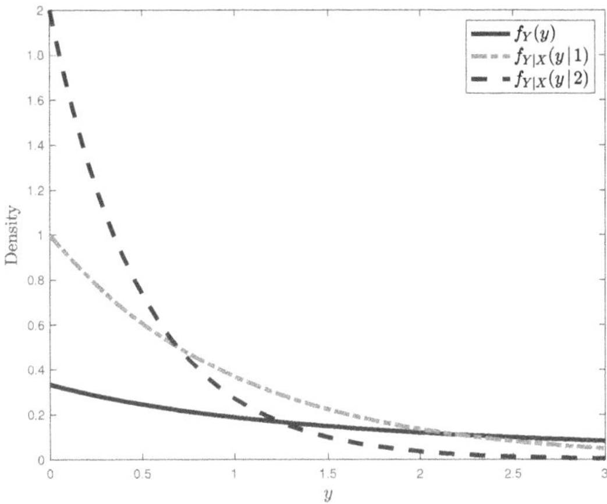

FIGURE 4.2
Marginal PDF $f_Y(y)$ of Y (solid line) and conditional PDFs $f_{Y|X}(y|x)$ of Y given X for two different values of x, $x = 1$ (dashed and dotted line) and $x = 2$ (dashed line), in Example 4.4.

4.3.3 Two random variables of mixed type

In this section, we consider how to model conditional distributions with random variables of mixed type. To do so, we define a *conditional cumulative distribution function* (conditional CDF). The reason is that the CDF exists for any type of random variable and, thus, can be used when the random variables involved are of the mixed type. Indeed, the same approach can be used if the two random variables are of different types, e.g. one is discrete and the other is continuous, or one is discrete and the other is mixed, etc. When the two random variables are either both discrete or both continuous, the approach presented here is equivalent to what we saw in the previous two sections, the difference being an emphasis on the conditional CDF instead of the conditional PMF or conditional PDF.

The conditional CDF is defined based on the joint CDF.

Definition 4.4 (Conditional distribution using CDFs) *Assume X and Y are two random variables with joint CDF $F_{X,Y}(x,y)$, and let $x \in \mathbb{R}$ be a real number such that $F_X(x) > 0$, where $F_X(x)$ is the marginal CDF of X. The **conditional cumulative distribution function (conditional CDF)** of Y, conditional on $X \leq x$, is defined as*

$$F_{Y|X}(y|x) = \Pr(Y \leq y | X \leq x) = \frac{F_{X,Y}(x,y)}{F_X(x)},$$

for all $y \in \mathbb{R}$.

The conditional CDF is defined with conditioning on $X \leq x$. This insures that the conditional CDF exists no matter the type of random variable of X.

Rather than conditioning on $X \leq x$, the interest may be the conditional CDF of Y conditional on $X = x$. The next two formulas can be used to derive the conditional CDF of Y given $X = x$ in the cases where X is either discrete or continuous. If X is discrete with PMF $f_X(x)$ and $x \in \mathbb{R}$ is such that $f_X(x) > 0$, then

$$\Pr(Y \leq y | X = x) = \frac{F_{X,Y}(x,y) - \lim_{a \to x^-} F_{X,Y}(a,y)}{f_X(x)},$$

for all y. If X is continuous with PDF $f_X(x)$ and $x \in \mathbb{R}$ is such that $f_X(x) > 0$, then

$$\Pr(Y \le y | X = x) = \frac{\frac{\partial F_{X,Y}(x,y)}{\partial x}}{f_X(x)},$$

for all y, where $\frac{\partial F_{X,Y}(x,y)}{\partial x}$ is the partial derivative of F with respect to x. In practice, when modeling random variables of mixed types, we often specify the conditional CDF conditional on $X = x$ directly without starting with a joint CDF.

4.4 Conditional distributions with three or more random variables

In case there are three random variables X, Y and Z, we can consider the distribution of one of them, say Y, given the other two, X and Z. We can also consider the joint distribution of two of them, say, X and Y, conditional on the third of them, Z. It is the application at hand that dictates which of these considerations is relevant. In case there are more than three variables, there are a host of different combinations of random variables on which we can condition. Below we consider some of the combinations which are most often used in practice.

We first extend the definition of conditional probability to three random variables. The definitions are similar for discrete and continuous random variables, except that in the former case the results are stated using PMFs whereas in the latter the results are stated using PDFs. For brevity, we collect the two cases in a single definition.

Definition 4.5 (Conditional distributions, three random variables) *Assume that X, Y, and Z are either discrete random variables with $f()$ denoting PMFs, or continuous random variables with $f()$ denoting PDFs. Let $z \in \mathbb{R}$ be a real number such that $f_Z(z) > 0$. Then*

$$f_{X,Y|Z}(x,y|z) = \frac{f(x,y,z)}{f_Z(z)},$$

*is the **conditional probability mass function** (**conditional PMF**) of X and Y, conditional on $Z = z$ if the random variables are discrete, and the **conditional density function** (**conditional PDF**) of X and Y conditional on $Z = z$ if the random variables are continuous.*

Let $x, z \in \mathbb{R}$ be real numbers such that $f_{X,Z}(x,z) > 0$. Then

$$f_{Y|X,Z}(y|x,z) = \frac{f(x,y,z)}{f_{X,Z}(x,z)}$$

*is the **conditional PMF of Y**, **conditional on $X = x$ and $Z = z$** if the random variables are discrete, and the **conditional PDF of Y**, **conditional on $X = x$ and $Z = z$** if the random variables are continuous.*

Example 4.5 (Example 4.2, continued) *We extend the example with an additional random variable Z, which indicates preschool language stimulation of students:*

$$Z = \begin{cases} 0 & \text{if preschool language stimulation low,} \\ 1 & \text{if preschool language stimulation high.} \end{cases}$$

The joint PMF $f_{X,Y,Z}(x,y,z)$ of X, Y and Z is given in the following two tables.

$f_{X,Y,Z}(x,y,0)$		*Y grade*	
		0	*1*
Ownership X	*0*	*0.06*	*0.09*
	1	*0.14*	*0.21*

$f_{X,Y,Z}(x,y,1)$		*Y grade*	
		0	*1*
Ownership X	*0*	*0.04*	*0.16*
	1	*0.06*	*0.24*

For example, a randomly selected school in this population has a 16% probability of being a private school with high grades and high language stimulation.

To calculate the conditional distribution of GPA (Y) conditional on ownership (X) and language stimulation (Z), the joint distribution of X and Z is needed. It can be calculated using Theorem 3.6. For example

$$f_{X,Z}(0,1) = f_{X,Y,Z}(0,0,1) + f_{X,Y,Z}(0,1,1) = 0.04 + 0.16 = 0.20.$$

Then, using Definition 4.5, we have, e.g. conditional on the school being private ($X = 0$) and having high preschool language stimulation ($Z = 1$), the probability that the school has a high GPA ($Y = 1$) is

$$f_{Y|X,Z}(1|0,1) = \frac{f_{X,Y,Z}(0,1,1)}{f_{X,Z}(0,1)} = \frac{0.16}{0.20} = 0.80 = 80\%.$$

The next two tables contain all values of the conditional probabilities of Y given X and Z

Ownership private ($X = 0$)			
$f_{Y|X,Z}(y|0,z)$		*Y grade*	
		0	*1*
language Z	*0*	*0.40*	*0.60*
	1	*0.20*	*0.80*

Ownership public ($X = 1$)			
$f_{Y|X,Z}(y|1,z)$		*Y grade*	
		0	*1*
language Z	*0*	*0.40*	*0.60*
	1	*0.20*	*0.80*

It is seen that the distribution of grades conditional on ownership and language stimulation is the same whether ownership is private or public, e.g. if $Z = 0$, then the probability of a high GPA is the same whether or not the school is public or private. The language stimulation Z, however, does change the conditional distribution of grades e.g. the probability of high GPA is 0.8 for high language stimulation compared to 0.6 for low language stimulation.

In general, we can divide random variables into those for which we want the joint distribution, and those upon which we want to condition. This leads to a general definition of conditional probability. We again state the definition for discrete and random variables together, since the only difference between the two cases is that in the former case we use PMFs and in the latter case we use PDFs.

Definition 4.6 (Conditional distributions, multiple random variables) *Let $k > m$ and assume $X_1, \ldots, X_m, X_{m+1}, \ldots, X_k$ are either discrete random variables with $f()$ being PMFs, or continuous random variables with $f()$ being PDFs. Let $x_{m+1}, \ldots, x_k \in \mathbb{R}$ be real numbers such that $f_{X_{m+1},\ldots,X_k}(x_{m+1}, \ldots, x_k) > 0$, where $f_{X_{m+1},\ldots,X_k}$ is the joint PMF/PDF of $X_{m+1}, \ldots, X_k$. Then*

$$f_{X_1,\ldots,X_m|X_{m+1},\ldots,X_k}(x_1, \ldots, x_m|x_{m+1}, \ldots, x_k) = \frac{f_{X_1,\ldots,X_m,\ldots,X_k}(x_1, \ldots, x_m, x_{m+1}, \ldots, x_k)}{f_{X_{m+1},\ldots,X_k}(x_{m+1}, \ldots, x_k)}$$

*is the **conditional PMF of** $X_1, \ldots, X_m$ **conditional on** $X_{m+1} = x_{m+1}, \ldots, X_k = x_k$ if the random variables are discrete, or the **conditional PDF of** $X_1, \ldots, X_m$ **conditional on** $X_{m+1} = x_{m+1}, \ldots, X_k = x_k$ if the random variables are continuous.*

Most often, we are interested in modeling the distribution of one random variable conditional on a host of other random variables. These conditioning random variables are often included such that we can focus on various subgroups of the population. For instance, in the context of Example 4.5, we might be interested in the two subgroups of the population determined by either low ($Z = 0$) or high ($Z = 1$) language stimulation.

A joint distribution can be decomposed into a product of conditional distributions. Such a decomposition turns out to be useful when modeling complex relationships. The result is presented in the next theorem.

Theorem 4.1 (Chain rule for probabilities) *Let X, Y, and Z be random variables with $f_{X,Y,Z}(x, y, z)$ being either a joint PMF or a joint PDF. Then*

$$f_{X,Y,Z}(x, y, z) = f_{X|Y,Z}(x|y, z) \cdot f_{Y|Z}(y|z) \cdot f_Z(z),$$

for all $x, y, z \in \mathbb{R}$.

In general, for k discrete random variables $X_1, \ldots, X_k$ with $f_{X_1,\ldots,X_k}(x_1, \ldots, x_k)$ being either a joint PMF or a joint PDF, then

$$f_{X_1,\ldots,X_k}(x_1, \ldots, x_k) = \left(\prod_{j=1}^{k-1} f_{X_j|X_{j+1},\ldots,X_k}(x_j|x_{j+1}, \ldots, x_k) \right) \cdot f_{X_k}(x_k),$$

for all $x_1, \ldots, x_k \in \mathbb{R}$, where $\prod_{j=1}^{k-1}$ means the product of all terms from $j = 1$ to $j = k - 1$.

The chain rule for probabilities is also known as the general product rule for probabilities.

4.5 Relation between long and short conditioning sets

In this section, show how conditional distributions with different conditioning sets of random variables are related. Knowing this can be important in applications, because we may only be able to condition on a subset of the variables upon which we ideally would have liked to condition, e.g. if some variables in practice are not observed by the statistician.

Suppose we want to calculate the conditional distribution of Y given X from the conditional distribution of Y given X and Z. We will here call conditioning on both X and Z for the "long conditioning set", and conditioning on only X for the "short conditioning set". In this terminology, the task is to find how the distribution of Y conditional on the long conditioning set relates to the distribution of Y conditional on the short conditioning set. The next theorem provides the answer.

Theorem 4.2 (Long and short conditioning sets)

1. *(Three variable case.) Let the long conditioning set be X and Z, and the short conditioning set be X. If X, Y, and Z are discrete random variables, then*

$$f_{Y|X}(y|x) = \sum_z f_{Y|X,Z}(y|x, z) f_{Z|X}(z|x),$$

for all $y, x \in \mathbb{R}$, where the sum is taken over all possible outcomes z of Z. If X, Y, and Z are continuous random variable, then

$$f_{Y|X}(y|x) = \int_{-\infty}^{\infty} f_{Y|X,Z}(y|x, z) f_{Z|X}(z|x) dz,$$

for all $y, x \in \mathbb{R}$.

2. *(General case.) Let $X_1, \ldots, X_m$ be the short conditioning set and let $X_1, \ldots, X_k$, $k > m$, be the long conditioning set. If $X_1, \ldots, X_k$ are discrete random variables, then*

$$f_{Y|X_1,\ldots,X_m}(y|x_1,\ldots,x_m) = \sum_{x_{m+1}} \cdots \sum_{x_k} \left[f_{Y|X_1,\ldots,X_k}(y|x_1,\ldots,x_k) \right.$$
$$\left. \cdot f_{X_{m+1},\ldots,X_k|X_1,\ldots,X_m}(x_{m+1},\ldots,x_k|x_1,\ldots,x_m) \right].$$

for all $y, x_1, \ldots, x_m \in \mathbb{R}$. If $X_1, \ldots, X_k$ are continuous random variables, then

$$f_{Y|X_1,\ldots,X_m}(y|x_1,\ldots,x_m) = \int_{-\infty}^{\infty} \cdots \int_{-\infty}^{\infty} \left[f_{Y|X_1,\ldots,X_k}(y|x_1,\ldots,x_k) \right.$$
$$\left. \cdot f_{X_{m+1},\ldots,X_K|X_1,\ldots,X_m}(x_{m+1},\ldots,x_k|x_1,\ldots,x_m) \right]$$
$$dx_{m+1} \ldots dx_k,$$

for all $y, x_1, \ldots, x_m \in \mathbb{R}$.

Example 4.6 (Example 4.2, continued) *The conditional PMF $f_{Y|X}(y|x)$ was calculated to be as follows:*

$f_{Y|X}$		Y grade	
		0	1
Ownership X	0	0.29	0.71
	1	0.31	0.69

This result shows that different ownership X is related to different distributions of grades: It is more likely to have a school with high GPA if the school is private.

In Example 4.5, the conditional PMF $f_{Y|X,Z}(y|x,z)$ of Y conditional on X and Z was calculated; we reproduce the results in the next table.

Ownership private ($X = 0$)			
$f_{Y|X,Z}(y|0,z)$		Y grade	
		0	1
language Z	0	0.40	0.60
	1	0.20	0.80

Ownership public ($X = 1$)			
$f_{Y|X,Z}(y|1,z)$		Y grade	
		0	1
language Z	0	0.40	0.60
	1	0.20	0.80

We see that, when conditioning on Z, conditioning also on X does not change the conditional distribution, i.e. $f_{Y|X,Z}(y|0,z) = f_{Y|X,Z}(y|1,z)$ for all y and z. This means that students with same language ability have the same GPA distribution no matter the type of school ownership.

According to Theorem 4.2, $f_{Y|X}$ can be calculated from $f_{Y|X,Z}$ and $f_{Z|X}$. Hence, by calculating $f_{Z|X}$ we can see why ownership X matters differently in $f_{Y|X}$ and $f_{Y|X,Z}$. The conditional PMF $f_{Z|X}$ is (see Problem 4.8.3)

$f_{Z|X}$		Z language	
		0	1
Ownership X	0	0.43	0.57
	1	0.54	0.46

It is seen that in privately owned schools, it is more likely that students have high preschool language stimulation. Hence, this would support a claim that the reason ownership relates to GPA in $f_{Y|X}$ is that students with high preschool language stimulation are more likely to go to privately owned schools.

4.6　Bayes' theorem

Bayes' theorem is useful for turning conditional probabilities which we know, e.g. because they are easy to assess in practice, into conditional probabilities which are of interest. First, to get some intuition for the theorem, we consider Bayes' theorem for events, and then follow with Bayes' theorem for random variables.

4.6.1　Bayes' theorem for events

In some situations we know $P(B|A)$, $P(B|A^c)$, $P(A)$, and $P(B)$ from past experience, and we want to know $P(A|B)$. Bayes' theorem is well suited for such situations.

Theorem 4.3 (Bayes' theorem for events) *Assume A and B are two events such that $P(A), P(B) > 0$. Then*

$$P(A|B) = \frac{P(B|A)P(A)}{P(B)} = \frac{P(B|A)P(A)}{P(B|A)P(A) + P(B|A^c)P(A^c)}.$$

Bayes' theorem can be very useful if we are interested in a conditional probability, say $P(A|B)$, and we have a way of assessing the other probabilities in Bayes' theorem. Problem 4.8.8 at the end of this section contains an example of this. The following example illustrates the usefulness of Bayes' theorem in the case where the object of interest is the conditional probability $P(A|B^c)$.

Example 4.7 *When screening for a disease, a concern is the quality of the screening. Let A be presence of a disease and let B be a screening indicating the disease is present. Denote A occurring as "positive" and A^c occurring as "negative". If the screening is positive, then one would proceed with more elaborate testing.*

The worry is when the screening is negative but the individual has the disease. In that situation one would not proceed with further testing so it is not detected that the individual has the disease. This corresponds to the probability of A given B^c. To calculate this probability, we can use Bayes' theorem,

$$P(A|B^c) = \frac{P(B^c|A)P(A)}{P(B^c|A)P(A) + P(B^c|A^c)P(A^c)},$$

because we can assess the various quantities on the right hand side by experiments and history.

Suppose we know from history that a certain proportion of the population gets the disease. Then we know $P(A)$ and $P(A^c)$. From medical experiments, we may also know the following quantities:

- *$B^c|A^c$: The screening is negative when there is no disease (called "true negative").*

- *$B|A^c$: The screening is positive when there is no disease (called "false positive").*

- *$B^c|A$: The screening is negative when there is a disease (called "false negative").*

- *$B|A$: The screening is positive when there is a disease (called "true positive").*

All these quantities can be assessed by applying the screening to individuals which we have examined extensively such that we know for certain if they have the disease or not.

Notice, false positives are false warnings, whereas false negatives correspond to no warning despite the disease being present.

In practice the interest is whether or not the individual has the disease, and to figure this out in a quick and inexpensive manner. Therefore screening is used and, thus, the probability of $P(A|B^c)$ is key to assess the quality of the screening.

4.6.2 Bayes' theorem for random variables

Bayes' theorem for random variables shows how to write the conditional distribution of X given Y using knowledge of the conditional distribution of Y given X. Bayes' theorem for random variables is stated next.

Theorem 4.4 (Bayes' theorem for random variables) *If X and Y are discrete random variables, then*

$$f_{X|Y}(x|y) = \frac{f_{Y|X}(y|x) \cdot f_X(x)}{f_Y(y)} = \frac{f_{Y|X}(y|x) \cdot f_X(x)}{\sum_{x_j} f_{Y|X}(y|x_j) \cdot f_X(x_j)},$$

for all $x, y \in \mathbb{R}$ such that $f_X(x), f_Y(y) > 0$, where the f's refer to PMFs. If X and Y are continuous random variables, then

$$f_{X|Y}(x|y) = \frac{f_{Y|X}(y|x) \cdot f_X(x)}{f_Y(y)} = \frac{f_{Y|X}(y|x) \cdot f_X(x)}{\int_{-\infty}^{\infty} f_{Y|X}(y|t) \cdot f_X(t)dt},$$

for all $x, y \in \mathbb{R}$ such that $f_X(x), f_Y(y) > 0$, where the f's refer to PDFs.

Assume we know how X influences Y. For example, this knowledge could be in terms of a model. This implies that we know $f_{Y|X}(y|x)$. Suppose our interest is the distribution of X. As a starting point, we have an assessment of the distribution of X, expressed as $f_X(x)$. If we are given new information in terms of Y, say $Y = y$ for a value $y \in \mathbb{R}$, we would like to know the distribution of X given this new information. This distribution can be calculated by Bayes' theorem. In other words, Bayes' theorem tells us how to update our knowledge about (the distribution of) X, given new information about Y.

4.7 Proofs

Proof of Theorem 4.1 *This result follows from the definition of the conditional probability. We show the result for three discrete random variables X, Y, Z. The other statements in the theorem can be proved using analogous arguments. Recall that, from Definition 4.5,*

$$f_{X,Y|Z}(x,y|z) = \frac{f(x,y,z)}{f_Z(z)}.$$

We may re-write this as

$$f(x,y,z) = f_{X,Y|Z}(x,y|z) \cdot f_Z(z).$$

Similarly, we have that

$$f_{X,Y|Z}(x,y|z) = f_{X|Y,Z}(x|y,z) \cdot f_{Y|Z}(y|z).$$

Taking this together yields

$$f(x, y, z) = f_{X|Y,Z}(x|y, z) \cdot f_{Y|Z}(y|z) \cdot f_Z(z),$$

as we wanted to show.

Proof of Theorem 4.2 *We again show this for the case of three discrete random variables X, Y, Z. The other statements in the theorem can be proved using analogous arguments. Note first that, according to (a conditional version of) Theorem 3.2, we have that*

$$f_{Y|X}(y|x) \sum_z f_{Y,Z|X}(y, z|x).$$

Further, by the definition of the conditional probability,

$$f_{Y,Z|X}(y, z|x) = f_{Y|X,Z}(y|x, z) f_{Z|X}(z|x).$$

Taking these together yields the desired result,

$$f_{Y|X}(y|x) = \sum_z f_{Y|X,Z}(y|x, z) f_{Z|X}(z|x).$$

Proof of Theorem 4.3 *The conditional probability of event A given event B can be expressed by the probability of event B given event A using the definition of the conditional probability of events (Definition 4.1) twice:*

$$P(A|B) = \frac{P(A \cap B)}{P(B)} = \frac{P(B|A) \cdot P(A)}{P(B)}. \tag{4.1}$$

Now, using first the law of total probability (Theorem 1.2) and then again Definition 4.1 twice, we may express $P(B)$ as

$$\begin{aligned} P(B) &= P(B \cap A) + P(B \cap A^c) \\ &= P(B|A)P(A) + P(B|A^c)P(A^c). \end{aligned}$$

Finally, insert this into (4.1) to get

$$P(A|B) = \frac{P(B|A)P(A)}{P(B|A)P(A) + P(B|A^c)P(A^c)},$$

which is what we wanted to show.

Proof of Theorem 4.4 *This follows along similar lines as the proof of Theorem 4.3. In particular, we have, from the definition of conditional probability, that*

$$f_{X|Y}(x|y) = \frac{f_{X,Y}(x, y)}{f_Y(y)} \tag{4.2}$$

and

$$f_{Y|X}(y|x) = \frac{f_{X,Y}(x, y)}{f_X(x)}.$$

Re-arrange the latter result to get

$$f_{X,Y}(x,y) = f_{Y|X}(y|x) \cdot f_X(x).$$

Insert this into (4.2) to get

$$f_{X|Y}(x|y) = \frac{f_{Y|X}(y|x) \cdot f_X(x)}{f_Y(y)}. \tag{4.3}$$

Suppose that X and Y are discrete random variables (the continuous case follows similar arguments). By (a conditional version of) Theorem 3.2, we have that

$$f_Y(y) = \sum_{x_j} f_{X,Y}(x_j, y).$$

Further, by the definition of conditional probability,

$$f_{X,Y}(x,y) = f_{Y|X}(y|x) f_X(x).$$

Inserting these expressions into (4.3) yields the result,

$$f_{X|Y}(x|y) = \frac{f_{Y|X}(y|x) \cdot f_X(x)}{f_Y(y)} = \frac{f_{Y|X}(y|x) \cdot f_X(x)}{\sum_{x_j} f_{X,Y}(x_j, y)} = \frac{f_{Y|X}(y|x) \cdot f_X(x)}{\sum_{x_j} f_{Y|X}(y|x_j) f_X(x_j)}.$$

4.8   Exercises

Problem 4.8.1 *The first of these problems is known as the "two-child problem".*

1. *A family has two children. You know that one of them is a boy. What is the probability that the other child is a girl?*

2. *A family has two children. You know that the first of them is a boy. What is the probability that the other child is a girl?*

Problem 4.8.2 *Consider the two random variables describing the behavior of a randomly selected car driver: Speeding violations (Y) and drivers who use mobile phones (X). Their distribution is given by their joint probability function $f_{X,Y}(x,y)$ displayed in the table below.*

Table: Car phone use and speeding violations

	Speeding violation, $Y = 1$	No speeding violation, $Y = 0$
Car phone user, $X = 1$	0.033	0.371
Not car phone user, $X = 0$	0.060	0.536

Calculate the conditional probability functions, $f_{X|Y}(x|y)$ and $f_{Y|X}(y|x)$, and interpret your results.

Problem 4.8.3 *Use the information in Examples 4.5 to calculate the conditional PMF $f_{Z|X}$, given in a table in Example 4.6.*

Problem 4.8.4 *Let X and Y be two continuous random variables and consider their joint PDF from Example 3.12,*

$$f_{X,Y}(x,y) = 3xe^{-xy}e^{-3x}, \quad x, y > 0.$$

1. Show that the marginal PDF of Y is given by $f_Y(y) = \frac{3}{(3+y)^2}$ for $y > 0$. (Hint: Use Theorem 3.9 and do integration by parts.)

2. Show that $f_Y(y) = \frac{3}{(3+y)^2}$, $y > 0$, is a valid PDF.

3. Derive the conditional PDF of X given Y, $f_{X|Y}(x|y)$. Sketch the marginal PDF of X and compare with the conditional PDFs of X given $Y = 1$ and $Y = 2$. (See Example 4.4 for a similar treatment of the conditional PDF of Y given X.)

Problem 4.8.5 *Let the conditional density function of X given Z be*

$$f_{X|Z}(x|z) = \begin{cases} \frac{1}{4}(x-z) & \text{if } 0 < (x-z) < 2, \\ -\frac{1}{4}(x-z)+1 & \text{if } 2 \le (x-z) \le 4, \\ 0 & \text{otherwise.} \end{cases}$$

1. Sketch the conditional density functions of X given Z for $z = 0$, $z = 1$, and $z = 2$. Discuss how the conditional density function of X given Z depends on z.

2. What is the conditional probability that $X \le 2$ given $Z = 0$? Compare to the conditional probability that $X \le 2$ given $Z = 1$.

Problem 4.8.6 *Let the conditional density function of Y given X be*

$$f_{Y|X}(y|x) = \begin{cases} \frac{1}{3}\frac{1}{x}y & \text{if } 0 < x < 6 \text{ and } 0 < y < x, \\ \frac{1}{3}\frac{1}{6-x}(6-y) & \text{if } 0 < x < 6 \text{ and } x \le y < 6, \\ 0 & \text{otherwise,} \end{cases}$$

and the marginal density function of X be

$$f_X(x) = \begin{cases} \frac{1}{18}x & \text{if } 0 < x < 6, \\ 0 & \text{otherwise.} \end{cases}$$

1. Sketch the conditional density functions of Y given X for $x = 1$, $x = 3$, and $x = 5$. Comment on how the conditional density function of Y given X depends on x.

2. Derive the joint density function $f_{X,Y}(x,y)$ of X and Y.

Problem 4.8.7 *Consider different areas of cities characterized by the crime rate (X), the presence of police (Y) and the economic wellbeing (Z). Each of these variables are defined as follows:*

$$X = \begin{cases} 0 & \text{low crime rate,} \\ 1 & \text{high crime rate,} \end{cases}$$

$$Y = \begin{cases} 0 & \text{little presence,} \\ 1 & \text{much presence,} \end{cases}$$

$$Z = \begin{cases} 0 & \text{not depressed,} \\ 1 & \text{depressed.} \end{cases}$$

Let the joint probability function $f_{X,Y,Z}(x,y,z)$ for X, Y, and Z be given by

$f_{X,Y,Z}(x,y,0)$

	X: 0	1
Y: 0	0.30	0.05
1	0.15	0.00

$f_{X,Y,Z}(x,y,1)$

	X: 0	1
Y: 0	0.00	0.10
1	0.10	0.30

The purpose is to investigate the association between crime and presence of police, and how this association depends on the economic wellbeing of an area.

1. *How is crime rate and police presence related? Interpret your result.*

2. *How is crime rate and police presence related when controlling for the economic wellbeing of an area? Interpret your result and compare it to your result in the previous question.*

Problem 4.8.8 *You are worried that you might be suffering from illness X. You go to the doctor and she performs a medical test; the test is indeed "positive" for illness X. The doctor informs you that in medical trials, the test has been found to have the following properties:*

(i) *The sensitivity of the test is 99%: This means that if the test is performed on a patient with the illness, then the test will indicate "positive" in 99% of cases. In other words, the "false negative" rate is 1%.*

(ii) *The specificity of the test is 99%: This means that if the test is performed on a healthy patient, then the test will indicate "negative" in 99% of cases. In other words, the "false positive" rate is 1%.*

Further, you know that illness X is uncommon in the population where you are from. In fact you know that:

(iii) *In the population where you are from, 0.5% have the illness.*

Use the information in (i)–(iii), to calculate the conditional probability that you are suffering from illness X, given the fact that you already tested "positive" for illness X. (Hint: Use Bayes' theorem for events, Theorem 4.3.)

5

Aspects of a univariate distribution

5.1 Introduction

In many circumstances, we are interested in certain aspects of a distribution. For instance, in the case of the distribution of incomes in a population, we might be interested in things like the mean or median income level, or a measure of the inequality of the incomes in the distribution. In general, being able to pinpoint particular aspects of a distribution can be useful to answer substantial questions.

In this chapter, we consider various ways, or measures, to describe a distribution. To characterize the distribution of a random variable, or the population that the random variable represents, we can define various descriptive measures that capture aspects of the distribution. Consider, for example, the distribution of personal incomes in Denmark and Germany. We may be interested in comparing features of the income distribution in these two countries. Rather than comparing a very long list of the incomes of the Danish population with an even longer list of incomes of the German population, we can instead compare descriptive measures of the income distributions, such as the mean income of the Danish and German populations, respectively.

It is important to note that the distribution itself contains all the information available. Indeed, the descriptive measures we will introduce below will be derived directly from the distribution. However, although these measures do not present new information over and above what is in the distribution, they can nonetheless help get a good overview of the descriptive aspects, or properties, of the distribution. Further, as we will see in later chapters, in cases where we do not know the distribution, we can use data to estimate these descriptive measures without having to estimate the distribution itself.

5.2 The expected value and mean of a random variable

The *expected value*, or *mean*, of a random variable is a measure of the central tendency of the distribution of the random variable. The technique to calculate the expected value depends on the type of random variable. Next, we formally define the expected value of a random variable.

Definition 5.1 (Expected value, E(X)) *The **expected value** $E(X)$ of a discrete random variable X with PMF $f_X(x)$ is*

$$E(X) = \sum_x x \cdot f_X(x),$$

where $\sum_x$ is the sum over all the possible outcomes x of X.

DOI: 10.1201/9781003591191-5

*The **expected value** $E(X)$ of a continuous random variable X with PDF $f_X(x)$ is*

$$E(X) = \int_{-\infty}^{\infty} x \cdot f_X(x)dx.$$

We will often denote the expectation of a random variable X by μ_X, that is, $\mu_X = E(X)$, and we call this value the mean of X. The two terms, "mean" and "expected value" of a random variable, will be used interchangeably in this book.

Example 5.1 *Let the random variable X be coding the attitude of a randomly selected individual toward a referendum, where 1 is "for" and 0 is "against". If the PMF of X is*

$$f_X(x) = \begin{cases} 0.52 & if \quad x = 1, \\ 0.48 & if \quad x = 0, \end{cases}$$

then, using Definition 5.1, the expected value of X is

$$E(X) = 0 \cdot f_X(0) + 1 \cdot f_X(1) = 0 \cdot 0.48 + 1 \cdot 0.52 = 0.52.$$

The mean of a random variable is informative about the location of the possible outcomes of the random variable. Physically it can be interpreted as a balanced point. In the context of Example 5.1, suppose we have a seesaw and we line up all the individuals who are "against" (0) at the left end and all the individuals who are "for" (1) at the right end of the seesaw. Assuming they have the same weight and the seesaw is 1 meter (0 to 1), the balance point of the seesaw is 0.52 meter from the left.

Example 5.2 *Let the random variable X be the outcome of one throw with a six-sided die. Assuming that the die is fair, the PMF of X is*

$$f_X(x) = \frac{1}{6}, \quad x = 1, 2, \ldots, 6.$$

Using Definition 5.1, the expected value of X is found to be

$$E(X) = \sum_{j=1}^{6} j \cdot f_X(j) = \frac{1}{6} \sum_{j=1}^{6} j = \frac{21}{6} = 3.5.$$

Example 5.2 illustrates that the expected value of a random variable (in the example, $\mu_X = E[X] = 3.5$) need not be a possible outcome for the random variable (in the example, the sample space is $\{1, 2, 3, 4, 5, 6\}$). Instead, the expected value can be thought of as the average value you would get if you could sample the random variable, i.e. perform the statistical experiment, infinitely many times. In Example 5.2, this means that if you could throw the die infinitely many times and average the results, you would get 3.5.

Example 5.3 (Example 2.9, continued) *Let the CDF and PDF of a continuous random variable X be given by*

$$F_X(x) = \begin{cases} 0 & if \quad x < 0, \\ 0.5 \cdot x & if \quad 0 \le x < 2, \\ 1 & if \quad x \ge 2, \end{cases}$$

and

$$f_X(x) = \begin{cases} 0 & if \quad x < 0, \\ 0.5 & if \quad 0 \le x < 2, \\ 0 & if \quad x \ge 2. \end{cases}$$

They are both illustrated in Figure 2.2. The mean of X is

$$E(X) = \int_{-\infty}^{\infty} x \cdot f_X(x)dx \;=\; \int_{-\infty}^{0} x \cdot 0 dx + \int_{0}^{2} x \cdot 0.5 dx + \int_{2}^{\infty} x \cdot 0 dx$$

$$= \left[0.5 \cdot 0.5 \cdot x^2\right]_0^2$$

$$= 0.5 \cdot 0.5 \cdot 2^2 - 0.5 \cdot 0.5 \cdot 0^2$$

$$= 1.$$

Figuratively speaking, if we imagine that the graph of the PDF is a body, here in the shape of a box, then the balance point is at $x = 1$.

There are distributions for which the expected value does not exist. Formally, we say that the expected value $E(X)$ *exists* if $E(|X|) < \infty$, where "$|\cdot|$" denotes the absolute value. An expected value may fail to exist if the sum or integral in the definition is not defined. The following example illustrates this.

Example 5.4 *Let X be a continuous random variable with PDF*

$$f_X(x) = \begin{cases} 0 & \text{if } x < 1, \\ 0.5x^{-1.5} & \text{if } x \geq 1. \end{cases}$$

The expected value is

$$E(X) = \int_{-\infty}^{\infty} x f_X(x)dx \;=\; \int_{-\infty}^{1} x \cdot 0 dx + \int_{1}^{\infty} x \cdot 0.5 x^{-1.5} dx$$

$$= 0.5 \int_{1}^{\infty} x^{-0.5} dx$$

$$= 0.5 \left[2x^{0.5}\right]_1^{\infty}$$

$$= \underbrace{\lim_{b \to \infty} \left(0.5 \cdot 2b^{0.5}\right)}_{=\infty} - 0.5 \left[2 \cdot 1^{0.5}\right]$$

$$= \infty,$$

where the second-to-last equality sign follows from the definition of an integral with infinite upper boundary, see Definition B.19 in the appendix. Hence, the integral does not exist (equals ∞) nor does the expected value. The reason is that the "tail" of the distribution, i.e. of $f_X(x)$, is too thick: very large outcomes for X are relatively likely, in a way such that the function $x f_X(x)$ is not integrable.

The expected value of a random variable of mixed type can be calculated by considering the expected values of the discrete part and the continuous part separately. Then the expected value is the weighted sum of those two. The formal definition is given next.

Definition 5.2 (Expected value of mixed random variable) *Let X be a mixed type random variable with CDF*

$$F_X(x) = \alpha F_X^d(x) + (1 - \alpha)F_X^c(x),$$

with $\alpha \in (0,1)$, see Theorem 2.15 for this decomposition. Let $X^{(d)}$ be a discrete random variable with CDF $F_X^d(x)$ and $X^{(c)}$ be a continuous random variable with CDF $F_X^c(x)$. Then, the expected value of the mixed type random variable X is

$$E(X) = \alpha E(X^{(d)}) + (1 - \alpha)E(X^{(c)}),$$

where $E(X^{(d)})$ and $E(X^{(c)})$ are defined in Definition 5.1.

The next example illustrates how to use Definition 5.2 to calculate the expected value of a mixed type random variable.

Example 5.5 (Example 2.14, continued) *Consider the CDF $F_X(x)$ of the mixed type random variable X, given by*

$$F_X(x) = 0.4 \cdot F_X^d(x) + 0.6 \cdot F_X^c(x),$$

where the continuous part is

$$F_X^c(x) = \begin{cases} 0 & \text{if} & x < 0, \\ 0.5 \cdot x & \text{if} & 0 \le x < 2, \\ 1 & \text{if} & x \ge 2, \end{cases}$$

and the discrete part is

$$F_X^d(x) = \begin{cases} 0 & \text{if} & x < 0, \\ 0.25 & \text{if} & 0 \le x < 2, \\ 1 & \text{if} & x \ge 2. \end{cases}$$

The expected value of a continuous random variable X^c with CDF $F_X^c(x)$ is

$$E(X^c) = \int_{-\infty}^{\infty} \frac{dF_X^c(x)}{dx} \, dx = \int_0^2 0.5 \, dx = [0.5x]_0^2 = 1 - 0 = 1.$$

The expected value of a discrete random variable with CDF $F_X^d(x)$ is

$$E(X^d) = \sum x \cdot f_X^d(x) = 0 \cdot 0.25 + 2 \cdot 0.75 = 1.5,$$

where $f_X^d(x)$ is the PMF corresponding to the CDF $F_X^d(x)$.
By Definition 5.2, the expected value of X can now be calculated by weighing together the expected values of the two parts:

$$E(X) = 0.4 \cdot E(X^d) + 0.6 \cdot E(X^c) = 0.4 \cdot 1.5 + 0.6 \cdot 1 = 1.2.$$

5.3 Rules of calculation for expected value

Consider a random variable X and a function of one variable, say $g(x)$. We can obtain a new random variable by evaluating the function at the random variable, i.e. $Y = g(X)$. When we need the expected value of the function $g(X)$ of a random variable X, it can be calculated without first having to derive the distribution of $g(X)$. The next theorem shows how.

Theorem 5.1 (Expectation of function of one random variable) *Let $g(x)$ be a function.*
If X is a discrete random variable with PMF $f_X(x)$, then

$$E(g(X)) = \sum_x g(x) \cdot f_X(x),$$

where $\sum_x$ denotes a summation over all possible values of X.
If X is a continuous random variable with PDF $f_X(x)$, then

$$E(g(X)) = \int_{-\infty}^{\infty} g(x) \cdot f_X(x) dx.$$

Example 5.6 (Example 5.2, continued) *Let again the random variable X denote the outcome of one throw of a fair six-sided die. Consider the random variable Y that equals the square of the outcome of the die. That is, we may write Y as*

$$Y = X^2.$$

We can calculate the mean of Y, i.e. the mean of X^2, using Theorem 5.1 and the function $g(x) = x^2$,

$$E(Y) = E(X^2) = \sum_{j=1}^{6} j^2 \cdot f_X(j) = \frac{1}{6} \sum_{j=1}^{6} j^2 = \frac{91}{6} = 15.17.$$

Based on Theorem 5.1, we can derive various rules of calculation. Some of these rules are collected in the next theorem.

Theorem 5.2 (Rules of calculation of mean) *Let $a \in \mathbb{R}$ be a constant, and X a random variable such that $E(X)$ exists. Then the following hold.*

1. *$E(a) = a$.*

2. *$E(a + X) = a + E(X)$.*

3. *$E(aX) = aE(X)$.*

The rules of calculations in Theorem 5.2 imply that, for constants $a, b \in \mathbb{R}$, we have (Problem 5.9.1)

$$E(a + bX) = a + bE(X).$$

Because of this result, we sometimes say that "taking expectations is a linear operation", meaning that additive and multiplicative constants can be taken outside of the expectation. Recall that a general definition of a function $g(x)$ being linear is that $g(ax) = ag(x)$ for all x, and $g(x_1 + x_2) = g(x_1) + g(x_2)$ for all x_1 and x_2.

It is worth stressing that, in general,

$$E(g(X)) \neq g(E(X)).$$

For instance, $E\left(X^2\right)$ is not equal to $(E(X))^2$ except under special circumstances.

Example 5.7 (Example 5.6, continued) *Let again the random variable X denote the outcome of one throw of a fair six-sided die. Recall from Examples 5.2 and 5.6 that*

$$E(X) = \frac{21}{6} = 3.5$$

and

$$E(X^2) = \frac{91}{6} = 15.17.$$

We see that

$$E(X)^2 = 3.5^2 = 12.25 \neq 15.17 = E(X^2).$$

5.4 Variance

Whereas the mean describes an aspect of the central tendency, or location, of the distribution, it may also be of interest to quantify how much the distribution varies around the mean. For example, the mean income of two countries may be the same, but in one of the countries the income may almost be the same for everyone and in the other country the income may be quite different over individuals. One measure of variation is the *variance* of a random variable. It is defined next.

Definition 5.3 (Variance, $Var(X)$) *The **variance** of a random variable X, denoted $Var(X)$, is defined as*

$$Var(X) = E\left((X - E(X))^2\right).$$

We often denote the variance of X with σ_X^2, that is, $\sigma_X^2 = Var(X)$. The unit of measurement of the variance is the squared unit of measurement of X, e.g. if X is in m, then $Var(X)$ is in m^2. The variance can be transformed into a measure of variation with the same unit of measurement as X by taking the square root of the variance. This measure of variation is called the *standard deviation* of X, σ_X, and it is defined next.

Definition 5.4 (Standard deviation) *Let $\sigma_X^2 = Var(X)$ be the variance of X. Then*

$$\sigma_X = \sqrt{\sigma_X^2} = \sqrt{Var(X)}$$

*is called the **standard deviation** of X.*

It can be seen that the definition of the variance is based on the Definition 5.1 of the expected value. It can be written as the expected value of the function $g()$ of X given by

$$g(X) = (X - E(X))^2.$$

Theorem 5.1 provides a way to calculate the variance: If X is discrete, then

$$Var(X) = \sum_x (x - E(X))^2 \cdot f_X(x),$$

whereas if X is continuous, then

$$Var(X) = \int_{-\infty}^{\infty} (x - E(X))^2 \cdot f_X(x)dx.$$

Example 5.8 (Example 5.6, continued) *Let again the random variable X denote the outcome of one throw of a fair six-sided die. In Example 5.2, the mean of X was found to be $\mu_X = 3.5$. The variance of X can be calculated using Theorem 5.1:*

$$
\begin{aligned}
Var(X) = E\left((X - \mu_X)^2\right) &= E\left((X - 3.5)^2\right) \\
&= \sum_{j=1}^{6} (j - 3.5)^2 \cdot f_X(j) \\
&= (1 - 3.5)^2 \cdot \frac{1}{6} + (2 - 3.5)^2 \cdot \frac{1}{6} + \ldots + (6 - 3.5)^2 \cdot \frac{1}{6} \\
&= \frac{17.5}{6} \\
&= 2.92.
\end{aligned}
$$

The following theorem collects a number of useful rules concerning calculation with variances. The rules can be derived using Theorem 5.1.

Theorem 5.3 (Rules of calculation of variance) *Let $a \in \mathbb{R}$ be a constant and X a random variable such that $Var(X) < \infty$. Then the following hold.*

1. $Var(X) = E\left(X^2\right) - E(X)^2.$

2. $Var(a + X) = Var(X).$

3. $Var(aX) = a^2 Var(X).$

The rules of calculations in Theorem 5.3 imply that, for constants $a, b \in \mathbb{R}$, we have (Problem 5.9.2)

$$Var(a + bX) = b^2 Var(X).$$

Hence, in contrast to the case of expectations, we see that taking the variance of a random variable is not a linear operation. Further, it is worth stressing that, in general,

$$Var\left(g(X)\right) \neq g\left(Var(X)\right).$$

Rule 1 in Theorem 5.3 can often be very useful when calculating the variance of a random variable, since it allows us to derive the variance by calculating $E(X^2)$ and $E(X)$. The next two examples illustrate this concept.

Example 5.9 (Example 5.8, continued) *Let again the random variable X denote the outcome of one throw of a fair six-sided die. Recall from Examples 5.2 and 5.6 that*

$$E(X) = \frac{21}{6} = 3.5$$

and

$$E(X^2) = \frac{91}{6} = 15.17.$$

Using rule 1 in Theorem 5.3, we have that

$$Var(X) = E(X^2) - E(X)^2 = 15.17 - 3.5^2 = 2.92,$$

corroborating the result found in Example 5.8.

Example 5.10 *Let X be a continuous random variable with PDF*

$$f_X(x) = 3x^{-4}, \quad x \geq 1.$$

This random variable is similar to the one in Example 5.4, except that the exponent is now -4 instead of -1.5. Using rule 1 in Theorem 5.3, we have that

$$Var(X) = E(X^2) - E(X)^2.$$

Hence, to calculate the variance of X, we can calculate $E(X)$ and $E(X^2)$. The former can be calculated using Definition 5.1, cf. also the calculation in Example 5.4,

$$
\begin{aligned}
E(X) = \int_{-\infty}^{\infty} x f_X(x)dx &= \int_{1}^{\infty} x \cdot 3x^{-4}dx \\
&= 3 \int_{1}^{\infty} x^{-3}dx \\
&= 3 \left[\frac{1}{-2}x^{-2}\right]_{1}^{\infty} \\
&= \frac{3}{2}.
\end{aligned}
$$

The expectation $E(X^2)$ can be calculated via Theorem 5.1,

$$
\begin{aligned}
E(X^2) = \int_{-\infty}^{\infty} x^2 f_X(x)dx &= \int_1^{\infty} x^2 \cdot 3x^{-4}dx \\
&= 3\int_1^{\infty} x^{-2}dx \\
&= 3\left[-1x^{-1}\right]_1^{\infty} \\
&= 3.
\end{aligned}
$$

Hence, we conclude that

$$
Var(X) = E(X^2) - E(X)^2 = 3 - \left(\frac{3}{2}\right)^2 = 3 - \frac{9}{4} = \frac{3}{4}.
$$

5.5   Moments

There are other aspects of a distribution that can be investigated by taking expectations. One class of such aspects are called *moments*. They are defined next.

Definition 5.5 (Moment, central moment) *Let X be a random variable. The kth **moment** m_k of X is defined as*

$$
m_k = E\left(X^k\right).
$$

*The kth **central moment** m_k^* of X is defined as*

$$
m_k^* = E\left((X - E(X))^k\right).
$$

It is clear from Definition 5.5 that the mean is the first moment and the variance is the second central moment. Measures of the asymmetry and the heaviness of the tails of a distribution may be expressed using third and fourth (central) moments, respectively, leading to the concepts of *skewness* and *kurtosis* of a distribution. The definition is next.

Definition 5.6 (Skewness and kurtosis) *Let X be a random variable. The coefficient of **skewness** γ_3 is*

$$
\gamma_3 = \frac{E\left((X - E(X))^3\right)}{\left(\sqrt{Var(X)}\right)^3}.
$$

*The coefficient of **kurtosis** γ_4 is*

$$
\gamma_4 = \frac{E\left((X - E(X))^4\right)}{\left(\sqrt{Var(X)}\right)^4}.
$$

As seen from the definition, the coefficients of skewness and kurtosis are given by the third and fourth central moment, respectively, divided by a power of the standard deviation. This "standardization" implies that the coefficient of skewness and the coefficient of kurtosis are without unit, regardless of the units in which the random variable is measured.

In case the skewness equals 0, the distribution is symmetric (around the mean). Depending on the sign of the skewness coefficient, a distribution is called *left skewed* or *right skewed*, or, alternatively, *negatively skewed* or *positively skewed*. Sometimes, kurtosis is measured as deviation from 3, called *excess kurtosis*. The reason is, as we shall see in Chapter 8, that the normal distribution has kurtosis equal to 3. Distributions with negative excess kurtosis are called *platykurtic*, distributions with zero kurtosis are called *mesokurtic*, and distributions with positive excess kurtosis are called *leptokurtic*. The tables below summarize the names and their definitions. The left panel of Figure 5.1 illustrates the three kinds of skewness (negative, symmetric, and positive) and the right panel of Figure 5.1 illustrates the three kinds of kurtosis (platykurtic, mesokurtic, and leptokurtic). As can be seen from the figure, skewness refers to which way the distribution is "tilting". Conversely, kurtosis refers to the "fatness" of the tails of the distributions, with "fat" tails, i.e. tails with a lot of mass, implying a higher kurtosis.

Skewness	Condition		Kurtosis	Condition
left / negative	$\gamma_3 < 0$		platykurtic	$\gamma_4 - 3 < 0$
symmetric	$\gamma_3 = 0$		mesokurtic	$\gamma_4 - 3 = 0$
right / positive	$\gamma_3 > 0$		leptokurtic	$\gamma_4 - 3 > 0$

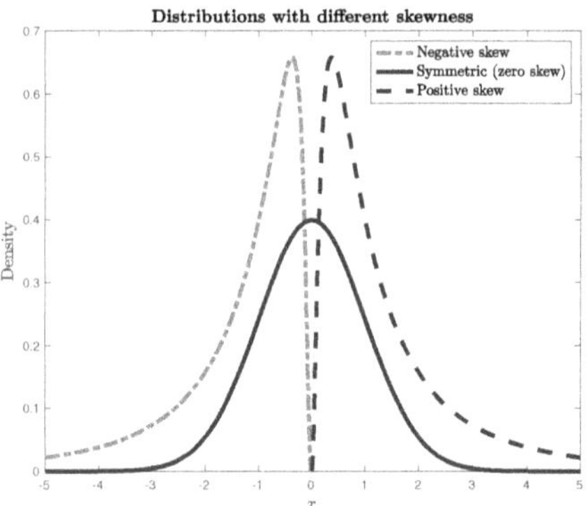

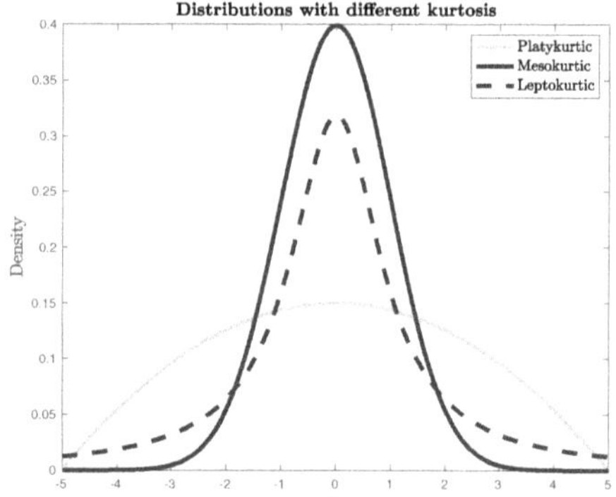

FIGURE 5.1

Distributions with different skewness and kurtosis. Left: Distributions with negative skew, symmetric (zero skew), and positive skew. Right: Platykurtic distribution (negative excess kurtosis), mesokurtic distribution (zero excess kurtosis), and leptokurtic distribution (positive excess kurtosis).

As was shown for the expected value, certain moments may not exist for some distributions. It can be shown that if a moment does not exist, then neither do any higher-order moments. This is formally stated in the next theorem.

Theorem 5.4 (Existence of moments) *Let X be a random variable and let m, k be any numbers such that $m > k > 0$. If $E\left(X^k\right)$ does not exist, then $E\left(X^m\right)$ does not exist. If $E\left(X^m\right)$ exists, then so does $E\left(X^k\right)$.*

The theorem implies that if the variance exists, then so does the mean.

5.6 Quantiles

A quantile is a value that divides a distribution, or population, into two parts: The first part is where outcomes have probability p of occurring, and a second part where outcomes have probability $(1 - p)$ of occurring. This value is in essence a *p-quantile*, and it is often written q_p. The formal definition is next.

Definition 5.7 (*p*-quantile, quantile function) *Let $p \in [0, 1]$ be a number and X a random variable. The value q_p is called a p-**quantile** of X if and only if*

$$\Pr\left(X < q_p\right) \leq p \ \text{and} \ \Pr\left(X \leq q_p\right) \geq p.$$

*If the p-quantile is considered as a function of p, it is sometimes called the **quantile function**.*

The condition in Definition 5.7 can equivalently be stated as

$$\Pr\left(X < q_p\right) \leq p \ \text{and} \ \Pr\left(X > q_p\right) \leq 1 - p.$$

It can also be formulated using the CDF of X:

$$F_X(q_p) - \Pr\left(X = q_p\right) \leq p \ \text{and} \ F_X(q_p) \geq p.$$

If X is a continuous random variable, then a p-quantile satisfies that

$$F_X(q_p) = p.$$

In case $F_X(x)$ is strictly increasing, it is an invertible function, and, hence,

$$q_p = F_X^{-1}(p),$$

where $F_X^{-1}(p)$ denotes the inverse function of $F_X(x)$. For this reason, the quantile function is sometimes referred to as the *inverse CDF*.

Some quantiles have their own names. The following table contains a collection of such names.

p	Name of q_p
0.50	Median
0.25, 0.50, 0.75	Quartiles
0.10, 0.20, 0.30, ...	Deciles
0.01, 0.02, 0.03, ...	Percentiles

For a continuous random variable X, the median, i.e. 0.5-quantile, is the value that divides the distribution in two equally large portions. That is, the median $q_{0.5}$, is the value that makes it equally likely that X takes an outcome above $q_{0.5}$, as an outcome below $q_{0.5}$. The first quartile, i.e. $q_{0.25}$, also divides the distribution in two, but now 25% is in the portion below $q_{0.25}$, and 75% is in the portion above q_{25}. That is, X has a 25% probability of being below $q_{0.25}$ and a 75% probability of being above $q_{0.25}$, and so on.

The following two examples show how we can calculate the quantiles of a continuous random variable by inverting the CDF.

Example 5.11 (Example 5.3, continued) *Let the CDF of the continuous random variable X be given by*

$$F_X(x) = \begin{cases} 0 & if \quad x < 0, \\ 0.5 \cdot x & if \quad 0 \le x < 2, \\ 1 & if \quad x \ge 2. \end{cases}$$

Let $p \in [0,1]$. Since $F_X(x)$ is strictly increasing, it is invertible. The p-quantile q_p can therefore be found by inverting $F_X(x)$, i.e. by solving the equation

$$F_X(q_p) = p.$$

We get

$$\begin{aligned} F_X(q_p) &= p \Rightarrow \\ 0.5 \cdot q_p &= p \Rightarrow \\ q_p &= F_X^{-1}(p) = 2 \cdot p. \end{aligned}$$

For instance, the median is $q_{0.5} = 1$ and the first quartile, that is, $q_{0.25}$, equals 0.5. For this distribution, the mean $E(X)$ and the median are the same, cf. Example 5.3.

Example 5.12 (Example 2.10, continued) *Assume that the CDF and PDF for the continuous random variable X are, respectively,*

$$F_X(x) = \begin{cases} 0 & if \quad x < 0, \\ 1 - e^{-2x} & if \quad x \ge 0, \end{cases}$$

and

$$f_X(x) = \begin{cases} 0 & if \quad x < 0, \\ 2e^{-2x} & if \quad x \ge 0. \end{cases}$$

The CDF $F_X(x)$ is strictly increasing and therefore invertible. The median $q_{0.5}$ can thus be found by inverting the CDF, i.e. by solving the equation

$$F_X(q_{0.5}) = 0.5.$$

We get

$$\begin{aligned} F_X(q_{0.5}) &= 0.5 \Rightarrow \\ 1 - e^{-2q_{0.5}} &= 0.5 \Rightarrow \\ q_{0.5} &= -0.5 \ln(0.5). \end{aligned}$$

This is illustrated in the left panel of Figure 5.2. The right panel of the figure shows how the median splits the area under the graph of the PDF with 0.5 in the left half and 0.5 in the right half.

When a random variable is discrete, its CDF is necessarily not invertible. Hence, to find the quantiles of a discrete random variable, we resort to checking the definition. The next example illustrates.

Example 5.13 *Consider the discrete random variable X with CDF $F_X(x)$ given by*

$$F_X(x) = \begin{cases} 0 & if \quad x < 10, \\ 0.2 & if \quad 10 \le x < 12, \\ 0.6 & if \quad 12 \le x < 14, \\ 1 & if \quad x \ge 14. \end{cases}$$

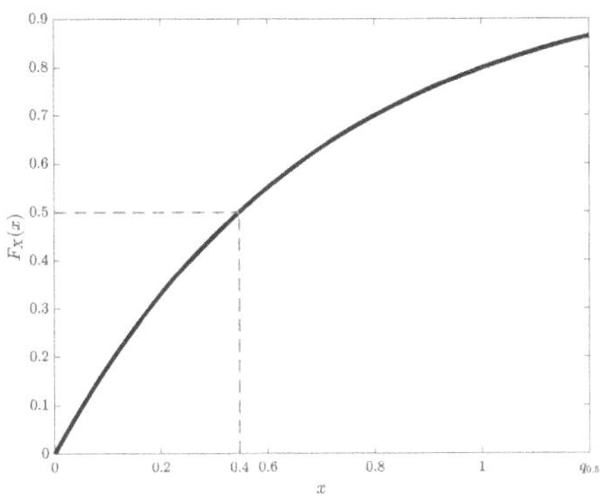 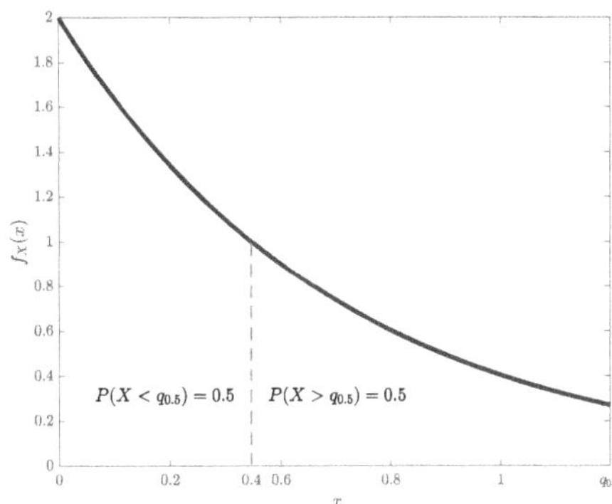

FIGURE 5.2

Left: Median $q_{0.5}$ found from CDF $F_X()$ via the inverse $q_{0.5} = F_X^{-1}(0.5)$. Right: The median $q_{0.5}$ divides the PDF $f_X()$ into two portions, such that the area under each portion is $p = 0.5$.

To calculate the median $q_{0.5}$, find candidates for the median. One candidate is 12. Then check if the conditions in Definition 5.7 are satisfied. First

$$\Pr(X < 12) = 0.2$$

and since $0.2 \leq 0.5$, the first part of the conditions is satisfied. The second part involves

$$\Pr(X \leq 12) = 0.6.$$

Since $0.6 \geq 0.5$, the second part of the conditions is satisfied. Thus, $q_{0.5} = 12$ is a median. We can check another candidate, say, 11 for median. This candidate does not work because

$$\Pr(X \leq 11) = 0.2,$$

and $0.2 \ngeq 0.6$ so the second condition is not satisfied. It can be checked that in this example, there is only one valid value for the median, namely 12.

To calculate the 0.2-quantile $q_{0.2}$, many values satisfy the conditions. For example, for the candidate 11, get

$$\Pr(X < 11) = 0.2 \leq 0.2$$

and

$$\Pr(X \leq 11) = 0.2 \geq 0.2.$$

Hence, both conditions are satisfied. In this example, it can be checked that all values in the interval $[10, 12]$ are valid 0.2-quantiles.

For a discrete random variable, one way to check if there is a unique or many values for a p-quantile is to check if there is a value of x such that $F(x) = p$. If there is such a value, then there are infinitely many solutions, namely, all the values of x for which $F(x) = p$. Those values form an half-open interval $[a, b)$. The endpoint b is also valid as a solution. In Example 5.13, for the 0.2-quantile, the half-open interval $[a, b)$ is $[10, 12)$. If there is no x for which $F(x) = p$, then there is a unique solution. Figuratively speaking, the solution is where $F(x)$ jumps over p. In Example 5.13, this happens at $x = 12$ for the median $q_{0.5}$.

5.7   Other descriptive measures of a distribution

Besides the variance, the spread of a distribution can be measured as the distance between two quantiles. For example, the *interquartile range* (IQR) is

$$IQR = q_{0.75} - q_{0.25}.$$

By definition of the quantiles, the IQR will contain the middle 50% of the distribution, and it measures the distance between the 75'th and 25'th percentiles in the distribution. For instance, if the distribution represents the income distribution in some population, the IQR measures the difference in income between the 25% wealthiest person ($q_{0.75}$) and the 25% poorest person ($q_{0.25}$) in the population. One could consider the spread of other quantiles, e.g. $q_{0.99} - q_{0.01}$, but the IQR is the most commonly used quantile spread in practice.

The *range* of a random variable is the difference between the largest possible and smallest possible outcomes of the random variable. It can be defined using the *support* of the distribution. The support of a distribution are the set of values of x for which the probability, or the density, are different from 0. It is given as stated as

$$supp(X) = \{x : f_X(x) > 0\},$$

where f is either a PMF or PDF. The range of a random variable can be defined as the difference between the largest and smallest values in the support of X, $supp(X)$. For example, if X can take the values 0, 1, 2 and 3, the range of X is $3 - 0 = 3$.

On a technical note, the largest or smallest values in the support of a random variable may not be well defined. For example, suppose the possible outcomes of a random variable are in the open interval $(2, 10)$. Then the largest possible number must be "just below" 10, but we cannot write up a specific number. To avoid this problem, we can define the smallest value for which no outcome of X is larger. That number is 10. Technically, this is called the *supremum*. Similarly, the largest number for which no outcome is smaller, is well defined and here it is 2. This is called *infimum*. Further mathematical details regarding the supremum and infimum can be found in the Chapter B of the appendix. Using supremum and infimum instead of minimum and maximum imply that the range of X is $10 - 2 = 8$. Formally, we define

$$\begin{aligned} range \;\; &= \;\; \sup supp(X) - \inf supp(X) \\ &= \;\; \sup_x \{x : f_X(x) > 0\} - \inf_x \{x : f_X(x) > 0\}. \end{aligned}$$

Sometimes it may be of interest to know the most likely outcome of a random variable. Such a value is called the *modal value*, or *mode*, of a distribution. For a discrete random variable, the modal value is the value most likely to occur. For a continuous random variable, all outcomes have probability equal to 0. Instead, we use the density function to define a model value. It is the value for which the density function is maximized. Formally, if f is either a PMF or a PDF, we define

$$mode = \arg\max_x f(x),$$

where "$\arg\max_x f(x)$" returns the x that maximizes (or attains the supremum) of $f(x)$.

Example 5.14 (Example 5.1, continued) *In the example where X denotes the attitude of a randomly selected individual toward a referendum, the PMF of X was given by*

$$f_X(x) = \begin{cases} 0.52 & if \;\; x = 1, \\ 0.48 & if \;\; x = 0. \end{cases}$$

The mode of X is $\arg\max_x f_X(x) = 1$, that is, $x = 1$ maximizes the function $f_X(x)$.

In a distribution, there may be several values for which the density function has a local maximum. We can describe an aspect of a distribution according to the number of local maxima, it has. If there is one local maximum, a distribution is called *unimodal* and if there are two local maxima, a distribution is called *bimodal.* In general, if a distribution has two or more local maxima, it may be called *multimodal.* The left and right panels of Figure 5.3 show examples of a unimodal and bimodal distribution, respectively.

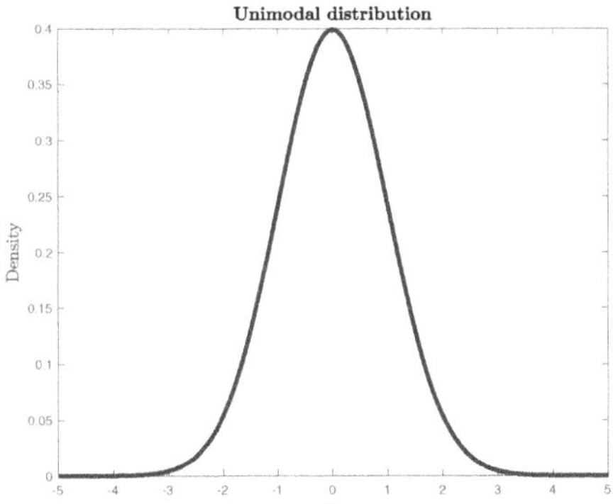
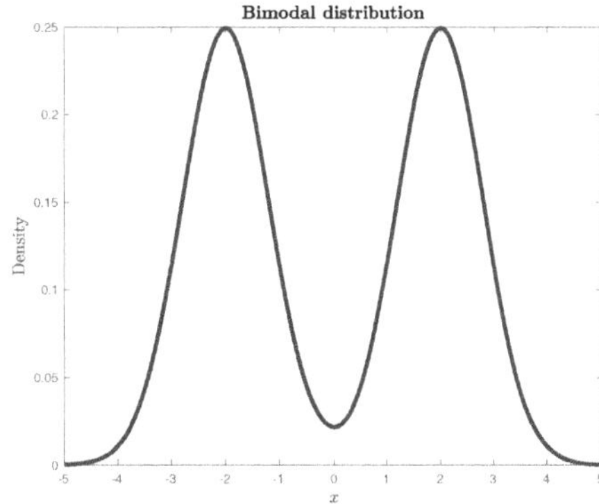

FIGURE 5.3
Examples of distributions. Left: PDF of a unimodal distribution. Right: PDF of a bimodal distribution.

5.8   Proofs

Proof of Theorem 5.1 *Suppose first that X is a discrete random variable. Define $Y = g(X)$. Then Y is a discrete random variable with outcomes $y = g(x)$, where x denotes the outcomes of X, and PMF $f_Y(y) = \sum_{x:g(x)=y} f_X(x)$. Hence, by the definition of the expected value of Y, we have*

$$E(Y) = \sum_y y f_Y(y) = \sum_y y \sum_{x:g(x)=y} f_X(x) = \sum_x g(x) f_X(x),$$

as we wanted to show.

The proof in the case where Y is a continuous random variable is slightly more involved. We give the result in the case where $g()$ is an invertible function. The proof can then be extended to include non-invertible functions by dividing them up into segments of invertible functions and treating each segment separately. For instance, the case $Y = g(X) = X^2$ can be handled by handling the segments where $x \leq 0$ and $x \geq 0$ separately. Suppose then, that $g(x)$ is invertible. Suppose for simplicity that $g(x)$ is strictly increasing (the decreasing case is handled analogously). Now, $Y = g(X)$ is a new continuous random variable with PDF (Theorem 2.14)

$$f_Y(y) = f_X\left(g^{-1}(y)\right) \cdot \frac{dg^{-1}(y)}{dy}. \tag{5.1}$$

Note that, from $y = g(x)$, $g^{-1}(y) = x$, and the chain rule, we have

$$1 = \frac{dy}{dy} = \frac{d}{dy}g(x) = g'(x)\frac{d}{dy}x = g'(x)\frac{d}{dy}g^{-1}(y),$$

which we can re-arrange to get

$$\frac{d}{dy}g^{-1}(y) = \frac{1}{g'(x)}. \tag{5.2}$$

By the definition of the expected value of Y, we have

$$E(Y) = \int_{-\infty}^{\infty} y f_Y(y)dy.$$

Using the change-of-variables $y = g(x)$ which implies $dy = g'(x)dx$, along with (5.1) and (5.2), we have

$$E(Y) = \int_{-\infty}^{\infty} y f_Y(y)dy = \int_{-\infty}^{\infty} g(x)f_X\left(g^{-1}(y)\right) \cdot \frac{1}{g'(x)}g'(x)dx = \int_{-\infty}^{\infty} g(x)f_X(x)\,dx,$$

as we wanted to show.

Proof of Theorem 5.2 *The proof of Theorem 5.2 is left as an exercise to the reader (Problem 5.9.1).*

Proof of Theorem 5.3 *The proof of Theorem 5.3 is left as an exercise to the reader (Problem 5.9.2).*

Proof of Theorem 5.4 *Recall that we say that the expectation $E(g(X))$ exists if $E(|g(X)|)$ is a finite number. Suppose $m > k > 0$ and that X is a continuous random variable (the discrete case follows using similar arguments). Suppose for simplicity that X is a positive random variable (the case where X can take negative values follows using similar arguments, by treating the positive and negative values separately). Then, using Theorem 5.1,*

$$E(X^k) = \int_0^{\infty} x^k f(x)dx = \int_0^1 x^k f(x)dx + \int_1^{\infty} x^k f(x)dx.$$

Note that the first term is always non-negative and finite. Using the above expression, we have

$$\begin{aligned}
E(X^k) &\leq \int_0^1 x^k f(x)dx + \int_1^{\infty} x^m f(x)dx \\
&\leq \int_0^1 x^k f(x)dx + \int_0^1 x^m f(x)dx + \int_1^{\infty} x^m f(x)dx \\
&= \int_0^1 x^k f(x)dx + E(X^m),
\end{aligned}$$

where the first inequality follows since $m > k$ and the second follows because $\int_0^1 x^m f(x)dx$ is a non-negative constant. Now, if $E(X^k)$ does not exist, i.e. if $E(X^k) = \infty$, then the *inequality above implies that $E(X^m) = \infty$, i.e. $E(X^m)$ does not exist. This completes the first part of the proof. To prove the second part of the proof, re-arrange the above to get*

$$E(X^k) - \int_0^1 x^k f(x)dx \leq E(X^m).$$

Using this, the assumption that $E(X^m)$ exists implies that $E(X^k)$ exists. This completes the second part of the proof.

5.9   Exercises

Problem 5.9.1

1. *Prove Theorem 5.2.*

2. *Let $a, b \in \mathbb{R}$ be constants and X a random variable such that $E(X)$ exists. Conclude from Theorem 5.2 that the following identity holds:*

$$E(a + bX) = a + bE(X).$$

Problem 5.9.2

1. *Prove Theorem 5.3.*

2. *Let $a, b \in \mathbb{R}$ be constants and X a random variable such that $Var(X)$ exists. Conclude from Theorem 5.3 that the following identity holds:*

$$Var(a + bX) = b^2 Var(X).$$

Problem 5.9.3 *Consider the following gamble: A coin is tossed repeatedly until the first time it shows "heads". If "heads" comes up in the first throw, you gain 2 units of money; if heads comes up the first time in the second throw, you gain 4 units of money; if heads comes up the first time in the third throw, you gain 8 units of money; etc. Let X be the discrete random variable representing the winnings from this gamble. X is given as*

$$X = 2^Y,$$

where Y is the number of throws until the first "heads" comes up.

1. *How much money would you be willing to pay to enter into this gamble? (Surely, you would pay at least 2 units of money, since this is the minimum amount of money you will win. But how much more than 2 units would you pay?)*

Maybe the answer to the above question is related to how much you would win on average, if you could the repeat the gamble many times, i.e. maybe it is related to the mean of X. Using the fact that the outcomes of the coin throws are "independent" (see Chapter 7 for a treatment of this concept), it can be shown that, if the coin is fair, the PMF of Y is

$$f_Y(y) = \frac{1}{2^y}, \quad y = 1, 2, \ldots.$$

2. *Show that the mean of X does not exist, i.e. show that $E(X) = \infty$.*

The setting studied in this problem is known as the St. Petersburg Paradox. *The "paradox" is that most people are willing to pay only a small amount (and definitely a finite amount) of money to enter this gamble, although, formally, the expected gain from the gamble is infinite. The "solution" to the paradox is the realization that people are not so much concerned about the amount of money that they win from the gamble, but instead are more concerned about how much well-being, or* utility, *they get from the gamble. According to this line of reasoning, it is not the expected value, but the expected utility, that is relevant when considering whether to enter into the gamble.*

Let the function $U()$ be a "utility function" such that $U(X)$ represents the utility a person gets from the gamble above. Suppose the function $U()$ is the square root function, i.e. $U(x) = \sqrt{x}$.

3. *Calculate the expected utility of the gamble, i.e. calculate $E(U(X))$.*

Suppose you have wealth $w > 0$ and suppose the cost of entering into the gamble is $c > 0$. Then the expected utility of your wealth from entering into the gamble is $E(U(w + X - c))$, while your expected utility from not entering into the gamble is $E(U(w)) = U(w)$. Suppose $w = 100$ units and your utility function is again $U(x) = \sqrt{x}$.

4. *Assuming your goal is to maximize your expected utility, how much would you maximally be willing to pay to enter the gamble? (Hint 1: You need to decide the value of c such that you are indifferent between gambling and not gambling, i.e. such that $E(U(w+X-c)) = E(U(w))$. This problem does not have a closed-form solution, so you may have to use computer software to approximate the value of c solving this equation. Hint 2: The answer is close to 10 units.)*

Problem 5.9.4 *You can choose to invest in either of two different assets. Let X be the random variable describing the profit of asset 1 and Y be the random variable describing the profit of asset 2. Assume that the PMFs for X and Y are given by*

$$f_X(x) = \begin{cases} 0.50 & x = -5000, \\ 0.45 & x = 6000, \\ 0.05 & x = 100000, \end{cases}$$

and

$$f_Y(y) = \begin{cases} 0.75 & y = 2000, \\ 0.25 & y = 4000. \end{cases}$$

Assume that you have $w = 25000$ in your bank account and that your utility function is logarithmic,

$$U(x) = \ln(x),$$

where $x > 0$ represents your wealth.

1. *Calculate your expected wealth if you choose asset 1, i.e. compute $E(w + X)$.*

2. *Calculate your expected wealth if you choose asset 2, i.e. compute $E(w + Y)$.*

3. *Calculate your expected utility if you choose asset 1, i.e. compute $E(U(w + X))$.*

4. *Calculate your expected utility if you choose asset 2, i.e. compute $E(U(w + Y))$.*

5. *Which asset provides you with the highest expected wealth? Which asset provides you with the highest expected utility?*

6. *Recalculate $E(U(w+X))$ and $E(U(w+Y))$, but now assume that your utility function is $U(x) = \sqrt{x}$. Which asset provides you with the highest expected utility in this case? Compare with the result using the logarithmic utility function.*

Problem 5.9.5 *Let X denote the income of a randomly selected person in one country and Y denote the income of a randomly selected person in another country (both measured in $10,000$ USD). Assume that X and Y are continuous random variables with density functions*

$$f_X(x) = (1+x)^{-2}, \quad x > 0,$$

and

$$f_Y(y) = 4 \cdot (1+y)^{-5}, \quad y > 0.$$

The PMFs $f_X(x)$ and $f_Y(y)$ are shown in Figure 5.4. Assume that individuals in both countries have utility given by the utility function

$$U(x) = \sqrt{1+x},$$

where $x \geq 0$ represents an individual's income.

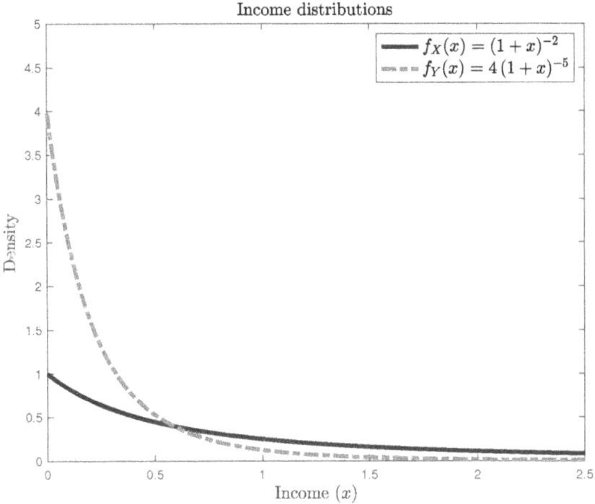

FIGURE 5.4
PDFs $f_X(x)$ and $f_Y(x)$ of the random variables X and Y, describing the income distributions in Problem 5.9.5. The PDFs are defined for alle $x \in (0, \infty)$, but are shown here for $x \in (0, 2.5]$.

1. *Show that $f_X(x)$ and $f_Y(y)$ satisfy the conditions for being valid PDFs.*

2. *Determine the probability that a randomly selected person in the first country has an income above $10,000$ USD, i.e. compute $P(X > 1)$. (Note: $P(X > 1)$ can also be interpreted as the proportion of the population in the first country that has an income above $10,000$ USD.)*

3. *Determine the probability that a randomly selected person in the second country has an income above $10,000$ USD, i.e. compute $P(Y > 1)$.*

4. Referring to Figure 5.4, explain what $P(X > 1)$ and $P(Y > 1)$ correspond to graphically.

5. Compute the expected utility for a randomly selected person in the first country, i.e. compute $E(U(X))$. (Note: $E(U(X))$ can also be interpreted as the average utility for this population.)

6. Compute the expected utility for a randomly selected person in the second country, i.e. compute $E(U(Y))$.

7. Based on the expected utility, which country is preferable?

Problem 5.9.6 *Consider again the case where X is the outcome of a single throw with a fair six-sided die. In Example 5.2 we saw that the mean of X is $\mu_X = E(X) = 3.5$. Calculate the median of X.*

Problem 5.9.7 *Let the PMF of X be*

$$
f_X(x) = \begin{cases}
0.1 & \text{if} & x = 0, \\
0.2 & \text{if} & x = 1, \\
0.4 & \text{if} & x = 2, \\
0.3 & \text{if} & x = 3, \\
0 & & \text{otherwise.}
\end{cases}
$$

1. Calculate the mean of X.

2. Calculate the mean of $g(X) = 3 \cdot X + 5$.

3. Calculate the median of X.

4. Calculate the 0.3-quantile of X.

5. Calculate the variance of X.

6. Calculate the variance of $g(X) = 3 \cdot X + 5$.

7. Calculate the coefficient of skewness of X.

8. Calculate the coefficient of kurtosis of X.

Problem 5.9.8 *Let X be a continuous random variable with density function*

$$
f_X(x) = \begin{cases}
\frac{1}{4}x & \text{if} \ \ 0 < x < 2, \\
-\frac{1}{4}x + 1 & \text{if} \ \ 2 \le x \le 4, \\
0 & \text{otherwise,}
\end{cases}
$$

and CDF

$$
F_X(x) = \begin{cases}
0 & \text{if} & x < 0, \\
\frac{1}{8}x^2 & \text{if} & 0 \le x < 2, \\
-\frac{1}{8}x^2 + x - 1 & \text{if} & 2 \le x < 4, \\
1 & \text{if} & x \ge 4.
\end{cases}
$$

1. Calculate the mean $E(X)$ of X.

   Hint, for this density function $f_X(x)$, the integral to calculate $E(X)$ can be written as

$$
\int_{-\infty}^{\infty} x \cdot f_X(x) \, dx = \int_0^2 x \cdot f_X(x) \, dx + \int_2^4 x \cdot f_X(x) \, dx.
$$

2. *Calculate the 0.2-quantile, the median, and the 0.8-quantile.*

Problem 5.9.9 *Let X be a discrete random variable with probability function*

$$f_X(x) = \begin{cases} \frac{1}{x} & \text{if } x = 2, 4, 8, 16, 32, \ldots, \\ 0 & \text{otherwise.} \end{cases}$$

1. *Show that the random variable X does not have a mean $E(X)$.*

2. *Argue, using Theorem 5.4, that X does not have a variance either.*

Problem 5.9.10 *Let X be a continuous random variable with PDF*

$$f_X(x) = \alpha x^{-(1+\alpha)}, \quad x \geq 1,$$

where $\alpha > 0$ is a constant. This distribution for X is called a Pareto distribution, *and it is an example of a* power law distribution. *It is often used to model the incomes in a population.*

1. *For what values of α does the mean of X exist? For these values of α, calculate the mean of X, $\mu_X = E(X)$.*

2. *Calculate the median of X. (Hint: Derive the CDF $F_X(x)$ of X and solve the equation $F_X(x) = 0.5$.)*

3. *For what values of α does the variance of X exist, i.e. for what values of α is $Var(X) < \infty$? For these values of α, calculate the variance of X, $\sigma_X^2 = Var(X)$. (Hint: You may want to use that $Var(X) = E(X^2) - E(X)^2$.)*

4. *Let $n \geq 1$ and consider the n'th order moment of X, $m_n = E(X^n)$. For a given value of n, for what values of α does $m_n = E(X^n)$ exist? For these values of α, calculate the n'th order moment of X, $m_n = E(X^n)$.*

5. *Suppose X represents the monthly income (measured in tens of thousands of Danish kr.) of a randomly chosen individual in some population and that we for this population have the value $\alpha = 1.15$. Calculate the mean, median, and variance of this income distribution. Comment on what these numbers tell you about the inequality of incomes in this population.*

6

Aspects of a multivariate distribution

6.1 Introduction

Certain aspects of a multivariate distribution can be used to characterize relationships between random variables. Examples include the *covariance* and *correlation* of two random variables. Of particular importance are aspects of conditional distributions, especially the aspect called *conditional expectation*. Conditional expectation characterizes the expectation, i.e. mean, of a random variable conditionally on the outcomes of other random variables. Conditional expectation is the foundation of regression analysis, covered in Volume II of this book.

Another aspect of a conditional distribution is a *conditional p-quantile*. A conditional p-quantile characterizes the p-quantile of a random variable conditionally on the outcomes of other random variables. Conditional p-quantiles are the foundation of quantile regression analysis, also covered in Volume II of this book.

Before discussing relationships between many random variables, we consider co-variation between two random variables. Measures of co-variation are useful in practice to make a first characterization of the relationship between two random variables.

6.2 Covariance

6.2.1 Definition of covariance

A measure of co-variation between two random variables is their *covariance*. In the context of a statistical experiment, co-variation can be interpreted as a tendency of the statistical experiment to have outcomes where a high value of one random variable, say X, is likely to be accompanied by a high value of another random variable, say Y, and vice versa. To make this interpretation operational, we have to define what we mean by high values of random variables, co-variation, and tendency. We interpret a high value of a random variable as a value which is larger than the mean of the random variable. That is, we consider deviations of the means of the random variables, $(X - E(X))$ and $(Y - E(Y))$. We may make the concept co-variation operational by considering the product of those two deviations

$$(X - E(X)) \cdot (Y - E(Y)).$$

This product is positive if a high value of X, relative to its mean $E(X)$, is accompanied by a high value of Y, relative to its mean $E(Y)$, and a low value of X, relative to its mean $E(X)$, is accompanied by a low value of Y, relative to its mean $E(Y)$. Conversely, if a high

DOI: 10.1201/9781003591191-6

value of X is accompanied by a low value of Y, or vice versa, the product will be negative. Since X and Y may have many different outcomes, we calculate the expected value of the product to assess the tendency of X and Y to "move together" in this sense. This all leads to the following definition of covariance.

Definition 6.1 (Covariance) *Let X and Y be two random variables. The **covariance** $Cov(X,Y)$ between X and Y is*

$$Cov(X,Y) = E\left((X - E(X)) \cdot (Y - E(Y))\right). \tag{6.1}$$

With this definition, the covariance between X and Y is positive if X tends to be larger than its mean when Y is larger than its mean and vice versa. A statistical experiment may also result in the covariance between X and Y being negative. This can be interpreted as X tends to be larger than its mean when Y is smaller than its mean and vice versa.

As we saw in Chapter 5, the mean or variance of a random variable may not exist. Similarly, the covariance between two random variables may not exist. The next theorem links the existence of the variance of the random variables X and Y to the existence of the covariance between them.

Theorem 6.1 (Existence of covariance) *If $Var(X)$ and $Var(Y)$ exist, then so does $Cov(X,Y)$.*

The joint distribution between two random variables X and Y can be used to calculate their covariance (6.1). To see how, let $\mu_X = E(X)$ and $\mu_Y = E(Y)$ denote the means of X and Y, and note that the term $(X - \mu_X) \cdot (Y - \mu_Y)$ defines a new random variable, say, Z given by

$$Z = (X - \mu_X) \cdot (Y - \mu_Y).$$

Note that, by Definition 6.1, $E(Z) = Cov(X,Y)$. If we knew the distribution of Z, we could calculate $E(Z)$ using Definition 5.1. Since Z is a function of two random variables X and Y, the distribution of Z can be derived using, for instance, Theorems 3.4 and 3.11. It turns out, however, that it is possible to calculate the mean of $Z = (X - \mu_X) \cdot (Y - \mu_Y)$, and thus the covariance between X and Y, without having to derive the distribution of Z. The details are given in the following section.

6.2.2 Expectation of a function of random variables

We next state a theorem which allows the expected value of a function of random variables to be calculated without having to derive the distribution of this function. This theorem provides a convenient method to calculate, for instance, the covariance. The theorem is an extension of Theorem 5.1 for one random variable, and here we state it for the case of two random variables.

Theorem 6.2 (Expectation of function of two random variables) *Let $g(x,y)$ be a function. If X and Y are discrete random variables with a joint PMF $f_{X,Y}(x,y)$, then*

$$E(g(X,Y)) = \sum_x \sum_y g(x,y) \cdot f_{X,Y}(x,y),$$

where $\sum_x$ is the sum over all possible outcomes of X and similarly for $\sum_y$.

If X and Y are continuous random variables with a joint PDF $f_{X,Y}(x,y)$, then

$$E(g(X,Y)) = \int_{-\infty}^{\infty} \int_{-\infty}^{\infty} g(x,y) \cdot f_{X,Y}(x,y)dxdy.$$

The benefit of Theorem 6.2 is that the expected value of $g(X, Y)$ can be calculated directly from the joint distribution $f_{X,Y}(x, y)$ of X and Y, without having to first derive the distribution of the random variable $Z = g(X, Y)$. The theorem may be straightforwardly generalized to more than two variables. For instance, suppose $X_1, \ldots, X_n$ are discrete random variables. Then

$$E(g(X_1, \ldots, X_n)) = \sum_{x_1} \cdots \sum_{x_n} g(x_1, \ldots, x_n) \cdot f_{X_1, \ldots, X_n}(x_1, \ldots, x_n), \tag{6.2}$$

where $f_{X_1, \ldots, X_n}(x_1, \ldots, x_n)$ denotes the joint PMF of $X_1, \ldots, X_n$.

The following result shows how to calculate the covariance of two random variables X and Y, based on their joint distribution. To this end, we define the function

$$g(x, y) = (x - \mu_X) \cdot (y - \mu_Y)$$

and apply Theorem 6.2 to calculate $E(g(X, Y))$.

Theorem 6.3 (Calculation of covariance) *If X and Y are discrete random variables with joint PMF $f_{X,Y}(x, y)$, then $Cov(X, Y)$ can be calculated as*

$$Cov(X, Y) = \sum_x \sum_y (x - \mu_X) \cdot (y - \mu_Y) \cdot f_{X,Y}(x, y). \tag{6.3}$$

Similarly, if X and Y are continuous random variables with joint PDF $f_{X,Y}(x, y)$, then $Cov(X, Y)$ can be calculated as

$$Cov(X, Y) = \int_{-\infty}^{\infty} \int_{-\infty}^{\infty} (x - \mu_X) \cdot (y - \mu_Y) \cdot f_{X,Y}(x, y) \, dxdy.$$

Example 6.1 *Consider two discrete random variables X and Y with joint PMF $f_{X,Y}(x, y)$ given in the table below.*

$f_{X,Y}(x, y)$		X	
		5	10
Y	0	0.4	0.3
	1	0.2	0.1

The means of X and Y are

$$\begin{aligned} \mu_X &= E(X) = 5 \cdot 0.6 + 10 \cdot 0.4 = 7, \\ \mu_Y &= E(Y) = 0 \cdot 0.7 + 1 \cdot 0.3 = 0.3. \end{aligned}$$

The covariance can be calculated using Equation (6.3),

$$\begin{aligned} Cov(X, Y) &= \sum_x \sum_y (x - \mu_X)(y - \mu_Y) f_{X,Y}(x, y) \\ &= (5 - 7) \cdot (0 - 0.3) \cdot 0.4 + (5 - 7) \cdot (1 - 0.3) \cdot 0.2 \\ &\quad + (10 - 7) \cdot (0 - 0.3) \cdot 0.3 + (10 - 7) \cdot (1 - 0.3) \cdot 0.1 \\ &= -0.1. \end{aligned}$$

Theorem 6.2 can also be used to derive a useful rule of calculation for the expectation of a sum of random variables. The next theorem states this result.

Theorem 6.4 (Expectation of a sum) *Let X and Y be random variables and $a, b \in \mathbb{R}$ constants. If $E(X)$ and $E(Y)$ exist, then*

$$E(aX + bY) = aE(X) + bE(Y).$$

Theorem 6.2 can also be used to derive the following rules of calculation for covariances.

Theorem 6.5 (Rules of calculation for covariance) *Let X and Y be random variables and $a, b \in \mathbb{R}$ constants. Then*

i) $Cov(X, Y) = E(X \cdot Y) - E(X) \cdot E(Y).$

ii) $Cov(X, X) = Var(X).$

iii) $Cov(a, X) = 0.$

iv) $Cov(aX, bY) = a \cdot b \cdot Cov(X, Y).$

v) $Cov(X, Y + Z) = Cov(X, Y) + Cov(X, Z).$

Property iv), $Cov(aX, bY) = a \cdot b \cdot Cov(X, Y)$, shows that the value of the covariance is sensitive to scaling of the random variables. For example, if X and Y are measured in 1000 kr's and they are rescaled to kr's, then the covariance increases by the factor 1,000,000. Therefore, one needs to be cautious in interpreting the value of the covariance. Below, in Section 6.3, a correlation coefficient, which is not sensitive to scaling in this way, is defined to address this issue.

Theorem 6.5i) provides an alternative formulation of the covariance, namely

$$Cov(X, Y) = E(X \cdot Y) - E(X) \cdot E(Y).$$

Combining this formula with Theorem 6.2 yields an alternative way of calculating the covariance. The following example illustrates.

Example 6.2 (Example 6.1, continued) *Theorem 6.2 implies that*

$$
\begin{aligned}
E(X \cdot Y) &= \sum_x \sum_y x \cdot y \cdot f_{X,Y}(x, y) \\
&= 5 \cdot 0 \cdot 0.4 + 5 \cdot 1 \cdot 0.2 + 10 \cdot 0 \cdot 0.3 + 10 \cdot 1 \cdot 0.1 \\
&= 2.
\end{aligned}
$$

Further, in Example 6.1, we found that $E(X) = 7$ and $E(Y) = 0.3$. Therefore, by property i) of Theorem 6.5,

$$Cov(X, Y) = E(XY) - E(X) \cdot E(Y) = 2 - 7 \cdot 0.3 = -0.1.$$

6.2.3 Covariance and the variance of a sum

The variance of the sum of two random variables can be written in terms of the variances of the random variables and their covariance. This link is important and therefore stated in the next theorem.

Theorem 6.6 (Variance of a sum) *Let X and Y be random variables . If $Var(X)$ and $Var(Y)$ exist, then*

$$Var(X + Y) = Var(X) + Var(Y) + 2Cov(X, Y).$$

Theorem 6.6 shows that the variance of a sum not only depends on how each of the two random variables varies, it also depends on how they co-varies, where co-variability is measured by the covariance.

Theorem 6.6 can be extended by combining it with rules of calculation for variance and covariance. Let $a, b \in \mathbb{R}$ be constants. Then

$$Var(aX + bY) = a^2 Var(X) + b^2 Var(Y) + 2ab Cov(X, Y).$$

Example 6.3 *Based on Example 6.1, the variance of X and Y are found to be*

$$\begin{aligned} Var(X) &= (5 - 7)^2 \cdot 0.6 + (10 - 7)^2 \cdot 0.4 = 6, \\ Var(Y) &= (0 - 0.3)^2 \cdot 0.7 + (1 - 0.3)^2 \cdot 0.3 = 0.21. \end{aligned}$$

Thus,
$$Var(X + Y) = 6 + 0.21 + 2 \cdot (-0.1) = 6.01.$$

In this case, the variance of the sum is less than the sum of variances. The reason is that X and Y have a negative covariance.

Theorem 6.6 shows that, for two random variables X and Y, the variance of the sum $X + Y$ is the sum of the variances and twice the covariance. In Problem 6.9.4 at the end of this chapter, you are asked to prove a generalization of this result, valid for $n \geq 2$ random variables $X_1, \ldots, X_n$.

6.3   The correlation coefficient

As the covariance between two random variables depends on the unit of measurement of each random variable, it can be practical to measure co-variation between two random variables in a manner which is not depending on the scaling of the two random variables. One such "scale-invariant" measure of co-variation is the *correlation coefficient*. It is defined next.

Definition 6.2 (Correlation coefficient) *Let X and Y be two random variables with variances $Var(X) > 0$ and $Var(Y) > 0$, and covariance $Cov(X, Y)$. Then the* **correlation coefficient** $\rho_{X,Y}$ *is*

$$\rho_{X,Y} = \frac{Cov(X, Y)}{\sqrt{Var(X) \cdot Var(Y)}}.$$

The correlation coefficient can be interpreted as the degree to which two random variables co-varies. The sign of the correlation coefficient is determined by the sign of the covariance. Thus, the interpretation of the covariance can similarly be used for the correlation coefficient. For instance, if the correlation coefficient $\rho_{X,Y}$ is positive, it means that large values of X (relative to its mean) tend to be associated with large values of Y (relative to its mean), and similarly for small values.

The next theorem shows that the correlation coefficient is always a number between -1 and 1.

Theorem 6.7 (Correlation coefficient is a number between -1 and 1) *Let X and Y be two random variables with correlation coefficient $\rho_{X,Y}$. It holds that $\rho_{X,Y} \in [-1, 1]$.*

The table below lists some of the terminology regarding two random variables X and Y when discussing their correlation coefficient.

Terminology	Value of correlation coefficient
Perfectly negatively correlated	$\rho_{X,Y} = -1$
Negatively correlated	$\rho_{X,Y} < 0$
Uncorrelated	$\rho_{X,Y} = 0$
Positively correlated	$\rho_{X,Y} > 0$
Perfectly positively correlated	$\rho_{X,Y} = 1$
Perfectly correlated	$\rho_{X,Y} = -1$ or $\rho_{X,Y} = 1$

The correlation coefficient does not have a unit of measurement, i.e. it is a dimensionless number. This can be seen from the result that the correlation coefficient is invariant to affine transformations of X and Y. Affine transformations of X and Y are $a + bX$ and $c + dY$, where $a, b, c, d \in \mathbb{R}$ are scalars. The correlation between X and Y is the same as the correlation between $a + bX$ and $c + dY$ provided b and d have the same sign, i.e. $\rho_{a+bX,c+dY} = \rho_{X,Y}$. You are asked to prove this in Problem 6.9.5. In words, changing the unit of measurement for a random variable, e.g. from kilos to tons or from Euros to Dollars, does not change its correlation with another random variable. It will, however, change its variance and the covariance.

The correlation coefficient can be given a more detailed interpretation. It is the degree of *linear* association between the two random variables involved. The next example illustrates that a deterministic linear relationship between two random variables implies a perfect negative or positive correlation coefficient. This statement is proved in full generality in Theorem 6.8 below.

Example 6.4 *Let X be a discrete random variable with PMF*

$$f_X(x) = \begin{cases} 0.3 & if \quad x = 1, \\ 0.5 & if \quad x = 2, \\ 0.2 & if \quad x = 3. \end{cases}$$

Define a new random variable Y via a deterministic positive linear relationship with X:

$$Y = 1 + 2 \cdot X.$$

Now, the joint PMF of X and Y is $f_{X,Y}(x,y)$ as given in the following table.

$f_{X,Y}(x,y)$		X		
		1	*2*	*3*
	3	*0.3*	*0.0*	*0.0*
Y	*5*	*0.0*	*0.5*	*0.0*
	7	*0.0*	*0.0*	*0.2*

Notice that, because of the deterministic relationship between X and Y, for each value of X, only one value of Y has probability different from 0. By inserting this into the expression for the correlation coefficient, we get $\rho_{X,Y} = 1$.

Alternatively, we can define Y via a deterministic negative linear relationship with X:

$$Y = 8 - 2 \cdot X.$$

In this case, their joint PMF $f_{X,Y}(x,y)$ becomes:

$f_{X,Y}(x,y)$			X	
		1	*2*	*3*
	2	*0.0*	*0.0*	*0.2*
Y	*4*	*0.0*	*0.5*	*0.0*
	6	*0.3*	*0.0*	*0.0*

By inserting this into the expression for the correlation coefficient, we get $\rho_{X,Y} = -1$. Alternatively, we can define Y via a deterministic non-linear relationship with X:

$$Y = 1 + X^2.$$

This is indeed a deterministic relationship between Y and X: If we know X, then we know Y with certainty; if we know Y, then we know X with certainty (note that this is because, in this example, we always have $X > 0$). Their joint PMF is

$f_{X,Y}(x,y)$			X	
		1	*2*	*3*
	2	*0.3*	*0.0*	*0.0*
Y	*5*	*0.0*	*0.5*	*0.0*
	10	*0.0*	*0.0*	*0.2*

Now $E(X) = 1.9$, $E(Y) = 5.1$, $Var(X) = 0.49$, $Var(Y) = 7.69$, and $Cov(X,Y) = 1.91$. By inserting this into the expression for the correlation coefficient, we get $\rho_{X,Y} = 0.98$. Though we know Y if we know X and vice versa, they are not perfectly correlated because their relationship is non-linear and the correlation coefficient is measuring the degree of linear dependence.

Summing up, we have seen that, in the first two examples, it is not because Y is a deterministic function of X that the correlation coefficient is 1 or -1, respectively. Rather, the perfect correlation in those two examples comes about because the deterministic relationship is linear. *Indeed, as we saw in the third example, if the deterministic relationship is* non-linear, *we cannot expect that the random variables are perfectly correlated.*

The next example shows that even in the case of a deterministic relationship between two random variables, their correlation coefficient might be zero when the relationship is non-linear. You are asked to prove the details of the example in Problem 6.9.6.

Example 6.5 *Let X be a continuous random variable with PDF*

$$f_X(x) = \frac{1}{2}, \quad x \in [-1, 1],$$

and define the new random variable Y as

$$Y = X^2.$$

It can be shown that (Problem 6.9.6)

$$Cov(X,Y) = 0,$$

which implies that

$$\rho_{X,Y} = 0.$$

The next result shows that if there is a deterministic linear relationship between X and Y, then they are perfectly correlated.

Theorem 6.8 *Let $a, b \in \mathbb{R}$ be constants and X a random variable. Define*

$$Y = a + bX.$$

It holds that

$$\rho_{X,Y} = \begin{cases} 0 & \text{if} \quad b = 0, \\ -1 & \text{if} \quad b < 0, \\ 1 & \text{if} \quad b > 0. \end{cases}$$

Linear relationships and the correlation coefficient play an important role in linear regression analysis, which is covered in Volume II of this book.

6.4 Conditional expectation

The *conditional expectation* of Y, given some conditioning random variables, is the mean of Y in the conditional distribution of Y given those conditioning random variables. Thus, the conditional expectation is an aspect of the conditional distribution of Y given these conditioning variables. See Chapter 4 for a discussion of conditional distributions.

Consider the case of two random variables X and Y. The conditional expectation of Y given $X = x$ is the expected value of a random variable Y in the conditional distribution of Y given $X = x$. Formally, this is written as

$$E(Y \mid X = x),$$

where "|" is read "conditional on" or "given". Informally, we think of $E(Y|X = x)$ as the expected value of Y in the case where it is known that the outcome of the stochastic variable X is $X = x$. Sometimes a subscript is used on E to help indicate the conditional distribution in mind, e.g. $E_{Y|X}(Y|X = x)$. We also refer to the conditional expectation of Y given $X = x$ as the conditional mean of Y given $X = x$ in a similar fashion as we sometimes refer to the expected value of X as the mean of X.

Next, the conditional expectation of Y given X is formally defined.

Definition 6.3 (Conditional expectation and CEF, two random variables)
*Let $x \in \mathbb{R}$ be a number. If Y is a discrete random variable and the conditional PMF of Y given $X = x$ is $f_{Y|X}(y|x)$, then the **conditional expectation of** Y **given** $X = x$ is defined as*

$$E(Y|X = x) = \sum_y y \cdot f_{Y|X}(y|x).$$

*If Y is a continuous random variable and the conditional PDF of Y given $X = x$ is $f_{Y|X}(y|x)$, then the **conditional expectation of** Y **given** $X = x$ is defined as*

$$E(Y|X = x) = \int_{-\infty}^{\infty} y \cdot f_{Y|X}(y|x)dy.$$

*When considering $E(Y|X = x)$ as a function of x, it is denoted the **conditional expectation function** (CEF) of Y given $X = x$.*

Notice, the only technical difference to the definition of the expectation of Y (Definition 5.1) is that $f_Y(y)$ is replaced by $f_{Y|X}(y|x)$ in the definition of the conditional expectation.

Also notice, there is no assumptions as to the distribution of X nor to the type of random variable of X.

The conditional expectation of Y given $X = x$ can be interpreted as the mean in the subpopulation defined by having $X = x$. For example, if Y is earnings of a randomly selected individual from a population and X is length of education of that individual, then $E(Y \mid X = 12)$ is the mean earnings in the subpopulation of individuals with 12 years of education.

Example 6.6 *Consider the following conditional PDF of X given $Z = z$*

$$f_{X|Z}(x|z) = \begin{cases} 0 & \text{if} & x < z, \\ 0.5 & \text{if} & z \leq x \leq z+2, \\ 0 & \text{if} & x > z+2. \end{cases}$$

Figure 6.1 illustrates the graph of this conditional density function. Notice, for $z = 0$, the conditional density function $f_{X|Z}(x|0)$ is the same as the density function $f_X(x)$ in Example 2.9.

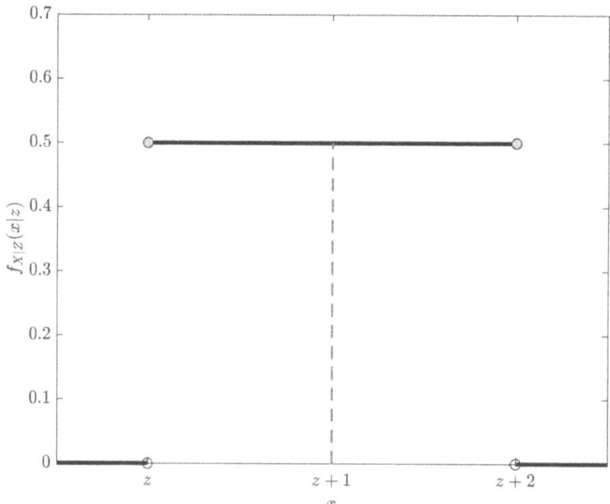

FIGURE 6.1

Conditional PDF $f_{X|Z}(x|z)$ of X given $Z = z$ from Example 6.6.

The conditional expectation of X given $Z = z$ can be calculated via Definition 6.3:

$$\begin{aligned} E(X \mid Z = z) &= \int_{-\infty}^{\infty} x \cdot f_{X|Z}(x \mid z)dx \\ &= \int_{-\infty}^{z} x \cdot 0 dx + \int_{z}^{z+2} x \cdot 0.5 dx + \int_{z+2}^{\infty} x \cdot 0 dx \\ &= \left[0.5 \cdot 0.5 \cdot x^2\right]_{z}^{z+2} \\ &= 0.5 \cdot 0.5 \cdot (z+2)^2 - 0.5 \cdot 0.5 \cdot z^2 \\ &= 0.25 \cdot (z^2 + 2^2 + 2 \cdot 2 \cdot z) - 0.25 \cdot z^2 \\ &= 1 + z. \end{aligned}$$

Hence, in this case, the conditional mean of X given $Z = z$ is a linear function of z.

In practice, we often specify the conditional expectation directly without first specifying the conditional distribution of Y given X. This is sufficient as long as our interest is the conditional expectation. For any specification of the conditional expectation, we can always find distributions, if we wanted to, that implies that specification of the conditional expectation. In fact, there are infinitely many different conditional distributions which have the same conditional expectation. This is no different than there are infinitely many distributions which have the same mean.

The rules of calculation of conditional expectations are similar to the rules of calculation for expectations, except that all the expressions are conditional statements. The next theorem collects some of the rules.

Theorem 6.9 (Rules of calculation for conditional expectation) *Let $a \in \mathbb{R}$ be a constant and $x \in \mathbb{R}$ a number. Assume the mean of the random variables Y and Z exist. Regardless of the type of random variables X, Y, and Z, then for any outcome x of X,*

i) $E(a + Y \mid X = x) = a + E(Y \mid X = x).$

ii) $E(a \cdot Y \mid X = x) = a \cdot E(Y \mid X = x).$

iii) $E(Y + Z \mid X = x) = E(Y \mid X = x) + E(Z \mid X = x).$

iv) $E(X \cdot Y \mid X = x) = x \cdot E(Y \mid X = x).$

Until now, we have considered the conditional expectation of the random variable Y, given $X = x$, i.e. given the knowledge that the random variable X has an outcome of x. As discussed above, this allows us to consider the conditional expectation function, $E(Y|X = x)$, as a function of x, and interpret $E(Y|X = x)$ as the mean in the subpopulation defined by having $X = x$. By replacing the number x by the random variable X, we can make it random which subpopulation we consider. That is, we consider the conditional expectation of Y given X, as a function of the random variable X. We denote the conditional expectation function as a function of the random variable X by $E(Y \mid X)$. To calculate the conditional expectation $E(Y \mid X)$, we simply calculate the conditional expectation function $E(Y \mid X = x)$, and then replace x by X. Note that, whereas $E(Y \mid X = x)$ is a (non-random) number, $E(Y \mid X)$ is a random variable. It is the problem at hand that determines when to use $E(Y \mid X)$ and when to use $E(Y \mid X = x)$.

Example 6.7 *Suppose*
$$E(Y \mid X = x) = 3 + 2 \cdot x.$$

Since the input x is a number, then the output of the function $E(Y \mid X = x)$ is a number. Now, replace x with the random variable X in the formula for $E(Y \mid X = x)$ to get

$$E(Y \mid X) = 3 + 2 \cdot X.$$

Since the input is a random variable X, then the output is also a random variable, namely, $3 + 2 \cdot X.$

Finally, we extend the definition of conditional expectation of Y given X to the case with many conditioning random variables.

Definition 6.4 (Conditional expectation and CEF, many random variables) *Let* $x_1, x_2, \ldots, x_k \in \mathbb{R}$. *The* **conditional expectation** *of a discrete random variable* Y **conditional on** $X_1 = x_1, \ldots, X_k = x_k$ *is*

$$E(Y|X_1 = x_1, \ldots, X_k = x_k) = \sum_y y \cdot f_{Y|X_1,\ldots,X_k}(y|x_1, \ldots, x_k),$$

where $f_{Y|X_1,\ldots,X_k}$ *is the conditional PMF of* Y *given* $X_1 = x_1, \ldots, X_k = x_k$.

The **conditional expectation** *of a continuous random variable* Y **conditional on** $X_1 = x_1, \ldots, X_k = x_k$ *is*

$$E(Y|X_1 = x_1, \ldots, X_k = x_k) = \int_{-\infty}^{\infty} y f_{Y|X_1,\ldots,X_k}(y|x_1, \ldots, x_k) dy,$$

where $f_{Y|X_1,\ldots,X_k}$ *is the conditional PDF of* Y *given* $X_1 = x_1, \ldots, X_k = x_k$.

When considering $E(Y|X_1 = x_1, \ldots X_k = x_k)$ *as a function of* $x_1, \ldots, x_k$, *it is denoted the* **conditional expectation function** *(CEF) of* Y *given* $X_1 = x_1, \ldots, X_k = x_k$.

6.5   Conditional expectation on different sets of random variables

In this section, we consider the difference between the conditional expectation of, say, Y given X, compared to the conditional expectation of Y given X and Z. For example, if Y is earnings of a randomly selected individual from a population, X is length of education, and Z is cognitive abilities of that individual, then $E(Y \mid X = 12)$ is the mean earnings in the subpopulation of individuals with 12 years of education. This is different from $E(Y \mid X = 12, Z = 110)$, which is the mean earnings in the subpopulation of individuals with 12 years of education and a cognitive ability of 110. The last subpopulation is a subpopulation of the first subpopulation. In this section, we consider how to calculate one conditional expectation from the other.

6.5.1   The Law of Iterated Expectations

To compare conditional expectations with different conditioning sets, the Law of Iterated Expectations is useful. It is also useful in many other situations we encounter later on. There are different variants of the Law of Iterated Expectations, three of which are shown in the next theorem.

Theorem 6.10 (Law of Iterated Expectations, LIE) *Assume the expectations of the random variables exist.*

1. For two random variables X *and* Y,

$$E(Y) = E\left(E(Y \mid X)\right). \tag{6.4}$$

2. For any function $g(x)$,

$$E(g(X) \cdot Y) = E\left(g(X) \cdot E(Y \mid X)\right). \tag{6.5}$$

3. For three random variables X, Y *and* Z,

$$E(Y \mid X) = E\left(E(Y \mid X, Z) \mid X\right). \tag{6.6}$$

Example 6.8 *Suppose there are three groups coded as $X = 1$, $X = 2$, and $X = 3$. Let the probabilities of a randomly selected individual being in a particular group be*

$$f_X(x) = \begin{cases} p_1 & if & x = 1, \\ p_2 & if & x = 2, \\ p_3 & if & x = 3, \\ 0 & & otherwise, \end{cases}$$

where $p_1, p_2, p_3 \geq 0$ and $p_1 + p_2 + p_3 = 1$.

Let Y be another random variable and suppose the mean of Y in each group is known. That is, $E(Y|X = x)$ is known. By Equation (6.4) in the Law of Iterated Expectations (Theorem 6.10), the mean of Y in the population is

$$E(Y) = E\left(E(Y|X)\right).$$

Using the definition of $E()$, we get

$$\begin{aligned} E(Y) &= \sum_x E(Y|X = x) \cdot f_X(x) \\ &= E(Y|X = 1) \cdot p_1 + E(Y|X = 2) \cdot p_2 + E(Y|X = 3) \cdot p_3. \end{aligned}$$

In words, the mean of Y is the mean of Y in each of the three groups weighted together with the probability of each group. In a population interpretation, the probability of each group is the fraction that each group has in the population.

Example 6.9 *Suppose we need to calculate the expected value of $X \cdot U$ and it is known that $E(U|X = x) = 0$ for all x. Then by Equation (6.5) in the Law of Iterated Expectations (Theorem 6.10), with $g(X) = X$,*

$$E(X \cdot U) = E\left(X \cdot E(U|X)\right) = E\left(X \cdot 0\right) = 0.$$

Note that, if it is only known that $E(U) = 0$, then it does not follow that $E(X \cdot U) = 0$. Indeed, the following provides a simple counter-example.

Example 6.10 *Let X be a discrete random variable with PMF*

$$f_X(x) = \begin{cases} 1/2 & if & x = -1, \\ 1/2 & if & x = 1. \end{cases}$$

Then,

$$E(X) = \sum_x x \cdot f(x) = -1 \cdot \frac{1}{2} + 1 \cdot \frac{1}{2} = 0.$$

Define $U = X$. Then, clearly, $E(U) = 0$, but

$$E(X \cdot U) = E(X^2) = \sum_x x^2 \cdot f(x) = (-1)^2 \cdot \frac{1}{2} + 1^2 \cdot \frac{1}{2} = 1,$$

which shows that the condition $E(U) = 0$ is not sufficient to conclude that $E(X \cdot U) = 0$.

6.5.2 Relationship between conditional expectation and covariance

Restrictions on the conditional expectation of Y given X typically have consequences for the covariance between Y and X. Below, we show some restrictions that are useful for interpretation of regression models discussed in Volume II of this book.

Importantly, if the conditional expectation $E(Y|X) = 0$, then the covariance $Cov(Y, X)$ between Y and X equals 0. The opposite conclusion does not necessarily hold. We state this as a slightly more general result in the next theorem.

Theorem 6.11 (Conditional expectation and covariance) *Let Y and X be random variables such that*

$$E(Y|X) = c,$$

where $c \in \mathbb{R}$ is a constant. Then

$$Cov(Y, X) = 0.$$

Notice, the condition $E(Y|X) = c$ is the same as requiring that $E(Y|X = x) = c$ for any outcome x of X.

6.5.3 Long and short conditioning sets

When doing causal analysis, covered in Volume II of this book, we often have to condition on a smaller set of random variables than we would have liked due to certain random variables not being observable. In this section, we consider theoretically the relationship between the conditional expectation given a large set of conditioning variables (called the "long conditioning set") to conditioning on a smaller set of random variables (called the "short conditioning set"). The relationship can be derived using the Law of Iterated Expectations (Theorem 6.10).

First, we consider the relationship between conditioning on two random variables and conditioning on only one of them. By Equation (6.6) in the Law of Iterated Expectations, the relationship between the conditional expectation of Y given X and the conditional expectation of Y given X and Z is

$$E(Y \mid X) = E\left(E(Y \mid X, Z) \mid X\right).$$

Here $E(Y \mid X, Z)$ is the conditional expectation with a long conditioning set (two stochastic variables, X and Z), and $E(Y \mid X)$ is the conditional expectation with a short conditioning set (one stochastic variable, X). This terminology is similar to the one used in Section 4.5.

To see more clearly how the conditional expectation with long and short conditioning sets are related, suppose the conditional expectation with a long conditioning set is a linear function, given by

$$E(Y|X, Z) = \beta_0 + \beta_1 X + \beta_2 Z,$$

where $\beta_0, \beta_1, \beta_2 \in \mathbb{R}$ are constants. Using the Law of Iterated Expectations (6.6), the conditional expectation with a short conditioning set is

$$
\begin{aligned}
E(Y|X) &= E\left(E(Y|X, Z) \mid X\right) \\
&= E\left((\beta_0 + \beta_1 X + \beta_2 Z) \mid X\right) \\
&= \beta_0 + \beta_1 X + \beta_2 E(Z|X),
\end{aligned}
\tag{6.7}
$$

where both parts iii) and iv) of Theorem 6.9 were used for the last equality sign.

Suppose further that the conditional expectation function of Z on X is linear and given by

$$E(Z|X) = \gamma_0 + \gamma_1 X,$$

where $\gamma_0, \gamma_1 \in \mathbb{R}$ are constants. Insert this into (6.7) to get

$$E(Y|X) = \underbrace{(\beta_0 + \beta_2\gamma_0)}_{=\delta_0} + \underbrace{(\beta_1 + \beta_2\gamma_1)}_{=\delta_1}X = \delta_0 + \delta_1 X. \tag{6.8}$$

The coefficient δ_1 on X in the conditional expectation with the short conditioning set is a function of coefficients partly from the conditional expectation with the long conditioning set, partly from the conditional expectation of Z on X. Thus, even though Z is not included in the conditional expectation of Y on X, it does influence the result through its relationship with X captured by the coefficient γ_1. Only if Z is not related to X, implying $\gamma_1 = 0$, or if Z is not related to Y, implying $\beta_2 = 0$, then Z does not influence the relationship between Y and X in the conditional expectation without conditioning on Z.

Example 6.11 *Suppose*

$$E(Y|X, Z) = 3X + 2Z$$

and

$$E(Z|X) = -4X.$$

Then using formula (6.8),

$$E(Y|X) = 3X + 2E(Z|X) = -5X.$$

This implies that the mean of Y is positively related to X for each subpopulation $Z = z$, whereas the mean of Y is negatively related to X over the whole population. In other words, when discussing the relationship between two random variables X and Y, it is crucial to be clear about what the population is, and if the relationship is assessed on subpopulations, here specified by Z.

In general, a conditional expectation function with a short conditioning set can be obtained from a conditional expectation function with a long conditioning set by the following extension of the Law of Iterated Expectations. Let the long conditioning set be $(X_1, \ldots, X_K)$ and the short conditioning set be $(X_1, \ldots, X_m)$, where $m < K$. Then

$$E(Y \mid X_1, \ldots, X_m) = E\left(E(Y \mid X_1, \ldots, X_K) \mid X_1, \ldots, X_m\right).$$

6.6 Conditional variance

Sometimes it is interesting to consider the relationship between X and the variance of Y either because of its substantive meaning, or because it may influence how inference can be made. For example, the consumption of low income individuals may vary less than the consumption of high income individuals. The relationship may be described as the variance of Y conditional on X. This is denoted $Var(Y \mid X = x)$. The definition is next.

Definition 6.5 (Conditional variance) *The **conditional variance of Y given $X = x$** is*

$$Var(Y|X = x) = E\left((Y - E(Y|X = x))^2 \mid X = x\right).$$

Since $Var(Y|X = x)$ is an expectation of a function of Y, it can be evaluated according to Theorem 5.1. Therefore, if Y is a discrete random variable, then $Var(Y|X = x)$ can be calculated as

$$Var(Y|X = x) = \sum_y (y - E(Y|X = x))^2 \cdot f_{Y|X}(y|x),$$

where

$$E(Y|X = x) = \sum_y y \cdot f_{Y|X}(y|x).$$

If Y is a continuous random variable, then $Var(Y|X = x)$ can be calculated as

$$Var(Y|X = x) = \int_{-\infty}^{\infty} (y - E(Y|X = x))^2 \cdot f_{Y|X}(y|x)dy,$$

where

$$E(Y|X = x) = \int_{-\infty}^{\infty} y \cdot f_{Y|X}(y|x)dy.$$

Example 6.12 (Example 6.6, continued) *Consider the conditional probability function of X given Z*

$$f_{X|Z}(x|z) = \begin{cases} 0 & \text{if} & x < z, \\ 0.5 & \text{if} & z \leq x < z + 2, \\ 0 & \text{if} & x \geq z + 2. \end{cases}$$

In Example 6.6, it was found that

$$E(X|Z = z) = 1 + z.$$

The conditional variance of X given z is

$$
\begin{aligned}
Var(X|Z = z) &= \int_{-\infty}^{\infty} (x - E(X \mid Z = z))^2 \cdot f_{X|Z}(x \mid z)dx \\
&= \int_{z}^{z+2} (x - (1+z))^2 \cdot 0.5dx \\
&= \left[0.5 \cdot \frac{1}{3}(x - 1 - z)^3 \right]_{z}^{z+2} \\
&= \frac{1}{6} \left[(z + 2 - 1 - z)^3 - (z - 1 - z)^3 \right] \\
&= \frac{1}{6} \left[1^3 - (-1)^3 \right] \\
&= \frac{1}{6} \cdot 2 \\
&= \frac{1}{3}.
\end{aligned}
$$

We see that in this example, the conditional variance of X given $Z = z$ does not depend on z. As illustrated in Figure 6.1, the conditioning value $Z = z$ influences the location of the conditional density of X given $Z = z$ but not the shape of the conditional density. For this reason, the conditional variance of X given $Z = z$ does not depend on z.

Example 6.13 (Example 4.3, continued) *Let the random variables X and Y denote, respectively, Public Service Motivation and Performance of a randomly chosen individual, i.e.*

$$X = \begin{cases} 1 & \text{if} & \text{low}, \\ 2 & \text{if} & \text{middle}, \\ 3 & \text{if} & \text{high}, \end{cases} \qquad Y = \begin{cases} 0 & \text{if} & \text{poor}, \\ 1 & \text{if} & \text{good}. \end{cases}$$

Suppose the conditional probability function $f_{Y|X}(y|x)$ of Y given X is given as in Example 4.3, i.e. by the numbers in the following table.

$f_{Y\mid X}(y\mid x)$		Y	
		0	1
X	1	0.80	0.20
	2	0.40	0.60
	3	0.40	0.60

To calculate the conditional variance, first calculate the conditional expectation function of Y given X using Definition 6.3

$$
\begin{aligned}
E(Y|X=1) &= 0\cdot 0.8 + 1\cdot 0.2 = 0.2,\\
E(Y|X=2) &= 0\cdot 0.4 + 1\cdot 0.6 = 0.6,\\
E(Y|X=3) &= 0\cdot 0.4 + 1\cdot 0.6 = 0.6.
\end{aligned}
$$

Then the conditional variance function can be found using Definition 6.5,

$$
\begin{aligned}
Var(Y|X=1) &= (0-0.2)^2\cdot 0.8 + (1-0.2)^2\cdot 0.2 = 0.16,\\
Var(Y|X=2) &= (0-0.6)^2\cdot 0.4 + (1-0.6)^2\cdot 0.6 = 0.24,\\
Var(Y|X=3) &= (0-0.6)^2\cdot 0.4 + (1-0.6)^2\cdot 0.6 = 0.24.
\end{aligned}
$$

It is seen that the conditional variance of Performance Y for the subpopulation with middle and high Public Service Motivation $(X=2,3)$ is higher than in the subpopulation with low Public Service Motivation $(X=1)$.

Some of the rules of calculation for conditional variances are listed in the next theorem.

Theorem 6.12 (Rules of calculation, conditional variance) *Let $a, b \in \mathbb{R}$ be constants and assume the variances exist. Then*

i) $Var(Y|X=x) = E\left(Y^2|X=x\right) - E(Y|X=x)^2.$

ii) $Var(a+Y|X=x) = Var(Y|X=x).$

iii) $Var(b\cdot Y|X=x) = b^2\cdot Var(Y|X=x).$

The following result can often be useful when calculating the variance of a random variable Y, which depends in a known way on another random variable, X.

Theorem 6.13 (Law of total variance) *Let X and Y be two random variables and assume their variances exist. Then*

$$
Var(Y) = E\left(Var(Y\mid X)\right) + Var\left(E(Y\mid X)\right). \tag{6.9}
$$

It is worth stressing that the variance of Y is not just the expected value over the variances for each subpopulation defined by X since this is the first term in Equation (6.9). The variance of Y also depends on how the mean of each subpopulation varies over the population. This is the second term in Equation (6.9). We sometimes denote the first term $E\left(Var(Y\mid X)\right)$ as the "within" variance and the second term $Var\left(E(Y\mid X)\right)$ as the "between" variance. For example, if X is age and Y is activity level, then the activity level of an age group may be quite different, that is, a high within variance, but the mean activity level over age may not be very different, that is, a low between variance.

As a side remark, the field of *Analysis of Variance* (commonly abbreviated "ANOVA") is based on the decomposition of a variance as given in Equation (6.9).

6.7 Conditional quantiles

A *conditional p-quantile* of Y given $X = x$ is the p-quantile of the conditional distribution of Y given $X = x$. For example, if Y is earnings of a randomly selected individual from a population and X is length of education of that individual, then the conditional 0.1-quantile of earnings, given $X = 12$, gives the value of income $q_{0.1}(12)$ for which 10% of the subpopulation with 12 years of education earns less than $q_{0.1}(12)$ and 90% of that subpopulation earns more than $q_{0.1}(12)$.

A p-quantile function is defined next. The definition is similar to the Definition 5.7 of a p-quantile except that the distribution used here is a conditional distribution.

Definition 6.6 (Conditional p-quantile function) *Let X and Y be random variables and $p \in [0,1]$. The function $q_p(x)$ is a **conditional p-quantile function of Y given** $X = x$ if and only if*

$$\Pr\left(Y < q_p(x) \mid X = x\right) \leq p \text{ and } \Pr\left(Y \leq q_p(x) \mid X = x\right) \geq p \text{ for all } x,$$

where $\Pr\left(Y < q_p(x) \mid X = x\right)$ and $\Pr\left(Y \leq q_p(x) \mid X = x\right)$ are probability statements on the conditional distribution of Y given $X = x$.

Example 6.14 *Let X and Y be two discrete random variables, and consider the discrete conditional PMF given in the following table.*

$f_{Y\mid X}$		Y	
		2	4	6
X	1	0.6	0.2	0.2
	2	0.0	0.4	0.6
	3	0.2	0.5	0.3

The conditional 0.5-quantile (conditional median) function is

$$q_{0.5}(x) = \begin{cases} 2 & \text{if } x = 1, \\ 6 & \text{if } x = 2, \\ 4 & \text{if } x = 3. \end{cases}$$

As comparison, the conditional mean function is

$$E(Y\mid X = x) = \begin{cases} 3.2 & \text{if } x = 1, \\ 5.2 & \text{if } x = 2, \\ 4.2 & \text{if } x = 3. \end{cases}$$

Note that the conditional median function is below the conditional mean function for some x and above for others.

Just as in the case of the definition of the (unconditional) p-quantile (Definition 5.7), the definition of the conditional quantile function simplifies in the case of continuous random variables. That is, suppose that Y is a continuous random variable and X is a random variable of any type. Then the conditional quantile function $q_p(x)$ satisfies that

$$F_{Y\mid X}(q_p(x)\mid x) = p,$$

where $F_{Y\mid X}$ is the conditional CDF of Y given X.

Example 6.15 *Consider the conditional CDF of X given Z*

$$F_{X|Z}(x|z) = \begin{cases} 0 & if & x < z, \\ 0.5(x-z) & if & z \le x < z+2, \\ 1 & if & x \ge z+2. \end{cases}$$

This is the CDF of the conditional PDF in Example 6.6 and if $z = 0$, then $F_{X|Z}(x|0)$ is the same CDF as $F_X(x)$ in Example 2.9.

The conditional p-quantile $q_p(z)$ of X given Z can be derived as follows

$$\begin{aligned} F_{X|Z}(q_p(z) \mid z) &= p \Rightarrow \\ 0.5(q_p(z) - z) &= p \Rightarrow \\ q_p(z) &= 2p + z. \end{aligned}$$

It is seen that the conditional median, denoted $Med(X \mid Z = z)$, is

$$Med(X \mid Z = z) = q_{0.5}(z) = 1 + z.$$

This is the same as the conditional expectation function derived in Example 6.6. This is no coincidence. If all the conditional distributions are symmetric, and their means exist, then the conditional median function and the conditional mean function coincide.

The conditional p-quantile is the basis for the method called quantile regression, covered in Volume II of this book.

6.8   Proofs

Proof of Theorem 6.1 *We use a version of the so-called Cauchy-Schwarz inequality, which states that*

$$|E(XY)|^2 \le E(X^2)E(Y^2). \tag{6.10}$$

Define the random variables $\tilde{X} = X - E(X)$ and $\tilde{Y} = Y - E(Y)$. By the Cauchy-Schwarz inequality (6.10) and the definition of covariance (Definition 6.1), we have

$$\begin{aligned} |Cov(X,Y)|^2 &= |E((X - E(X))(Y - E(Y)))|^2 \\ &= |E(\tilde{X}\tilde{Y})|^2 \\ &\le E(\tilde{X}^2)E(\tilde{Y}^2) \\ &= E((X - E(X))^2)E((X - E(X))^2) \\ &= Var(X)Var(Y), \end{aligned}$$

showing that

$$Var(X), Var(Y) < \infty \Rightarrow |Cov(X,Y)| < \infty,$$

which is what we wanted to prove.

Proof of Theorem 6.2 *Consider the "two-dimensional" random variable*

$$(X,Y) : \Omega \to \mathbb{R}^2.$$

The proof now follows along similar lines as the proof of Theorem 5.1 in Chapter 5, applied to (X, Y). We omit the details.

Proof of Theorem 6.3 *This follows by applying Theorem 6.2 to the function $g(x, y) = (x - \mu_X)(y - \mu_Y)$.*

Proof of Theorem 6.4 *Apply Theorem 6.2 by setting $g(x, y) = x + y$. We provide the proof in the case of two discrete random variables; the case of two continuous random variables is left as an exercise for the reader in Problem 6.9.2. First, note that*

$$
\begin{aligned}
E(X + Y) &= \sum_x \sum_y (x + y) \cdot f_{X,Y}(x, y) \\
&= \sum_x \sum_y x \cdot f_{X,Y}(x, y) + \sum_x \sum_y y \cdot f_{X,Y}(x, y) \\
&= \sum_x x \cdot \left(\sum_y f_{X,Y}(x, y) \right) + \sum_y y \cdot \left(\sum_x f_{X,Y}(x, y) \right),
\end{aligned}
$$

where the equality signs follow from rearranging sums. The two terms in parentheses are joint PMFs where one of the variables is "summed out", thus yielding a marginal PMF (Theorem 3.2). Therefore, insert for the marginal PMFs to get

$$
\begin{aligned}
E(X + Y) &= \sum_x x \cdot f_X(x) + \sum_y y \cdot f_Y(y) \\
&= E(X) + E(Y).
\end{aligned}
$$

Proof of Theorem 6.5 *The proof of Theorem 6.5 is left as an exercise to the reader (Problem 6.9.3).*

Proof of Theorem 6.6 *The derivation of the variance of a sum is as follows. Start by using the definition of the variance (Definition 5.3)*

$$
\begin{aligned}
Var(X + Y) &= E\left(((X + Y) - E(X + Y))^2 \right) \\
&= E\left(((X + Y) - (\mu_X + \mu_Y))^2 \right) \\
&= E\left(((X - \mu_X) + (Y - \mu_Y))^2 \right) \\
&= E\left((X - \mu_X)^2 + (Y - \mu_Y)^2 + 2(X - \mu_X)(Y - \mu_Y) \right),
\end{aligned}
$$

where we in the second line applied Theorem 6.4. Now, apply Theorem 6.4 again to get

$$
\begin{aligned}
Var(X + Y) &= E\left((X - \mu_X)^2 \right) + E\left((Y - \mu_Y)^2 \right) + E(2(X - \mu_X)(Y - \mu_Y)) \\
&= Var(X) + Var(Y) + 2\underbrace{E((X - \mu_X)(Y - \mu_Y))}_{=Cov(X,Y)}.
\end{aligned}
$$

Proof of Theorem 6.7 *Apply the definition of covariance (Definition 6.1) and the Cauchy-Schwarz inequality (6.10) to get*

$$|(Cov(X,Y))|^2 = |E\left((X-\mu_X)(Y-\mu_Y)\right)|^2 \ \leq \ E\left((X-\mu_X)^2\right)E\left((Y-\mu_Y)^2\right)$$
$$= \ Var(X)Var(Y).$$

Now, divide through by $Var(X)Var(Y)$ and use the definition of the correlation coefficient to get

$$1 \geq \frac{|(Cov(X,Y))|^2}{Var(X)Var(Y)} = \left(\frac{Cov(X,Y)}{\sqrt{Var(X)\cdot Var(Y)}}.\right)^2 = (\rho_{X,Y})^2.$$

Taking square roots yields $-1 \leq \rho_{X,Y} \leq 1$, as we wanted to show.

Proof of Theorem 6.8 *From property iii) of Theorem 6.5, we have that $\beta_1 = 0$ implies that $Cov(X,Y) = Cov(X,\beta_0) = 0$ and therefore $\rho_{X,Y} = 0$. Assume $\beta_1 \neq 0$. Then*

$$\rho_{X,Y} = \frac{Cov(X,Y)}{\sqrt{Var(X)Var(Y)}}.$$

The covariance is

$$\begin{aligned}
Cov(X,Y) &= Cov(X,\beta_0 + \beta_1 X) \\
&= Cov(X,\beta_0) + Cov(X,\beta_1 X) \\
&= \beta_1 Cov(X,X) \\
&= \beta_1 Var(X).
\end{aligned}$$

The variance of Y is

$$Var(\beta_0 + \beta_1 X) = \beta_1^2 Var(X).$$

Insert into the expression for correlation

$$\begin{aligned}
\rho_{X,Y} &= \frac{\beta_1 Var(X)}{\sqrt{Var(X)\beta_1^2 Var(X)}} = \frac{\beta_1 Var(X)}{\sqrt{\beta_1^2\left(Var(X)\right)^2}} \\
&= \frac{\beta_1 Var(X)}{\sqrt{\beta_1^2}\sqrt{\left(Var(X)\right)^2}} = \frac{\beta_1 Var(X)}{|\beta_1|\cdot|Var(X)|} \\
&= \frac{\beta_1 Var(X)}{|\beta_1|\cdot Var(X)} = \frac{\beta_1}{|\beta_1|} = \left\{\begin{array}{ll} -1 & \text{if} \ \ \beta_1 < 0, \\ 1 & \text{if} \ \ \beta_1 > 0, \end{array}\right.
\end{aligned}$$

where it has been used that $\sqrt{\beta_1^2} = |\beta_1|$, and that $Var(X)$ by definition is non-negative implying $|Var(X)| = Var(X)$. In conclusion, when $\beta_1 \neq 0$, X and Y are perfectly correlated and the sign of the correlation is determined by the sign of β_1.

Proof of Theorem 6.9 *The proof of properties i)–iii) are similar to their unconditional counterparts, but with conditional distributions instead of marginal distributions, and we hence omit the proofs of these. Property iv) can be seen by*

$$E(X \cdot Y | X = x) = E(x \cdot Y | X = x) = x \cdot E(Y | X = x),$$

where the last equality comes from x being a constant, that is, not a random variable.

Proof of Theorem 6.10   *We prove (6.4) for the case where X and Y are discrete random variables; the case of continuous random variables is similar, with PDFs replacing PMFs. Let $f_Y(y)$ be the marginal PMF for Y. By definition,*

$$E(Y) = \sum_y y \cdot f_Y(y).$$

Let $f_{X,Y}(x,y)$ be the joint PMF for X and Y. Then the marginal PMF of Y can be written as (Theorem 3.2)

$$f_Y(y) = \sum_x f_{X,Y}(x,y) = \sum_x f_{Y|X}(y|x) \cdot f_X(x),$$

where we used the definition of conditional probability, $f_{X,Y}(x,y) = f_{Y|X}(y|x) \cdot f_X(x)$. Insert this into the expression for $E(Y)$ to get

$$E(Y) = \sum_y y \cdot \left(\sum_x f_{Y|X}(y|x) \cdot f_X(x) \right).$$

Rearrange the two sums as

$$
\begin{aligned}
E(Y) &= \sum_x \sum_y y \cdot f_{Y|X}(y|x) \cdot f_X(x) \\
&= \sum_x \left(\sum_y y \cdot f_{Y|X}(y|x) \right) \cdot f_X(x) \\
&= \sum_x \left(E(Y|X = x) \right) \cdot f_X(x) \\
&= E\left(E(Y|X) \right).
\end{aligned}
$$

The last equality sign follows by regarding $E(Y|X = x)$ as a function of x according to Theorem 5.1. When that function is averaged at all the possible inputs x with the probability of each of those inputs, then this is the expected value of that function.

*   The remaining parts of the theorem can be proved in a similar manner.*

Proof of Theorem 6.11   *Recall that, by property i) in Theorem 6.5, we can write the covariance as*

$$Cov(Y, X) = E(YX) - E(Y)E(X)$$

*   The assumption that $E(Y|X) = c$, together with Equation (6.4) in the Law of Iterated Expectations (Theorem 6.10), implies that*

$$E(Y) = E\left(E(Y \mid X) \right) = E\left(c \right) = c.$$

Similarly, using Equation (6.5) in the Law of Iterated Expectations, we get

$$E(X \cdot Y) = E(X \cdot E(Y|X)) = E(X \cdot c) = c \cdot E(X).$$

Insert these two results into expression for the covariance above to get

$$Cov(Y, X) = c \cdot E(X) - c \cdot E(X) = 0.$$

Proof of Theorem 6.12 *The proof is similar to the proof of Theorem 5.3, except that we use conditional expectations and conditional distributions, and we hence omit the proof.*

Proof of Theorem 6.13 *We first note that, by the definition of the conditional variance (Definition 6.5)*

$$Var(Y|X) = E((Y - E(Y|X))^2) \tag{6.11}$$

and the variance (Definition 5.3)

$$Var(E(Y|X)) = E((E(Y|X) - E(Y))^2), \tag{6.12}$$

where we used that $E(E(Y|X)) = E(Y)$ by the Law of Iterated Expectations (Theorem 6.10). Now, by using the definition of the variance and adding and subtracting $E(Y|X)$, we get

$$Var(Y) = E((Y - E(Y))^2) = E\left((Y - E(Y|X) + E(Y|X) - E(Y))^2\right).$$

Expanding the square yields

$$\begin{aligned}
Var(Y) \quad = \quad & E\left((Y - E(Y|X))^2\right) + E\left((E(Y|X) - E(Y))^2\right) \\
& + 2E\left((Y - E(Y|X))(E(Y|X) - E(Y))\right).
\end{aligned}$$

By the Law of Iterated Expectations, the first term equals

$$E\left((Y - E(Y|X))^2\right) = E\left(E\left((Y - E(Y|X))^2|X\right)\right) = E\left(Var(Y|X)\right),$$

cf. Equation (6.11). As for the second term

$$E\left((E(Y|X) - E(Y))^2\right) = Var(E(Y|X))$$

cf. Equation (6.12). The third term equals zero by the Law of Iterated Expectations,

$$\begin{aligned}
E\left((Y - E(Y|X))(E(Y|X) - E(Y))\right) \quad = \quad & E\left(E\left((Y - E(Y|X))(E(Y|X) - E(Y))|X\right)\right) \\
= \quad & E\left((E(Y|X) - E(Y|X))(E(Y|X) - E(Y))\right) \\
= \quad & E(0 \cdot (E(Y|X) - E(Y))) \\
= \quad & 0.
\end{aligned}$$

We conclude that

$$Var(Y) = E\left(Var(Y|X)\right) + Var(E(Y|X)),$$

as we wanted to show.

6.9   Exercises

Problem 6.9.1 *The table below contains the joint probability function $f_{X,Y}(x,y)$ for the two random variables X and Y.*

		Y	
		0	*1*
	4	*0.1*	*0.3*
X	*7*	*0.2*	*0.1*
	10	*0.1*	*0.2*

1. Calculate the marginal probability functions for X and Y, $f_X(x)$ and $f_Y(y)$.

2. Calculate the conditional probability function for Y given $X = 10$, $f_{Y|X}(y|X = 10)$.

3. Calculate the mean of Y.

4. Calculate the conditional mean of Y given $X = 10$, $E(Y|X = 10)$.

5. Calculate the variance of Y.

6. Calculate the conditional variance of Y given $X = 10$, $Var(Y|X = 10)$.

7. Calculate the $p = 0.4$ quantile for Y.

8. Calculate the $p = 0.4$ conditional quantile for Y given $X = 10$, $q_{0.4}(10)$.

Problem 6.9.2 *Prove Theorem 6.4 in the case where X and Y are continuous random variables with joint PDF $f_{X,Y}(x, y)$.*

Problem 6.9.3 *Prove Theorem 6.5.*

Problem 6.9.4 *Theorem 6.6 shows that, for two random variables X and Y, the variance of the sum $X + Y$ is the sum of the variances and twice the covariance, i.e.*

$$Var(X + Y) = Var(X) + Var(Y) + 2Cov(X, Y).$$

This property can be generalized. Let $n \geq 2$ be an integer and consider the n random variables $X_1, \ldots, X_n$. Prove the following string of equalities

$$Var\left(\sum_{i=1}^{n} X_i\right) = \sum_{i=1}^{n}\sum_{j=1}^{n} Cov(X_i, X_j)$$

$$= \sum_{i=1}^{n} Var(X_i) + \sum_{i \neq j} Cov(X_i, X_j)$$

$$= \sum_{i=1}^{n} Var(X_i) + 2\sum_{i < j} Cov(X_i, X_j),$$

where "$\sum_{i \neq j}$" means the sum over all $i, j = 1, \ldots, n$ such that $i \neq j$ and "$\sum_{i < j}$" means the sum over all $i, j = 1, \ldots, n$ such that $i < j$.

Problem 6.9.5 *Let X and Y be two random variables with $Var(X), Var(Y) > 0$ and let $a, b, c, d \in \mathbb{R}$ be scalars.*

1. Show that $Var(aX + bY) = a^2 Var(X) + b^2 Var(Y) + 2ab Cov(X, Y)$.

2. Show that the correlation of $a + bX$ and $c + dY$ is equal to the correlation between X and Y, i.e. show that $\rho_{a+bX,c+dY} = \rho_{X,Y}$.

Problem 6.9.6 *Prove the statements in Example 6.5, i.e. prove that $Cov(X, Y) = 0$ and $\rho_{X,Y} = 0$.*

Problem 6.9.7 *Let X and ϵ be random variables such that $\sigma_X^2 = Var(X) < \infty$, $\sigma_\epsilon^2 = Var(\epsilon) < \infty$, and $Cov(X, \epsilon) = 0$. Define the new random variable*

$$Y = \beta_0 + \beta_1 X + \epsilon,$$

where $\beta_0, \beta_1 \in \mathbb{R}$ are constants.

1. *Calculate* $Var(Y)$.

2. *Calculate the correlation coefficient between* X *and* Y, $\rho_{X,Y}$.

Problem 6.9.8 *Consider the conditional density function* $f_{Y|X}(y|x)$ *of* Y *given* $X = x$ *for* $x > 0$,

$$f_{Y|X}(y|x) = \begin{cases} \frac{1}{x} & if \ \ 0 \leq y \leq x, \\ 0 & otherwise. \end{cases}$$

(Note: The conditional distribution of Y *given* $X = x$ *is a so-called "uniform distribution" with the endpoints 0 and* x*, see Definition 8.6 in Chapter 8 for further details on the uniform distribution.)*

1. *Derive the conditional expectation function* $E(Y|X = x)$.

2. *Derive the conditional variance* $Var(Y|X = x)$.

Problem 6.9.9 *Let the joint probability function* $f_{X,Y}(x,y)$ *of* X *and* Y *be*

$f_{X,Y}(x,y)$		Y	
		0	*1*
X	*2*	*0.30*	*0.50*
	4	*0.20*	*0.00*

1. *Calculate the mean of* X *and mean of* Y.

2. *Calculate the variance of* X *and variance of* Y.

3. *Calculate the correlation coefficient between* X *and* Y, *and comment on the sign of the correlation.*

Problem 6.9.10 *Let* X *and* U *be two random variables. Assume* $E(U|X) = 0$.

1. *Prove that* $E\left(X^2 \cdot U\right) = 0$. *(Hint: Use the Law of Iterated Expectations.)*

2. *Prove that the correlation coefficient* $\rho_{X,U} = 0$.

Problem 6.9.11 *Let* X *be gender (male* $= 0$; *female* $= 1$) *with probability function given by*

$$f_X(x) = \begin{cases} 0.80 & if \ \ \ x = 0, \\ 0.20 & if \ \ \ x = 1, \\ 0 & otherwise. \end{cases}$$

Denote by Y *yearly income (in 1000 kr). Let* $E(Y|X = 0) = 610$ *and* $E(Y|X = 1) = 640$.

Calculate the mean yearly income $E(Y)$ *in the population. Interpret your result in view of the distribution of gender. (Hint: You can use the Law of Iterated Expectations.)*

Problem 6.9.12 *A Danish company sells its products in the USA. Therefore, its earnings in Danish kroner (DKK) depend both on its earnings in US dollars (USD) and the exchange rate between USD and DKK. Let* X *be the random variable representing the earnings next year in USD, and let* Y *be the exchange rate next year between DKK and USD. That is, for 1 USD, you can get* Y *DKK. Let the random variable* Z *represent the earnings measured in DKK.*

Assume that the joint probability function between X *and* Y *is given as in the following table:*

$f_{X,Y}(x,y)$	$X = 10000$	$X = 50000$
$Y = 6$	0.20	0.05
$Y = 7$	0.10	0.30
$Y = 8$	0.05	0.30

It can be seen that the earnings in USD can only take on two values, namely "low earnings" ($X = 10000$ USD) and "high earnings" ($X = 50000$ USD). Similarly, it is assumed that the exchange rate can only take on three values, namely "low" ($Y = 6$ DKK/USD), "medium" ($Y = 7$ DKK/USD), and "high" ($Y = 8$ DKK/USD).

1. *Calculate the marginal probability functions $f_X(x)$ and $f_Y(y)$.*

2. *Calculate $E(X)$, $E(Y)$, $Var(X)$, $Var(Y)$, $Std(X)$, and $Std(Y)$.*

3. *Calculate $Cov(X,Y)$ and $Corr(X,Y)$.*

4. *Argue that the random variable Z describing the earnings in DKK is given by $Z = X \cdot Y$.*

5. *Use this to calculate the expected earnings in DKK, i.e. $E(Z) = E(X \cdot Y)$.*

6. *Calculate $E(X) \cdot E(Y)$ and compare with your answer from the previous question, i.e. with $E(X \cdot Y)$. Is $E(X) \cdot E(Y)$ equal to $E(X \cdot Y)$ in this case?*

7. *Specify the sample space for Z and derive the marginal probability function for Z, $f_Z(z)$.*

8. *Use the marginal probability function $f_Z(z)$ found in the previous question to once again calculate $E(Z)$ and also to calculate $Var(Z)$.*

7

Independence between random variables

7.1 Introduction

In Chapter 3, we saw that the dependence between random variables are encoded in their joint distribution. An important special case arises in the situation where the random variables possess no dependence. Indeed, the concept of *independence* between random variables plays an important role in establishing properties of statistical methods, and when investigating causal relationships. Independence is a property that a joint distribution of random variables may or may not possess. An implication of independence between, say, two random variables is that knowledge about the outcome of one of them does not change the assessment of the probabilities of the outcomes of the other random variable. Because it is a property of the joint distribution, independence can be investigated empirically since it may be assessed with a sufficient number of observations from the population. In Volume II of this book, we will investigate the relationship between the existence of a causal effect and independence of random variables.

We first consider independence between events. Studying events gives a basic interpretation of the concept of independence. In practical applications, we are usually interested in settings which cannot easily be described by single events, and therefore we use random variables for the subsequent analysis. Since different techniques must be used dependent on the type of random variables, we discuss each type of random variables separately. Though the techniques are different, the interpretations are the same for each type of random variables. We therefore develop the intuition and interpretation of independence of random variables in the section on discrete random variables.

There are several types of independence between random variables. These include *statistical independence* and *conditional statistical independence*, which are used for modeling of samples, assessing the quality of statistical methods, and for causal analysis. Other types of independence are *mean independence* and *conditional mean independence*, which are widely used in causal analysis.

Below, we go through various independence concepts together with a number of examples, such that the concepts become familiar. The concepts will be important later in the book, e.g. when deriving properties of statistical methods, and in Volume II, where we discuss prediction and causal relationships.

DOI: 10.1201/9781003591191-7

7.2   Statistical independence

In this section, we consider the property of statistical independence, often simply abbreviated to "independence". As mentioned above, we start by considering events to develop an understanding of statistical independence. After considering events, we proceed to discuss statistical independence between random variables.

7.2.1   Independence between events

The spirit of statistical independence is to capture that two events, A and B, say, are not related in a statistically relevant sense. Consider two events A and B. Suppose A happens with probability $1/2$ and B happens with probability $1/3$. If A and B are unrelated, we would expect that irrespective of whether B has happened, A will happen with probability $1/2$. And similarly for B. This line of reasoning implies that the probability that both A and B will happen is $1/2 \cdot 1/3 = 1/6$. This leads us to the definition of statistical independence.

Definition 7.1 (Statistical independence between two events) *The two events A and B are **statistically independent** if and only if*

$$P(A \cap B) = P(A) \cdot P(B). \tag{7.1}$$

An implication of the definition is that statistical independence can be expressed using conditional probability. The next theorem shows that two events A and B are statistically independent if the probability of A occurring does not depend on whether B has occurred.

Theorem 7.1 (Statistical independence between two events, conditional formulation) *Assume $P(B) > 0$. Then A and B are statistically independent if and only if*

$$P(A|B) = P(A).$$

Theorem 7.1 provides an alternative interpretation of statistical independence: two events A and B are (statistically) independent if the knowledge of whether B has happened does not influence our assessment of the probability of whether A will happen.

Example 7.1 *Let A be the event that the local football team wins the next game and let B be the event that the local ice hockey team wins the next game. Suppose $P(A) = 0.6$ and $P(B) = 0.7$. If there is statistical independence between the football team wins and the ice hockey team wins, then the probability that both teams win their next game is*

$$P(A \cap B) = P(A) \cdot P(B) = 0.6 \cdot 0.7 = 0.42.$$

If, instead, we can argue that there are some e.g. psychological factors that has a positive influence on one of the teams winning if the other team wins, then we would assert that

$$P(A|B) > P(A) \quad and \quad P(B|A) > P(B).$$

Since $P(A \cap B) = P(A|B) \cdot P(B)$, then $P(A \cap B) = P(A|B) \cdot P(B) > P(A) \cdot P(B) = 0.42$ in case there is a positive influence between the teams winning.

7.2.2   Statistical independence between two random variables

We extend the concept of statistical independence of two events to two random variables. We first define statistical independence using CDFs since these exist no matter the type of random variables under consideration.

Definition 7.2 (Statistical independence between two random variables) *Let X and Y be two random variables and let F denote CDFs. Then X and Y are **statistically independent** if and only if*

$$F_{X,Y}(x,y) = F_X(x) \cdot F_Y(y)$$

for all values of x and y.

If both random variables are discrete, then their joint PMF can be used to determine statistical independence. Similarly, if both random variables are continuous, then their joint PDF can be used. The next theorem contains these results.

Theorem 7.2 (Statistical independence, two discrete or continuous random variables) *Assume either X and Y are discrete random variables with PMFs denoted by f, or X and Y are continuous random variables with PDFs denoted by f. Then X and Y are statistically independent if and only if*

$$f_{X,Y}(x,y) = f_X(x) \cdot f_Y(y)$$

for all values of x and y.

A key requirement in Definition 7.2, and consequently also in Theorem 7.2, is that the defining equation of statistical independence hold for *all* x and y. The next example illustrates.

Example 7.2 *Let X and Y be discrete random variables with marginal PMFs*

$$f_X(x) = \begin{cases} 1/2 & \text{if} \quad x = 0, \\ 1/2 & \text{if} \quad x = 1, \end{cases} \qquad f_Y(y) = \begin{cases} 0.2 & \text{if} \quad y = -1, \\ 0.3 & \text{if} \quad y = 0, \\ 0.5 & \text{if} \quad y = 1. \end{cases}$$

Assume their joint PMF is given by the numbers in the following table.

		Y		
	$f_{X,Y}(x,y)$	-1	0	1
X	0	0.1	0.15	0.25
	1	0.1	0.15	0.25

Then, by Theorem 7.2, X and Y are statistically independent because $f_{X,Y}(x,y) = f_X(x) \cdot f_Y(y)$ for all x and y. If instead the joint PMF of X and Y is, say:

		Y		
	$f_{X,Y}(x,y)$	-1	0	1
X	0	0.15	0.15	0.20
	1	0.05	0.15	0.30

Then, by Theorem 7.2, X and Y are no longer statistically independent because there exists at least one pair of x and y where $f_{X,Y}(x,y) \neq f_X(x) \cdot f_Y(y)$. In particular, for $x = 0$ and $y = 1$ we have

$$f_{X,Y}(0,1) = 0.2 \neq 0.25 = \frac{1}{2} \cdot 0.5 = f_X(0) \cdot f_Y(1).$$

Note that, for $y = 0$, we have that $f_{X,Y}(x, y) = f_X(x) \cdot f_Y(y)$ for all x, but this is not enough for independence, since the condition has to hold for all *x and y. And, as shown above, it does not hold true for $(x, y) = (0, 1)$ (and neither does it hold for $(x, y) = (0, -1), (1, -1)$ and $(1, 1)$).*

A more intuitive understanding of statistical independence between two random variables can be obtained by looking at the consequences for the conditional distributions.

Theorem 7.3 (Statistical independence, conditional formulation) *Let F denote CDFs. Then the random variables X and Y are statistically independent if and only if*

$$F_{Y|X}(y|x) = F_Y(y)$$

for all x, y with $F_X(x) > 0$.

If Y is either a discrete random variable with PMFs denoted by f, or if Y is a continuous random variable with PDFs denoted by f, then X and Y are statistically independent if and only if

$$f_{Y|X}(y|x) = f_Y(y)$$

for all x, y such that $f_X(x) > 0$.

The interpretation of the theorem is that X and Y are independent if knowing the outcome of, say, X, does not change the assessment of the probability of Y, i.e. the conditional distribution of Y given X is equal to the marginal distribution of Y.

Example 7.3 *Consider tossing two coins with outcomes represented by the random variables X and Y, respectively. Start by tossing the coin represented by X. A question is if the outcome, say, tails, should change the assessment of the probability of a tails for the coin represented by Y.*

Suppose the person flipping the coin Y knows the result of X, and accordingly just drops coin Y with the same face up without flipping it. Then we probably will make a different assessment of the outcome of Y than had we not known the outcome of X. In the language of conditional probabilities, $f_{Y|X}(y|x) \neq f_Y(y)$.

On the other hand, if the person flipping the coin Y makes sure it rotates many times, then knowing the outcome of coin X probably does not change our assessment of the outcome of coin Y. In the language of conditional probabilities, $f_{Y|X}(y|x) = f_Y(y)$.

In general, we cannot calculate a joint distribution from only the marginal distributions, because the marginal distributions only encode information on each random variable separately, whereas the joint distribution also encodes information on the dependence between the random variables. If, however, we know that two random variables are statistically independent, then we can calculate their joint distribution from their marginal distributions using Theorem 7.3. The next example illustrates.

Example 7.4 *Consider tossing the same coin twice and code the outcomes by*

$$Y_j = \begin{cases} 0 & \text{if} \quad \text{tails in } j^{th} \text{ toss}, \\ 1 & \text{if} \quad \text{heads in } j^{th} \text{ toss}, \end{cases} \quad j = 1, 2.$$

Suppose the coin is bend such that the probabilities of tails ($Y_j = 0$) and heads ($Y_j = 1$) are:

$$f_{Y_j}(y_j) = \begin{cases} 0.4 & \text{if} \quad y_j = 0, \\ 0.6 & \text{if} \quad y_j = 1, \end{cases} \quad j = 1, 2.$$

Assume the two coin flips are performed such that the outcome of one of them does not influence the outcome of the other. Then we may assume Y_1 and Y_2 are statistically independent. Using Theorem 7.3, this implies that their joint PMF is

$$f_{Y_1,Y_2}(y_1,y_2) = f_{Y_1}(y_1) \cdot f_{Y_2}(y_2) = \begin{cases} 0.16 & if \quad y_1 = 0, y_2 = 0, \\ 0.24 & if \quad y_1 = 0, y_2 = 1, \\ 0.24 & if \quad y_1 = 1, y_2 = 0, \\ 0.36 & if \quad y_1 = 1, y_2 = 1. \end{cases}$$

An important consequence of statistical independence between two random variables is that it implies that their covariance, if it exists, equals zero. This is formally stated in the next result.

Theorem 7.4 (Statistical independence implies zero covariance) *Let X and Y be random variables such that their variances exist. If X and Y are statistically independent, then*

$$Cov(X,Y) = 0.$$

Since the correlation coefficient is given as the covariance divided by the standard deviation (Definition 6.2), Theorem 7.4 similarly implies that the correlation coefficient of two statistically independent random variables is necessarily zero. It is important to note that the converse of Theorem 7.4 is not true in general. That is, zero covariance (or correlation) does not necessarily imply statistical independence. The following example illustrates this.

Example 7.5 (Counter-example to the converse of Theorem 7.4) *Let X be a discrete random variable with PMF*

$$f_X(x) = \begin{cases} 1/4 & if & x = -1, \\ 1/2 & if & x = 0, \\ 1/4 & if & x = 1, \\ 0 & otherwise. \end{cases}$$

Since the distribution of X is symmetric around zero, it holds that $E(X) = 0$ and $E(X^3) = 0$. Define the random variable $Y = X^2$. Then X and Y have covariance

$$\begin{aligned} Cov(X,Y) = Cov(X, X^2) = E(X \cdot X^2) - E(X) \cdot E(X^2) &= E(X^3) - E(X) \cdot E(X^2) \\ &= 0 + 0 \cdot E(X^2) \\ &= 0, \end{aligned}$$

where we in the second equality used property 1 of Theorem 6.5. Hence, X and Y are uncorrelated. They are not, however, statistically independent. For instance, knowing that $Y = 1$ means that the probability that X equals zero is zero. That is,

$$f_{X|Y}(0|1) = 0 \neq \frac{1}{2} = f_X(0).$$

Finally, we note that the formula (Theorem 6.5, property 1)

$$Cov(X,Y) = E(X \cdot Y) - E(X) \cdot E(Y)$$

implies that

$$Cov(X,Y) = 0 \iff E(X \cdot Y) = E(X) \cdot E(Y).$$

Together with Theorem 7.4, this also implies the following useful property

$$X \text{ and } Y \text{ statistically independent} \Rightarrow E(X \cdot Y) = E(X) \cdot E(Y). \tag{7.2}$$

Similarly to what is discussed below Theorem 7.4, we note that the converse of (7.2) does not hold. That is, knowing that $E(X \cdot Y) = E(X) \cdot E(Y)$ does not imply that X and Y are statistically independent.

7.2.3   Statistical independence between many random variables

When considering more random variables than two, different concepts for statistical independence may be defined. The relevant concept to use depends on context of the analysis. First, we give the definition of the concept of *mutual statistical independence.*

Definition 7.3 (Mutual statistical independence among many random variables)
Let $X_1, \ldots, X_K$ *be* K *random variables with joint CDF* $F_{X_1,\ldots,X_K}(x_1,\ldots,x_K)$. *Then* $X_1, \ldots, X_K$ *are **mutual statistically independent** if and only if*

$$F_{X_1,\ldots,X_K}(x_1,\ldots,x_K) = F_{X_1}(x_1)\cdots F_{X_K}(x_K)$$

for all $x_1, \ldots, x_K$.

If the random variables are all either discrete or all continuous, then we typically use the PMFs or PDFs, respectively, to check for mutual statistical independence. The next theorem contains these results.

Theorem 7.5 (Mutual statistical independence with discrete or continuous random variables) *Let Assume that either* $X_1, \ldots, X_K$ *are discrete random variables with* $f()$ *denoting PMFs, or continuous random variables with* $f()$ *denoting PDFs. Then* $X_1, \ldots, X_K$ *are mutual statistically independent if and only if*

$$f_{X_1,\ldots,X_K}(x_1,\ldots,x_K) = f_{X_1}(x_1)\cdots f_{X_K}(x_K)$$

for all $x_1, \ldots, x_K$.

Mutual statistical independence implies that, say, knowing the outcomes of $X_2, \ldots, X_K$ do not change the assessment of the probability of X_1. The next theorem states this result.

Theorem 7.6 (Mutual statistical independence, conditional formulation) *Assume that either* $X_1, \ldots, X_K$ *are discrete random variables with* $f()$ *denoting PMFs, or continuous random variables with* $f()$ *denoting PDFs. Then the random variables* $X_1, \ldots, X_K$ *are mutual statistically independent if and only if for all* $j = 1, \ldots, K$,

$$f(x_j | x_1, \ldots, x_{j-1}, x_{j+1}, \ldots, x_K) = f_{X_j}(x_j)$$

for all $x_1, \ldots, x_K$ *such that* $f(x_1, \ldots, x_{j-1}, x_{j+1}, \ldots, x_K) > 0$.

In the case of more than two random variables, we next consider a kind of statistical independence based directly on the definition of statistical independence between two random variables, applied to all pairs of random variables. This leads to the concept of *pairwise statistical independence* defined next.

Definition 7.4 (Pairwise statistical independence) *The random variables* $X_1, \ldots, X_K$ *are **pairwise statistically independent** if and only if* X_i *and* X_j *are statistically independent for all* i *and* j *such that* $i \neq j$.

Mutual statistical independence is a more restrictive property than pairwise statistical independence, in the sense that mutual statistical independence implies pairwise statistical independence. This is stated in the following theorem.

Theorem 7.7 (Mutual statistical independence implies pairwise statistical independence) *Suppose the* K *random variables* $X_1, X_2, \ldots, X_K$ *are mutual statistically independent. Then they are also pairwise statistically independent.*

The following example illustrates pairwise and mutual statistical independence and shows that the converse of Theorem 7.7 does not hold, that is, pairwise statistical independence does not imply mutual statistical independence.

Example 7.6 *Let X and Y represent outcomes of two coin tosses with marginal probability functions*

$$f_X(x) \;=\; \begin{cases} 0.5 & if \quad x = 0, \\ 0.5 & if \quad x = 1, \end{cases}$$

$$f_Y(y) \;=\; \begin{cases} 0.5 & if \quad y = 0, \\ 0.5 & if \quad y = 1, \end{cases}$$

where 0 is coded for tails and 1 is coded for heads. Assume X and Y are statistically independent, that is,

$$f_{X,Y}(x, y) = f_X(x) f_Y(y).$$

Let Z be a random variable that equals 0 if the two coin tosses resulted in the same face (two heads or two tails) and 1 if they are different. This can be written as

$$Z = \begin{cases} 0 & if \quad X = 0, Y = 0 \ or \ X = 1, Y = 1, \\ 1 & if \quad X = 0, Y = 1 \ or \ X = 1, Y = 0. \end{cases}$$

Each of the combinations of outcomes of X and Y, for example $X = 1, Y = 1$, happens with probability $1/4$ because X and Y are statistically independent. This implies that the probability function $f_Z(z)$ of Z is

$$f_Z(z) = \begin{cases} 0.5 & if \quad z = 0, \\ 0.5 & if \quad z = 1. \end{cases}$$

Next we show that Z and X are statistically independent. Given $X = 0$, the outcome of Z is determined solely by the outcome of Y. If $Y = 0$, then $Z = 0$ and if $Y = 1$, then $Z = 1$. Thus, the conditional probability function $f_{Z|X}(z|x)$ of Z given $X = 0$ is

$$f_{Z|X}(z|0) = \begin{cases} 0.5 & if \quad z = 0, \\ 0.5 & if \quad z = 1. \end{cases}$$

By similar arguments we have, given $X = 1$,

$$f_{Z|X}(z|1) = \begin{cases} 0.5 & if \quad z = 0, \\ 0.5 & if \quad z = 1. \end{cases}$$

Since $f_{Z|X}(z|0) = f_{Z|X}(z|1) = f_Z(z)$, then X and Z are statistically independent (Theorem 7.3).

The same arguments for Z and X being statistically independent can be invoked to show that Z and Y are statistically independent. Since X and Y are statistically independent by assumption, then, all together, X, Y, and Z are pairwise statistically independent according to Definition 7.4.

The random variables X, Y, and Z are not mutual statistically independent, however. This can be seen by considering the conditional probability of Z given X and Y. Since Z is a deterministic function of X and Y, we can perfectly assess the outcome of Z when X and Y are known. For instance, if $Y = X = 0$, then $Z = 0$, that is

$$f_{Z|X,Y}(y|0,0) = \begin{cases} 1 & if \quad z = 0, \\ 0 & if \quad z = 1. \end{cases}$$

Since e.g. $f_{Z|X,Y}(0|0,0) = 1 \neq f_Z(0) = 0.5$, then it follows from Theorem 7.6 that X, Y, and Z are not mutual statistically independent.

Example 7.6 illustrates an important point when studying a system of phenomena. We can be seriously mislead about the whole system if we only consider smaller parts in isolation. This is illustrated in the example, where we observed that examining the dependence of random variables in pairs suggested independence, while analyzing all variables together revealed dependence.

7.3   Conditional statistical independence

A core concept in modeling empirical relationships is *conditional statistical independence*. Conditional statistical independence is statistical independence between some of the random variables when conditioning on, or "controlling for", the remaining random variables. It is a property that a conditional distribution may, or may not, possess. We first state the definition using conditional CDFs (Definition 4.4), ensuring that the definition applies to random variables of all types. For ease of presentation, we formulate the following concepts in a context of three random variables, but it can be straightforwardly extended to more than three random variables by replacing the conditioning random variable Z by $Z_1, Z_2, \ldots, Z_m$ for some $m \geq 1$.

Definition 7.5 (Conditional statistical independence) *Let X, Y, Z be random variables with $F()$ denoting CDFs. Then X and Y are* **conditional statistically independent** *given Z if and only if*

$$F_{X,Y|Z}(x,y|z) = F_{X|Z}(x|z) \cdot F_{Y|Z}(y|z)$$

for all values of x, y, and z, such that $F_Z(z) > 0$.

In case X and Y are not conditional statistically independent on Z, we may also say that X and Y are *conditional statistically dependent* given Z.

Similarly to Theorem 7.2, when the random variables X and Y are either both discrete or both continuous, conditional statistical dependence can also be formulated using PMFs or PDFs, respectively. The next result gives the details.

Theorem 7.8 (Statistical conditional independence, two discrete or continuous random variables) *Let either X and Y be discrete random variables with $f()$ denoting PMFs, or continuous random variables with $f()$ denoting PDFs. Let Z be a random variable of any type. Then X and Y are conditional statistically independent given Z if and only if*

$$f_{X,Y|Z}(x,y|z) = f_{X|Z}(x|z) \cdot f_{Y|Z}(y|z)$$

for all values of x, y, and z, such that $f_Z(z) > 0$.

Similarly to the conditional formulation of independence between two random variables X and Y (Theorem 7.3), there is also a conditional formulation of conditional statistical independence. The next theorem has this result.

Theorem 7.9 (Conditional statistical independence, conditional formulation) *Let X, Y, Z be random variables with $F()$ denoting CDFs. Then X and Y are conditional statistically independent given Z if and only if*

$$F_{Y|X,Z}(y|x,z) = F_{Y|Z}(y|z)$$

for all values of x, y, and z, such that $F_{X,Z}(x,z) > 0$. If the random variables are either discrete with $f()$ denoting PMFs, or if they are continuous with $f()$ denoting PDFs, then X and Y are conditional statistically independent given Z if and only if

$$f_{Y|X,Z}(y|x,z) = f_{Y|Z}(y|z)$$

for all values of x, y, and z, such that $f_{X,Z}(x,z) > 0$.

The theorem shows that conditional independence can be interpreted as the fact that knowledge of the outcome of X does not change the assessment of the distribution of Y, if we also use knowledge of the outcome of Z.

Example 7.7 *In Example 4.5 about school performance in terms of GPA, Y, school ownership, X, and preschool language stimulation, Z, it turned out that the conditional distribution of GPA given ownership and preschool language stimulation only depends on preschool language stimulation but not on ownership. Thus, GPA and ownership are conditional statistically independent given preschool language stimulation. In other words, when we control for preschool language stimulation, then there is no statistical dependence between GPA and ownership.*

The next example shows how conditional statistical independence may come about.

Example 7.8 *Consider tossing the same coin twice. Let X and Y be the outcome of the first and second toss of the coin with 0 representing "tails" and 1 representing "heads". Let the probability of heads be $p \in [0,1]$ and assume the marginal distributions of X and Y are*

$$f_X(x) = \begin{cases} 1-p & \text{if } x = 0, \\ p & \text{if } x = 1, \end{cases}$$

$$f_Y(y) = \begin{cases} 1-p & \text{if } y = 0, \\ p & \text{if } y = 1. \end{cases}$$

Suppose the coin to be tossed twice is selected among two coins, denoted "A" and "B", respectively. Let Z be a random variable indicating the coin used to select the coin to be tossed twice,

$$Z = \begin{cases} 0 & \text{if } \text{coin A tossed}, \\ 1 & \text{if } \text{coin B tossed}. \end{cases}$$

Assume the two coins have different probability of coming up heads. Let coin A have probability of heads equal to 0.7 and coin B have probability of heads equal to 0.5. This can be represented by letting p be a function of z in the following manner

$$p(z) = \begin{cases} 0.7 & \text{if } z = 0, \\ 0.5 & \text{if } z = 1. \end{cases}$$

For example, if $z = 0$, then the probability of X being heads ($X = 1$) is 0.7 whereas for $z = 1$ it is 0.5.

We can now calculate the joint probability of X and Y conditional on the coin being tossed. Assuming the outcome of a coin is statistically independent over tosses, this gives

Coin A

$f_{X,Y|Z}(x,y|0)$

	X		$f_{Y|Z}(y|0)$
	0	1	
Y 0	0.09	0.21	0.30
1	0.21	0.49	0.70
$f_{X|Z}(x|0)$	0.30	0.70	1

Coin B

$f_{X,Y|Z}(x,y|1)$

	X		$f_{Y|Z}(y|1)$
	0	1	
Y 0	0.25	0.25	0.50
1	0.25	0.25	0.50
$f_{X|Z}(x|1)$	0.50	0.50	1

By examining the entries in the tables, we see that $f_{X,Y|Z}(x,y|z) = f_{X|Z}(x|z) \cdot f_{Y|Z}(y|z)$ for all x, y, z. Thus, X and Y are conditionally statistical independent given Z (Theorem 7.8).

Now, suppose the selection of the coin to be tossed is done by another coin toss with 0.5 probability of coin A being chosen for the two tosses, and similarly for coin B. In other words, Z is now considered a random variable with $f_Z(0) = f_Z(1) = 0.5$. Then the joint probability function of X and Y is

$$f_{X,Y}(x,y) = \sum_{z=0}^{1} f_{X,Y,Z}(x,y,z) = \sum_{z=0}^{1} f_{X,Y|Z}(x,y|z) \cdot f_Z(z).$$

Using that $f_Z(0) = f_Z(1) = 0.5$ and the values for $f_{X,Y|Z}(x,y|z)$ given in the tables above, we can calculate the joint distribution of X and Y, as well as their marginal distributions:

$f_{X,Y}(x,y)$		X		$f_Y(y)$
		0	1	
Y	0	0.17	0.23	0.40
	1	0.23	0.37	0.60
$f_X(x)$		0.40	0.60	

From the table is can be seen that X and Y are not statistically independent (Theorem 7.2). For example,

$$\begin{aligned} f_{X,Y}(0,0) &= 0.17, \\ f_X(0) \cdot f_Y(0) &= 0.40 \cdot 0.40 = 0.16, \end{aligned}$$

and, thus, $f_{X,Y}(0,0) \neq f_X(0) \cdot f_Y(0)$.

Though it may seem odd that X and Y are statistically dependent just because the coin tossed is selected by another coin toss, the reason is that after the first toss (i.e. after observing the outcome of X), we are more informed about the outcome of Y than the marginal distribution $f_Y(y)$. Indeed, the first toss (X) is informative about whether the coin tossed is coin A or coin B, and this information can be utilized to assess the probability of the outcome of Y. If, on the other hand, we know which coin is tossed, that is, we know the outcome of Z, then there is no information in the outcome of X that is not already included in $f_{Y|Z}(y|z)$.

In conclusion, this is an example of X and Y being statistically dependent but conditional statistically independent on Z.

Whether or not X and Y are conditional statistically independent may depend upon which variables we condition. Two opposite cases may happen:

1. X and Y are conditional statistically independent when conditioning on a long conditioning set, but conditional statistically dependent when conditioning on a short conditioning set.

2. X and Y are conditional statistically dependent when conditioning on a long conditioning set, but conditional statistically independent when conditioning on a short conditioning set.

Example 7.7 is an example of case 1. In that example, GPA and school ownership are conditional statistically independent given preschool language stimulation (the long conditioning set) whereas they are statistically dependent when not conditioning on preschool language stimulation (the short conditioning set). Example 7.8 is also an example of case

1. Example 7.6 is an example of case 2. In that example, two coin tosses are represented by X and Y, and Z is a random variable being 0 if the two coin tosses give different outcomes and 1 if they give the same outcome. Then X and Y are conditional statistically dependent given Z (the long conditioning set), but X and Y are statistically independent when not conditioning on Z (the short conditioning set).

Of the properties of statistical independence considered so far, mutual statistical independence is the most restrictive. Mutual statistical independence of X, Y, and Z is more restrictive than conditional statistical independence of X and Y given Z, since mutual statistical independence implies $f_{Y|X,Z}(y|x,z) = f_Y(y)$ whereas conditional statistical independence of X and Y given Z only implies $f_{Y|X,Z}(y|x,z) = f_{Y|Z}(y|z)$.

In the definitions and theorems in this section, we only consider three random variables. As mentioned above, this can be easily extended for example by having X and Y being conditional statistical independent given $Z_1, \ldots, Z_m$. Formally, this can be done by replacing Z with $Z_1, \ldots, Z_m$ in the above expressions.

7.4 Mean independence and conditional mean independence

Depending on the circumstances, it can sometimes be sufficient to investigate dependence and independence between random variables using only particular aspects of the joint distribution. In this section, we will focus on dependence in the framework of conditional expectation. Considering the conditional expectation rather than the conditional distribution may be an advantage when we have to estimate the relationship in that we can typically estimate a conditional expectation with higher precision than a conditional distribution. We may also prefer to consider a conditional expectation for substantive reasons in that we may only be interested in the expected behavior of the population rather than the entire distribution of the population.

Similarly to statistical independence between two random variables, we now consider *mean independence* between two random variables. The definition is next.

Definition 7.6 (Mean independence) *Let Y and X be random variables. Then Y is* ***mean independent*** *of X if and only if*

$$E(Y|X = x) = E(Y)$$

for all values of x.

Note that the definition of mean independence is not symmetric in the random variables X and Y. Indeed, it is possible for Y to be mean independent of X, while X is mean dependent on Y and vice versa.

Statistical independence is a more restrictive property than mean independence, in the sense that statistical independence implies mean independence. This is stated in the following theorem.

Theorem 7.10 (Statistical independence implies mean independence) *Let X and Y be statistically independent random variables. Then X is mean independent of Y and Y is mean independent of X.*

We note that the converse of Theorem 7.10 is not true in general. That is, mean independence does not necessarily imply statistical independence.

Similarly to conditional statistical independence, we can consider conditional mean independence. The definition for the case of three random variables is next.

Definition 7.7 (Conditional mean independence) *Let Y, X, and Z be random variables. Then Y is **conditional mean independent** of X given Z if and only if*

$$E(Y|X = x, \ Z = z) = E(Y|Z = z)$$

for all values of x and z.

We can illustrate the concept of conditional mean independence by specifying a linear conditional expectation function. Suppose

$$E(Y|X = x, \ Z = z) = \beta_0 + \beta_1 x + \beta_2 z,$$

where $\beta_0, \beta_1, \beta_2 \in \mathbb{R}$ are constants. If $\beta_1 = 0$, then Y is conditional mean independent of X given Z.

Conditional mean independence between X and Y may depend upon the other conditioning variables. Consider again a linear conditional expectation function. Suppose Y is conditional mean independent of X given Z:

$$E(Y|X = x, \ Z = z) = \beta_0 + 0 \cdot x + \beta_2 z = \beta_0 + \beta_2 z.$$

Compared to a smaller conditioning set than with Z, which here would be no random variable in the conditioning set, we can calculate $E(Y|X = x)$ from $E(Y|X = x, \ Z = z)$ using the Law of Iterated Expectations (Theorem 6.10)

$$
\begin{aligned}
E(Y|X = x) &= E\left(E(Y|X = x, Z) \mid X = x\right) \\
&= E(\beta_0 + \beta_2 Z \mid X = x) \\
&= \beta_0 + \beta_2 E(Z \mid X = x).
\end{aligned}
$$

Suppose

$$E(Z|X = x) = \gamma_0 + \gamma_1 x,$$

where $\gamma_0, \gamma_1 \in \mathbb{R}$ are constants. Then

$$
\begin{aligned}
E(Y|X = x) &= \beta_0 + \beta_2(\gamma_0 + \gamma_1 x) \\
&= \underbrace{(\beta_0 + \beta_2\gamma_0)}_{=\delta_0} + \underbrace{\beta_2\gamma_1 x}_{=\delta_1} \\
&= \delta_0 + \delta_1 x.
\end{aligned}
$$

This result shows that Y and X are not conditional mean independent in the smaller conditioning set where Z is not included.

7.5 Proofs

Proof of Theorem 7.1 *First, use the definition of conditional probability of events (Definition 4.1) to rewrite (7.1):*

$$P(A|B) = \frac{P(A \cap B)}{P(B)} \Rightarrow P(A \cap B) = P(A|B)P(B). \tag{7.3}$$

To prove the "if"-part, assume that $P(A|B) = P(A)$. Use this in Equation (7.3) to arrive at

$$P(A \cap B) = P(A) \cdot P(B),$$

which, by Definition 7.1 means that A and B are statistically independent. To prove the "only if"-part, assume that A and B are statistically independent, i.e. $P(A \cap B) = P(A) \cdot P(B)$, and insert into Equation (7.3) to get

$$P(A) \cdot P(B) = P(A \cap B) = P(A|B)P(B),$$

which, using the assumption that $P(B) \neq 0$, implies that (cancelling $P(B)$)

$$P(A|B) = P(A).$$

Proof of Theorem 7.2 *The proof of Theorem 7.2 is left as an exercise to the reader (Problem 7.6.10).*

Proof of Theorem 7.3 *The proof of Theorem 7.3 is left as an exercise to the reader (Problem 7.6.11).*

Proof of Theorem 7.4 *The proof of Theorem 7.4 is left as an exercise to the reader (Problem 7.6.12).*

Proof of Theorem 7.5 *The proof is similar to the proof of Theorem 7.2 and is therefore omitted.*

Proof of Theorem 7.6 *The proof is similar to the proof of Theorem 7.3 and is therefore omitted.*

Proof of Theorem 7.7 *Consider the case with $K = 3$ random variables X_1, X_2, and X_3. Assume they are mutual statistically independent. Then, according to Definition 7.3, we have*

$$F_{X_1, X_2, X_3}(x_1, x_2, x_3) = F_{X_1}(x_1)F_{X_2}(x_2)F_{X_3}(x_3) \quad for \ all \ x_1, x_2, x_3. \tag{7.4}$$

We first show that X_1 and X_2 are statistically independent. First derive the CDF of X_1 and X_2. It can be done by letting $x_3 \to \infty$. This gives

$$F_{X_1, X_2}(x_1, x_2) = \lim_{x_3 \to \infty} F_{X_1, X_2, X_3}(x_1, x_2, x_3).$$

Now, insert the expression (7.4) for $F_{X_1, X_2, X_3}(x_1, x_2, x_3)$ provided by the mutual statistical independence. This implies

$$\begin{aligned}
F_{X_1, X_2}(x_1, x_2) &= \lim_{x_3 \to \infty} F_{X_1, X_2, X_3}(x_1, x_2, x_3) \\
&= \lim_{x_3 \to \infty} \left(F_{X_1}(x_1)F_{X_2}(x_2)F_{X_3}(x_3) \right) \\
&= F_{X_1}(x_1)F_{X_2}(x_2) \lim_{x_3 \to \infty} F_{X_3}(x_3) \\
&= F_{X_1}(x_1)F_{X_2}(x_2).
\end{aligned}$$

So, by Definition 7.2, X_1 and X_2 are statistically independent. Similar arguments can be used if there are more than $K = 3$ random variables by considering every pair of random variables. Thus, by Definition 7.4, $X_1, \ldots, X_K$ are pairwise statistically independent, which is what we wanted to show.

Proof of Theorem 7.8   *The proof is similar to the proof of Theorem 7.2 and is therefore omitted.*

Proof of Theorem 7.9   *The proof is similar to the proof of Theorem 7.3 and is therefore omitted.*

Proof of Theorem 7.10   *The proof of Theorem 7.10 is left as an exercise to the reader (Problem 7.6.13).*

7.6   Exercises

Problem 7.6.1 *The following two problems are classic problems in probability theory. They were proposed by Antoine Gombaud (aka. the Chevalier de Méré (1607–1684), a gambler), to Blaise Pascal (1623–1662), one of the founding fathers of probability theory. The solutions to the problems, by Blaise Pascal and Pierre de Fermat (1601–1665), helped spawn and develop probability theory.*

In the following, you may assume that the random variables (or, equivalently, the events) describing the dice throws are statistically independent.

1. *Consider four throws of a fair six-sided die. The Chevalier de Méré guessed that the probability of throwing at least one "6" in these four throws was greater than 50%. Calculate the probability and assess whether the Chavalier was right.*

2. *Consider 24 throws of two fair six-sided dice. The Chevalier de Méré guessed that the probability of throwing at least one "double 6" in these 24 throws was greater than 50%. Calculate the probability and assess whether the Chavalier was right.*

Problem 7.6.2 *This problem is a variant of the so-called "birthday paradox". You may ignore leap years and assume that the day a randomly selected person has a birthday is evenly distributed across the year (i.e. the probability of being born on any one given day is $1/365$).*

1. *You are at a cocktail party with 253 other guests. You may assume that these guests are randomly selected (so that there is independence between their birthdays and yours). What is the probability that at least one person shares your birthday?*

Problem 7.6.3 *Consider two coins, Coin 1 and Coin 2. Coin 1 is a regular "fair" coin with "heads" on one side and "tails" on the other. Coin 2 has "heads" on both sides. Now, one of the coins is tossed four times: with a 50% probability, it is Coin 1 that is tossed, and with a 50% probability, it is Coin 2 that is tossed. You do not know whether it is Coin 1 or Coin 2 being tossed. You may assume that the three tosses are statistically independent.*

1. *Suppose the outcome of the first toss is "heads". What is the probability that it is Coin 2 being tossed?*

2. *Suppose the outcome of the second toss is also "heads". What is the probability that it is Coin 2 being tossed?*

3. *Suppose the outcome of the third toss is also "heads". What is the probability that it is Coin 2 being tossed?*

4. *Now, suppose the outcome of the fourth toss is "tails". What is the probability that it is Coin 2 being tossed?*

Problem 7.6.4 *Let the random variable X indicate the gender of an individual and Y indicate whether the individual has changed job within the past year $(Y = 1)$ or not $(Y = 0)$. Suppose the joint probability function $f_{X,Y}(x, y)$ of X and Y is given by*

$f_{X,Y}(x, y)$	$Y = 1$	$Y = 0$
$X = 1$ *(man)*	0.35	0.23
$X = 0$ *(woman)*	0.15	0.27

1. *Calculate the marginal probability functions $f_X(x)$ and $f_Y(y)$.*

2. *What is the probability of a change of jobs if the individual is a woman? And if it is a man?*

3. *Are the random variables X and Y statistically independent?*

Problem 7.6.5 *Let*
$$E(Y|X, Z) = \beta_0 + \beta_1 X + \beta_2 Z,$$

and $E(Z) = \mu_Z$. Assume that X and Z are mean independent. Derive the conditional expectation function $E(Y|X)$ of Y given X.

Problem 7.6.6 *Let the joint probability function $f_{X,Y,Z}(x, y, z)$ of X, Y and Z be*

$f_{X,Y,Z}(x, y, 0)$ Y

X	3	5
2	0.25	0
4	0	0.25

and

$f_{X,Y,Z}(x, y, 1)$ Y

X	3	5
2	0	0.25
4	0.25	0

1. *Are X, Y, and Z mutual statistically independent?*

2. *Are X and Y conditional statistically independent given Z?*

3. *Are X and Y statistically independent?*

Problem 7.6.7

Let X be a continuous random variable with PDF

$$f(x) = \frac{3}{8}x^2, \qquad x \in (0, 2]$$

and let Y be a continuous random variable with PDF

$$f_Y(y) = \frac{1}{12}y, \qquad y \in [1, 5].$$

Assume that X and Y are statistically independent.

1. *Calculate $E(YX)$, i.e. the mean of the product $X \cdot Y$.*

Problem 7.6.8 *Example 7.5 showed that zero covariance does not imply statistical independence by giving a counter-example involving discrete random variables. Here, you are asked to provide another counter-example, this time with continuously distributed random variables.*

Let X be a continuous random variable with PDF

$$f_X(x) = \begin{cases} 1/2 & \text{if} \quad x \in [-1,1], \\ 0 & \text{otherwise.} \end{cases}$$

Let $Y = X^2$. Show that the covariance of X and Y equals zero. Argue that X and Y are not statistically independent.

Problem 7.6.9 *In Equation (7.2) it was stated that:*

$$X \text{ and } Y \text{ statistically independent } \Rightarrow E(X \cdot Y) = E(X) \cdot E(Y).$$

You are now asked to generalize this result. Let $X_1, \ldots, X_n$ be random variables.

1. *Assume that $X_1, \ldots, X_n$ are mutual statistically independent and discrete (the continuous case follows similarly). Prove that*

$$E(X_1 \cdot X_2 \cdots X_n) = E(X_1) \cdot E(X_2) \cdots E(X_n).$$

   (Hint: You may apply the result in (6.2) to $E(X_1 \cdot X_2 \cdots X_n)$.)

2. *Does a similar result hold if $X_1, \ldots, X_n$ are only assumed to be pairwise statistically independent?*

Problem 7.6.10 *Prove Theorem 7.2. (Hint: Apply the results of going from the CDF to the PMF/PDF.)*

Problem 7.6.11 *Prove Theorem 7.3.*

Problem 7.6.12 *Prove Theorem 7.4.*

Problem 7.6.13 *Prove Theorem 7.10.*

8

Commonly used univariate distributions

DOI: 10.1201/9781003591191-8

8.1 Introduction

In the previous chapters, we have studied random variables with distributions given by their cumulative distribution function (CDF), their probability mass function (PMF), or their probability density function (PDF). There are a number of distributions that are especially useful for modeling the outcome of statistical experiments, or, as we shall see in the later chapters of this book, for describing estimators and test statistics. In the next two chapters, we present a selection of these distributions. First, in this chapter, we will look at distributions describing a single random variable, so-called *univariate distributions*. Then, in the next chapter, we will look at a particular instance of a distribution describing two random variables, a so-called *bivariate distribution*. In Volume II of the book, we will consider distributions describing two or more random variables, so-called *multivariate distributions*.

An important aspect of a distribution of a random variable is the mean and variance implied by the distribution, and we will pay special attention to these aspects. In addition, we also consider distributions of the sum of random variables having one of these distributions since sums of random variables are often encountered when deriving distributions of estimators.

The distributions in this chapter are specified by one or two parameters. They can be viewed as members of families of distributions, where one set of values of the parameters correspond to one family member. Considering a family of distributions makes it easier to apply the distributions in practice. In chapters to come, we will apply these concepts and when characterizing the distributions of various estimators. In Volume II of the book, the concepts will be used for modeling causal relationships.

8.2 Bernoulli and Binomial distributions

8.2.1 Bernoulli distributions

Let X be a random variable that has only two possible outcomes. Unless otherwise stated, we assume the two outcomes are coded 0 and 1. For example, the random variables may be coded 0 for "no" and 1 for "yes". The probability of the outcome 1 is denoted by a parameter p. That is,

$$\Pr(X = 1) = p.$$

Hence, the parameter p must be a number in the interval $p \in [0, 1]$. A random variable with this distribution is said to have the *Bernoulli distribution*. The formal definition is as follows.

Definition 8.1 (Bernoulli distribution) *Let $p \in [0, 1]$. A discrete random variable X is* **Bernoulli distributed** *with parameter p if and only if the PMF of X is*

$$f_X(x) = \begin{cases} 1 - p & if & x = 0, \\ p & if & x = 1, \\ 0 & otherwise. \end{cases} \tag{8.1}$$

Since the Bernoulli distribution is widely used, it has its own notation:

$$X \sim Ber(p),$$

which should be read as "X is Bernoulli distributed with parameter p" and means that X has the PMF $f_X(x)$ given in (8.1). Each value of $p \in [0, 1]$ corresponds to a member of the family of Bernoulli distributions. Figure 8.1 illustrates the PMF and CDF of a Bernoulli distribution.

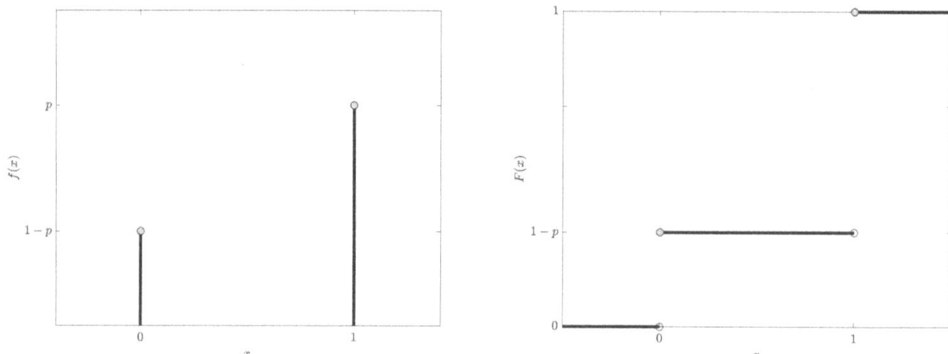

FIGURE 8.1
The Bernoulli distribution, $Ber(p)$. Left: PMF $f(x)$. Right: CDF $F(x)$.

The mean and variance of a Bernoulli distribution with parameter value p are given in the next theorem.

Theorem 8.1 (Mean and variance of the Bernoulli distribution) *Assume that $X \sim Ber(p)$. Then*

$$\begin{aligned} E(X) &= p, \\ Var(X) &= p(1 - p). \end{aligned}$$

Example 8.1 *Let the outcome of a coin toss be represented by the random variable X, where*

$$X = \begin{cases} 0 & if & \text{"tails"}, \\ 1 & if & \text{"heads"}. \end{cases}$$

Then $X \sim Ber(p)$, where $p \in [0, 1]$ is the probability of "heads". If the coin is fair, that is, has equal probability of heads and tails, and no other outcomes are possible, then $p = 0.5$, i.e. X has the Bernoulli distribution $Ber(0.5)$.

A population of elements, where each element only can be one of two values, is called a *Bernoulli population*. Hence, we can think about a random variable $X \sim Ber(p)$ as a random draw from a Bernoulli population, where the fraction of "1"s is p.

Example 8.2 *We are interested in assessing whether the members of a given population are in favor of a particular political proposal. Let the random variable X describe whether a randomly selected individual from the population is "for" the proposal (coded as $X = 1$) or whether the individual is "against" the proposal (coded as $X = 0$). If the fraction of members of the population who are "for" the proposal is $p \in [0, 1]$, then $X \sim Ber(p)$.*

8.2.2 Binomial distributions

Suppose we are interested in the number of "1"s, e.g. "1" coded for "yes", obtained by drawing n times from a Bernoulli distribution. Let X_i be the outcome of draw i and let S_n be the number, or sum, of "1"s in the n draws. Then

$$S_n = X_1 + X_2 + \ldots + X_n = \sum_{i=1}^{n} X_i.$$

To find the distribution of the sum S_n, it is necessary to make assumptions on X_i and how the X_i's relate to each other. Consider the following two assumptions:

1. $X_1, X_2, \ldots, X_n$ are mutual statistically independent.

2. All X_i's have identical distribution: That is, $X_i \sim Ber(p)$, where $p \in [0, 1]$ is fixed.

Under these two assumptions, the distribution of S_n can be derived. The result is a distribution called the *Binomial distribution* and it is formally defined next.

Definition 8.2 (Binomial distribution) *Let $p \in [0, 1]$ and $n \geq 1$. The discrete random variable S_n has a **Binomial distribution** with parameters n and p if and only if it has the PMF*

$$f_{S_n}(s) = \begin{cases} \binom{n}{s} p^s (1 - p)^{n-s} & \text{if} \quad s = 0, 1, \ldots, n, \\ 0 & \text{otherwise,} \end{cases}$$

where $\binom{n}{s}$ is the binomial coefficient, defined as

$$\binom{n}{s} = \frac{n!}{(n - s)! \cdot s!},$$

and $n!$ is n factorial, defined as

$$n! = 1 \cdot 2 \cdots n.$$

Note that, by convention, $0! = 1$. If S_n is a Binomially distributed random variable with parameters n and p, we write $S_n \sim Bin(n, p)$.

To summarize, a sum S_n of n independent and identically distributed Bernoulli $Ber(p)$ random variables has the Binomial distribution $Bin(n, p)$. Figure 8.2 illustrates the PMF and CDF of a Binomial distribution.

The terms in the formula for the PMF can each be given an interpretation. The binomial coefficient $\binom{n}{s} = \frac{n!}{(n-s)! \cdot s!}$ counts the number of different outcomes of $X_1, \ldots, X_n$ for $S_n = s$, while the term $p^s (1 - p)^{n-s}$ is the probability of each of these ways. There are two parameters, n and p, specifying the distribution. Thus, the Binomial distributions are a family, where each member is specified by an n (n being a natural number, $n \in \mathbb{N}$) and a p (p being a number between zero and one, $p \in [0, 1]$).

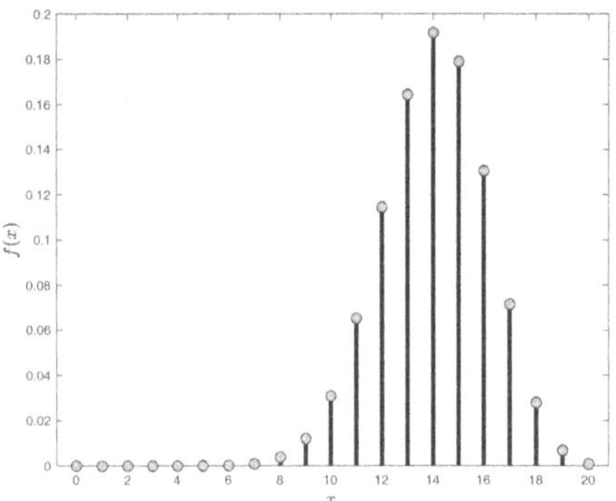 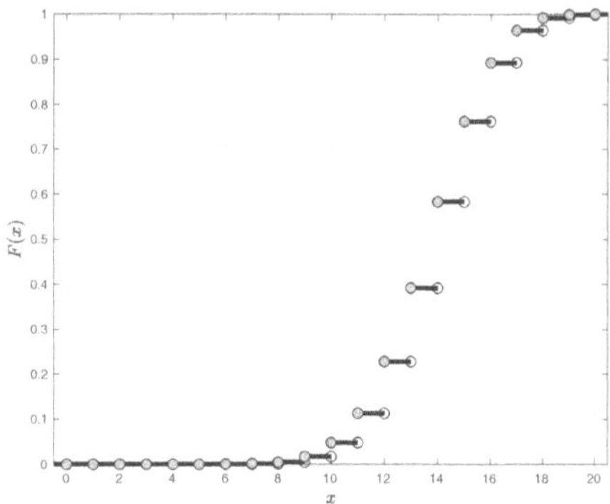

FIGURE 8.2
The Binomial distribution, $Bin(n, p)$, with $n = 20$ and $p = 0.7$. Left: PMF $f(x)$. Right: CDF $F(x)$.

Example 8.3 *Suppose we toss a coin $n = 3$ times and that we are interested in getting exactly $s = 2$ heads. We have that*

$$\frac{n!}{(n-s)! \cdot s!} = \frac{3!}{(3-2)! \cdot 2!} = \frac{3 \cdot 2 \cdot 1}{(1) \cdot (2 \cdot 1)} = 3,$$

is the number of ways to obtain a total of 2 heads, namely, the combinations $(1, 1, 0)$, $(1, 0, 1)$, and $(0, 1, 1)$. The number $p^2(1-p)^{3-2} = p^2(1-p)^1$ is the probability of getting two heads in a specific sequence, e.g. $(1, 1, 0)$. Notice, the power s of p^s is the number of 1's (heads) and the power $n - s$ of $(1-p)^{n-s}$ is the number of 0's (tails).

The mean and variance of the Binomial distribution are given in the next theorem.

Theorem 8.2 (Mean and variance of the Binomial distribution) *Assume that $S_n \sim Bin(n, p)$. Then*

$$\begin{aligned} E(S_n) &= n \cdot p, \\ Var(S_n) &= n \cdot p(1 - p). \end{aligned}$$

8.3 Exponential, Erlang, and Poisson distributions

8.3.1 Exponential distributions

The *exponential distribution* describes a positive continuous random variable and is often used to model waiting times until an event occurs, for instance, the time until a light bulb burns out, a customer arrives, or a radioactive particle decays. This distribution is particularly suitable when the waiting time until the event is independent of how much time has already passed without the event occurring. The definition of the distribution is given next.

Definition 8.3 (Exponential distribution) *Let $\lambda > 0$. The continuous random variable X has an **exponential distribution** with parameter λ if and only if it has the PDF*

$$f(x) = \begin{cases} 0 & \text{if } x < 0, \\ \lambda \exp(-\lambda x) & \text{if } x \geq 0. \end{cases}$$

If X is an exponentially distributed random variable with parameter λ, we write $X \sim Exp(\lambda)$.

The exponential distribution could equivalently have been defined using its CDF instead of its PDF. The CDF of the exponential distribution is

$$F(x) = \begin{cases} 0 & \text{if } x < 0, \\ 1 - \exp(-\lambda x) & \text{if } x \geq 0, \end{cases} \tag{8.2}$$

see Problem 8.8.4 at the end of this chapter. The mean and variance of the exponential distribution are given next.

Theorem 8.3 (Mean and variance of the exponential distribution) *Assume that $X \sim Exp(\lambda)$. Then*

$$\begin{aligned} E(X) &= \frac{1}{\lambda}, \\ Var(X) &= \frac{1}{\lambda^2}. \end{aligned}$$

In a context where X is the time until an event happens, then λ can be interpreted as the average number of events happening per time unit. For example, if the average number of events happening per hour is $\lambda = 2$, then the expected time $E(X)$ until an event happens is $1/2$ hour. In Figure 2.5 on page 28, the graphs of the CDF and PDF for the exponential distribution with $\lambda = 2$ are shown.

Example 8.4 *A professor has office hours where individual students can come and discuss exercises in their statistics course. The students arrive at random times and independently of each other in such a way that, on average, two students will arrive at the professor's office each hour. Let X be the random variable describing the waiting time until the next student arrives. We can model X as a random variable with the exponential distribution, that is $X \sim Exp(2)$. By the properties of the exponential distribution, the average waiting time until the next student arrives is*

$$E(X) = \frac{1}{2}.$$

Similarly, using the CDF of the exponential distribution, we can calculate the probability of, say, waiting at most 1 hour for the next student to arrive:

$$\Pr(X \leq 1) = F(1) = 1 - \exp(-2 \cdot 1) = 1 - \exp(-2) = 0.86.$$

Figure 8.3 illustrates the PDF and CDF of the exponential distribution $Exp(2)$.

8.3.2 Erlang distributions

The sum of independent and identically distributed random variables with the exponential distribution is Erlang distributed. Let $\lambda > 0$ and consider n mutual statistically independent

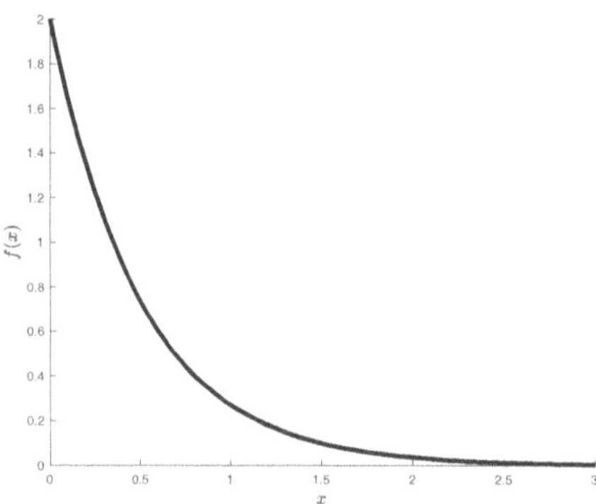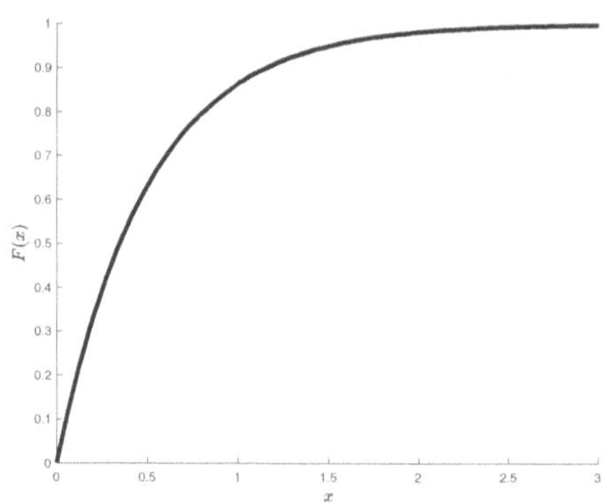

FIGURE 8.3
The exponential distribution, $Exp(\lambda)$, with $\lambda = 2$. Left: PDF $f(x)$. Right: CDF $F(x)$.

exponential random variables, $X_i \sim Exp(\lambda)$ for $i = 1, 2, \ldots, n$, and define a new random variable S_n to be their sum,

$$S_n = X_1 + X_2 + \ldots + X_n = \sum_{i=1}^{n} X_i.$$

The distribution of S_n is called the *Erlang distribution* with parameters λ and n. The random variable S_n can be interpreted as the time it takes for n events to happen, when the time between each event is independent and exponentially distributed with parameter λ. The Erlang distribution is defined next.

Definition 8.4 (Erlang distribution) *Let $\lambda > 0$ and $n \geq 1$. The continuous random variable S_n has an **Erlang distribution** with parameters λ and n if and only if it has the PDF*

$$f_{S_n}(s) = \frac{\lambda^n \cdot s^{n-1}}{(n-1)!} \exp(-\lambda \cdot s), \quad s > 0.$$

If S_n is an Erlang distributed random variable with parameters λ and n, we write $S_n \sim Erlang(\lambda, n)$.

The Erlang distribution could equivalently have been defined using its CDF instead of its PDF. The CDF of the Erlang distribution is

$$F_{S_n}(s) = 1 - \sum_{i=0}^{n-1} \frac{(\lambda \cdot s)^i}{i!} \exp(-\lambda \cdot s), \quad s > 0.$$

The mean and variance of the Erlang distribution are given next.

Theorem 8.4 (Mean and variance of the Erlang distribution) *Assume that $S_n \sim Erlang(\lambda, n)$. Then*

$$\begin{aligned}
E(S_n) &= \frac{n}{\lambda}, \\
Var(S_n) &= \frac{n}{\lambda^2}.
\end{aligned}$$

Example 8.5 (Example 8.4, continued) *The waiting time for, say, two students to arrive is $S_2 = X_1 + X_2$, where $X_1 \sim Exp(\lambda)$ is the waiting time for the first student and $X_2 \sim Exp(\lambda)$ is the waiting time from the first to the second student. Under the assumption that the waiting times X_1 and X_2 are independent, $S_2 \sim Erlang(\lambda, 2)$. Since $\lambda = 2$, then, by Theorem 8.4, the average waiting time for two students to arrive is*

$$E(S_2) = \frac{2}{2} = 1.$$

Similarly, using the CDF of the Erlang distribution, we can calculate the probability of, say, waiting at most 1 hour for the next two students to arrive:

$$\Pr(S_2 \leq 1) = F_{S_2}(1) = 1 - \sum_{i=0}^{2-1} \frac{(2 \cdot 1)^i}{i!} \exp(-2 \cdot 1) = 1 - (\exp(-2) + 2 \cdot \exp(-2)) = 0.59.$$

Figure 8.4 illustrates the PDF and CDF of the Erlang distribution $Erlang(2, 2)$.

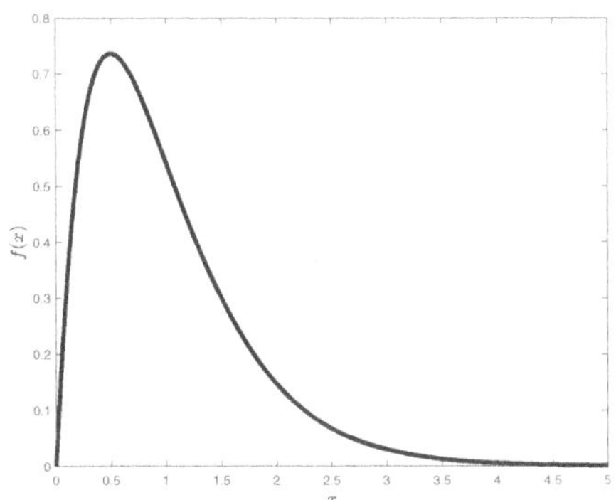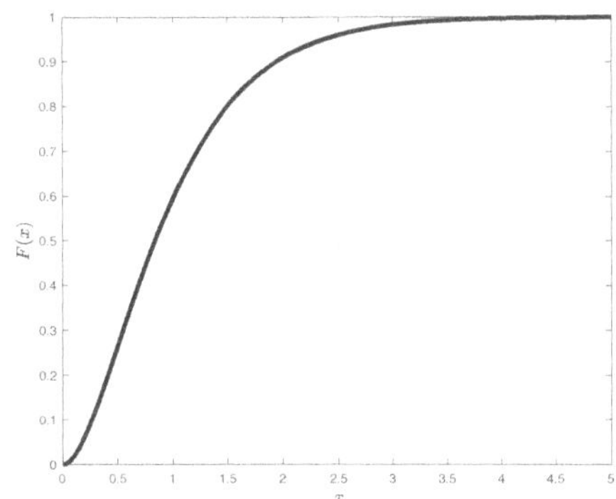

FIGURE 8.4
The Erlang distribution, $Erlang(\lambda, n)$, with $\lambda = 2$ and $n = 2$. Left: PDF $f(x)$. Right: CDF $F(x)$.

We note that the Erlang PDF and CDF can be generalized to the situation where the parameter n is a positive real number, i.e. it is allowed to not be an integer. The resulting distribution is known as the *gamma distribution*.

8.3.3 Poisson distributions

The *Poisson distribution* describes the number of events that happens within a fixed interval of time or space, when the intervals between events are independent and exponentially distributed with a constant parameter λ. The Poisson distribution is thus closely related to the exponential and Erlang distributions. The parameter λ is often referred to as the intensity, or rate, of arrivals of the events. The formal definition of the Poisson distribution is given next.

Definition 8.5 (Poisson distribution) *Let $\lambda > 0$. The discrete random variable X has a **Poisson distribution** with parameter λ if and only if it has the PMF*

$$f(x) = \frac{\lambda^x}{x!} \exp(-\lambda), \quad x = 0, 1, 2, 3, \ldots.$$

If X is a Poisson distributed random variable with parameter λ, we write $X \sim Poi(\lambda)$.

The Poisson distribution could equivalently have been defined using its CDF instead of its PDF. The CDF of the Poisson distribution is

$$F(x) = \sum_{k=0}^{x} \frac{\lambda^k}{k!} \exp(-\lambda), \quad x = 0, 1, 2, 3, \ldots.$$

The mean and variance of the Poisson distribution are given next.

Theorem 8.5 (Mean and variance of the Poisson distribution) *Assume that $X \sim Poi(\lambda)$. Then*

$$\begin{aligned} E(X) &= \lambda, \\ Var(X) &= \lambda. \end{aligned}$$

Example 8.6 (Example 8.5, continued) *Let the discrete random variable Y describe the number of students arriving in a one hour interval. In this case, we have $Y \sim Poi(\lambda)$ since λ is measured in the time unit hours. Since $\lambda = 2$, then, by Theorem 8.5, the expected number of students arriving in a one hour interval is*

$$E(Y) = 2.$$

Using the PMF of the Poisson distribution, we can also find the probabilities of, say, 0, 1, 2, and 3 students arriving in a one hour interval

$$\begin{aligned} f(0) &= \frac{2^0}{0!} \exp(-2) = 0.14, \\[6pt] f(1) &= \frac{2^1}{1!} \exp(-2) = 0.27, \\[6pt] f(2) &= \frac{2^2}{2!} \exp(-2) = 0.27, \\[6pt] f(3) &= \frac{2^3}{3!} \exp(-2) = 0.18. \end{aligned}$$

It is seen that the probability of exactly 2 students arriving during a one hour interval is 0.27, whereas the probability of waiting at most 1 hour for two students to arrive is 0.59 according to the Erlang distribution in Example 8.5. Figure 8.5 illustrates the PMF and CDF of the Poisson distribution $Poi(2)$.

We may also define the random variable Y_t, which describes the number of students arriving in a time interval of length $t > 0$. For instance, if $t = \frac{1}{2}$ hours, then $Y_{1/2}$ describes the number of students arriving in a half-hour interval. We have that Y_t is Poisson distributed and $E(Y_t) = t\lambda$, which implies $Y_t \sim Poi(\lambda t)$. In the case where $t = \frac{1}{2}$, we can use this to calculate the probabilities that 0, 1, 2, and 3 students arrive in a half-hour interval

$$\begin{aligned} f_{Y_{1/2}}(0) &= \frac{\left(2 \cdot \frac{1}{2}\right)^0}{0!} \exp\left(-2 \cdot \frac{1}{2}\right) = 0.37, \\[6pt] f_{Y_{1/2}}(1) &= \frac{\left(2 \cdot \frac{1}{2}\right)^1}{1!} \exp\left(-2 \cdot \frac{1}{2}\right) = 0.37, \\[6pt] f_{Y_{1/2}}(2) &= \frac{\left(2 \cdot \frac{1}{2}\right)^2}{2!} \exp\left(-2 \cdot \frac{1}{2}\right) = 0.28, \\[6pt] f_{Y_{1/2}}(3) &= \frac{\left(2 \cdot \frac{1}{2}\right)^3}{3!} \exp\left(-2 \cdot \frac{1}{2}\right) = 0.061. \end{aligned}$$

The collection of random variables Y_t as a function of $t > 0$ is called a Poisson process.

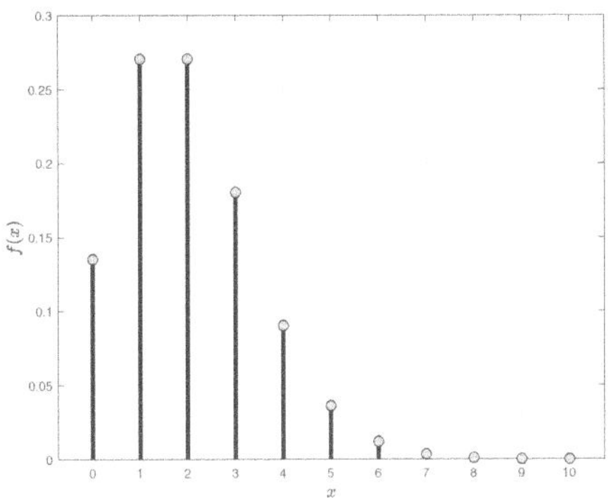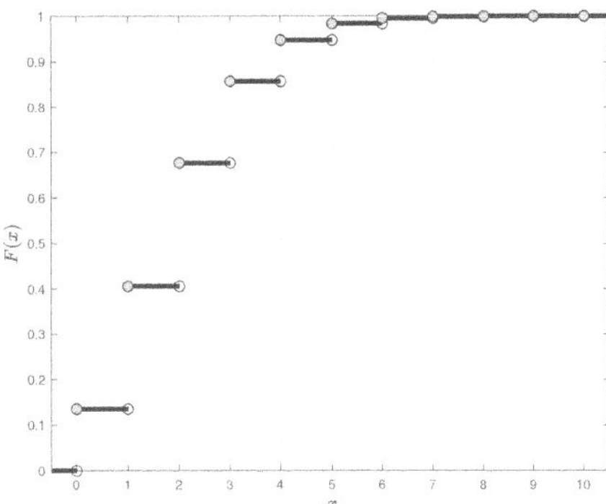

FIGURE 8.5
The Poisson distribution, $Poi(\lambda)$, with $\lambda = 2$. Left: PMF $f(x)$. Right: CDF $F(x)$.

8.4 Uniform and beta distributions

8.4.1 Uniform distributions

The *uniform distribution* is a continuous distribution which assigns equal probability to intervals of equal length. The support of the uniform distribution is an interval from a to b with $b > a$, i.e. $[a, b]$. The definition of the uniform distribution is given next.

Definition 8.6 (Uniform distribution) *Let $a, b \in \mathbb{R}$ such that $b > a$. The continuous random variable X has a **uniform distribution** if and only if it has the PDF*

$$f(x) = \begin{cases} \frac{1}{b-a} & if \quad a \le x \le b, \\ 0 & otherwise. \end{cases}$$

If X is a uniformly distributed random variable with support $[a, b]$, we write $X \sim U(a, b)$.

The uniform distribution could equivalently have been defined using its CDF instead of its PDF. The CDF of the uniform distribution is

$$F(x) = \begin{cases} 0 & if \quad x < a, \\ \frac{x-a}{b-a} & if \quad a \le x \le b, \\ 1 & if \quad x > b, \end{cases} \tag{8.3}$$

see Problem 8.8.8 at the end of this chapter. Figure 8.6 illustrates the PDF and CDF for a uniform distribution $U(a, b)$.

The mean and variance of the uniform distribution are given next.

Theorem 8.6 (Mean and variance of the uniform distribution) *Assume that $X \sim U(a, b)$. Then*

$$E(X) = \frac{a+b}{2},$$

$$Var(X) = \frac{1}{12}(b - a)^2.$$

Finally, we note that if $X \sim U(a, b)$ and $c \in \mathbb{R}$ and $d > 0$ are constants, then (Problem 8.8.8)

$$Y = c + dX \sim U(c + da, c + db). \tag{8.4}$$

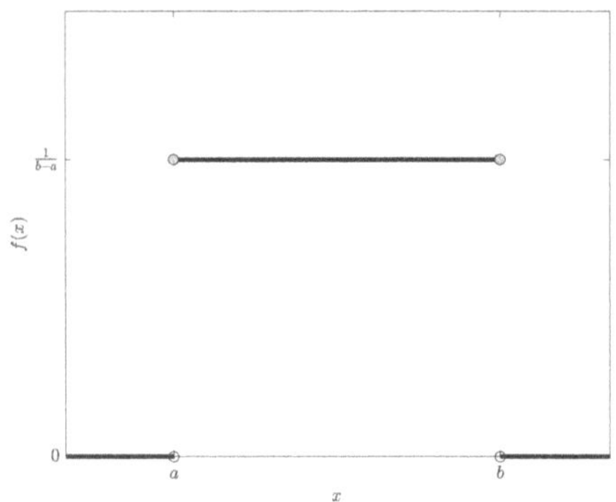 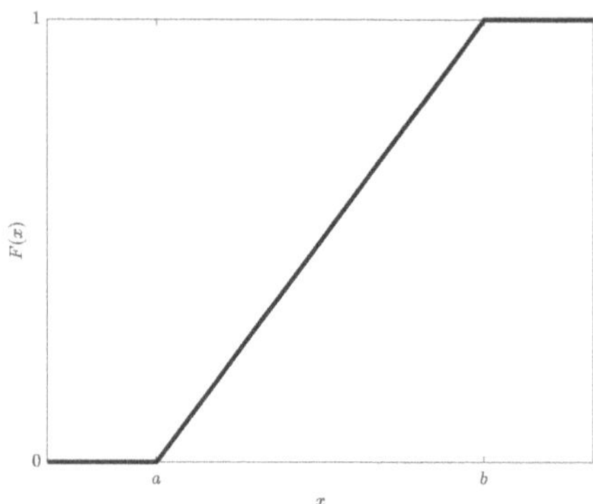

FIGURE 8.6
The uniform distribution, $U(a, b)$. Left: PDF $f(x)$. Right: CDF $F(x)$.

8.4.2   Beta distributions

Like the uniform distribution, the *beta distribution* has support in an interval from a to b with $b > a$, i.e. $[a, b]$. Unlike the uniform distribution, beta distributions are a family of distributions which can be symmetric, asymmetric, left-skewed, and right-skewed. They can be bell-shaped and U-shaped, describing situations where extreme values have low and high probabilities, respectively. The definition of the beta distribution is given next.

Definition 8.7 (Beta distribution) *Let $a, b \in \mathbb{R}$ such that $b > a$ and $\alpha, \beta > 0$. The continuous random variable X has a **beta distribution** with support $[a, b]$ and parameters α, β if and only if it has the PDF*

$$f(x) = \frac{\Gamma(\alpha + \beta)}{\Gamma(\alpha)\Gamma(\beta)}(x - a)^{\alpha - 1}(b - x)^{\beta - 1}(b - a)^{1 - \alpha - \beta}, \quad x \in [a, b], \tag{8.5}$$

where $\Gamma(z) = \int_0^\infty s^{z-1}e^{-s}ds$ is called the Gamma function. *If X is a beta distributed random variable with support $[a, b]$ and parameters α, β, we write $X \sim Beta(\alpha, \beta; a, b)$. In the case $a = 0$ and $b = 1$, i.e. where X is beta distributed over $[0, 1]$, this is usually shortened to $X \sim Beta(\alpha, \beta)$.*

The Gamma function $\Gamma()$ used in (8.5) can be calculated by many statistical software programs using numerical procedures.

Figure 8.7 plots the PDF and CDF of the beta distribution $Beta(\alpha, \beta)$ for various values of α, β in the case where $b = 1$ and $a = 0$, i.e. for a beta distribution over the unit interval $[0, 1]$. We see how choosing $\alpha = \beta$ leads to symmetric distributions, making small and large values of X equally likely, whereas letting, e.g. $\alpha < \beta$ will result in an asymmetric distribution where small values of X are more likely than large values. When $\alpha = \beta = 1$, the beta distribution $Beta(1, 1; a, b)$ reduces to the uniform distribution $U(a, b)$.

We note that the function

$$\mathcal{B}(\alpha, \beta) = \frac{\Gamma(\alpha)\Gamma(\beta)}{\Gamma(\alpha + \beta)}, \quad \alpha, \beta > 0, \tag{8.6}$$

is sometimes called the *beta function*. Using the beta function, we may write the PDF of the beta distribution (8.5) as

$$f(x) = \frac{1}{\mathcal{B}(\alpha, \beta)}(x - a)^{\alpha - 1}(b - x)^{\beta - 1}(b - a)^{1 - \alpha - \beta}, \quad x \in [a, b].$$

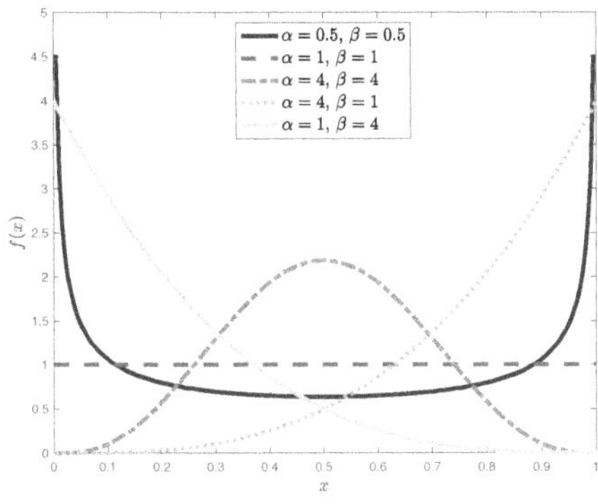 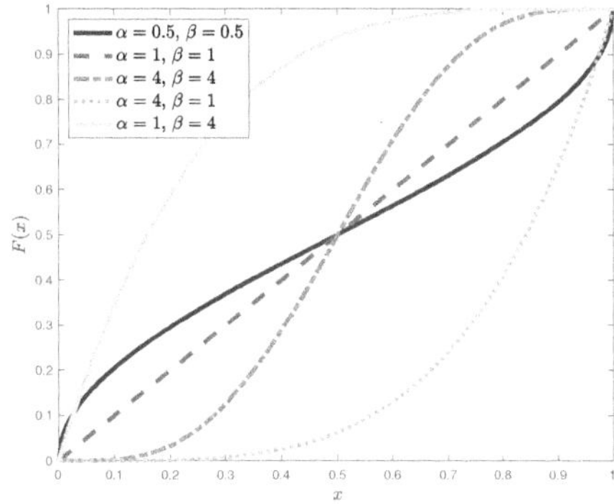

FIGURE 8.7
The beta distribution $Beta(\alpha, \beta; a, b)$ with $a = 0$, $b = 1$ and α, β as given in the plots. Left: PDF $f(x)$. Right: CDF $F(x)$.

Letting $a = 0$ and $b = 1$, and using that a PDF integrates to 1, this also shows that we may represent the beta function as

$$\mathcal{B}(\alpha, \beta) = \int_0^1 x^{\alpha-1}(1-x)^{\beta-1}dx.$$

The CDF of the beta distribution can be represented using the so-called *regularized incomplete beta function*.

The mean and variance of the beta distribution are given next.

Theorem 8.7 (Mean and variance of the beta distribution) *Assume that $X \sim Beta(\alpha, \beta; a, b)$. Then*

$$E(X) = a + (b-a)\frac{\alpha}{\alpha+\beta},$$
$$Var(X) = (b-a)^2\frac{\alpha\beta}{(\alpha+\beta)^2(\alpha+\beta+1)}.$$

In Chapter 19, and in particular in Example 19.4, we will see how the beta distribution is useful for representing an individual's beliefs regarding unknown probabilities and proportions.

8.5 Normal distributions

The *normal distribution* is widely used for two reasons: many natural phenomena are approximately normally distributed and many estimators and test statistics are approximately normally distributed. The latter fact is a consequence of the so-called *Central Limit Theorem*, which we will discuss in detail in Chapter 13. The normal distribution is sometimes called the *Gaussian distribution* in honor of the German mathematician Carl Friedrich Gauss (1777–1855), who made significant contributions to our understanding of the normal distribution.

There is a whole family of normal distributions, depending on two parameters, describing the mean and the variance of the distribution, respectively. We often use the notation μ for the mean and σ^2 for the variance. The definition of the normal, or Gaussian, distribution is given next.

Definition 8.8 (Normal distribution, Gaussian distribution) *Let $\mu_X \in \mathbb{R}$ and $\sigma_X^2 > 0$ be constants. The continuous random variable X has a **normal distribution** (or **Gaussian distribution**) with mean μ_X and variance σ_X^2 if and only if it has PDF*

$$f_X(x) = \frac{1}{\sqrt{2\pi\sigma_X^2}} \exp\left(-\frac{1}{2}\left(\frac{x - \mu_X}{\sigma_X}\right)^2\right),$$

for all $x \in \mathbb{R}$. If X is a normally distributed random variable with parameters μ_X and σ_X^2, we write $X \sim N(\mu_X, \sigma_X^2)$.

From the PDF, it is possible to verify that the mean and variance of X is indeed μ_X and σ_X^2, respectively. This is stated in the following theorem.

Theorem 8.8 (Mean and variance of the normal distribution) *Assume that $X \sim N(\mu, \sigma^2)$. Then*

$$\begin{aligned} E(X) &= \mu_X, \\ Var(X) &= \sigma_X^2. \end{aligned}$$

Further, using Definitions 8.8 and 2.5, we may infer that the CDF of $X \sim N(\mu_X, \sigma_X^2)$ is given by

$$F_X(x) = \frac{1}{\sqrt{2\pi\sigma_X^2}} \int_{-\infty}^{x} \exp\left(-\frac{1}{2}\left(\frac{y - \mu_X}{\sigma_X}\right)^2\right) dy,$$

for all $x \in \mathbb{R}$. This integral does not have a solution with well-known functions and, hence, numerical procedures are needed to evaluate the CDF of a normally distributed random variable. Figure 8.8 illustrates the PDF and CDF of the normal distribution. As can be seen from the figure, the PDF resembles a bell, which is why the PDF of a normal random variable is sometimes referred to as "the bell curve".

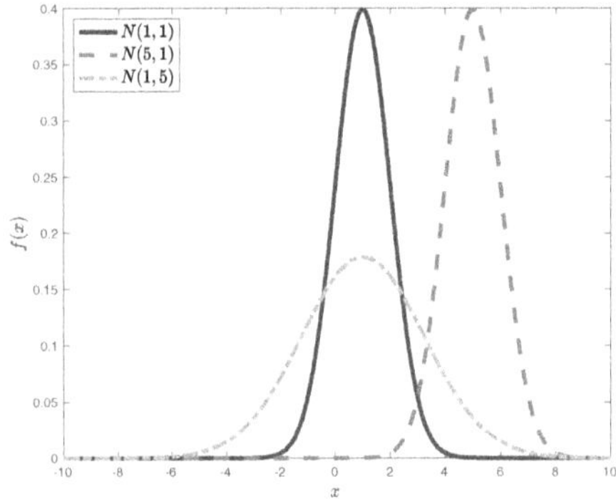
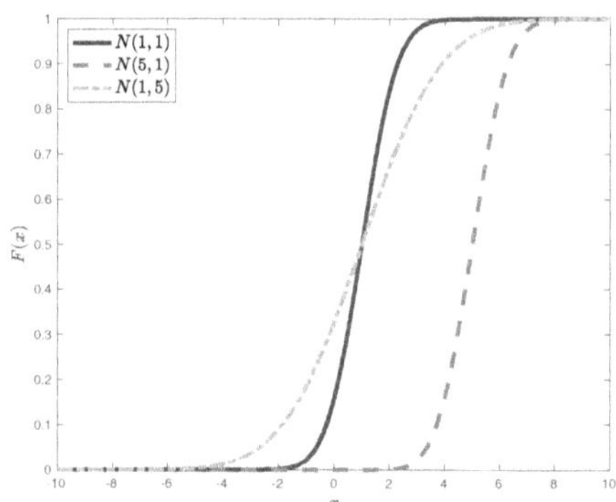

FIGURE 8.8
The normal distribution, $N(\mu, \sigma^2)$, with $\mu = 1, 5$ and $\sigma^2 = 1, 5$. Left: PDF $f(x)$. Right: CDF $F(x)$.

It turns out that linear transformations of a normally distributed random variable is again normally distributed. The details of this important result is given next.

Theorem 8.9 (Linear transformation of normal random variable) *Let* $\mu_X \in \mathbb{R}$, $\sigma_X^2 > 0$, *and* $X \sim N(\mu_X, \sigma_X^2)$. *Consider the numbers* $a, b \in \mathbb{R}$ *such that* $b \neq 0$. *It holds that*

$$a + bX \sim N(a + b\mu_X, b^2 \sigma_X^2).$$

A consequence of Theorem 8.9 is that, starting from one normal random variable, we can construct all other normal random variables by suitable linear transformations. In other words, we can define the whole family of normal distributions from linear combinations of a single family member. The family member chosen as the baseline normal distribution is the one with mean zero and variance one, which we call the *standard normal distribution*. It is defined next.

Definition 8.9 (Standard normal distribution) *The continuous random variable* Z *has a **standard normal distribution** if and only if it is normally distributed with mean* $\mu_Z = 0$ *and variance* $\sigma_Z^2 = 1$, *that is* $Z \sim N(0, 1)$. *The PDF of the standard normal distribution is*

$$f_Z(z) = \frac{1}{\sqrt{2\pi}} \exp\left(-\frac{1}{2}z^2\right),$$

for all $z \in \mathbb{R}$.

In Figure 8.9, the graphs of the PDF and CDF of the standard normal distribution are illustrated. We note that a tradition has developed that the CDF of a standard normal distribution is denoted $\Phi(z)$ and the PDF of a standard normal distribution is denoted $\phi(z)$.

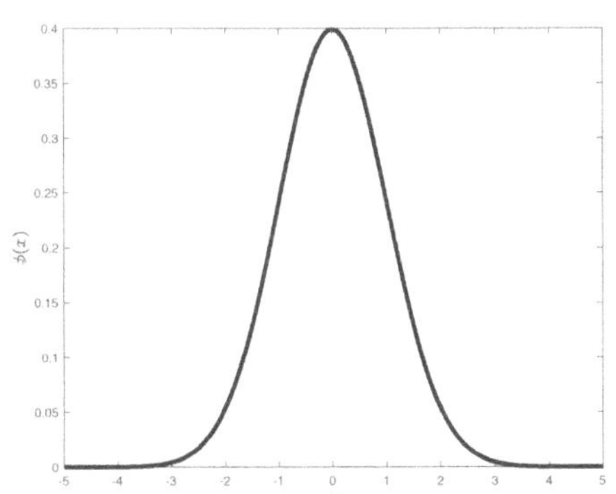
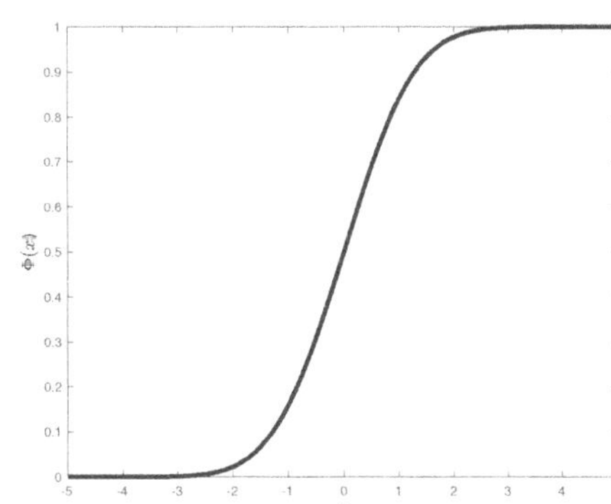

FIGURE 8.9
The standard normal distribution, $N(0, 1)$. Left: PDF, $\phi(x)$. Right: CDF, $\Phi(x)$.

As discussed above, Theorem 8.9 implies that the whole family of normal distributions can be defined starting from the standard normal distribution. To see this, let $Z \sim N(0, 1)$, $\mu_X \in \mathbb{R}$, $\sigma_X^2 > 0$, and define the random variable X via

$$X = \mu_X + \sigma_X Z.$$

Then Theorem 8.9 implies that

$$X \sim N(\mu_X, \sigma_X^2).$$

As mentioned above, the CDF of a normal random variable does not have a closed form solution, that is, we cannot write it as a known function without the integral. Therefore, to evaluate the CDF, it is necessary to make a numerical approximation of the integral using specialized computer software. Since there are infinitely many different normal distributions, i.e. infinitely many possibilities of (μ, σ^2), it would be necessary to have a computer program for each, had it not been for the fact that the CDF of the standard normal distribution, $\Phi(z)$, can be used to evaluate all of them via Theorem 8.9. To evaluate the CDF F_X of $X \sim N(\mu_X, \sigma_X^2)$, rewrite F_X to obtain it as a function of the standard normal CDF $F_Z(z) = \Phi(z)$, as follows:

$$
\begin{aligned}
F_X(x) &= \Pr(X \le x) \\
&= \Pr\left(\frac{X - \mu_X}{\sigma_X} \le \frac{x - \mu_X}{\sigma_X} \right) \\
&= \Pr\left(Z \le \frac{x - \mu_X}{\sigma_X} \right) \\
&= F_Z\left(\frac{x - \mu_X}{\sigma_X} \right),
\end{aligned}
$$

implying that

$$
F_X(x) = \Phi\left(\frac{x - \mu_X}{\sigma_X} \right). \tag{8.7}
$$

In other words, if we wish to calculate the CDF of $X \sim N(\mu_X, \sigma_X^2)$ at a number $x \in \mathbb{R}$, i.e. $F_X(x)$, we can instead evaluate the CDF of the standard normal distribution at $\frac{x-\mu_X}{\sigma_X}$, i.e. $\Phi\left(\frac{x-\mu_X}{\sigma_X} \right)$. The following examples illustrate this.

Example 8.7 *Let $X \sim N(10, 16)$. To calculate the probability of X being no higher than 8, use the expression in (8.7)*

$$
\Pr(X \le 8) = F_X(8) = F_Z\left(\frac{8 - 10}{\sqrt{16}} \right) = \Phi(-0.5) = 0.31,
$$

where a statistical software package has been used to calculate the probability $\Phi(-0.5)$.

Example 8.8 *For any normal distribution, the probability of an outcome within a neighborhood of 2 times the standard deviation σ from the mean μ is approximately 0.95. This can be derived as follows. Let $X \sim N(\mu, \sigma^2)$. Then*

$$
\begin{aligned}
\Pr\left(\mu - 2\sigma \le X \le \mu + 2\sigma \right) &= \Pr\left(X \le \mu + 2\sigma \right) - \Pr\left(X \le \mu - 2\sigma \right) \\
&= \Pr\left(\frac{X - \mu}{\sigma} \le 2 \right) - \Pr\left(\frac{X - \mu}{\sigma} \le -2 \right) \\
&= \Phi(2) - \Phi(-2) \\
&= 0.9772 - 0.0228 \\
&= 0.9544,
\end{aligned}
$$

where a statistical software package has been used to calculate the probabilities $\Phi(2)$ and $\Phi(-2)$.

8.5.1   Sums of independent normally distributed random variables

In later chapters, it will prove useful to be able to characterize the distribution of a sum of normally distributed random variables. It turns out that, when the random variables are statistically independent, their sum is also normally distributed with a mean and variance equal to the sum of the mean and variances of the random variables. When the random variables are not statistically independent, things are slightly more complicated. The case of dependent random variables will be considered in Chapter 9, when we discuss the bivariate normal distribution. The next theorem contains the results in the case of independent random variables.

Theorem 8.10 (Distribution of a sum of independent normal random variables)
Let $n \geq 1$ and $\mu_i \in \mathbb{R}$, $\sigma_i^2 > 0$, for $i = 1, 2, \ldots, n$, and consider the mutual statistically independent random variables $X_i \sim N(\mu_i, \sigma_i^2)$ for $i = 1, 2, \ldots, n$. It holds that

$$S_n = X_1 + X_2 + \ldots + X_n = \sum_{i=1}^{n} X_i \sim N(\mu, \sigma^2),$$

where

$$\mu \;=\; \mu_1 + \mu_2 + \ldots + \mu_n = \sum_{i=1}^{n} \mu_i,$$

$$\sigma^2 \;=\; \sigma_1^2 + \sigma_2^2 + \ldots + \sigma_n^2 = \sum_{i=1}^{n} \sigma_i^2.$$

In the special case where the X_i's have identical distributions, $X_i \sim N(\mu_X, \sigma_X^2)$ for all i, and X_i, $i = 1, \ldots, n$ are mutual statistically independent, then

$$S_n \sim N(n \cdot \mu_X, n \cdot \sigma_X^2), \tag{8.8}$$

while, by Theorem 8.9, the average of the X_i's has the distribution (Problem 8.8.13)

$$\frac{1}{n} S_n = \frac{1}{n} \sum_{i=1}^{n} X_i \sim N\left(\mu_X, \frac{\sigma_X^2}{n}\right). \tag{8.9}$$

8.6 Distributions derived from the normal distribution

As mentioned above, the normal distribution is important because the Central Limit Theorem (Theorem 13.1 in Chapter 13) implies that many estimators and test statistics are (approximately) normally distributed. For the same reason, particular transformations of the normal distribution are often encountered in statistics. The following subsections discuss some of the distributions arising from such transformations, which will be useful in the later chapters.

8.6.1 χ^2 distributions

Let $n \geq 1$ and $Z_1, Z_2, \ldots, Z_n$ be mutual statistically independent $N(0, 1)$ random variables. Define the continuous random variable

$$Y = Z_1^2 + Z_2^2 + \ldots + Z_n^2.$$

The distribution of Y is called the χ^2 *distribution* (pronounced "chi squared distribution") with n degrees of freedom, and we write $Y \sim \chi^2(n)$.

The PDF and CDF of the χ^2 distribution have complicated forms and are not stated here. The χ^2 CDF and its quantile function can be found in tables or by using statistical software programs. The support of the distribution are the positive real numbers. Figure 8.10 illustrates the PDF and CDF of the χ^2 distribution.

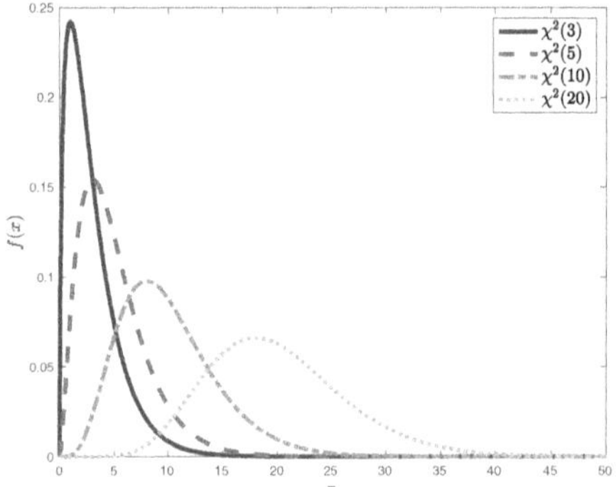
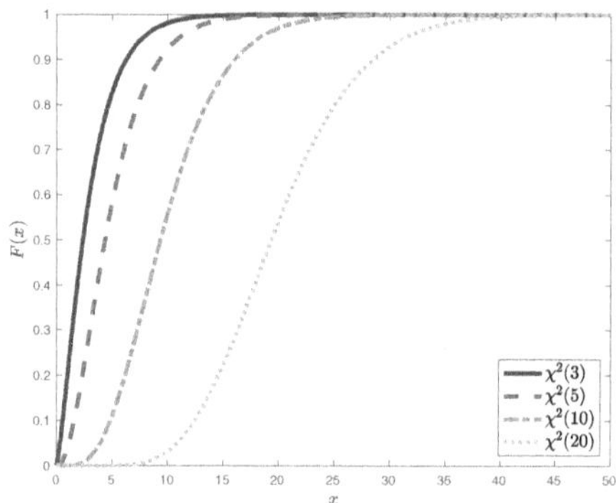

FIGURE 8.10
The χ^2 distribution, $\chi^2(n)$, with $n = 3, 5, 10, 20$. Left: PDF $f(x)$. Right: CDF $F(x)$.

The mean and variance of a χ^2 distribution with n degrees of freedom are (Problem 8.8.18)

$$\begin{aligned} E(Y) &= n, \\ Var(Y) &= 2n. \end{aligned}$$

Let $n \geq 1$, but assume now that $X_1, X_2, \ldots, X_n$ are mutual statistically independent random variables, distributed according to

$$X_i \sim N(\mu_i, 1),$$

where $\mu_i \in \mathbb{R}$ for $i = 1, 2, \ldots, n$. Define the continuous random variable

$$W = X_1^2 + X_2^2 + \ldots + X_n^2.$$

The distribution of W is called the *non-central χ^2 distribution* with n degrees of freedom and non-centrality parameter λ given by

$$\lambda^2 = \mu_1^2 + \mu_2^2 + \ldots + \mu_n^2.$$

We write $W \sim \chi^2(n, \lambda)$. Thus, the non-central χ^2 distribution with $\mu_1 = \ldots = \mu_n = 0$ is equal to the χ^2 distribution, i.e. $\chi^2(n, 0) = \chi^2(n)$. The means and variances of the non-central χ^2 distribution are

$$\begin{aligned} E(W) &= n + \lambda^2, \\ Var(W) &= 2(n + 2\lambda^2). \end{aligned}$$

Similarly to the χ^2 distribution, the analytical forms of the PDF and CDF of the non-central χ^2 distribution are quite complicated and are therefore not presented here. Evaluating the CDF and calculating quantiles of the non-central χ^2 distribution is typically done using statistical software programs.

8.6.2 t-distributions

Let $n \geq 1$ and $Z \sim N(0,1)$, $Y \sim \chi^2(n)$ be two statistically independent random variables. Define the continuous random variable

$$X = \frac{Z}{\sqrt{Y/n}}.$$

The distribution of X is called the *t-distribution* (or *student-t distribution*) with n degrees of freedom, and we write $X \sim t(n)$.

The PDF and CDF of the t-distribution have complicated forms and are not given here. The CDF and quantile functions can be found in tables or using statistical software programs. The support of the distribution is the real number line, i.e. $(-\infty, \infty)$.

It can be shown that the t-distribution with n degrees of freedom only have moments which are strictly lower than n. Thus, the mean does not exist for a t-distribution with one degree of freedom, whereas the mean exists for a t-distribution with two degrees of freedom, but the variance does not. In particular, The mean and variance of a t-distribution with n degrees of freedom are

$$
\begin{aligned}
E(X) &= 0, && \text{if } n \geq 2, \text{ otherwise does not exist,} \\
Var(X) &= \frac{n}{n-2}, && \text{if } n \geq 3, \text{ otherwise does not exist.}
\end{aligned}
$$

As n increases, the t-distribution will become closer to the standard normal distribution. Indeed, in the limit, as $n \to \infty$, the $t(n)$ distribution will be equal to the $N(0,1)$ distribution. Figure 8.11 compares the $N(0,1)$ distribution with the $t(n)$ distribution for $n = 1, 2, 5, 10, 20$.

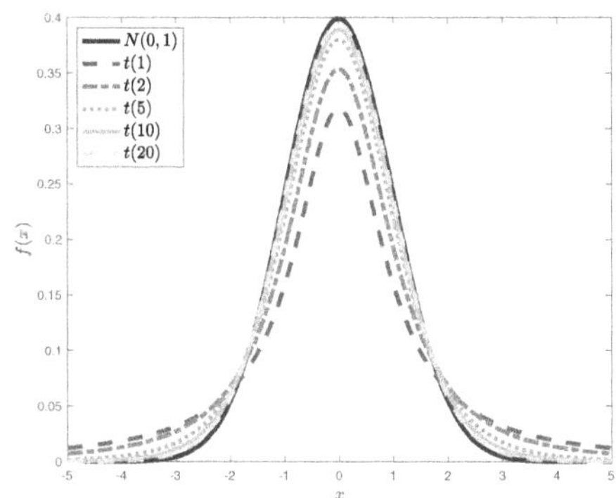
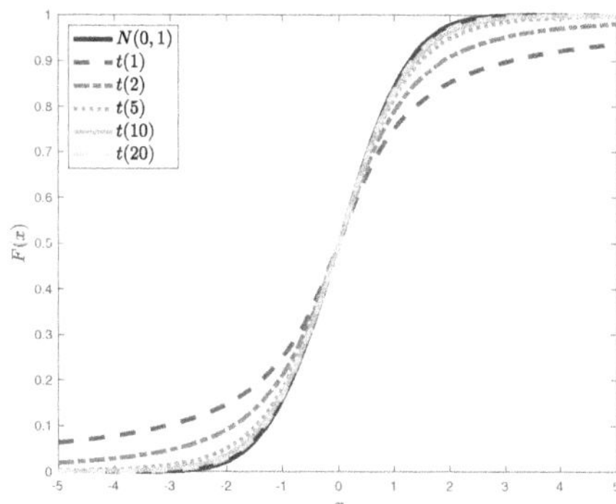

FIGURE 8.11
The standard normal and $t(n)$-distributions for $n = 1, 2, 5, 10, 20$. Left: PDF $f(x)$. Right: CDF $F(x)$.

A useful generalization of the t-distribution changes the location of the mode of the t-distribution. Let $\mu \in \mathbb{R}$ and let Z and Y be as above. Consider the random variable

$$W = \frac{Z + \mu}{\sqrt{Y/n}}.$$

The distribution of W is called the *non-central t-distribution* with n degrees of freedom, and we write $X \sim t(n, \lambda)$. The non-central t-distribution with $\mu = 0$ is equal to the t-distribution, i.e. $t(n, 0) = t(n)$. Similarly to the t-distribution, the non-central t-distribution with n degrees of freedom only have moments strictly smaller than n. The analytical forms of the PDF, CDF, mean, and variance of the non-central t-distribution are quite complicated and are therefore not presented here. Evaluating the CDF and calculating quantiles of the non-central t-distribution is typically done using statistical software programs.

8.6.3 F distributions

Let $r, s \geq 1$ and $Y \sim \chi^2(r)$, $W \sim \chi^2(s)$ be two statistically independent random variables. Define the continuous random variable

$$X = \frac{Y/r}{W/s}.$$

The distribution of X is called the *F-distribution* with (r, s) degrees of freedom, and we write $X \sim F(r, s)$. Note that, from the definition of X above, we see that if $X \sim F(r, s)$, then $X^{-1} \sim F(s, r)$.

The PDF and CDF of the F-distribution have complicated forms and are not given here. The CDF and quantile functions can be found in tables or in statistical software programs. The support of the distribution is the positive real numbers. Figure 8.12 illustrates the PDF and CDF of the F-distribution.

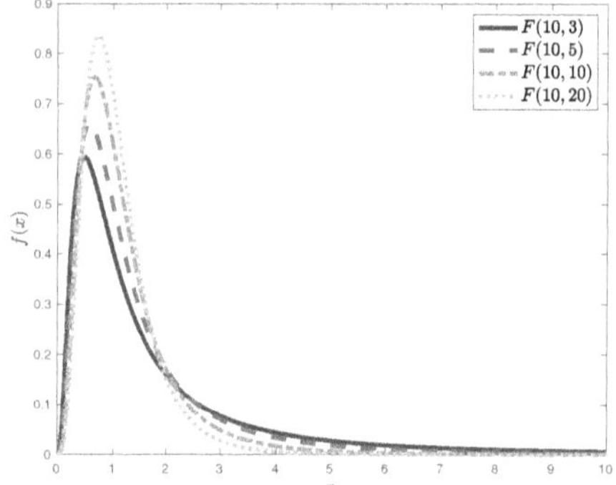
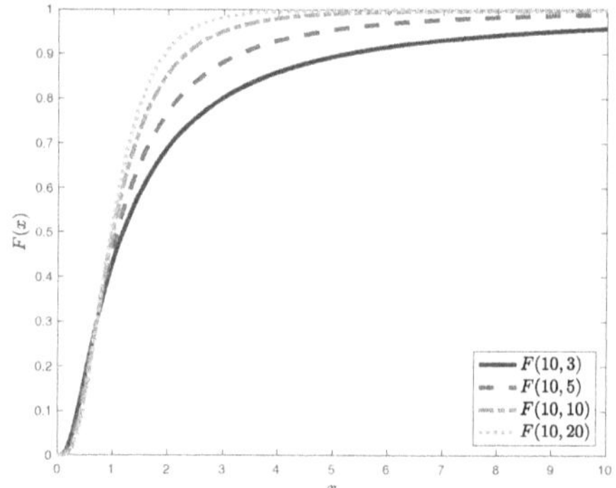

FIGURE 8.12
The F-distribution, $F(r, s)$, with $r = 10$ and $s = 3, 5, 10, 20$. Left: PDF $f(x)$. Right: CDF $F(x)$.

It can be shown that the F-distribution with (r, s) degrees of freedom only have moments which are strictly lower than $\frac{s}{2}$. Similarly, it can be shown that the mean and variance of the F-distribution with (r, s) degrees of freedom are

$$E(X) = \frac{s}{s-2} \quad \text{if } s \geq 3, \text{ otherwise does not exist,}$$

$$Var(X) = \frac{2s^2(r+s-2)}{r(s-2)^2(s-4)} \quad \text{if } s \geq 5, \text{ otherwise does not exist.}$$

Let $\alpha \in (0, 1)$. The quantiles of the F-distribution have the following property

$$F_{1-\alpha}(r, s) = \frac{1}{F_\alpha(s, r)}.$$

The F-distribution is linked to the χ^2 distribution in that the $F(r, s)$ distribution converges to the $\chi^2(r)$ distribution when $s \to \infty$. The F-distribution is also linked to the t-distribution via the following relation: If $X \sim t(s)$, then $X^2 \sim F(1, s)$.

8.7 Proofs

Proof of Theorem 8.1 *The proof of Theorem 8.1 is left as an exercise to the reader (Problem 8.8.1).*

Proof of Theorem 8.2 *The proof of Theorem 8.2 is left as an exercise to the reader (Problem 8.8.2).*

Proof of Theorem 8.3 *The proof of Theorem 8.3 is left as an exercise to the reader (Problem 8.8.4).*

Proof of Theorem 8.4 *The proof of Theorem 8.4 is left as an exercise to the reader (Problem 8.8.5).*

Proof of Theorem 8.5 *The proof of Theorem 8.5 is left as an exercise to the reader (Problem 8.8.6).*

Proof of Theorem 8.6 *The proof of Theorem 8.6 is left as an exercise to the reader (Problem 8.8.8).*

Proof of Theorem 8.7 *The proof of Theorem 8.7 is left as an exercise to the reader (Problem 8.8.11).*

Proof of Theorem 8.8 *The proof of Theorem 8.8 is left as an exercise to the reader (Problem 8.8.12).*

Proof of Theorem 8.9 *Define the random variable $Y = g(X) = a + bX$ and note that $g^{-1}(y) = \frac{y-a}{b}$. Recall from Theorem 2.14 that the PDF of Y is given by*

$$f_Y(y) = f_X\left(g^{-1}(y)\right)\left|\frac{dg^{-1}(y)}{dy}\right| = f_X\left(\frac{y-a}{b}\right)\left|\frac{1}{b}\right|.$$

Since $X \sim N(\mu_X, \sigma_X^2)$ we can plug in its PDF (Definition 8.8) to arrive at

$$
\begin{aligned}
f_Y(y) &= \frac{1}{\sqrt{2\pi\sigma_X^2}}\exp\left(-\frac{1}{2}\left(\frac{\frac{y-a}{b}-\mu_X}{\sigma_X}\right)^2\right)\cdot\frac{1}{|b|} \\
&= \frac{1}{\sqrt{2\pi\sigma_X^2 b^2}}\exp\left(-\frac{1}{2}\left(\frac{y-(a+b\mu_X)}{b\sigma_X}\right)^2\right).
\end{aligned}
$$

By Definition 8.8, Y is a normally distributed random variable with mean $a + b\mu_X$ and variance $b^2\sigma_X^2$, that is $Y = a + bX \sim N(a + b\mu_X, b^2\sigma_X^2)$, as we wanted to show.

Proof of Theorem 8.10 *The proof that S_n is normally distributed is beyond the scope of this book. The interested reader may find a proof of this fact in most, if not all, textbooks on advanced probability theory. Once we know that S_n is normally distributed, however, the proof of the theorem follows easily. Indeed, since the normal distribution only relies on two parameters, the mean μ and the variance σ^2, all that is left is to calculate these. From the properties of the expectation, we get*

$$\mu = E(S_n) = E\left(\sum_{i=1}^{n} X_i\right) = \sum_{i=1}^{n} E(X_i) = \sum_{i=1}^{n} \mu_i.$$

Similarly, from the properties of the variance, along with the fact that the X_i's are mutual statistically independent (which implies $Cov(X_i, X_j) = 0$ for $i \neq j$), we get

$$\sigma^2 = Var(S_n) = Var\left(\sum_{i=1}^{n} X_i\right) = \sum_{i=1}^{n} Var(X_i) = \sum_{i=1}^{n} \sigma_i^2,$$

which concludes the proof.

8.8　Exercises

Problem 8.8.1 *Prove that the mean and variance of $X \sim Ber(p)$ are*

$$\begin{aligned} E(X) &= p, \\ Var(X) &= p(1-p). \end{aligned}$$

Problem 8.8.2 *Prove that the mean and variance of $S_n \sim Bin(n, p)$ are*

$$\begin{aligned} E(X) &= n \cdot p, \\ Var(X) &= n \cdot p(1-p). \end{aligned}$$

Problem 8.8.3 *You might want to use a calculator or statistical software to solve this exercise.*

1. *Find $P(X = 10)$, when $X \sim Bin(n = 50, p = 0.4)$.*

2. *Find $P(X = 50)$, when $X \sim Bin(n = 170, p = 0.4)$.*

3. *Find $P(X \leq 50)$, when $X \sim Bin(n = 170, p = 0.4)$.*

4. *Find $P(X > 4)$, when $X \sim Bin(n = 10, p = 0.7)$.*

5. *Find $P(2 < X \leq 5)$, when $X \sim Bin(n = 10, p = 0.7)$.*

Problem 8.8.4 *Let $X \sim Exp(\lambda)$, where $\lambda > 0$.*

1. *Use the PDF of the exponential distribution to prove that the CDF is given by Equation (8.2).*

2. *Prove that*

$$E(X) \;=\; \frac{1}{\lambda},$$
$$Var(X) \;=\; \frac{1}{\lambda^2}.$$

Problem 8.8.5 *Let* $X \sim Erlang(\lambda, n)$, *where* $\lambda > 0$ *and* $n \geq 1$. *Prove that*

$$E(X) \;=\; \frac{n}{\lambda},$$
$$Var(X) \;=\; \frac{n}{\lambda^2}.$$

Hint: Use the fact that X *can be represented as a sum of iid exponentially distributed random variables.*

Problem 8.8.6 *Let* $X \sim Poi(\lambda)$, *where* $\lambda > 0$. *Prove that*

$$E(X) \;=\; \lambda,$$
$$Var(X) \;=\; \lambda.$$

Problem 8.8.7 *Two teams, Team A and Team B, are set to play a football match. Let the random variables* X *and* Y *denote the number of goals that Team A and Team B will score during the match, respectively. You may assume that* X *and* Y *are independent. (Note: This assumption is likely too strong in practice.) From a preliminary statistical analysis, you know that*

$$X \sim Poi(\lambda_A), \quad Y \sim Poi(\lambda_B),$$

where $\lambda_A = 2$ *and* $\lambda_B = 1.2$.

1. *What is the expected number of goals that Team A will score during the match?*

2. *What is the expected number of total goals that will be scored during the match?*

3. *What is the probability that Team A will score at least 3 goals during the match?*

4. *What is the probability that the match will end* $1 - 1$?

5. *What is the probability that Team A will win the match? What is the probability of a tie? And what is the probability that Team B will win the match?*

   Hint 1: You might want to assume that the number of goals a team may score is capped, e.g. assume that the maximum number of goals a team will score is, say, 10. Strictly speaking, this assumption will imply that the answers you give to the final question are approximations; but since it is exceedingly rare that more than 10 goals are scored by a team, the approximation will in turn be exceedingly accurate.

   Hint 2: Since the final question requires the calculations of many probabilities, you might want to use statistical software to answer this question.

Problem 8.8.8 *Let* $X \sim U(a, b)$, *where* $b > a$.

1. *Use the PDF of the uniform distribution to prove that the CDF is given by Equation (8.3).*

2. *Prove that*

$$E(X) = \frac{a+b}{2},$$

$$Var(X) = \frac{1}{12}(b-a)^2.$$

3. *Prove the relationship in Equation (8.4). Hint: Use Theorem 2.14*

Problem 8.8.9 *Let $p \in [0,1]$. Derive an expression for the p-quantile of the uniform distribution $X \sim U(a,b)$.*

Problem 8.8.10 *Consider the gamma function, given by*

$$\Gamma(z) = \int_0^\infty s^{z-1}e^{-s}ds, \quad z > 0.$$

Let $z > 0$. Prove that

$$\Gamma(z+1) = z\Gamma(z).$$

Hint: Use integration by parts.

Problem 8.8.11 *Let $Y \sim Beta(\alpha,\beta)$, where $\alpha,\beta > 0$, i.e. $Y \sim Beta(\alpha,\beta;0,1)$.*

1. *Show that $E(Y) = \frac{\alpha}{\alpha+\beta}$. (Hint: You may express $E(Y) = \int_0^1 yf(y)dy$ using the gamma and beta functions, then use (8.6), and the fact that $\Gamma(z+1) = z\Gamma(z)$, see Problem 8.8.10.)*

2. *Show that $E(Y^2) = \frac{\alpha(\alpha+1)}{(\alpha+\beta)(\alpha+\beta+1)}$. (Hint: Same hint as above.)*

3. *Use your expressions of $E(Y)$ and $E(Y^2)$, found above, to conclude that $Var(Y) = \frac{\alpha\beta}{(\alpha+\beta)^2(\alpha+\beta+1)}$.*

Let $a,b \in \mathbb{R}$ such that $b > a$ and consider the random variable $X \sim Beta(\alpha,\beta;a,b)$. It can be shown that

$$X \overset{d}{=} a + (b-a) \cdot Y,$$

where $Y \sim Beta(\alpha,\beta)$ and where "$\overset{d}{=}$" means that the distribution of random variable on the left hand side (i.e. X) is the same as the distribution of the random variable on the right hand side (i.e. $a + (b-a) \cdot Y$).

4. *Use your expressions for $E(Y)$ and $Var(Y)$, together with the fact that $X \overset{d}{=} a+(b-a)\cdot Y$, to prove Theorem 8.7.*

Problem 8.8.12

1. *Let $Z \sim N(0,1)$. Prove that $E(Z) = 0$ and $Var(Z) = 1$. (Hint: To calculate $Var(Z)$ you may calculate $E(Z^2)$ using integration by parts and the value of the so-called Gaussian integral, $\int_{-\infty}^\infty e^{-z^2}dz = \sqrt{\pi}$.)*

2. *Let $X \sim N(\mu,\sigma^2)$, where $\mu \in \mathbb{R}$ and $\sigma^2 > 0$. Prove that $E(X) = \mu$ and $Var(X) = \sigma^2$. (Hint: Use your answer from the first part, along with Theorem 8.9.)*

Problem 8.8.13 *Use Theorems 8.9 and 8.10 to prove the relation in Equation (8.9). Discuss what happens to the distribution of S_n as n increases. (This is related to the Law of Large Numbers, which will be discussed in Chapter 11, in particular in Theorem 11.4.)*

Problem 8.8.14 *Let $X \sim N(-2, 1)$ and $Y \sim N(2, 16)$ be statistically independent normally distributed random variables.*

1. *Calculate the probabilities $P(X < -2)$ and $P(Y \geq 0)$.*

2. *Consider the random variable $Q = 2X + 2Y$. Derive the distribution of Q.*

3. *Calculate $Cov(X, Q)$ and $Cov(Y, Q)$.*

4. *Calculate the probabilities $P(Q = 0)$ and $P(Q > 1)$.*

Problem 8.8.15 *An agency distributes relief payments due to natural disasters for two separate regions, A and B, of a country. The relief payments X_A (in mill kr) to region A in a given year is modeled as $X_A \sim N(15, 4)$ and the relief payments X_B to region B as $X_B \sim N(10, 5)$.*

1. *Calculate the probability that no more than 12 mill kr are distributed to region B in a given year.*

2. *Assume the relief payments to the two regions are statistically independent. Calculate the probability that the agency distributes more than 28 mill kr in a given year.*

Problem 8.8.16 *A risk manager oversees the risk of her company's portfolio of projects. The company can invest in three projects. The analyst department has informed the risk manager that the future returns of the projects are uncertain, but can be described by random variables with a normal distribution. In particular, the returns are given by:*

- *Project 1: $X \sim N(0.05, 0.025)$.*

- *Project 2: $Y \sim N(0.15, 0.10)$.*

- *Project 3: $Z \sim N(0.50, 1.00)$.*

1. *What is the probability that each of the projects incur a loss (have a negative return)?*

2. *What is the probability that each of the projects has a return of more than 10%?*

3. *Assume that the returns on the three projects are independent. The company has invested 10M DKK in the first project, 5M DKK in the second project, and 1M DKK in the third project. Let W be the random variable denoting the total profit from this portfolio. What is the distribution of W?*

4. *What is the probability that the company incurs a loss on the entire portfolio?*

5. *What is the probability that the company makes more than 2M DKK on the portfolio?*

Problem 8.8.17 *A factory is using a machine to produce a small piece of equipment. The piece is used as a part in the final product of a larger piece of equipment, produced by the same factory. It is a requirement for the small piece that its diameter is in the interval 2.30 ± 0.301 millimeters. Pieces that are below the lower limit $(2.30 - 0.301)$ are discarded, while pieces that are above the upper limit $(2.30 + 0.301)$ are sent to be grinded down in another section of the factory. Experience shows that the diameter of the pieces thus produced is normally distributed with mean 2.26 and standard deviation 0.20.*

1. *What percentage of pieces will be discarded?*

2. *If* 30 *pieces are picked at random what is the probability that at most two pieces will have to be discarded?*

3. *If* 1000 *pieces are picked at random, calculate the expected number of pieces that will be sent to be grinded down.*

Problem 8.8.18 *Let* $Y \sim \chi^2(n)$, *where* $n \geq 1$. *Prove that*

$$\begin{aligned} E(Y) &= n, \\ Var(Y) &= 2n. \end{aligned}$$

Hint: Use the fact that Y *can be represented as a sum of iid squared normally distributed random variables.*

9

Bivariate normal distributions

9.1 Introduction

In the previous chapter, we introduced the concept of a normal distribution for a random variable, also called the *univariate normal distribution* (Section 8.5). In this chapter, we extend the discussion of the normal distribution to the joint distribution of two random variables, which gives rise to the *bivariate normal distributions.*

The bivariate normal distributions are a family of distributions indexed by the means, variances, and covariance between the two random variables. To fix ideas, let X_1 and X_2 be two random variables. Denote the means of X_1 and X_2 by $\mu_1 = E(X_1)$ and $\mu_2 = E(X_2)$, denote their variances as $\sigma_1^2 = Var(X_1)$ and $\sigma_2^2 = Var(X_2)$, and denote their covariance as $\sigma_{12} = Cov(X_1, X_2)$. In the following, we describe the bivariate normal distribution, which is a function of the five parameters μ_1, μ_2, σ_1^2, σ_2^2, and $\sigma_{1,2}$. Note that, because the correlation $\rho = Corr(X_1, X_2)$ between X_1 and X_2 can be expressed via the standard deviations and the covariance by the relation $\rho = \frac{\sigma_{12}}{\sigma_1 \sigma_2}$, we may equivalently express the bivariate normal distribution as a function of the five parameters μ_1, μ_2, σ_1^2, σ_2^2, and ρ.

9.2 Density of a bivariate normal distribution

Each bivariate normal distribution can be defined by a specific functional form of the joint PDF of X_1 and X_2 as shown next.

Definition 9.1 (Bivariate normal distribution) *Let* $\mu_1, \mu_2 \in \mathbb{R}$, $\sigma_1^2, \sigma_2^2 > 0$, *and* $\sigma_{12} \in \mathbb{R}$ *such that* $\sigma_{12} \in (-\sigma_1\sigma_2, \sigma_1\sigma_2)$. *The collection* $X = (X_1, X_2)$ *of two continuous random variables* X_1 *and* X_2 *has a **bivariate normal distribution** with parameters* $(\mu_1, \mu_2, \sigma_1^2, \sigma_2^2, \sigma_{12})$ *if and only if the joint PDF* $f_{X_1, X_2}(x_1, x_2)$ *of* X_1 *and* X_2 *equals*

$$f_{X_1, X_2}(x_1, x_2) \;=\; \frac{1}{\sqrt{(2\pi)^2 \cdot (\sigma_1^2 \sigma_2^2 - \sigma_{12}^2)}} \tag{9.1}$$
$$\cdot \exp\left(-\frac{1}{2} \frac{(x_1 - \mu_1)^2 \, \sigma_2^2 + (x_2 - \mu_2)^2 \, \sigma_1^2 - 2\,(x_1 - \mu_1)\,(x_2 - \mu_2)\,\sigma_{1,2}}{(\sigma_1^2 \sigma_2^2 - \sigma_{12}^2)} \right),$$

for all $x_1, x_2 \in \mathbb{R}$. *If* $X = (X_1, X_2)$ *has a bivariate normal distribution with parameters* $(\mu_1, \mu_2, \sigma_1^2, \sigma_2^2, \sigma_{12})$, *we write* $X = (X_1, X_2) \sim N(\mu_1, \mu_2, \sigma_1^2, \sigma_2^2, \sigma_{12})$.

The condition in Theorem 9.1 that $\sigma_{12} \in (-\sigma_1\sigma_2, \sigma_1\sigma_2)$ is equivalent to assuming that $|Corr(X_1, X_2)| < 1$. This, in particular, rules out the possibility that X_1 and X_2 are perfectly correlated. It also ensures that $\sigma_{1,2}$ is a valid value for a covariance, in the sense that it implies a correlation coefficient with absolute value that is not greater than one.

DOI: 10.1201/9781003591191-9

Similarly to the univariate case (Section 8.5), all bivariate normal distributions can be understood as members of a bivariate normal family, where each member is characterized by unique values of the five parameters, namely the means μ_1, μ_2, the variances σ_1^2, σ_2^2, and the covariance σ_{12}. The left panels of Figures 9.1 and 9.2 illustrate the graphs of the PDFs for two members of the family of bivariate normal distributions, namely, the members with zero means ($\mu_1 = 0$, $\mu_2 = 0$) and unit variances ($\sigma_1^2 = 1$, $\sigma_2^2 = 1$), but different covariance, namely $\sigma_{1,2} = 0$ and $\sigma_{1,2} = 0.7$, respectively. The right panels of the figures show the corresponding contours, or level curves, of the PDFs $f(x_1, x_2)$. A level curve of the function $f(x, y)$ shows all combinations of x and y that imply the same value of the function $f()$, i.e. the level curve of the function $f(x, y)$ associated to the value $c \in \mathbb{R}$ are the pairs (x, y) that satisfy the equation $f(x, y) = c$. The level curves in the right panel of Figure 9.1 show that the contours of the PDF of the $N(0, 0, 1, 1, 0)$ distribution are circles, reflecting the fact that $Cov(X_1, X_2) = 0$, i.e. X_1 and X_2 are uncorrelated. In contrast, the level curves in the right panel of Figure 9.2 show that the contours of the PDF of the $N(0, 0, 1, 1, 0.7)$ distribution are ellipses, reflecting the fact that $Cov(X_1, X_2) \neq 0$, i.e. that X_1 and X_2 are correlated. In general, the direction of the major axis of the ellipses, which is the direction the probability of an outcome is concentrated, is determined by the correlation between X_1 and X_2, that is, the parameter $\rho = Corr(X_1, X_2)$. If the correlation is positive, as is e.g. the case for the $N(0, 0, 1, 1, 0.7)$ distribution, then the major axis has a positive slope when viewed in the x_1 and x_2 plane, as also seen in the right panel of Figure 9.2.

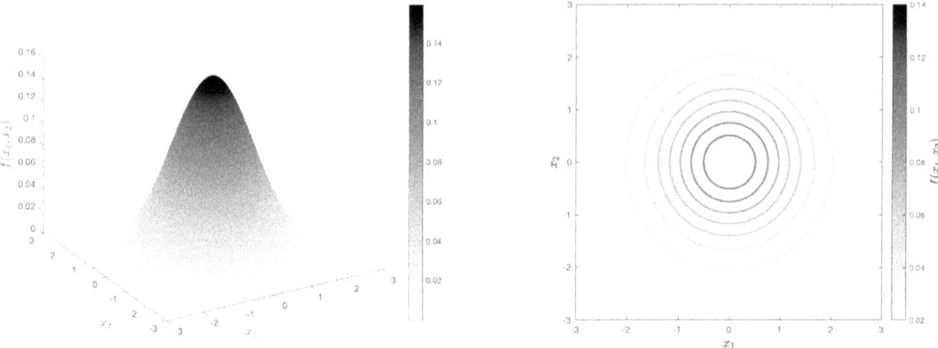

FIGURE 9.1
PDF of bivariate normal $N(0, 0, 1, 1, 0)$-distribution. Left: PDF $f(x_1, x_2)$. Right: Level curves of $f(x_1, x_2)$, i.e. those (x_1, x_2) such that $f(x_1, x_2) = c$ for $c = 0.02, 0.04, \ldots, 0.14$.

As mentioned above, it will sometimes be convenient to parametrize the bivariate normal distribution in terms of the correlation $\rho = Corr(X_1, X_2)$ instead of the covariance $\sigma_{12} = Cov(X_1, X_2)$. For this purpose, the PDF in (9.1) can equivalently be written as (Problem 9.7.1)

$$f_{X_1, X_2}(x_1, x_2) = \frac{1}{\sqrt{(2\pi)^2 \sigma_1^2 \sigma_2^2 (1 - \rho^2)}} \cdot \exp\left\{ -\frac{1}{2(1 - \rho^2)} \left(\left(\frac{x_1 - \mu_1}{\sigma_1} \right)^2 + \left(\frac{x_2 - \mu_2}{\sigma_2} \right)^2 \right. \right.$$
$$\left. \left. - 2\rho \left(\frac{x_1 - \mu_1}{\sigma_1} \right) \left(\frac{x_2 - \mu_2}{\sigma_2} \right) \right) \right\}, \tag{9.2}$$

for all $x_1, x_2 \in \mathbb{R}$.

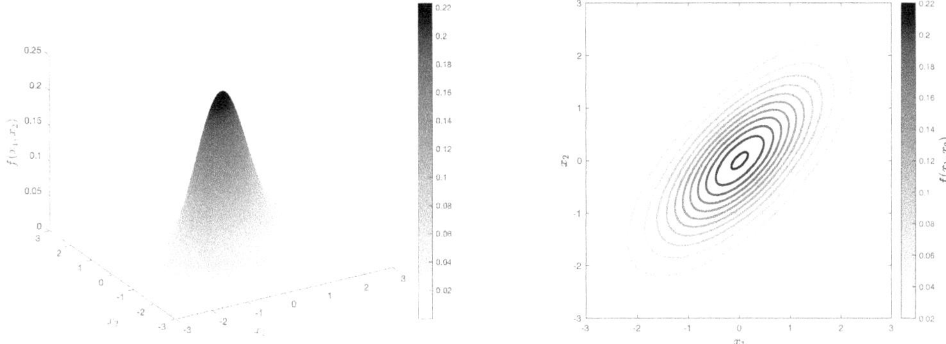

FIGURE 9.2
PDF of bivariate normal $N(0,0,1,1,0.7)$-distribution. Left: PDF $f(x_1, x_2)$. Right: Level curves of $f(x_1, x_2)$, i.e. those (x_1, x_2) such that $f(x_1, x_2) = c$ for $c = 0.02, 0.04, \ldots, 0.22$.

9.3 Marginal distributions and linear combination of bivariate normal random variables

We proceed to present a number of important results related to the bivariate normal distribution. The first result states that if $X = (X_1, X_2)$ is bivariate normal, then the marginal distributions of the random variables X_1 and X_2 are univariate normal.

Theorem 9.1 (Marginal distributions of a bivariate normal distribution) *Assume* $X \sim N(\mu_1, \mu_2, \sigma_1^2, \sigma_2^2; \sigma_{12})$. *Then the marginal distributions of* X_1 *and* X_2 *are*

$$X_1 \sim N\left(\mu_1, \sigma_1^2\right) \quad and \quad X_2 \sim N\left(\mu_2, \sigma_2^2\right).$$

The converse of Theorem 9.1 is, in general, not true. That is, from the knowledge that the two random variables X_1 and X_2 each have a univariate normal distribution, we cannot conclude that their joint distribution is bivariate normal. The following example contains a counterexample to the converse of Theorem 9.1. You are asked to prove the details of the example in Problem 9.7.6 at the end of this chapter.

Example 9.1 (Counterexample to the converse of Theorem 9.1) *Let X and Q be two statistically independent random variables where $X \sim N(0, 1)$ and where Q can have the two outcomes -1 or 1 with $\Pr(Q = -1) = \Pr(Q = 1) = 1/2$. Let $Y = X \cdot Q$. It can be shown that (Problem 9.7.6)*

$$Y \sim N(0, 1).$$

That is, $X \sim N(0, 1)$ and $Y \sim N(0, 1)$ both have univariate normal distributions. It can be shown, however, that (X, Y) does not have a bivariate normal distribution (Problem 9.7.6).

The next theorem states that if $X = (X_1, X_2)$ has a bivariate normal distribution, then two linear combinations of X_1 and X_2 will also have a bivariate normal distribution.

Theorem 9.2 (Linear combinations of bivariate normal random variables) *Assume $X \sim N(\mu_1, \mu_2, \sigma_1^2, \sigma_2^2; \sigma_{12})$ and let $a_1, a_2, b_1, b_2 \in \mathbb{R}$ be constants. Define the two* *continuous random variables W_1 and W_2 as linear combinations of X_1 and X_2,*

$$
\begin{aligned}
W_1 &= a_1 X_1 + a_2 X_2, \\
W_2 &= b_1 X_1 + b_2 X_2.
\end{aligned}
$$

Then $W = (W_1, W_2)$ has a bivariate normal distribution, $W = (W_1, W_2) \sim N(\mu_{W1}, \mu_{W2}, \sigma_{W1}^2, \sigma_{W2}^2, \sigma_{W12})$, where

$$
\begin{aligned}
\mu_{W2} &= a_1 \mu_1 + a_2 \mu_2, \\
\mu_{W2} &= b_1 \mu_1 + b_2 \mu_2, \\
\sigma_{W1}^2 &= a_1^2 \sigma_1^2 + a_2 \sigma_2^2 + 2 a_1 a_2 \sigma_{12}, \\
\sigma_{W2}^2 &= b_1^2 \sigma_1^2 + b_2 \sigma_2^2 + 2 b_1 b_2 \sigma_{12},
\end{aligned}
$$

and

$$
\sigma_{W12} = a_1 b_1 \sigma_1^2 + a_2 b_2 \sigma_2^2 + (a_1 b_2 + a_2 b_1) \sigma_{12}.
$$

If we combine Theorem 9.2 with Theorem 9.1, we see that if $X = (X_1, X_2)$ is bivariate normal as in the theorems, then a linear combination of X_1 and X_2 has a univariate normal distribution. That is, for constants $a_1, a_2 \in \mathbb{R}$, then

$$
a_1 X_1 + a_2 X_2 \sim N\left(a_1 \mu_1 + a_2 \mu_2, a_1^2 \sigma_1^2 + a_2 \sigma_2^2 + 2 a_1 a_2 \sigma_{12}\right).
$$

Example 9.2 *Consider two random variables, $X \sim N(2, 8)$ and $Y \sim N(1, 2)$, with $Cov(X, Y) = 3$. Define the new random variable*

$$
W = -X + 2Y.
$$

If the joint distribution of X and Y is bivariate normal, then Theorem 9.2 and Theorem 9.1 allow us to conclude that W has a normal distribution. Note that this conclusion only holds if the assumption of bivariate normality of (X, Y) holds; if this assumption is not fulfilled, then we cannot be sure that W is normally distributed, see Example 9.1 for a counterexample.

Assume that the joint distribution of X and Y is indeed bivariate normal and that we therefore know that W is (univariate) normally distributed. The normal distribution only depends on the mean and variance, so if we calculate these, then we have pinpointed the distribution of W. Using the rules of calculation for expectations, we get

$$
E(W) = E(-X + 2Y) = -E(X) + 2E(Y) = -2 + 2 = 0.
$$

Similarly, using the rules for variances, we get

$$
\begin{aligned}
Var(W) = Var(-X + 2Y) &= (-1)^2 Var(X) + 2^2 Var(Y) + 2 \cdot (-1) \cdot 2 \cdot Cov(X, Y) \\
&= (-1)^2 \cdot 8 + 2^2 \cdot 2 + 2 \cdot (-1) \cdot 2 \cdot 3 \\
&= 8 + 8 - 12 \\
&= 4.
\end{aligned}
$$

Therefore,

$$
W \sim N(0, 4).
$$

When two random variables are independent, we can, in general, derive their joint distribution based on their individual univariate distributions. The next theorem shows that two independent random variables, each normally distributed, jointly have a bivariate normal distribution.

Theorem 9.3 (Independent univariate normals to bivariate normal) *Assume that*

$$X_1 \sim N\left(\mu_1, \sigma_1^2\right) \quad and \quad X_2 \sim N\left(\mu_2, \sigma_2^2\right).$$

If X_1 and X_2 are statistically independent, then $X = (X_1, X_2)$ has the bivariate normal distribution

$$X = (X_1, X_2) \sim N(\mu_1, \mu_2, \sigma_1^2, \sigma_2^2, 0).$$

In contrast to many other types of bivariate distributions, if the two random variables with the bivariate normal distribution are uncorrelated, then they are also independent. This is formally stated in the next theorem.

Theorem 9.4 *Assume $(X_1, X_2) \sim N(\mu_1, \mu_2, \sigma_1^2, \sigma_2^2, 0)$. Then X_1 and X_2 are statistically independent.*

9.4 Conditional distributions of bivariate normal distribution

The next result states that if $X = (X_1, X_2)$ has a bivariate normal distribution, then the conditional distribution of X_1 given X_2 is univariate normal.

Theorem 9.5 (Properties of bivariate normal distributions) *Assume that $X = (X_1, X_2) \sim N(\mu_1, \mu_2, \sigma_1^2, \sigma_2^2, \sigma_{12})$. The conditional distribution of X_1 given $X_2 = x_2$ is*

$$X_1 | X_2 = x_2 \sim N\left(\gamma_0 + \gamma_1 x_2 \, , \, \tau^2\right),$$

where

$$\gamma_1 = \frac{\sigma_{12}}{\sigma_2^2},$$

$$\gamma_0 = \mu_1 - \gamma_1 \mu_2,$$

and

$$\tau^2 = \sigma_1^2 \left(1 - \frac{\sigma_{12}^2}{\sigma_1^2 \sigma_2^2}\right).$$

Theorem 9.5 shows that if Y and X are jointly bivariate normal, then the conditional expectation function of Y given X is linear, i.e.

$$E(Y|X = x) = \gamma_0 + \gamma_1 x.$$

Furthermore, in Theorem 9.5, the conditional variance of Y given $X = x$ equals the constant τ^2. Hence, the conditional variance does not depend on the value of x. Notice also that τ^2 can be expressed as

$$\tau^2 = \sigma_1^2 \left(1 - \rho^2\right),$$

where $\rho = Corr(X, Y)$ is the correlation coefficient between Y and X. If Y and X are highly correlated, then τ^2 is close to 0. This implies that the variance of Y, given the knowledge of the outcome of X, is small.

Example 9.3 *Let $(Y, X) \sim N(5, 10, 8, 12, -2)$. The conditional expectation function of Y given X is*

$$E(Y|X = x) = \gamma_0 + \gamma_1 x,$$

where

$$\gamma_1 = \frac{-2}{12} = -\frac{1}{6},$$

and

$$\gamma_0 = 5 - \gamma_1 \cdot 10 = \frac{40}{6}.$$

Thus, the conditional expectation function of Y given X is

$$E(Y|X = x) = \frac{40}{6} - \frac{1}{6}x.$$

9.5 Constructing bivariate normal distribution from univariate normal distributions

Similar to the univariate case, where all normal distributions can be constructed from any one normal distribution, Theorem 9.2 implies that the whole family of bivariate normal distributions can be constructed from any one bivariate normal distribution. Analogous to the univariate case, the canonical bivariate normal distribution we propose to use is the one with zero means and unit variances. This distribution is called the *standard bivariate normal distribution* and it is formally defined as follows.

Definition 9.2 (Standard bivariate normal distributions) *Let $X = (X_1, X_2)$ have a bivariate normal distribution according to*

$$X = (X_1, X_2) \sim N(0, 0, 1, 1, \rho),$$

*where $\rho \in (-1, 1)$. Then we say that $X = (X_1, X_2)$ has a **standard bivariate normal distribution** with correlation ρ.*

Theorem 9.1 implies that if $X = (X_1, X_2)$ has a standard bivariate normal distribution, then each random variable, X_1 and X_2, has a standard (univariate) normal distribution as their marginal distribution, i.e. $X_1 \sim N(0, 1)$ and $X_2 \sim N(0, 1)$. Note also that, since $\sigma_1^2 = \sigma_2^2 = 1$, then the correlation and the covariance coincide, i.e. $\rho = \sigma_{12}$. Two examples of the PDFs of the standard normal distribution, namely the $N(0, 0, 1, 1, 0)$ and $N(0, 0, 1, 1, 0.7)$ distributions, are shown in Figures 9.1 and 9.2. In the former case, we have $\rho = Corr(X_1, X_2) = 0$, implying that X_1 and X_2 are statistically independent (Theorem 9.4). In the latter case, we have $\rho = Corr(X_1, X_2) = 0.7$, implying that X_1 and X_2 are positively correlated.

We can use the standard bivariate normal distribution to construct any other bivariate normal distribution. In fact, there is a particularly convenient way of doing this construction, which can be useful for simulating random draws from the bivariate normal distribution. Problem 9.7.3 at the end of this chapter asks you to simulate random instances from the bivariate normal distribution using this approach. The result can be stated as follows.

Theorem 9.6 (Generating random instances from the bivariate normal distribution) *Let $Z_1, Z_2 \sim N(0, 1)$ be statistically independent random variables. Let also $\mu_1, \mu_2 \in \mathbb{R}$, $\sigma_1, \sigma_2 > 0$, and $\rho \in (-1, 1)$ be constants. Define the two random variables U_1 and U_2 according to*

$$\begin{aligned} U_1 &= Z_1, \\ U_2 &= \rho Z_1 + \sqrt{1 - \rho^2} Z_2. \end{aligned}$$

Then $U = (U_1, U_2)$ has a standard bivariate normal distribution according to

$$U = (U_1, U_2) \sim N(0, 0, 1, 1, \rho).$$

In addition, define the random variables X_1 and X_2 as follows

$$\begin{aligned} X_1 &= \mu_1 + \sigma_1 U_1, \\ X_2 &= \mu_2 + \sigma_2 U_2. \end{aligned}$$

Then, $X = (X_1, X_2)$ has a bivariate normal distribution according to

$$X = (X_1, X_2) \sim N(\mu_1, \mu_2, \sigma_1^2, \sigma_2^2, \sigma_{12}), \tag{9.3}$$

where $\sigma_{12} = \sigma_1 \sigma_2 \rho$.

As mentioned, Theorem 9.6 can be used to simulate a random draw from X with distribution given in (9.3). This can be done by making two independent draws from a standard normal distribution $N(0, 1)$, which is the distribution of Z_1 and Z_2, respectively. Then these two draws can be transformed into U_1 and U_2, and further into X_1 and X_2 according to the formulas in the theorem.

9.6 Proofs

Proof of Theorem 9.1 *We show that $X_1 \sim N(\mu_1, \sigma_1^2)$. The proof that $X_2 \sim N(\mu_2, \sigma_2^2)$ follows along similar lines. Let $f_{X_1, X_2}(x_1, x_2)$ be the joint PDF of X_1 and X_2, given in Equation (9.1). Recall that the PDF of the marginal distribution of X_1 is given by (Theorem 3.9)*

$$f_{X_1}(x_1) = \int_{-\infty}^{\infty} f_{X_1, X_2}(x_1, x_2) dx_2.$$

To show that $X_1 \sim N(\mu_1, \sigma_1^2)$, we need to show that $f_{X_1}(x_1)$ is equal to the PDF of a univariate normal distribution with mean μ_1 and variance σ_1^2, i.e. we want to show that

$$f_{X_1}(x_1) = \frac{1}{\sqrt{2\pi\sigma_1^2}} \exp\left(-\frac{1}{2}\left(\frac{x_1 - \mu_1}{\sigma_1}\right)^2\right)$$

for all $x_1 \in \mathbb{R}$.

Note first that, using the equivalent formulation of the joint PDF of X_1 and X_2, given in Equation (9.2), we can write

$$\begin{aligned} & f_{X_1, X_2}(x_1, x_2) \\ &= \frac{1}{\sqrt{(2\pi)^2 \sigma_1^2 \sigma_2^2 (1 - \rho^2)}} \cdot \exp\left\{ -\frac{1}{2(1-\rho^2)} \left(\left(\frac{x_1 - \mu_1}{\sigma_1}\right)^2 + \left(\frac{x_2 - \mu_2}{\sigma_2}\right)^2 \right. \right. \\ & \qquad \left. \left. -2\rho \left(\frac{x_1 - \mu_1}{\sigma_1}\right) \left(\frac{x_2 - \mu_2}{\sigma_2}\right) \right) \right\} \\ &= \frac{1}{\sqrt{2\pi\sigma_1^2}} \exp\left(-\frac{1}{2}\left(\frac{x_1 - \mu_1}{\sigma_1}\right)^2\right) \cdot A(x_1, x_2), \end{aligned}$$

where $A(x_1, x_2)$ can be written as

$$
\begin{aligned}
A(x_1, x_2) \;=\; & \frac{1}{\sqrt{2\pi\sigma_2^2\,(1-\rho^2)}} \cdot \exp\left\{-\frac{1}{2(1-\rho^2)}\left(\left(\frac{x_1-\mu_1}{\sigma_1}\right)^2 + \left(\frac{x_2-\mu_2}{\sigma_2}\right)^2\right.\right.\\
& \left.\left. -2\rho\left(\frac{x_1-\mu_1}{\sigma_1}\right)\left(\frac{x_2-\mu_2}{\sigma_2}\right) - (1-\rho^2)\left(\frac{x_1-\mu_1}{\sigma_1}\right)^2\right)\right\}\\[4pt]
=\; & \frac{1}{\sqrt{2\pi\sigma_2^2\,(1-\rho^2)}} \cdot \exp\left\{-\frac{1}{2(1-\rho^2)}\left(\rho^2\left(\frac{x_1-\mu_1}{\sigma_1}\right)^2 + \left(\frac{x_2-\mu_2}{\sigma_2}\right)^2\right.\right.\\
& \left.\left. -2\rho\left(\frac{x_1-\mu_1}{\sigma_1}\right)\left(\frac{x_2-\mu_2}{\sigma_2}\right)\right)\right\}\\[4pt]
=\; & \frac{1}{\sqrt{2\pi\sigma_2^2\,(1-\rho^2)}} \exp\left(-\frac{1}{2}\left(\frac{x_2-(\mu_2+\sigma_1^{-1}(x_1-\mu_1)\rho\sigma_2)}{\sqrt{(1-\rho^2)\sigma_2^2}}\right)^2\right).
\end{aligned}
$$

For fixed x_1, we recognize the function $g(x_2) = A(x_1, x_2)$ as the PDF of a normal distribution with mean $\mu_2 + \sigma_1^{-1}(x_1 - \mu_1)\rho\sigma_2$ and variance $(1-\rho^2)\sigma_2^2$, i.e. from the $N(\mu_2 + \sigma_1^{-1}(x_1 - \mu_1)\rho\sigma_2, (1-\rho^2)\sigma_2^2)$ distribution. Since a PDF must always integrates to one, this means that, for fixed x_1, then

$$
\int_{-\infty}^{\infty} A(x_1, x_2)\,dx_2 = 1.
$$

Let $x_1 \in \mathbb{R}$. The above shows that

$$
\begin{aligned}
f_{X_1}(x_1) \;=\; & \int_{-\infty}^{\infty} f_{X_1, X_2}(x_1, x_2)\,dx_2\\[4pt]
=\; & \frac{1}{\sqrt{2\pi\sigma_1^2}}\exp\left(-\frac{1}{2}\left(\frac{x_1-\mu_1}{\sigma_1}\right)^2\right) \cdot \int_{-\infty}^{\infty} A(x_1, x_2)\,dx_2\\[4pt]
=\; & \frac{1}{\sqrt{2\pi\sigma_1^2}}\exp\left(-\frac{1}{2}\left(\frac{x_1-\mu_1}{\sigma_1}\right)^2\right),
\end{aligned}
$$

the PDF of a $N(\mu_1, \sigma_1^2)$ distributed random variable. We conclude that $X_1 \sim N(\mu_1, \sigma_1^2)$, as we wanted to show.

Proof of Theorem 9.2   *The proof that $\widetilde{X}_1 = a_1 X_1 + a_2 X_2$ and $\widetilde{X}_2 = b_1 X_1 + b_2 X_2$ is bivariate normally distributed is beyond the scope of this book. The interested reader may find a proof of this fact in most textbooks on advanced probability theory.*

Once (joint) normality of $\widetilde{X}_1 = a_1 X_1 + a_2 X_2$ and $\widetilde{X}_2 = a_1 X_1 + a_2 X_2$ has been established, however, what remains is to calculate the means and variances of $\widetilde{X}_1$ and $\widetilde{X}_2$, as well as $Cov(\widetilde{X}_1, \widetilde{X}_2)$. From the rules of calculations with expectations, we find that

$$
E(\widetilde{X}_1) = E(a_1 X_1 + a_2 X_2) = a_1 E(X_1) + a_2 E(X_2) = a_1\mu_1 + a_2\mu_2,
$$

and likewise for $E(\widetilde{X}_2)$. Similarly, using the rules of variances, we find

$$
\begin{aligned}
Var(\widetilde{X}_1) \;=\; & Var\,(a_1 X_1 + a_2 X_2)\\
=\; & a_1^2 Var(X_1) + a_2^2 Var(X_2) + 2a_1 a_2 Cov(X_1, X_2)\\
=\; & a_1^2\sigma_1^2 + a_2^2\sigma_2^2 + 2a_1 a_2\sigma_{12},
\end{aligned}
$$

and, analogously,

$$Var(\widetilde{X}_2) = b_1^2 \sigma_1^2 + b_2^2 \sigma_2^2 + 2b_1 b_2 \sigma_{12}.$$

The rules for calculations of covariances yield

$$
\begin{aligned}
Cov(\widetilde{X}_1, \widetilde{X}_2) &= Cov\,(a_1 X_1 + a_2 X_2, b_1 X_1 + b_2 X_2) \\
&= a_1 b_1 Cov(X_1, X_1) + a_1 b_2 Cov(X_1, X_2) + a_2 b_1 Cov(X_2, X_1) \\
&\quad + a_2 b_2 Cov(X_2, X_2) \\
&= a_1 b_1 \sigma_1^2 + a_1 b_2 \sigma_{12} + a_2 b_1 \sigma_{12} + a_2 b_2 \sigma_2^2.
\end{aligned}
$$

Combining these calculations completes the proof.

Proof of Theorem 9.3 *Let $X_1 \sim N(\mu_1, \sigma_1^2)$, $X_2 \sim N(\mu_2, \sigma_2^2)$, and assume X_1 and X_2 are statistically independent. Statistical independence implies that $\sigma_{12} = Cov(X_1, X_2) = 0$ (Theorem 7.4). We want to show that the joint distribution of X_1 and X_2 is bivariate normal according to Definition 9.1, i.e. that their joint PDF is given by Equation (9.1) with $\sigma_{12} = 0$, which can be written as*

$$
\begin{aligned}
&\frac{1}{\sqrt{(2\pi)^2 \cdot (\sigma_1^2 \sigma_2^2)}} \cdot \exp\left(-\frac{1}{2} \frac{(x_1 - \mu_1)^2 \sigma_2^2 + (x_2 - \mu_2)^2 \sigma_1^2}{(\sigma_1^2 \sigma_2^2)} \right) \\
&= \frac{1}{\sqrt{2\pi \cdot \sigma_1^2}} \frac{1}{\sqrt{2\pi \cdot \sigma_2^2}} \exp\left(-\frac{1}{2} \left(\left(\frac{x_1 - \mu_1}{\sigma_1}\right)^2 + \left(\frac{x_2 - \mu_2}{\sigma_2}\right)^2 \right) \right).
\end{aligned}
$$

But since X_1 and X_2 are statistically independent, their joint PDF is equal to the product of their marginal PDFs. Using the form of the univariate normal distribution, the joint PDF of X_1 and X_2 is thus

$$
\begin{aligned}
f_{X_1, X_2}(x_1, x_2) &= f_{X_1}(x_1) f_{X_2}(x_2) \\
&= \frac{1}{\sqrt{2\pi \cdot \sigma_1^2}} \exp\left(-\frac{1}{2}\left(\left(\frac{x_1 - \mu_1}{\sigma_1}\right)^2 \right) \right) \\
&\quad \cdot \frac{1}{\sqrt{2\pi \cdot \sigma_2^2}} \exp\left(-\frac{1}{2}\left(\left(\frac{x_2 - \mu_2}{\sigma_2}\right)^2 \right) \right) \\
&= \frac{1}{\sqrt{2\pi \cdot \sigma_1^2}} \frac{1}{\sqrt{2\pi \cdot \sigma_2^2}} \exp\left(-\frac{1}{2}\left(\left(\frac{x_1 - \mu_1}{\sigma_1}\right)^2 + \left(\frac{x_2 - \mu_2}{\sigma_2}\right)^2 \right) \right).
\end{aligned}
$$

That is, the joint PDF $f_{X_1, X_2}(x_1, x_2)$ is equal to the PDF from the bivariate normal distribution with $\sigma_{12} = 0$, as derived above. We conclude that (X_1, X_2) has a bivariate normal distribution according to

$$X = (X_1, X_2) \sim N(\mu_1, \mu_2, \sigma_1^2, \sigma_2^2, 0),$$

as we wanted to show.

Proof of Theorem 9.4 *The proof of Theorem 9.4 is left as an exercise to the reader (Problem 9.7.2).*

Proof of Theorem 9.5 *Recall that the conditional distribution of X_1 given $X_2 = x_2$ can be represented by the conditional PDF*

$$f_{X_1|X_2}(x_1|x_2) = \frac{f_{X_1,X_2}(x_1,x_2)}{f_{X_2}(x_2)}.$$

Since (X_1, X_2) is bivariate normal, the PDF $f_{X_1,X_2}(x_1,x_2)$ is given in Equation (9.2). Theorem 9.1 implies that $X_2 \sim N(\mu_2, \sigma_2)$, and, therefore,

$$f_{X_2}(x_2) = \frac{1}{\sqrt{2\pi\sigma_2^2}} \exp\left(-\frac{1}{2}\left(\frac{x_2 - \mu_2}{\sigma_2}\right)^2\right).$$

We thus have that

$$
\begin{aligned}
f_{X_1|X_2}(x_1|x_2) &= \frac{f_{X_1,X_2}(x_1,x_2)}{f_{X_2}(x_2)} \\[2mm]
&= \frac{1}{\sqrt{(2\pi)^2\sigma_1^2\sigma_2^2(1-\rho^2)}}\sqrt{2\pi\sigma_2^2} \cdot A(x_1, x_2) \\[2mm]
&= \frac{1}{\sqrt{(2\pi)\sigma_1^2(1-\rho^2)}} \cdot A(x_1, x_2),
\end{aligned}
$$

where

$$
\begin{aligned}
A(x_1, x_2) &= \exp\left\{-\frac{1}{2(1-\rho^2)}\left(\left(\frac{x_1-\mu_1}{\sigma_1}\right)^2 + \left(\frac{x_2-\mu_2}{\sigma_2}\right)^2\right.\right. \\
&\qquad \left.\left. -2\rho\left(\frac{x_1-\mu_1}{\sigma_1}\right)\left(\frac{x_2-\mu_2}{\sigma_2}\right) - (1-\rho^2)\left(\frac{x_2-\mu_2}{\sigma_2}\right)^2\right)\right\} \\[2mm]
&= \exp\left\{-\frac{1}{2(1-\rho^2)}\left(\left(\frac{x_1-\mu_1}{\sigma_1}\right)^2 + \rho^2\left(\frac{x_2-\mu_2}{\sigma_2}\right)^2\right.\right. \\
&\qquad \left.\left. -2\rho\left(\frac{x_1-\mu_1}{\sigma_1}\right)\left(\frac{x_2-\mu_2}{\sigma_2}\right)\right)\right\} \\[2mm]
&= \exp\left(-\frac{1}{2(1-\rho^2)}\left(\left(\frac{x_1-\mu_1-\rho\sigma_1\sigma_2^{-1}(x_2-\mu_2)}{\sigma_1}\right)^2\right)\right) \\[2mm]
&= \exp\left(-\frac{1}{2}\left(\left(\frac{x_1-(\mu_1+\rho\sigma_1\sigma_2^{-1}(x_2-\mu_2))}{\sqrt{(1-\rho^2)\sigma_1^2}}\right)^2\right)\right).
\end{aligned}
$$

We recognize $f_{X_1|X_2}(x_1|x_2) = \frac{1}{\sqrt{(2\pi)\sigma_1^2(1-\rho^2)}} \cdot A(x_1, x_2)$ as the PDF of a normal distribution $N(\tilde{\mu}, \tilde{\sigma}^2)$ with mean

$$\tilde{\mu} = \mu_1 + \rho\sigma_1\sigma_2^{-1}(x_2-\mu_2) = \gamma_0 + \gamma_1 x_2$$

and variance

$$\tilde{\sigma}^2 = \sqrt{(1-\rho^2)\sigma_1^2} = \tau^2,$$

with γ_0, γ_1, and τ^2 as given in the theorem. This concludes the proof.

Proof of Theorem 9.6 *Since $Z_1, Z_2 \sim N(0, 1)$ are statistically independent, Theorem 9.3 implies that $Z = (Z_1, Z_2)$ has a bivariate normal distribution. Theorem 9.2, in turn, implies* *that two linear combinations of Z_1 and Z_2 is bivariate normally distributed. In particular, letting $U_1 = Z_1$ and $U_2 = \rho Z_1 + \sqrt{1 - \rho^2} Z_2$, Theorem 9.2 implies that $U = (U_1, U_2)$ has a bivariate normal distribution. Another application of Theorem 9.2 then implies that $X = (X_1, X_2)$, where $X_1 = \mu_1 + \sigma_1 U_1$ and $X_2 = \mu_2 + \sigma_1 U_2$, has a bivariate normal distribution. What remains is to calculate the means, variances, and covariances of X_1 and X_2. We find*

$$E(X_1) = E\left(\mu_1 + \sigma_1 U_1\right) = \mu_1 + \sigma_1 E(U_1) = \mu_1$$

and

$$E(X_2) = E\left(\mu_2 + \sigma_2 U_2\right) = \mu_2 + \sigma_2 E(U_2) = \mu_2,$$

where we used that $E(U_1) = E(Z_1) = 0$ and $E(U_2) = E(\rho Z_1 + \sqrt{1 - \rho^2} Z_2) = \rho E(Z_1) + \sqrt{1 - \rho^2} E(Z_2) = 0$. As for variances, we have

$$Var(X_1) = Var\left(\mu_1 + \sigma_1 U_1\right) = \sigma_1^2 Var(U_1) = \sigma_1^2 Var(Z_1) = \sigma_1^2,$$

and

$$
\begin{aligned}
Var(X_2) &= Var\left(\mu_2 + \sigma_2 U_2\right) \\
&= \sigma_2^2 Var(U_2) \\
&= \sigma_2^2 Var\left(\rho Z_1 + \sqrt{1 - \rho^2} Z_2\right) \\
&= \sigma_2^2 \left(\rho^2 Var(Z_1) + (1 - \rho^2) Var(Z_2) + \rho\sqrt{1 - \rho^2} Cov(Z_1, Z_2)\right) \\
&= \sigma_2^2 \left(\rho^2 \cdot 1 + (1 - \rho^2) \cdot 1 + \rho\sqrt{1 - \rho^2} \cdot 0\right) \\
&= \sigma_2^2.
\end{aligned}
$$

Lastly,

$$
\begin{aligned}
Cov(X_1, X_2) &= Cov\left(\mu_1 + \sigma_1 U_1, \mu_2 + \sigma_2 U_2\right) \\
&= \sigma_1 \sigma_2 Cov\left(U_1, U_2\right) \\
&= \sigma_1 \sigma_2 Cov\left(Z_1, \rho Z_1 + \sqrt{1 - \rho^2} Z_2\right) \\
&= \sigma_1 \sigma_2 \rho Cov(Z_1, Z_1) + \sigma_1 \sigma_2 \sqrt{1 - \rho^2} Cov(Z_1, Z_2) \\
&= \sigma_1 \sigma_2 \rho Var(Z_1) + \sigma_1 \sigma_2 \sqrt{1 - \rho^2} Cov(Z_1, Z_2) \\
&= \sigma_1 \sigma_2 \rho \cdot 1 + \sigma_1 \sigma_2 \sqrt{1 - \rho^2} \cdot 0 \\
&= \sigma_1 \sigma_2 \rho.
\end{aligned}
$$

In summary, we conclude that

$$X = (X_1, X_2) \sim N(\mu_1, \mu_2, \sigma_1^2, \sigma_2^2, \sigma_1 \sigma_2 \rho),$$

as we wanted to show.

9.7 Exercises

Problem 9.7.1 *Prove the equivalence between the two equalities (9.1) and (9.2).*

Problem 9.7.2 *We know that if two random variables X_1 and X_2 are independent then their covariance must necessarily be zero (Theorem 7.4). We can state this as*

$$X_1, X_2 \text{ statistically independent} \;\Rightarrow\; Cov(X_1, X_2) = 0.$$

We also know that the opposite is not the case in general (see, e.g. Example 7.5). However, it turns out that if X_1 and X_2 have a bivariate normal distribution, then the converse is true. That is

$$X = (X_1, X_2) \sim N(\mu_1, \mu_2, \sigma_1^2, \sigma_2^2, 0) \;\Rightarrow\; X_1, X_2 \text{ statistically independent.} \qquad (9.4)$$

Prove (9.4). (Hint: Show that $f_{X_1, X_2}(x_1, x_2) = f_{X_1}(x_1) f_{X_2}(x_2)$ for all $x_1, x_2 \in \mathbb{R}$. You may, for simplicity, assume that $\mu_1 = \mu_2 = 0$ and $\sigma_1^2 = \sigma_2^2 = 1$.)

Problem 9.7.3 *Let $\mu_1 = 0, \mu_2 = 1, \sigma_1 = 1, \sigma_2 = 4, \rho = 0.5$, and $\sigma_{12} = \rho \sigma_1 \sigma_2$. Consider $X = (X_1, X_2)$ distributed according to the bivariate normal distribution*

$$X = (X_1, X_2) \sim N(\mu_1, \mu_2, \sigma_1^2, \sigma_2^2, \sigma_{12}).$$

1. *Use Theorem 9.6 to simulate $n = 1000$ independent draws from X in a programming language of your choice.*

2. *Plot the n simulated random draws in a scatter plot with the realized values of X_1 on the x-axis and the corresponding realized values of X_2 on the y-axis.*

3. *Repeat the two steps for $\rho = -0.5$ and $\rho = 0$. Comment on how the scatter plot changes as the value of the correlation parameter ρ changes.*

Problem 9.7.4 *Let $X \sim N(1, 2)$ and $Y \sim N(-2, 4)$ and assume that $Corr(X, Y) = 0.25$. Assume further that X and Y are jointly distributed according to the bivariate normal distribution, i.e. $(X, Y) \sim N(\mu_1, \mu_2, \sigma_1^2, \sigma_2^2, \sigma_{12})$.*

1. *Calculate the parameters $\mu_1, \mu_2, \sigma_1^2, \sigma_2^2$, and σ_{12}.*

2. *Define a new random variable $Q = -3X + 2Y$. State the distribution of Q.*

Problem 9.7.5 *Let $(X, Y) \sim N(2, 3, 4, 9, 3)$.*

1. *Calculate the correlation between X and Y.*

2. *Consider the conditional expectation function given by $E(Y|X = x) = \beta_0 + \beta_1 x$. Derive the values of β_0 and β_1. (Hint: Use Theorem 9.5.)*

3. *Use a programming language of your choice to simulate $n = 500$ independent draws from (X, Y). (Hint: You may use Theorem 9.6.) Plot the n simulated random draws in a scatter plot with the realized values of X on the x-axis and the corresponding realized values of Y on the y-axis. In the plot, depict also the line $\beta_0 + \beta_1 x$ for relevant values of x.*

Problem 9.7.6 *Let X and Q be two statistically independent random variables where $X \sim N(0, 1)$ and where Q can have the two outcomes -1 or 1 with $\Pr(Q = -1) = \Pr(Q = 1) = 1/2$. Let $Y = X \cdot Q$.*

1. *Show that $Y \sim N(0,1)$ by showing that the CDF $F_Y(y) = \Pr(Y \leq y)$ is equal to the CDF of a standard normal random variable $\Phi(y)$.*

   Hints: Apply the Law of Total Probability

   $$\Pr(Y \leq y) = \Pr(Y \leq y \,|\, Q = -1)\,\Pr(Q = -1) + \Pr(Y \leq y \,|\, Q = 1)\,\Pr(Q = 1),$$

   insert expression for Y, and use that X and Q are assumed statistically independent.

2. *Show that (Y, X) does not have a bivariate normal distribution.*

   Hints: Show that the distribution of the sum of X and Y is not normal. Then invoke Theorem 9.2, which says that any linear combination of two bivariate normal random variables is normal.

10

Distribution of a sample

10.1 Introduction

In statistical analyses, the main source of information about a population is going to be a sample of observations drawn from the population. For a sample to be useful, it must reflect the aspects of the population in which we are interested. For example, if we are interested in the fraction of a population with a certain characteristic, then the sample must reflect that fraction.

In this chapter, we consider how to derive the distribution of a sample based on the distribution of the population and knowledge of the sampling mechanism. Figure 1.2 on page 3 illustrates how they are related. Our main focus will be on a type of sample denoted a *simple random sample* (Definition 10.3). Even when the sampling mechanism is such that the sample obtained is non-simple, the techniques considered here are often still useful. This is illustrated in Section 10.4, where we discuss how data collected over time may lead to non-simple random samples, and in Section 10.5, where we give two further examples of how non-simple random samples might arise.

10.2 Random samples

Let X be a random variable, the probability distribution of which describes the distribution of a particular characteristic in the population under study. The distribution of the random variable X is represented by, e.g. a CDF $F_X(x)$ or, when appropriate, a PMF or a PDF $f_X(x)$. Our goal is to infer certain statistical properties of the population or, equivalently, of the distribution of X. For this purpose, we assume that we have n observations from the population, $\widetilde{X}_1, \ldots, \widetilde{X}_n$, called a *sample*. We use a tilde on top to distinguish an observation from the population (represented by $\widetilde{X}_i$) from the random variable describing the distribution of the characteristic in the population (represented by X).

Example 10.1 (Coin tosses) *Suppose you are interested in analyzing the statistical properties of a particular coin, which may be bend such that the probability of heads and tails are different. Let X be the random variable describing the distribution of heads (coded as $X = 1$) and tails (coded as $X = 0$) of this coin when it is thrown in a perfectly "fair" way (e.g. it rotates many times when thrown). Assume that the probability of heads is given by the parameter $p \in [0, 1]$. This implies that $X \sim Ber(p)$ and thus $f_X(1) = p$ and $f_X(0) = 1 - p$. In this case, if the parameter p in the probability distribution of the random variable X is known, then the population distribution would be known as well.*

Now, consider actually throwing the coin $n = 100$ times and recording the outcomes in the observations $\widetilde{X}_1, \ldots, \widetilde{X}_{100}$. These outcomes could, e.g. be used to infer the value of the parameter p, i.e. the probability that the coin lands on heads when thrown.

DOI: 10.1201/9781003591191-10

Example 10.2 (Political polling) *Suppose we are interested in analyzing the political zeitgeist in a given country. The country has a multi-party system with 10 political parties. We are interested in the proportion of potential voters that would vote for the respective parties in case there is an election due the next day. Let X be the random variable describing the party that a completely randomly selected person from the population would vote for, in case there was an election the next day ($X = 1$ denotes that the person would vote for the first party, $X = 2$ denotes that the person would vote for the second party, etc.; $X = 11$ denotes that the person will not vote). Then the probability distribution of X describes the distribution of the voter preferences in the population, e.g. $f_X(1) = p_1 \in [0,1]$ where p_1 is the proportion of the population that would vote for the first party. In this case, if the parameters $p_1, \ldots, p_{11}$ in the probability distribution of the random variable X are known, then the population distribution would be known as well.*

Consider asking $n = 1000$ people the question "Which party would you vote for if there was an election tomorrow?" and recording the outcomes in the observations $\widetilde{X}_1, \ldots, \widetilde{X}_{1000}$. These outcomes are going to be the basis for inferring the values of the parameters $p_1, \ldots, p_{11}$, i.e. the current political landscape of the population.

A sample $\widetilde{X}_1, \ldots, \widetilde{X}_n$ may be obtained in various different manners, and the way a sample is obtained influences the subsequent analysis. According to the statistical experiment, the procedure by which a sample is drawn from the population is the sampling mechanism. If the sample is obtained by means of some kind of stochastic mechanism, i.e. if there is an element of chance when deciding which elements in the population is drawn, then we call the sample a *random sample*. It is the involvement of chance that leads to the use of random variables to represent a random sample. The definition is next.

Definition 10.1 (Random sample) *A **random sample** of n observations consists of n random draws from the population, represented by n random variables, written as*

$$(\widetilde{X}_1, \widetilde{X}_2, \ldots, \widetilde{X}_n),$$

where $\widetilde{X}_i$ denotes the ith draw from the population.

A realized sample is n outcomes $\widetilde{x}_1, \ldots, \widetilde{x}_n$ of the random variables $\widetilde{X}_1, \ldots, \widetilde{X}_n$. This is written

$$(\widetilde{X}_1, \widetilde{X}_2, \ldots, \widetilde{X}_n) = (\widetilde{x}_1, \widetilde{x}_2, \ldots, \widetilde{x}_n),$$

or, in short,

$$(\widetilde{x}_1, \widetilde{x}_2, \ldots, \widetilde{x}_n).$$

Example 10.3 (Example 10.2, continued) *In the example with 10 parties, let the random variable $\widetilde{X}_1$ represent the possible outcomes when drawing the first observation by the sampling mechanism available and let $F_{\widetilde{X}_1}()$ be the CDF of the distribution of those possible outcomes. Suppose the first observation is party number 6. Then the realized observation is $\widetilde{x}_1 = 6$. Notice, $\widetilde{X}_1$ is a random variable and $\widetilde{x}_1$ is a number.*

A random sample has a distribution called the *sample distribution*. It is an n-dimensional joint distribution of the n random variables $\widetilde{X}_1, \ldots, \widetilde{X}_n$. It is defined next.

Definition 10.2 (Sample distribution) *The distribution of a sample $(\widetilde{X}_1, \ldots, \widetilde{X}_n)$ is called a **sample distribution**. It can be represented by a joint CDF $F()$, denoted by*

$$F_{\widetilde{X}_1, \ldots, \widetilde{X}_n}(x_1, \ldots, x_n).$$

If all $\widetilde{X}_1, \ldots, \widetilde{X}_n$ are either discrete with joint PMF $f()$, or continuous with joint PDF $f()$, then the sample distribution can also be represented by

$$f_{\widetilde{X}_1, \ldots, \widetilde{X}_n}(x_1, \ldots, x_n).$$

As discussed above, it is the goal of a statistical analysis to use the random sample $\widetilde{X}_1, \ldots, \widetilde{X}_n$ to infer something about the underlying population, represented by the random variable X. To this end, it is important that the random sample is in some way *representative* of the underlying population, in the sense that there should be some connection between observations $\widetilde{X}_i$ and the random variable X. An important instance of this, which we will often use throughout the book, is when the sampling mechanism used for sampling the observations $\widetilde{X}_i$ results in the distribution of $\widetilde{X}_i$ being the same as the distribution of X, i.e. such that $f_{\widetilde{X}_i}(x) = f_X(x)$ for all x.

Example 10.4 (Example 10.1, continued) *Consider drawing the sample $\widetilde{X}_1, \ldots, \widetilde{X}_{100}$ of $n = 100$ coin tosses. One way of obtaining this sample is to use the sampling mechanism that simply places the coin with heads face up. In this case, $f_{\widetilde{X}_i}(1) = 1$ and $f_{\widetilde{X}_i}(0) = 0$ for all $i = 1, \ldots, 100$. Obviously, unless $p = 1$, this sample cannot be used to infer anything about the underlying population represented by X. In particular, the sample cannot be used to infer anything about the probability that the coin, were it to be thrown in a fair way, lands on heads. That is, with this sampling mechanism, the sample is not representative of the underlying population.*

Consider instead the sampling mechanism where we do our best to throw the coin in a manner such that it rotates many times. In this case, it is not unreasonable to assume that the resulting observations $\widetilde{X}_1, \ldots, \widetilde{X}_{100}$ are representing the underlying population accurately, i.e. that $f_{\widetilde{X}_i}(1) = p$ and $f_{\widetilde{X}_i}(0) = 1 - p$ for all $i = 1, \ldots, 100$. In other words, in this case, we are justified in assuming that $f_{\widetilde{X}_i}(x) = f_X(x)$ for all x and all $i = 1, \ldots, n$.

Example 10.5 (Example 10.2, continued) *Recall that the sample $\widetilde{X}_1, \ldots, \widetilde{X}_{1000}$ was obtained by asking $n = 1000$ people the question "Which party would you vote for if there was an election tomorrow?" and recording the outcomes. How to actually do this in practice such that the sample is representative of the underlying population is an extremely challenging task that polling institutes grabble with regularly.*

We could, for instance, call potential voters on the phone and ask the question "Which party would you vote for if there was an election tomorrow?". Alternatively, we could send out e-mails asking the same question or go out into the street and randomly ask by passers the question. Or a combination of these sampling mechanisms. However, each strategy comes with pitfalls. For instance, it might only be a particular demographic that is able to answer the phone during working hours. Likewise, it might only be a particular demographic that regularly checks their e-mails or that strolls around on the street. It is not clear that these demographics are representative of the entire population. Another problem is that certain political parties may be stigmatized in a way that would make a polled person, who would actually vote for such a party, hesitant to divulge this information when asked. Hence, any sampling mechanism using these approaches run the risk of being unrepresentative of the underlying population. In statistical terms, this would imply that $f_{\widetilde{X}_i}(x) \neq f_X(x)$ for some x and i. Polling institutes are acutely aware of this problem and may thus spend considerable effort in ensuring that their random samples are as representative as possible.

The following sections discuss various assumptions on random samples that ensure that they are representative. The assumptions will also specify the dependence structure between observations, which will influence subsequent statistical analyses using the sample. In empirical analyses, it is important to always keep in mind how the available sample is obtained and whether the sample is actually representative of the underlying population. This could, for instance, be done by asking whether the actual sampling mechanism used in practice may plausibly lead to a random sample satisfying the assumptions presented in the coming sections. If this is not the case, the subsequent statistical analysis may not be able to describe anything meaningful about the underlying population. Unless this is kept in mind, the statistician runs the risk of making faulty assessments, which could in turn result in making faulty decisions.

10.3 Simple random sample

A *simple random sample* is a sample where the observations are mutual statistically independent with marginal distribution equal to the population distribution. It is formally defined next.

Definition 10.3 (Simple random sample) *The random sample* $(\widetilde{X}_1, \widetilde{X}_2, \ldots, \widetilde{X}_n)$ *of the population characteristic X is called a **simple random sample**, if and only if the following two conditions hold.*

1. $\widetilde{X}_1, \widetilde{X}_2, \ldots, \widetilde{X}_n$ *are mutual statistically independent.*

2. *All $\widetilde{X}_i$ have marginal distributions identical to the distribution of the characteristic X in the population, i.e.*

$$F_{\widetilde{X}_i}(x) = F_X(x), \quad i = 1, \ldots, n \ \ \text{and for all } x,$$

   where F denotes CDFs.

Condition 2 of Definition 10.3 implies that the marginal distributions of $\widetilde{X}_1, \widetilde{X}_2, \ldots, \widetilde{X}_n$ are identical, i.e.

$$F_{\widetilde{X}_1}(x) = F_{\widetilde{X}_2}(x) = \ldots = F_{\widetilde{X}_n}(x),$$

for all x. Since the observations in a simple random sample are mutual statistically independent and identically distributed, we sometimes say that the members of the sample $\widetilde{X}_1, \widetilde{X}_2, \ldots, \widetilde{X}_n$ are *iid* ("iid" standing for "independent and identically distributed").

The sample distribution can be derived for a simple random sample if we assume a distribution of the population. This is due to the mutual statistical independence and the identical distribution (iid) assumptions. The result is stated in the next theorem.

Theorem 10.1 (Sample distribution with simple random sample) *Assume* $(\widetilde{X}_1, \widetilde{X}_2, \ldots, \widetilde{X}_n)$ *is a simple random sample of the population characteristic X. Then the CDF $F_{\widetilde{X}_1,\ldots,\widetilde{X}_n}$ of the sample $(\widetilde{X}_1, \ldots, \widetilde{X}_n)$ is*

$$F_{\widetilde{X}_1,\ldots,\widetilde{X}_n}(x_1, \ldots, x_n) = \prod_{i=1}^{n} F_X(x_i),$$

where F_X is the CDF of X.

If the population characteristic X is either a discrete random variable with PMF $f()$, or a continuous random variable with PDF $f()$, then the sample distribution can also be represented as

$$f_{\widetilde{X}_1,\ldots,\widetilde{X}_n}(x_1, \ldots, x_n) = \prod_{i=1}^{n} f_X(x_i),$$

where $f_{\widetilde{X}_1,\ldots,\widetilde{X}_n}$ is a joint PMF or PDF according to the random variable type of X.

Later, we are interested in estimating parameters, say θ, of the probability distribution $F_X(x)$ of a population random variable X. As we will see in the coming chapters, it is vital that the sample distribution $F_{\widetilde{X}_1,\ldots,\widetilde{X}_n}(x_1,\ldots,x_n)$ also depends on those parameters, otherwise the parameters cannot be estimated from the sample. If the random sample $\widetilde{X}_1,\ldots,\widetilde{X}_n$ is a simple random sample, then the second property of Definition 10.3 ensures that the sample distribution depends on the same parameters as the distribution of the population characteristic. The next example illustrates this implication.

Example 10.6 *Consider a population consisting of 10 individuals of which 8 is "for" a proposal and 2 are "against". Let this population be represented by the random variable X, i.e. X takes the value 1 if a randomly chosen individual is "for" and 0 if the individual is "against". Then X is Bernoulli distributed $X \sim Ber(p)$ with parameter value $p = 8/10$. The PMF of X has the form*

$$f_X(x) = \begin{cases} 1-p & \text{if } x = 0, \\ p & \text{if } x = 1. \end{cases} \tag{10.1}$$

For notational convenience, the "0 otherwise" part is understood and hence left out. To ease the derivations to come, we rewrite the PMF (10.1) as

$$f_X(x) = (1-p)^{1-x} \cdot p^x, \quad x = 0, 1,$$

which holds because $p^0 = 1$ and $p^1 = p$.

Suppose a simple random sample with $n = 3$ observations is drawn. Then, by Theorem 10.1, the joint PMF of the sample is

$$\begin{aligned} f_{\widetilde{X}_1,\widetilde{X}_2,\widetilde{X}_3}(\widetilde{x}_1,\widetilde{x}_2,\widetilde{x}_3) &= \left((1-p)^{1-\widetilde{x}_1} \cdot p^{\widetilde{x}_1}\right) \cdot \left((1-p)^{1-\widetilde{x}_2} \cdot p^{\widetilde{x}_2}\right) \cdot \left((1-p)^{1-\widetilde{x}_3} \cdot p^{\widetilde{x}_3}\right) \\ &= (1-p)^{n-(\widetilde{x}_1+\widetilde{x}_2+\widetilde{x}_3)} \cdot p^{(\widetilde{x}_1+\widetilde{x}_2+\widetilde{x}_3)}. \end{aligned}$$

It is seen that the sample distribution depends on the same parameter p as the population distribution does.

In Example 10.6, the sampling mechanism has to be in such a manner that the same element in the population can be drawn more than once. If that is not the case, then the observations cannot be mutual statistically independent, and the sample is therefore not a simple random sample. This can be seen if, say, the first two draws are the two "against". If these two elements cannot be drawn again, then the probability of a "for" in the third draw is 1, since "for" are the only elements left from which to draw. A sampling mechanism where an element can be drawn repeatedly is called sampling *with replacement*. The opposite is called sampling *without replacement* and leads to a non-simple random sample. We return to sampling without replacement in Example 10.14 below. First, we discuss a different sampling mechanism, which might also lead to a non-simple random sample, namely when the observations in the sample are collected over time.

10.4   Time series

In this section, we consider a sample, the observations of which are drawn over time, and where the time perspective influences the properties of the sample. Such a sample is called a *time series sample* or, simply, a *time series*. To emphasize the time perspective, we let the symbol T denote the number of observations in a time series, instead of the symbol n used for a generic random sample above. Thus, when discussing time series, we consider random samples of size T, denoted $(\widetilde{X}_1, \widetilde{X}_2, \ldots, \widetilde{X}_T)$. The observations are ordered according to time, such that $\widetilde{X}_1$ is the observation made in the first time period, $\widetilde{X}_2$ is the observation made in the second time period, and so on. Instead of labelling the observations in a time series from 1 to T, we may also use calendar time as labels, e.g. $\widetilde{X}_{2020}$ could refer to an observation obtained for the year 2020, $\widetilde{X}_{2021}$ to the observation obtained for the year 2021, and so on. Time series are commonly encountered in social and economic studies. Typically, a time series describes sequential observations of one or more characteristics on the same unit, where a unit can be, e.g. an individual, a firm, or a country. We give two simple examples to illustrate the concepts.

Example 10.7 (Blood pressure data as time series) *A person measures their blood pressure every morning over the course of a year. Here, the unit is the person and the observed characteristic is the blood pressure of that person. Let $\widetilde{X}_1$ be the random variable denoting the blood pressure of the person on the first morning, $\widetilde{X}_2$ be the random variable denoting the blood pressure of the person on the second morning, and so on. In this case, we would have $T = 365$ days and $\widetilde{X}_{365}$ would be the random variable denoting the blood pressure of the person on the last morning of the sample.*

Example 10.8 (Stock price data as time series) *Consider the price of a financial stock, e.g. the Apple Inc. stock, over the course of a trading day, where a trading day is usually defined as the hours when a stock exchange is open. For instance, the New York Stock Exchange is typically open from 9:30 to 16:00. Here, the unit would be the financial stock and the observed characteristic the price of the stock. If we observe the price every minute, say, then $\widetilde{X}_1$ would be the random variable denoting the price of the Apple Inc. stock one minute after the opening of the stock exchange, $\widetilde{X}_2$ would be the random variable denoting the price two minutes after the opening, and so on.*

An important characteristic of a time series is that the random variables in the sample $\widetilde{X}_1, \widetilde{X}_2, \ldots, \widetilde{X}_T$ tend *not* to be statistically independent. For this reason, a time series is seldom a simple random sample, since it, in general, violates the first condition of Definition 10.3. Further, due to the nature of the data, where sample points arrive one after the other, the dependence often has a special structure, where the outcome of, say, $\widetilde{X}_t$ depends on the previous values, i.e. $\widetilde{X}_{t-1}, \widetilde{X}_{t-2}, \ldots$. For instance, we expect that the blood pressure of an individual on a given day t will be related to the blood pressure on the previous day, $t-1$, e.g. because the person might be going through a stressful, or tranquil, period of time. Similarly, we might expect that the stock price of Apple Inc. at a given time t is closely related to the price a minute earlier at $t-1$. Such dependence is typical of time series, and it is denoted *serial dependence*.

To interpret uncertainty in a context with a time series, we extend the understanding of an element in the population. This turns out to be useful in order to focus on the dependence between observations in a time series. This extension implies that we will interpret an element in a time series population as a *path*. For example, whereas we so far have interpreted an element in a population as a particular characteristic (e.g. income), which can be represented by the random variable X, an element in a time series population is a collection of characteristics over time (e.g. the trajectory of a stock price over a day), which can be represented by the path $(X_1, \ldots, X_T)$.

A path $(X_1, \ldots, X_T)$ representing a time series population can be expressed by a single symbol $X_{1:T}$, where $X_{1:T} = (X_1, \ldots, X_T)$. A realization of $X_{1:T}$ is a path and, thus, $X_{1:T}$ represents the distribution of all possible paths in the time series population. Notice, we use the word "path" to convey that the ordering of the random variables in $X_{1:T} = (X_1, \ldots, X_T)$

is important. In the following sections, we consider how to specify the distribution $F_{X_1,\ldots,X_T}$ of a time series population and, in particular, how this can be done with the dependence between $X_1, X_2, \ldots, X_T$, in focus.

10.4.1   Time series population modeled by conditional distributions

One approach to specifying a time series population is to use conditional distributions. Let the random variable X_1 have a distribution represented by the *CDF* $F_{X_1}(x_1)$, the random variable X_2 have a distribution represented by a distribution conditional on X_1, namely the conditional *CDF* $F_{X_2|X_1}(x_2|x_1)$, and the third random variable X_3 have a distribution represented by a distribution conditional on X_2 and X_1, namely $F_{X_3|X_2,X_1}(x_3|x_2,x_1)$. The random variable X_t has a distribution represented by a distribution conditional on $X_{t-1}, \ldots, X_1$, namely $F_{X_t|X_{t-1},\ldots,X_2,X_1}(x_t|x_{t-1},\ldots,x_1)$, for $t = 2, 3, \ldots, T$. From the conditional distributions, we can derive the joint distribution of the population path $(X_1, \ldots, X_T)$ using the chain rule for probabilities (Theorem 4.1). When employing the chain rule for probabilities, we order the random variables such that the conditioning variables on X_t refer back in time. This is motivated by the assumption that the present may be a result of the past, but not the other way round. This implies the following application of the chain rule for probabilities

$$F_{X_1,\ldots,X_T}(x_1,\ldots,x_T) = F_{X_1}(x_1) \prod_{t=2}^{T} F_{X_t|X_{t-1},X_{t-2},\ldots,X_1}(x_t|x_{t-1},x_{t-2},\ldots,x_1).$$

If the X_t's consists of either discrete random variables with PMFs $f()$, or continuous random variables with PDFs $f()$, then the distribution of the population path can also be represented as

$$f_{X_1,\ldots,X_T}(x_1,\ldots,x_T) = f_{X_1}(x_1) \prod_{t=2}^{T} f_{X_t|X_{t-1},X_{t-2},\ldots,X_1}(x_t|x_{t-1},x_{t-2},\ldots,x_1). \qquad (10.2)$$

The next example illustrates.

Example 10.9 (Time series population) *In this example, we model a stock price in $T = 2$ time periods. Let the possible stock prices at time t be represented by the random variables X_t, for $t = 1, 2$. For simplicity of presentation, assume that*

$$X_1 \sim N(100, 1).$$

That is, the stock price at $t = 1$ is drawn from a normal distribution with mean 100 (USD, say) and variance 1. Next, we model the stock price at time $t = 2$, X_2, conditional on the value at time $t = 1$, X_1. Assume that

$$X_2 | (X_1 = x_1) \sim N(x_1, 1).$$

This specification implies that if the stock price is high in the first period, then it is more likely it is high in the second period. We can now use the distribution of the initial value, $X_1 \sim N(100, 1)$, as well as the distribution of the second value, conditional on the first, $X_2|X_1 = x_1 \sim N(x_1, 1)$ to derive the joint distribution of X_1 and X_2. Using the properties of the normal distribution (Definition 8.8), we have

$$f_{X_1}(x_1) \;=\; \frac{1}{\sqrt{2\pi}} e^{-\frac{1}{2}(x_1 - 100)^2},$$

$$f_{X_2|X_1}(x_2|x_1) \;=\; \frac{1}{\sqrt{2\pi}} e^{-\frac{1}{2}(x_2 - x_1)^2}.$$

From (10.2), we deduce that the joint PDF of the population paths (X_1, X_2) is

$$f_{X_1,X_2}(x_1,x_2) = f_{X_2|X_1}(x_2|x_1) \cdot f_{X_1}(x_1) = \frac{1}{2\pi}e^{-\frac{1}{2}(x_2-x_1)^2}e^{-\frac{1}{2}(x_1-100)^2}.$$

10.4.2 Time series population modeled as a process

The collection of $X_1, \ldots, X_T$ can be viewed as a process, where each random variable X_t is a function of other random variables, in particular, past random variables $X_{t-1}, X_{t-2}, \ldots$. Viewing $X_1, \ldots, X_T$ as a process often makes modeling easier in practice. After modeling the process, the distribution of the time series population may be derived. This approach is different to modeling with conditional distributions, as for instance done in Example 10.9, where we directly specified the conditional distribution $f_{X_2|X_1}$ to arrive at a population distribution of the path (X_1, X_2).

A process in X_t can often be expressed with the aid of other random variables possessing certain regularities, e.g. they could be iid. Such conditions will allow us to use similar statistical methods to analyze a time series as we do with a simple random sample. We will see an example of this in Chapter 17 of this book, when discussing how to estimate population parameters using a time series sample, and several more examples in Volume II. The next example illustrates how the distribution of the time series population in Example 10.9 can alternatively be modeled by a process for X_2 as a function of X_1.

Example 10.10 (Example 10.9, continued) *Let again,*

$$X_1 \sim N(100, 1).$$

Instead of specifying the conditional distribution of X_2 given X_1 directly, we assume that the stock price at time $t = 2$ is equal to the price at time $t = 1$ plus a random term, i.e.

$$X_2 = X_1 + \varepsilon_2, \tag{10.3}$$

where ε_2 is a random variable, independent of X_1. In this model, the randomness in the stock price from time $t = 1$ to time $t = 2$ comes from ε_2. As a model of this randomness, assume

$$\varepsilon_2 \sim N(0, 1).$$

To see that this results in the same time series population distribution as in Example 10.9, note that the conditional density $f_{X_2|X_1}(x_2|x_1)$ can be calculated from (10.3), since, conditional on $X_1 = x_1$,

$$X_2 = x_1 + \varepsilon_2 \sim N(x_1, 1),$$

where only ε_2 is random, independent of X_1, and standard normally distributed. Thus, the conditional PDF is

$$f_{X_2|X_1}(x_2|x_1) = \frac{1}{\sqrt{2\pi}}e^{-\frac{1}{2}(x_2-x_1)^2},$$

showing that the distribution of (X_1, X_2), as derived here using (10.3), is the same as the distribution of (X_1, X_2), derived by specifying the conditional distribution directly in Example 10.9.

Specification of the paths by a process (Example 10.10) is often easier and more interpretable than the specification of conditional distribution functions (Example 10.9). For instance, in Example 10.10, ε_2 has an intuitive interpretation: It captures the effect of changes of circumstances between time $t = 1$ and time $t = 2$. If nothing of relevance happens for the stock between $t = 1$ and time $t = 2$, then $\varepsilon_2 = 0$ and the price X_2 is the same as X_1. If something positive happens to the stock between time $t = 1$ and time $t = 2$, then $\varepsilon_2 > 0$ and the price of the stock at time $t = 2$ is larger than the price at time $t = 1$, and vice versa.

Another advantage of specifying paths by a process is that the serial dependence, i.e. the dependence of X_t on its past, becomes clear. The next example illustrates.

Example 10.11 (Serial dependence in time series) *From Example 10.10, the process of the paths*

$$X_2 = X_1 + \varepsilon_2$$

implies correlation over time, i.e. serial dependence. To see this, insert $X_1 + \epsilon_2$ for X_2 in the expression for the correlation between X_1 and X_2, and use that X_1 and ϵ_2 are statistically independent. This gives

$$
\begin{aligned}
Corr(X_1, X_2) &= \frac{Cov(X_1, X_2)}{\sqrt{Var(X_1)Var(X_2)}} \\
&= \frac{Cov(X_1, X_1 + \varepsilon_2)}{\sqrt{Var(X_1)Var(X_1 + \varepsilon_2)}} \\
&= \frac{Var(X_1)}{\sqrt{Var(X_1)(Var(X_1) + Var(\varepsilon_2))}} \\
&= \frac{1}{\sqrt{1 \cdot (1 + 1)}} = \frac{1}{\sqrt{2}} = 0.71.
\end{aligned}
$$

In Examples 10.9, 10.10, and 10.11, we set the sample size $T = 2$ for presentation purposes. In practice, the sample size T will often be much larger than 2. In these cases, we may model the process of the paths using a sequence of iid random variables. As mentioned above, the upshot of this is that many of the statistical methods available for simple random samples will also be available for time series, possibly in slightly modified form. The next example extends Example 10.10 to show how a sequence of iid random variables can be used to specify a model for the paths $(X_1, X_2, \ldots, X_T)$.

Example 10.12 (Example 10.10, extended) *We extend Example 10.10 to multiple periods. Let $T \geq 2$ and assume again that $X_1 \sim N(100, 1)$. Consider a sequence $(\varepsilon_2, \varepsilon_3, \ldots, \varepsilon_T)$ of iid random variables, also independent of X_1, where $\varepsilon_t \sim N(0, 1)$ for all t. Define*

$$X_t = X_{t-1} + \varepsilon_t, \quad t = 2, 3, \ldots, T.$$

As is the case in Example 10.10, this specification completely determines the conditional distribution of X_t given its past, and, hence, the population distribution of $(X_1, X_2, \ldots, X_T)$. This is a so-called random walk model, *named so because the observations X_t will tend to meander around randomly according to the iid random variables $(\varepsilon_2, \varepsilon_3, \ldots, \varepsilon_T)$.*

10.4.3 Time series sample

To obtain a time series sample $(\widetilde{X}_1, \widetilde{X}_2, \ldots, \widetilde{X}_T)$ with the same distribution as the population paths $(X_1, X_2, \ldots, X_T)$, we assume that the random variables comprising the time series has been drawn according to the (conditional) distributions as specified by the process of the paths $(X_1, X_2, \ldots, X_T)$. That is, $\widetilde{X}_1$ is a random variable with distribution represented by $F_{X_1}(x_1)$, and $\widetilde{X}_2$ is a random variable with distribution represented by $F_{X_2|X_1}(x_2|\widetilde{x}_1)$, where $\widetilde{x}_1$ is the realization of $\widetilde{X}_1$. More generally, we assume that $\widetilde{X}_t$ has a distribution represented by $F_{X_t|X_{t-1},\ldots,X_1}(x_t|\widetilde{x}_{t-1}, \ldots, \widetilde{x}_1)$ for $t = 2, 3, \ldots, T$. Notice how the outcomes of the previous sample members $\widetilde{X}_{t-1}, \ldots, \widetilde{X}_1$ may influence the distribution of $\widetilde{X}_t$. This way of obtaining a time series sample can be contrasted to the simple random sample (Definition 10.3), where each observation $\widetilde{X}_i$ is simply sampled independently from a fixed population distribution $F_X(x)$. An important consequence of the time series sampling mechanism, where the random variable $\widetilde{X}_t$ is drawn conditional on the preceding observations $\widetilde{X}_{t-1}, \ldots, \widetilde{X}_1$, is that the time series sample $(\widetilde{X}_1, \widetilde{X}_2, \ldots, \widetilde{X}_T)$ inherits the time series properties (serial dependence) from the population paths $(X_1, X_2, \ldots, X_T)$.

Equivalently, we may think of the time series sample $(\widetilde{X}_1, \widetilde{X}_2, \ldots, \widetilde{X}_T)$ as arising from the same process as that describing the population paths $(X_1, X_2, \ldots, X_T)$. That is, the sampling mechanism for $\widetilde{X}_t$ is equivalent to the process describing the population path. The next example illustrates.

Example 10.13 (A random walk sample) *Consider the population* $(X_1, X_2, \ldots, X_T)$ *as implied by the setting in Example 10.12. Let $T \geq 2$ and assume that $\widetilde{X}_1 \sim N(100, 1)$. Consider an iid sequence $(\widetilde{\varepsilon}_2, \widetilde{\varepsilon}_3, \ldots, \widetilde{\varepsilon}_T)$, where $\widetilde{\varepsilon}_t \sim N(0, 1)$ for all t. Define the time series sample $(\widetilde{X}_1, \widetilde{X}_2, \ldots, \widetilde{X}_T)$ via*

$$\widetilde{X}_t = \widetilde{X}_{t-1} + \widetilde{\varepsilon}_t, \quad t = 2, 3, \ldots, T.$$

By construction, the time series sample $(\widetilde{X}_1, \widetilde{X}_2, \ldots, \widetilde{X}_T)$ will have the same distribution as the population time series $(X_1, X_2, \ldots, X_T)$.

10.5   Sampling methods used in practice

In practice, sampling of observations is done with a view toward the cost of sampling. This implies that random samples available typically are not simple random samples and the random sample may not be representative in terms for $F_{\widetilde{X}_i}(x) = F_X(x)$, where Fs are CDFs, X represents the population, and $\widetilde{X}_i$ represents the ith observation. This need not be a problem, in fact, it may be an advantage. The two necessary points for such random samples to be useful are that we know how the random sample is drawn and that the sample distribution depends on the parameters that we wish to infer.

In this book we often assume a simple random sample is available. We do so because the various concepts may easier be understood in the context of a simple random sample. Further, many of the sampling methods used in practice build on simple random samples and, therefore, understanding the properties of the methods in the context of simple random samples is essential. In this section, we show how a commonly used sampling method, *stratification*, builds on simple random samples, and how another often used method, *sampling without replacement*, results in a sample with approximately similar properties as a simple random sample, as long as the sample size is not too large compared to the population.

10.5.1   Sampling without replacement

In practice, we often avoid drawing the same element more than once. For instance, if we were conducting a poll for an upcoming election, we would make sure not to poll the same person twice. In the context of random samples, this can be modeled by specifying that if an element is drawn from the population, then that element is not returned to the population for potential subsequent drawings. Such a sampling scheme is called sampling *without replacement*. If, on the other hand, an element can be drawn again after having been drawn once, we call it sampling *with replacement*. The next example is similar to Example 10.6, except the sampling is without replacement. This implies that the observations are identically distributed but not mutual statistically independent. That is, although the random sample satisfies the second property of Definition 10.3, it does not satisfy the first property, and, hence, is not a simple random sample.

Example 10.14 *Consider a population consisting of 10 individuals of which 8 are "for" a proposal and 2 are "against". Suppose a random sample of 3 observations are drawn from this population. Assume that once an individual has been drawn, the same individual cannot be drawn (asked) again. That is, the sampling is without replacement. Before each draw, assume all individuals not previously drawn have the same probability of being drawn.*

To derive the sample distribution, we use the chain rule for probabilities (Theorem 4.1), which implies that the joint PMF of the sample can be expressed as

$$f_{\widetilde{X}_1,\widetilde{X}_2,\widetilde{X}_3}(\widetilde{x}_1,\widetilde{x}_2,\widetilde{x}_3) = f_{\widetilde{X}_3 \mid \widetilde{X}_1,\widetilde{X}_2}(\widetilde{x}_3 \mid \widetilde{x}_1,\widetilde{x}_2) \cdot f_{\widetilde{X}_2 \mid \widetilde{X}_1}(\widetilde{x}_2 \mid \widetilde{x}_1) \cdot f_{\widetilde{X}_1}(\widetilde{x}_1). \qquad (10.4)$$

The PMF of the first observation is $\widetilde{X}_1 \sim Ber(8/10)$. Hence,

$$f_{\widetilde{X}_1}(\widetilde{x}_1) = \left(1 - \frac{8}{10}\right)^{1-\widetilde{x}_1} \cdot \left(\frac{8}{10}\right)^{\widetilde{x}_1}.$$

The conditional PMF of the second observation given the first observation is

$$\widetilde{X}_2 \mid \widetilde{X}_1 = \widetilde{x}_1 \sim Ber\left(\frac{8 - \widetilde{x}_1}{9}\right),$$

since there are 9 elements left from which to draw, and amongst those, $8 - \widetilde{x}_1$ are "for". Hence,

$$f_{\widetilde{X}_2 \mid \widetilde{X}_1}(\widetilde{x}_2 \mid \widetilde{x}_1) = \left(1 - \frac{8 - \widetilde{x}_1}{9}\right)^{1-\widetilde{x}_2} \cdot \left(\frac{8 - \widetilde{x}_1}{9}\right)^{\widetilde{x}_2}.$$

It is seen that $f_{\widetilde{X}_2 \mid \widetilde{X}_1}(\widetilde{x}_2 \mid \widetilde{x}_1)$ depends on $\widetilde{x}_1$ and, thus, observations 1 and 2 are statistically dependent.

Similarly, the conditional PMF of the third observation given the first two observations is

$$\widetilde{X}_3 \mid \left(\widetilde{X}_1 = \widetilde{x}_1, \widetilde{X}_2 = \widetilde{x}_2\right) \sim Ber\left(\frac{8 - (\widetilde{x}_1 + \widetilde{x}_2)}{8}\right).$$

The reason is that there are $10 - 2 = 8$ elements left from which to draw observation 3, and amongst those, $8 - (\widetilde{x}_1 + \widetilde{x}_2)$ are "for". Hence,

$$f_{\widetilde{X}_3 \mid \widetilde{X}_1,\widetilde{X}_2}(\widetilde{x}_3 \mid \widetilde{x}_1,\widetilde{x}_2) = \left(1 - \frac{8 - (\widetilde{x}_1 + \widetilde{x}_2)}{8}\right)^{1-\widetilde{x}_3} \cdot \left(\frac{8 - (\widetilde{x}_1 + \widetilde{x}_2)}{8}\right)^{\widetilde{x}_3}.$$

It is seen that observation 3 is not statistically independent of the first two observations.

The sample distribution $f_{\widetilde{X}_1,\widetilde{X}_2,\widetilde{X}_3}(\widetilde{x}_1,\widetilde{x}_2,\widetilde{x}_3)$ can be derived by inserting the three PMFs into (10.4). By doing so, it can be seen that this sample distribution differs from the sample distribution obtained in Example 10.6, where the sampling is with replacement and the sample is a simple random sample.

Despite the fact that the observations are statistically dependent, due to sampling without replacement, they all have the same marginal distribution. Consider observation 2. The marginal distribution of observation 2 can be derived as

$$
\begin{aligned}
f_{\widetilde{X}_2}(\widetilde{x}_2) &= \sum_{\widetilde{x}_1=0}^{1} f_{\widetilde{X}_1,\widetilde{X}_2}(\widetilde{x}_1,\widetilde{x}_2) \\
&= \sum_{\widetilde{x}_1=0}^{1} f_{\widetilde{X}_2 \mid \widetilde{X}_1}(\widetilde{x}_2 \mid \widetilde{x}_1) \cdot f_{\widetilde{X}_1}(\widetilde{x}_1) \\
&= \sum_{\widetilde{x}_1=0}^{1} \left(1 - \frac{8-\widetilde{x}_1}{9}\right)^{1-\widetilde{x}_2} \cdot \left(\frac{8-\widetilde{x}_1}{9}\right)^{\widetilde{x}_2} \cdot \left(1 - \frac{8}{10}\right)^{1-\widetilde{x}_1} \cdot \left(\frac{8}{10}\right)^{\widetilde{x}_1} \\
&= \left(\frac{1}{9}\right)^{1-\widetilde{x}_2} \cdot \left(\frac{8}{9}\right)^{\widetilde{x}_2} \cdot \left(\frac{2}{10}\right) + \left(\frac{2}{9}\right)^{1-\widetilde{x}_2} \cdot \left(\frac{7}{9}\right)^{\widetilde{x}_2} \cdot \left(\frac{8}{10}\right).
\end{aligned}
$$

If $\widetilde{x}_2 = 0$, *then*

$$
f_{\widetilde{X}_2}(0) = \left(\frac{1}{9}\right) \cdot \left(\frac{2}{10}\right) + \left(\frac{2}{9}\right) \cdot \left(\frac{8}{10}\right) = \frac{2+16}{9\cdot 10} = \frac{2}{10},
$$

and, thus

$$
f_{\widetilde{X}_2}(1) = 1 - f_{\widetilde{X}_2}(0) = \frac{8}{10}.
$$

We see that $f_{\widetilde{X}_2}(x) = f_{\widetilde{X}_1}(x) = f_X(x)$. *A similar result can be shown for* $f_{\widetilde{X}_3}(x)$. *Thus, this random sample has identically distributed observations.*

Although the most used sampling scheme in practice is arguably sampling without replacement, we are often justified in assuming that the sampling has, in fact, been done with replacement. Indeed, in the limit where the population is infinitely large, it does not matter if sampling is done with or without replacement, since the probability that the same element in the population is drawn twice is zero. When the population is not infinite but merely very large compared to the sample size, the probability that an element in the population is drawn more than once is still very small. The upshot is that the statistical methods based on either sampling with or without replacement give almost identical results. Since the statistical analyses are often simpler under the assumption that the sampling is done with replacement, e.g. because this can lead to a simple random sample, it is therefore commonplace to ignore the fact that the sampling is in fact done without replacement, as long as the sample size is small compared to the size of the population.

10.5.2 Stratified sampling

There are sampling methods that may lead to higher statistical quality using the resulting random sample compared to a simple random sample. One such widely used sampling method is *stratified sampling*. Stratified sampling is characterized by deliberately sampling from certain subgroups of the population, e.g. 40% from urban areas and 60 % from rural areas. From each of these subgroups, we draw a simple random sample. In case the distributions for each of the subpopulations are different, then the observations in the sample are not identically distributed. That is, although the sample turns out to satisfy the first property of Definition 10.3, it does not satisfy the second property, and, hence, it is not a simple random sample, despite the fact that it is made up of parts of simple random samples. The next example illustrates.

Example 10.15 *Consider a population consisting of 10 individuals of which 8 are "for" a proposal and 2 are "against". Assume that the individuals live at two different locations A and B with 2 "against" and 5 "for" in location A, and 0 "against" and 3 "for" in location B.*

Suppose we decide to draw a random sample of 3 observations from this population such that the first 2 observations are drawn from location A and the last observation is drawn from location B. Assume sampling with replacement.

The observations are mutual statistically independent. The reason is that the probability of getting e.g. "for" in the second draw is not influenced by outcomes of the other draws due to sampling with replacement.

The observations are not identically distributed. To see this, let "for" be coded as 1. Since it is decided that the first two observations are drawn from location A, the first two observations each has a marginal distribution according to the Bernoulli distribution $\text{Ber}(5/7)$, whereas the last observation, decided to be drawn from location B, has a marginal distribution according to a Bernoulli distribution $\text{Ber}(1)$.

The sampling mechanism described in Example 10.15 is an example of stratified sampling. It can be a cost efficient method to draw a sample. For example, suppose there are two subpopulations, where one contains similar units and the other contains very different units. Then it is cost efficient to take most of the observations from the subpopulation with the different units since this subpopulation is the hardest to learn, whereas just a few observations from the subpopulation with similar units is enough to learn about that subpopulation.

Example 10.16 (Cost efficiency of stratified sampling) *Consider the extreme case that the population consists of two equally large subpopulations, say A and B, where everyone within a subpopulation is the same. Suppose it cost 500 kr to draw and survey an element. If we do stratified sampling, then we can learn the whole population by drawing one individual from subpopulation A and one from subpopulation B. That is, for 1000 kr we have learned the whole population.*

Alternatively, if we draw randomly from the population, e.g. because we cannot identify who is in group A and B before the observation is drawn, then with probability 0.5 we need more than 2 observations before we have an individual from each subpopulation. That is, suppose the first draw is a person from subpopulation A. Then there is 0.5 probability that the second person also is from subpopulation A and, thus, there is $0.5 \cdot 0.5 = 0.5^2 = 0.25$ probability that all three persons are from the same subpopulation. If we decide that we want to be at least 90% sure to have a person from each subpopulation, then we need to draw 5 observations, because the probability of all five being from the same subpopulation is $0.5^4 = 0.0625$. The cost is $5 \cdot 500 = 2500$ kr. With stratified sampling, the cost is minimized, namely 1000 kr.

10.6   Proofs

Proof of Theorem 10.1 *We prove the results for CDFs $F()$. The results for PMFs/PDFs $f()$ follow using similar arguments.*

By the first property of Definition 10.3, a simple random sample means that the observations are mutual statistically independent. By the definition of mutual statistical independence (Definition 7.3), this implies

$$F_{\widetilde{X}_1,\ldots,\widetilde{X}_n}(x_1,\ldots,x_n) = F_{\widetilde{X}_1}(x_1) \cdot \ldots \cdot F_{\widetilde{X}_n}(x_n) = \prod_{i=1}^{n} F_{\widetilde{X}_i}(x_i).$$

By the second property of Definition 10.3, a simple random sample also means that the observations have the same distribution as the population distribution, i.e. $F_{\widetilde{X}_i}(x_i) = F_X(x_i)$ for all $i = 1, 2, \ldots, n$. Insert to get the CDF of the sample distribution

$$F_{\widetilde{X}_1,\ldots,\widetilde{X}_n}(x_1,\ldots,x_n) = \prod_{i=1}^{n} F_X(x_i).$$

10.7 Exercises

Problem 10.7.1 *Consider a population of waiting times until a customer arrives. Let X be the waiting time and assume X is exponentially distributed with parameter λ, that is, $X \sim Exp(\lambda)$.*

Let $(\widetilde{X}_1, \ldots, \widetilde{X}_n)$ be a simple random sample from this population. Derive the sample distribution, i.e. the joint PDF, for this simple random sample.

Hint: Recall that $\exp(a) \cdot \exp(b) = \exp(a + b)$.

Problem 10.7.2 *Consider a population of blood pressures. Let X be the blood pressure and assume X is normally distributed with mean μ and variance σ^2, that is, $X \sim N(\mu, \sigma^2)$.*

Let $(\widetilde{X}_1, \ldots, \widetilde{X}_n)$ be a simple random sample from this population. Derive the sample distribution, i.e. the joint PDF, for this simple random sample.

Hint: Recall that $\exp(a) \cdot \exp(b) = \exp(a + b)$.

Problem 10.7.3 *Let X be the attitude toward a new road, where $X = 0$ is "against" the road and $X = 1$ is "for" the road. Suppose the population is split in their attitude such that $X \sim Ber(0.5)$.*

Consider the following sampling mechanism. Randomly draw one individual from the population. This is the first observation. Assume each individual has a partner, and let that partner be the second observation. Suppose there is a probability of 0.8 that partners have the same attitude. For simplicity of calculation, assume everyone has a partner.

Assume a random sample of $n = 2$ observations.

1. *Calculate the probability that the second observation is "for" given the first observation is "against".*

2. *Calculate the probability that the second observation is "for".*

3. *Are the observations independent?*

4. *Are the observations identically distributed?*

5. *Is the random sample a simple random sample?*

Problem 10.7.4 *Consider the time series model*

$$X_t = \alpha + \beta X_{t-1} + \varepsilon_t, \quad t = 2, \ldots, T.$$

Assume $\varepsilon_t's$ are iid, the distribution of ε_t is $f_\varepsilon()$ and the distribution of X_1 is $f_{X_1}()$. Derive the time series sample distribution $f_{\widetilde{X}_1,\ldots,\widetilde{X}_T}(x_1,\ldots,x_T)$.

Problem 10.7.5 *Consider the random walk process*

$$X_t = X_{t-1} + \varepsilon_t.$$

Assume that ε_t is statistically independent of ε_s for $t \neq s$, and statistically independent of X_1. Also, assume $E(X_1) = 5$, $Var(X_1) = \sigma^2$, $E(\varepsilon_t) = 0$, and $Var(\varepsilon_t) = \sigma^2$.

1. *Derive $E(X_t)$. (Hint: write X_t as a function of X_1 and the ε's.)*

2. *Derive $Var(X_t)$. (Hint: write X_t as a function of X_1 and the ε's.)*

3. *Compare $E(X_t)$ derived in 1 with $E(X_t|X_{t-1} = x_{t-1})$ and interpret the difference.*

4. *Compare $Var(X_t)$ derived in 2 with $Var(X_t|X_{t-1} = x_{t-1})$ and interpret the difference.*

5. *Derive the distribution of X_t under the additional assumptions that $X_1 \sim N(5, \sigma^2)$ and $\varepsilon_t \sim N(0, \sigma^2)$, $t = 2, \ldots, T$*

6. *Compare the distribution of X_t in 5 with the distribution $X_t|(X_{t-1} = x_{t-1})$ of X_t given X_{t-1}.*

Problem 10.7.6 *Consider the random walk process*

$$X_t = X_{t-1} + \varepsilon_t.$$

Assume that ε_t is independent of ε_s for $t \neq s$, and independent of X_1. Show that ε_t is statistically independent of X_{t-1} and statistically dependent on X_{t+1}.

11

Estimation theory

11.1 Introduction

The idea of an *estimator* is to synthesize the information contained in a random sample into a guess, denoted an *estimate*, of a population quantity of interest. To assess the various methods of converting the sample into an estimate, we define a number of concepts that measure the statistical quality of an estimator. Some of these measures of quality rely on the properties of the estimator for a given sample size, whereas others rely on the properties of the estimator for very large sample sizes. We often use the latter measures of quality because it may be hard to derive properties of the estimator for a small sample size, whereas certain regularities dominate when the sample size is large. An important result relating to very large sample sizes is the so-called *Law of Large Numbers* (Theorem 11.4).

As we shall see, an estimator is a function of a random sample consisting of random variables, which implies that an estimator is itself a random variable. In particular, the estimator has a distribution. Consequently, the (statistical) properties of an estimator can be assessed using aspects of the distribution of the estimator. In general, the distribution of an estimator will depend on the specifics of the construction of the estimator and on the distribution of the random sample. Recall that the distribution of the sample is, in turn, a result of the distribution of the population and the sampling mechanism, as discussed in Chapter 10. In theory, if the distribution of the population and the sampling mechanism are known, then the distribution of the estimator can be derived. Such a derivation, however, is typically difficult to perform. Instead, the distribution of the estimator may be simulated using a computer. This is done by so-called *Monte Carlo simulation*. Monte Carlo simulation is described in Section 11.7 at the end of this chapter.

11.2 Estimand, estimator, and estimate

Any random variable constructed as a function of observations from a random sample is called a *sample statistics* and the distribution of the sample statistics is called the *sampling distribution* of the statistics.

Definition 11.1 (Sample statistics, sampling distribution) *A **sample statistics** T is a random variable, which is a function $h()$ of a random sample $(\widetilde{X}_1, \ldots, \widetilde{X}_n)$,*

$$T = h(\widetilde{X}_1, \ldots, \widetilde{X}_n).$$

*The distribution of a sample statistics is called a **sampling distribution**.*

DOI: 10.1201/9781003591191-11

Since some of the properties of a sample statistics depend on the sample size n, we will sometimes indicate this by attaching a subscript n to the sample statistics, i.e. we will write T_n.

When a sample statistics is used for assessing the value of a parameter, the sample statistics is called an estimator of that parameter. Here, a parameter can be the parameter of a specific distribution, e.g. the parameter p of the Bernoulli distribution (Section 8.2), or it can be an aspect of an unknown distribution, such as the mean of a population or an effect of one variable on another variable. To make it explicit when a distribution depends on a parameter, say, θ, we may add a semicolon followed by the parameters in the expression for a distribution function. That is, $F_X(x; \theta)$ will denote the CDF of the random variable X, where the CDF depends on the parameter θ. The following example illustrates this.

Example 11.1 *Let $p \in [0, 1]$ and consider a population represented by the random variable X, where $X \sim Ber(p)$. The CDF of X is (Definition 8.1)*

$$F_X(x) = \begin{cases} 0 & \text{if} & x < 0, \\ 1 - p & \text{if} & x \in [0, 1), \\ 1 & \text{if} & x \geq 1. \end{cases}$$

The CDF $F_X(x)$ of X depends on the parameter p. To make this explicit, we will often write $F_X(x; p)$ instead of $F_X(x)$.

Next, the main concepts related to estimation is defined.

Definition 11.2 (Estimand, estimator, estimate) *Consider a population distribution $F_X(x)$, and let the parameter of interest be θ, where θ is an aspect of the population distribution $F_X(x)$.*

*The parameter of interest θ is called the **estimand** and it is the quantity to be estimated. An **estimator** $\widehat{\theta}$ of the estimand θ is a sample statistics (Definition 11.1)*

$$\widehat{\theta} = h(\widetilde{X}_1, \ldots, \widetilde{X}_n),$$

where $\widetilde{X}_1, \ldots, \widetilde{X}_n$ is a random sample.
*An **estimate** $\widehat{b}$ is a realized value of the estimator $\widehat{\theta}$*

$$\widehat{b} = h(\widetilde{x}_1, \ldots, \widetilde{x}_n),$$

where $\widetilde{x}_1, \ldots, \widetilde{x}_n$ is a realization of the random sample $\widetilde{X}_1, \ldots, \widetilde{X}_n$.

The estimand θ can be any aspect of a distribution, such as its mean or variance. In many cases, the estimand will be a parameter describing the distribution $F(x; \theta)$.

Example 11.2 (Example 11.1, continued) *Consider the population represented by the random variable X, where $X \sim Ber(p)$. The CDF of X, $F(X; p)$, was given in Example 11.1. If the parameter p is unknown, we may wish to estimate it using a sample, $\widetilde{X}_1, \ldots, \widetilde{X}_n$, from X. In this case, p is the estimand. An estimator of p may be denoted as $\widehat{p}$, where $\widehat{p}$ is a function of $\widetilde{X}_1, \ldots, \widetilde{X}_n$.*

It is worth emphasizing that since an estimator is a sample statistics, i.e. it is a function of the random sample $(\widetilde{X}_1, \ldots, \widetilde{X}_n)$, it is a random variable, and it therefore has a distribution, e.g. represented by the CDF $F_{\widehat{\theta}}$. In contrast, an estimand and an estimate are numbers, not random variables, and hence they do not have distributions. The estimand is a fixed, but unknown, number, and the estimate is a realization of the estimator.

An example of an estimator is the sample average. We will see shortly that the sample average is an estimator of the mean of a distribution.

Definition 11.3 (Sample average) *Let $(\widetilde{X}_1, \ldots, \widetilde{X}_n)$ be a random sample. Then the* **sample average** $\overline{X}$ *is defined as*

$$\overline{X} = \frac{1}{n}\left(\widetilde{X}_1 + \ldots + \widetilde{X}_n\right) = \frac{1}{n}\sum_{i=1}^{n}\widetilde{X}_i. \tag{11.1}$$

The sample average is a random variable with a CDF $F_{\overline{X}}$. In contrast, the realized sample average is a number given by

$$\overline{x} = \frac{1}{n}\left(\widetilde{x}_1 + \widetilde{x}_2 + \ldots + \widetilde{x}_n\right) = \frac{1}{n}\sum_{i=1}^{n}\widetilde{x}_i,$$

where $(\widetilde{x}_1, \ldots, \widetilde{x}_n)$ is the realized random sample.

Example 11.3 (Sample average) *You decide that you want to measure the height of the next five people you meet. Before encountering anyone, you can represent the height of the next five people with $\widetilde{X}_1, \ldots, \widetilde{X}_5$, where $\widetilde{X}_1$ denotes the height of the first person, $\widetilde{X}_2$ the height of the second person, and so on. Since you do not know which five people you will encounter next, you can think of the heights $\widetilde{X}_1, \ldots, \widetilde{X}_5$ as random variables. You are particularly interested in the average height of the five people, i.e. you consider the sample average (Definition 11.3)*

$$\overline{X} = \frac{1}{5}\left(\widetilde{X}_1 + \ldots + \widetilde{X}_5\right) = \frac{1}{5}\sum_{i=1}^{5}\widetilde{X}_i.$$

Since the individual heights, $\widetilde{X}_i$, are random variables, then so is their average $\overline{X}$.

Now you go about measuring the height of the next five people you meet. Say that the first person is 169cm, the second is 181cm, the third 172cm, the fourth 192cm, and the fifth 165cm. These represent the realizations of the random variables $\widetilde{X}_1, \widetilde{X}_2, \ldots, \widetilde{X}_5$, i.e. $\widetilde{x}_1 = 169, \widetilde{x}_2 = 181, \ldots, \widetilde{x}_5 = 165$. The realized sample average is

$$\overline{x} = \frac{1}{5}\left(\widetilde{x}_1 + \widetilde{x}_2 + \ldots + \widetilde{x}_5\right) = \frac{1}{5}\sum_{i=1}^{5}\widetilde{x}_i = \frac{879}{5} = 175.80.$$

Note how $\overline{X}$ is a random variable while $\overline{x}$ is a realization of $\overline{X}$, namely, a number.

11.3 Distribution of an estimator

The quality of an estimation method can be assessed using the sampling distribution of the estimator. The sampling distribution of an estimator can be derived if the distribution of the population and the sampling mechanism are known. In this section, we will assume that these are known. In practice, however, we will not know the distribution of the estimator since this will, in general, depend on the estimand, which is unknown (if the estimand was known, there would be no reason to estimate it). Hence, the results of this section are *infeasible* in practice, in the sense that they rely on knowing the population distribution.

In Chapter 13, we use the insights obtained in this section, to show how we can estimate the sampling distribution of an estimator in a *feasible* way, i.e. without relying on knowing the population distribution.

In the following, we will present two examples of distributions of populations that we considered in Chapter 8. For the sampling mechanism, we assume that the mechanism leads to the random sample being a simple random sample (Definition 10.3).

Example 11.4 (Sampling distribution of sample average from a normal population) *Let $\mu_0 \in \mathbb{R}$ and $\sigma_0^2 > 0$ be numbers. Suppose the distribution of the population is $X \sim N\left(\mu_0, \sigma_0^2\right)$ and the random sample $\left(\tilde{X}_1, \ldots, \tilde{X}_n\right)$ is a simple random sample from this population. Assume the estimand is the population mean μ_0 and the estimator is the sample average*

$$\overline{X} = \frac{1}{n} \sum_{i=1}^{n} \tilde{X}_i.$$

Using the result on a sum of mutual statistically independent normal random variables given in (8.8), we have

$$Y = \sum_{i=1}^{n} \tilde{X}_i \sim N\left(n \cdot \mu_0, n\sigma_0^2\right).$$

Using the result that the distribution of a linear transformation of a normally distributed random variable is again normal (Theorem 8.9), we have

$$\overline{X} = \frac{1}{n} \sum_{i=1}^{n} \tilde{X}_i = \frac{1}{n} Y \sim N\left(\left(\frac{1}{n}\right)(n \cdot \mu_0), \left(\frac{1}{n}\right)^2 \left(n\sigma_0^2\right)\right) = N\left(\mu_0, \frac{1}{n}\sigma_0^2\right).$$

To sum up, in the case where the population distribution is normal, the sample average of a simple random sample $\overline{X}$ has itself a normal distribution, namely $\overline{X} \sim N\left(\mu_0, \frac{1}{n}\sigma_0^2\right)$. In other words, this is the sampling distribution of the estimator $\overline{X}$. Note that the sampling distribution will be unknown in practice because the parameters μ_0 and σ_0 are unknown in practice.

Example 11.5 (Sampling distribution of sample average from a Bernoulli population) *Let $p_0 \in [0, 1]$ be a number and suppose the distribution of the population is a Bernoulli distribution, $X \sim Ber(p_0)$, and the sample $\left(\tilde{X}_1, \ldots, \tilde{X}_n\right)$ is a simple random sample from this population. Suppose the estimand is the population parameter p_0 and the estimator is the sample average*

$$\overline{X} = \frac{1}{n} \sum_{i=1}^{n} \tilde{X}_i.$$

Recall that the sum of mutual statistically independent Bernoulli random variables is binomially distributed, i.e.

$$Y = \sum_{i=1}^{n} \tilde{X}_i \sim Bin(n, p_0).$$

This implies that the CDF of $\overline{X}$ can be evaluated using the CDF of the $Bin(n, p_0)$ distribution since,

$$
\begin{aligned}
F_{\overline{X}}(x) &= \Pr\left(\overline{X} \le x\right) \\
&= \Pr\left(\frac{1}{n}\sum_{i=1}^{n}\widetilde{X}_i \le x\right) \\
&= \Pr\left(\sum_{i=1}^{n}\widetilde{X}_i \le n \cdot x\right) \\
&= \Pr\left(Y \le n \cdot x\right) \\
&= F_Y(n \cdot x),
\end{aligned}
$$

where $F_Y(y)$ is the CDF of the Binomial distribution $Bin(n, p_0)$. Again we see that the sampling distribution of $\overline{X}$ depends on the unknown parameter p_0 and, therefore, it is unknown in practice.

11.4 Precision of an estimator

11.4.1 Loss and risk functions

Based on the sampling distribution, we can define various measures of the precision of an estimator. The idea of a precision measure is to evaluate how good the estimator $\widehat{\theta}$ is at estimating the estimand θ_0. We will often use the subscript 0 on the estimand and refer to it as the *true value* of the parameter θ. In order to quantify precision, we need to specify how to measure "good". We will specify "good" by the cost, or *loss*, associated with not getting an estimate equal to the estimand. Therefore, we next define a *loss function* as a function of the estimate and the estimand.

Definition 11.4 (Loss function of estimator) *Let b be an estimate of the estimand θ_0. Then the function $L(b, \theta_0)$ is a **loss function of an estimator** if and only if it satisfies the following two properties.*

1. *$L(b, \theta_0) \ge 0$ for all b and θ_0.*

2. *$L(\theta_0, \theta_0) = 0$.*

In exploring the precision of an estimator, we will consider the loss as a function of an estimator $\widehat{\theta}$ instead of an estimate, that is, we consider $L\left(\widehat{\theta}, \theta_0\right)$. Since the estimator $\widehat{\theta}$ is a random variable, $L\left(\widehat{\theta}, \theta_0\right)$ is likewise a random variable. The distribution of this random variable is the distribution of losses associated with using $\widehat{\theta}$ as an estimator of the estimand θ_0. In view of the thought experiment of repeating a statistical experiment infinitely many times, the distribution of losses is a result of how losses are distributed over all the repetitions.

Typically we are interested in assigning a single number, and not a random quantity, for the quality of an estimator. This can be achieved by considering the *expected loss* of the estimator $\widehat{\theta}$, i.e. the expectation of the loss function $L\left(\widehat{\theta}, \theta_0\right)$. The expected value of a loss function is called the *risk* of an estimator and it is defined next.

Definition 11.5 (Risk function of estimator) *Let $\widehat{\theta}$ be an estimator of the estimand θ_0. The* **risk function** $R\left(\widehat{\theta}, \theta_0\right)$ *of the estimator $\widehat{\theta}$ is defined as*

$$R\left(\widehat{\theta}, \theta_0\right) = E\left(L\left(\widehat{\theta}, \theta_0\right)\right),$$

where the expectation $E()$ is taken with respect to the distribution of the random variable $\widehat{\theta}$, which may depend on θ_0.

The risk of an estimator can be thought of as the average loss the estimator would incur if we were to repeat the statistical experiment infinitely many times.

In Chapter 5, we discussed that an expected value need not exists. Since the risk of an estimator is defined via an expected value, this also means that the risk function may not exist for certain loss functions and estimators.

In terms of a risk function, the closer the risk of an estimator is to 0, the better. In particular, we will say that the *precision* of an estimator is higher the closer the risk of the estimator is to 0. That is, other things being equal, we will prefer one estimator over another if the former has lower risk than the latter.

11.4.2   Mean squared error, variance, and bias

The most commonly used loss function is the so-called *squared loss function*. The squared loss function for an estimator $\widehat{\theta}$ is given by

$$L\left(\widehat{\theta}, \theta_0\right) = \left(\widehat{\theta} - \theta_0\right)^2. \tag{11.2}$$

It can be seen that this function is indeed a loss function in the sense of Definition 11.4. Correspondingly, the squared risk function is

$$R\left(\widehat{\theta}, \theta_0\right) = E\left(\left(\widehat{\theta} - \theta_0\right)^2\right).$$

This risk function based on the squared error loss function is called the *mean squared error* (MSE) of the estimator $\widehat{\theta}$. Since this is a very commonly used risk function, it is formally defined next.

Definition 11.6 (Mean squared error (MSE) of estimator) *Let $\widehat{\theta}$ be an estimator of the estimand θ_0. The* **mean squared error** (**MSE**) *of the estimator $\widehat{\theta}$ is defined as*

$$MSE\left(\widehat{\theta}, \theta_0\right) = E\left(\left(\widehat{\theta} - \theta_0\right)^2\right).$$

The squared loss function behind the MSE assigns a relatively higher loss on differences between $\widehat{\theta}$ and θ_0 the further $\widehat{\theta}$ is from θ_0. For example, the loss of $b - \theta_0 = 6$ is more than twice as large as the loss of $b - \theta_0 = 3$, in fact, it is 4 times as large.

The MSE has the property that it can be decomposed into the variance of the estimator and the so-called *bias* of the estimator. The bias of an estimator is defined as the expected value of the difference between the estimator and the estimand.

Definition 11.7 (Bias of estimator) *Let $\widehat{\theta}$ be an estimator of the estimand θ_0. The* **bias** *of the estimator $\widehat{\theta}$ is defined as*

$$Bias\left(\widehat{\theta}, \theta_0\right) = E\left(\widehat{\theta}\right) - \theta_0,$$

where $E()$ is taken with respect to $\widehat{\theta}$, whose distribution may depend on θ_0. We say that an estimator is unbiased if $Bias\left(\widehat{\theta}, \theta_0\right) = 0$, upwards biased if $Bias\left(\widehat{\theta}, \theta_0\right) > 0$, and downwards biased if $Bias\left(\widehat{\theta}, \theta_0\right) < 0$.

One interpretation of bias of an estimator is in view of a statistical experiment. If we could consider the average estimate of infinitely many repetitions of the statistical experiment, then the estimator is unbiased when this average equals the true value.

Next, we state how the MSE of an estimator can be decomposed into a sum of the variance of $\widehat{\theta}$ and the squared bias of $\widehat{\theta}$.

Theorem 11.1 (MSE link to bias and variance) *Let $\widehat{\theta}$ be an estimator of the estimand θ_0. If $Var\left(\widehat{\theta}\right)$ exists, then*

$$MSE\left(\widehat{\theta}, \theta_0\right) = \left(Bias\left(\widehat{\theta}, \theta_0\right)\right)^2 + Var\left(\widehat{\theta}\right).$$

Notice that, like the bias $Bias\left(\widehat{\theta}, \theta_0\right)$, the variance $Var\left(\widehat{\theta}\right)$ may also depend on θ_0 because the distribution of $\widehat{\theta}$ may depend on θ_0.

As mentioned above, the sample average is an estimator of the population mean. The next theorem states the MSE, bias, and variance of this estimator.

Theorem 11.2 (Properties of the sample average) *Let X be a random variable and $\mu_0 = E(X)$, $\sigma_0^2 = Var(X)$, such that $\sigma_0^2 < \infty$. Assume a simple random sample $\left(\widetilde{X}_1, \ldots, \widetilde{X}_n\right)$ from X. Then the sample average has the following properties.*

1. $Bias\left(\overline{X}, \mu_0\right) = 0.$

2. $Var\left(\overline{X}\right) = \frac{\sigma_0^2}{n}.$

3. $MSE\left(\overline{X}, \mu_0\right) = \frac{\sigma_0^2}{n}.$

Since the MSE decreases with the sample size n, the precision of the sample average as an estimator of the population mean μ_0, increases as the sample size n increases.

11.4.3 Other measures of precision

Ideally, the researcher should choose a loss function that matches the loss that is in practice inflicted by obtaining an estimate different from the estimand. Thus, the empirical problem is supposed to guide us toward an appropriate choice of loss function. The squared loss is but one of many possible loss functions. One reason for the popularity of the squared loss function is that it is mathematical convenient. The MSE may not exist in some cases of interest since, e.g. the variance of a population may not exist. In this subsection, we present two alternative loss functions.

The *mean absolute loss function* is given by

$$L\left(\widehat{\theta}, \theta_0\right) = \left|\widehat{\theta} - \theta_0\right|, \tag{11.3}$$

where $|\cdot|$ denotes the absolute value. The corresponding risk function is

$$R\left(\widehat{\theta}, \theta_0\right) = E\left(\left|\widehat{\theta} - \theta_0\right|\right),$$

where the expectation $E()$ is taken with respect to the distribution of $\widehat{\theta}$. This risk function is also called the *mean absolute error* (MAE) of an estimator, and we often write $MAE(\widehat{\theta}, \theta_0)$ for this risk function. With this loss function, the loss is proportional to the difference between the estimate and the estimand.

The MAE exists if the mean of $\widehat{\theta}$ exists, and may thus exist even in cases where the variance of $\widehat{\theta}$ does not exist. An example would be if $\widehat{\theta}$ has a t-distribution with 2 degrees of freedom.

Next we consider a loss function which is dichotomous in that either the estimate is close to the estimand, or it is not close. Close to the estimand is defined as a neighborhood of θ_0, where the neighborhood is represented by the open interval

$$(\theta_0 - \varepsilon, \theta_0 + \varepsilon), \tag{11.4}$$

where $\varepsilon > 0$ determines the width, equal to $2 \cdot \varepsilon$, of the interval. The estimator has an outcome in that neighborhood if

$$(\theta_0 - \varepsilon) < \widehat{\theta} < (\theta_0 + \varepsilon),$$

or, written more compactly,

$$\left|\widehat{\theta} - \theta_0\right| < \varepsilon.$$

We formulate a loss function that is equal to one if the estimate $\widehat{\theta}$ is outside the interval (11.4), and 0 if the estimate is inside the interval. That is, we define the loss function

$$L\left(\widehat{\theta}, \theta_0\right) = I\left(|\widehat{\theta} - \theta_0| \geq \varepsilon\right),$$

where $I()$ denotes the indicator function. This is an example of a so-called "0-1 loss function", since it results in a loss of either 0 or 1, depending on whether the estimator $\widehat{\theta}$ is inside or outside the ε-neighborhood of the estimand θ_0, defined in (11.4). The 0-1 loss function models the view that either the estimate is close to θ_0, in which case there is no loss, or the estimate is not close to θ_0, in which case there is a loss of 1 no matter how far the estimate is from θ_0.

The risk of the 0-1 loss function is the expected value of the 0-1 loss function (Definition 11.5). Since $E(I(A)) = Pr(A)$ for any event A, the risk can be written as

$$\begin{aligned} R\left(\widehat{\theta}, \theta_0\right) = E\left(L\left(\widehat{\theta}, \theta_0\right)\right) = E\left(I\left(|\widehat{\theta} - \theta_0| \geq \varepsilon\right)\right) &= \Pr\left(\left|\widehat{\theta} - \theta_0\right| \geq \varepsilon\right) \\ &= 1 - \Pr\left(\left|\widehat{\theta} - \theta_0\right| < \varepsilon\right). \end{aligned}$$

That is, the risk is equal to the probability that $\widehat{\theta}$ has an outcome outside the ε neighborhood defined in (11.4). Note that, since the loss function always outputs a 0 or a 1, the risk will in this case be a number in the interval $[0, 1]$. Since the probability always exists, the risk of the 0-1 loss function also always exists.

Assuming the estimator $\widehat{\theta}$ is a continuous random variable, the probability of an outcome of $\widehat{\theta}$ in this neighborhood is

$$\Pr\left(\left|\widehat{\theta} - \theta_0\right| < \varepsilon\right) = \Pr\left(\theta_0 - \varepsilon < \widehat{\theta} < \theta_0 + \varepsilon\right) = F_{\widehat{\theta}}(\theta_0 + \varepsilon) - F_{\widehat{\theta}}(\theta_0 - \varepsilon), \tag{11.5}$$

where $F_{\widehat{\theta}}()$ denotes the CDF of $\widehat{\theta}$. The higher this probability, the lower the risk, and, therefore, the higher the precision of the estimator.

11.5 Consistency of an estimator

The concept of *consistency* evaluates whether an estimator can assess the estimand closely when the sample size is very large. To quantify what is meant by the term "closely" in this context, we use the probability of the estimator having an outcome in a neighborhood of the estimand, that is, we use the 0-1 loss function defined in the previous section. Consistency is concerned about the properties of the risk of the 0-1 loss function as a function of the sample size n.

To formalize this, we let $\varepsilon > 0$ be a number, and use the construction stated in (11.5), namely,

$$\Pr\left(\left|\widehat{\theta} - \theta_0\right| < \varepsilon\right).$$

The concept of consistency is based not only on the probability of $\widehat{\theta}$ being in the neighborhood of the estimand θ_0, but also on how this probability changes with larger sample sizes n. To make it explicit that the estimator $\widehat{\theta}$ is different for different sample sizes, we put a subscript n on the estimator $\widehat{\theta}_n$. To be precise, we study

$$\Pr\left(\left|\widehat{\theta}_n - \theta_0\right| < \varepsilon\right)$$

for growing sample sizes n, i.e. for $n \to \infty$. The study of the properties of estimators as the sample size grows infinitely large, i.e. as $n \to \infty$, is called *asymptotic analysis*.

The precision of the estimator may depend on the true value θ_0. For example, for a given sample size, estimation of a variance can be done more precisely if the true variance is small than if the true variance is large, see Section 12.4.4 below. We would like an estimator to have a good quality no matter the true value of θ_0. For that purpose, we define a set of values in which the parameter θ_0 is permitted. This set of values is called a *parameter space*, and it is defined next.

Definition 11.8 (Parameter space) *In a model specified by parameters, the set of permitted values of the parameters is called a **parameter space**.*

We will often denote a parameter space by an uppercase Greek letter, e.g. Θ. The parameter space Θ for the mean μ of a population might be all the real numbers, that is, $\Theta = \mathbb{R}$. If the population is height of people measured in meters, we could set $\Theta = (0, 4)$ if we are confident that the average height is below 4 meters. We will always exclude numbers in the parameter space which cannot occur; for instance, since the variance σ^2 of a random variable cannot be negative, the parameter space for σ^2 will always be a subset of $\Theta = [0, \infty)$.

If we want an estimator to have good quality no matter the value of θ_0, this can be obtained by requiring that we have a good quality for all θ_0 in the parameter space, i.e. for all $\theta_0 \in \Theta$. For example, if we study the 0-1 risk function, then we need to study $\Pr\left(\left|\widehat{\theta}_n - \theta_0\right| < \varepsilon\right)$ for all $\theta_0 \in \Theta$. Notice, this does not mean that there is more than one true value, but it means that we allow the true value to be any number in Θ.

The concept of consistency is defined as the probability of $\widehat{\theta}_n$ being in any neighborhood of the estimand θ_0 must approach 1 as the sample size n increases toward infinity. This has to happen no matter the value of θ_0 in the parameter space Θ and width of the neighborhood, specified by ε. The formal definition is next.

Definition 11.9 (Consistent estimator) *Let $\widehat{\theta}_n$ be an estimator of the estimand θ_0 when the sample size is n, and let Θ be the parameter space of θ_0. Then $\widehat{\theta}_n$ is a **consistent estimator** of θ_0 if and only if for any $\varepsilon > 0$ and for any $\theta_0 \in \Theta$,*

$$\Pr\left(\left|\widehat{\theta}_n - \theta_0\right| < \varepsilon\right) \to 1 \ as \ n \to \infty. \tag{11.6}$$

The definition states that the probability of the event $\{|\widehat{\theta}_n - \theta_0| < \varepsilon\} = \{\theta_0 - \varepsilon < \widehat{\theta}_n < \theta_0 + \varepsilon\}$ must approach 1 as the sample grows. This must happen no matter which neighborhood of θ_0 we consider, since it must hold true for any $\varepsilon > 0$, and ε determines the width of the neighborhood. That is, it must hold for arbitrarily small neighborhoods of θ_0. In other words, consistency ensures that the probability that the estimator $\widehat{\theta}_n$ has an outcome in a very small neighborhood of the estimand θ_0, must approach 1 as the sample size grows. Hence, if an estimator is consistent and the sample size n is large, we can have a high degree of certainty that the estimate $\widehat{\theta}_n$ will be close to the estimand θ_0, no matter the value of θ_0.

Note that, even though the probability is close to 1 for $\widehat{\theta}_n$ having an outcome in a neighborhood close to θ_0, typically $\Pr\left(\widehat{\theta}_n = \theta_0\right) = 0$, which is always the case if $\widehat{\theta}_n$ is a continuous random variable. This is the main reason for not using $\Pr\left(\widehat{\theta}_n = \theta_0\right)$ to define the concept of consistency but, instead, using small neighborhoods around θ_0, i.e. $\Pr\left(\left|\widehat{\theta}_n - \theta_0\right| < \varepsilon\right)$.

There are different, but equivalent, terminologies used to convey that an estimator is consistent. Sometimes it is said that "$\widehat{\theta}_n$ converges in probability to θ_0", or the "probability limit of $\widehat{\theta}_n$ is θ_0". There are also different, but equivalent, notations used to denote that an estimator is consistent. For instance, the following four statements are equivalent:

1. $\widehat{\theta}_n$ is a consistent estimator of θ_0.

2. $\widehat{\theta}_n \xrightarrow{p} \theta_0$ as $n \to \infty$.

3. $plim\ \widehat{\theta}_n = \theta_0$ as $n \to \infty$.

4. $\underset{n \to \infty}{plim}\ \widehat{\theta}_n = \theta_0$.

The "p" on top of the convergence arrow in 2, and in front of the "lim" in 3 and 4, indicates that the convergence is defined in terms of probabilities, cf. (11.6).

11.5.1 Relationship between MSE and consistency

The MSE and consistency of an estimator are related in that convergence of MSE to 0 as $n \to \infty$ implies the estimator is consistent. This result is stated next.

Theorem 11.3 (MSE and consistency) *Let $\widehat{\theta}_n$ be an estimator of the estimand θ_0 and let Θ be the parameter space for θ_0. If*

$$MSE\left(\widehat{\theta}_n, \theta_0\right) \to 0 \ as\ n \to \infty,$$

for any $\theta_0 \in \Theta$, then $\widehat{\theta}_n$ is a consistent estimator of θ_0.

Example 11.6 *Let $\widetilde{X}_1, \ldots, \widetilde{X}_n$ be a simple random sample from the random variable X, which is such that $\sigma_0^2 = Var(X) < \infty$. Then the sample average $\overline{X}$ is a consistent estimator of the population mean $E(X) = \mu_0$. This can be shown using the result in Theorem 11.2, i.e. that the MSE is*

$$MSE\left(\overline{X}, \mu_0\right) = \frac{1}{n} \cdot \sigma_0^2.$$

If we let $n \to \infty$, then $\frac{1}{n} \to 0$, and, hence, $MSE\left(\overline{X}, \mu_0\right) \to 0$. This happens for any value of μ_0, in fact, the MSE of $\overline{X}$ does not depend on μ_0. We conclude from Theorem 11.3 that the sample average $\overline{X}$ is a consistent estimator of the population mean μ_0.

The converse of Theorem 11.3 is not true. That is, an estimator may be consistent without $MSE\left(\widehat{\theta}_n, \theta_0\right) \to 0$ as $n \to \infty$. Further, as mentioned above, the $MSE\left(\widehat{\theta}_n, \theta_0\right)$ may not even exist, since it is defined using an expected value, and expected values may not exist. Indeed, an estimator may be consistent even if its MSE does not exist.

By combining Theorems 11.1 and 11.3, we may conclude that if the bias and variance of an estimator both approach 0 as the sample size grows, then the estimator is consistent. We state this as a corollary to Theorem 11.3.

Corollary 11.5.1 (Corollary to Theorem 11.3) *Let* $\widehat{\theta}_n$ *be an estimator of* θ_0. *If* $Bias\left(\widehat{\theta}_n, \theta_0\right) \to 0$ *and* $Var(\widehat{\theta}_n) \to 0$ *as* $n \to \infty$ *for any* $\theta_0 \in \Theta$, *then* $\widehat{\theta}_n$ *is a consistent estimator of* θ_0.

In most cased, we only consider estimators which are consistent. In other words, if an estimator does not get arbitrarily close to the true value of the estimand when the sample size increases toward infinity, we generally discard the estimator.

11.5.2 The Law of Large Numbers

Consistency of an estimator can often be proved using a so-called *law of large numbers*. The reason is that many estimators are constructed by sample averages, and laws of large numbers describe how such sample averages behave when the sample size becomes infinitely large. We will here present a law of large numbers, which is applicable with a simple random sample, that is, when the random variables are mutual statistically independent and identically distributed (iid).

Theorem 11.4 (Law of Large Numbers) *Assume* $(X_1, \ldots, X_n)$ *are mutual statistically independent and identically distributed random variables with population mean* $\mu_0 = E(X)$ *and population variance* $\sigma_0^2 = Var(X) < \infty$. *Let*

$$\overline{X}_n = \frac{1}{n} \sum_{i=1}^{n} X_i.$$

Then

$$\underset{n \to \infty}{plim} \, \overline{X}_n = \mu_0.$$

We note that the assumption of a finite population variance, $Var(X) = \sigma_0^2 < \infty$, made in Theorem 11.4 can be relaxed.

By setting $X_i = \widetilde{X}_i$, the Law of Large Numbers in Theorem 11.4 is directly applicable to show that the sample average $\overline{X}$ is a consistent estimator.

Example 11.7 (The Law of Large Numbers for dice rolls) *Consider throwing a fair six-sided die n times, yielding the random sample* $(\widetilde{X}_1, \widetilde{X}_2, \ldots, \widetilde{X}_n)$, *where* $\widetilde{X}_j = k$ *if the j'th roll is equal to k, where* $k = 1, 2, \ldots, 6$. *We assume that the die is thrown in such a way that the outcomes are mutual statistically independent and has the same distribution, i.e.* $(\widetilde{X}_1, \widetilde{X}_2, \ldots, \widetilde{X}_n)$ *constitutes a simple random sample from a population X with PMF* $f_X(1) = f_X(2) = \ldots = f_X(6) = \frac{1}{6}$. *Recall that the population mean of X with this PMF is* $\mu_0 = E(X) = 3.5$. *The left panel of Figure 11.1 shows one realization of the sample mean* $\overline{X}_n$ *as a function of the sample size n (solid line) for* $n = 1, 2, \ldots, 10000$. *That is, after sampling* $x_1, \ldots, x_{10000}$, *we plot* $(1, \overline{x}_1 = x_1), (2, \overline{x}_2 = (x_1 + x_2)/2), (3, \overline{x}_3 = (x_1 + x_2 + x_3)/3)$, *and so on. The horizontal dashed line denotes* $\mu_0 = 3.5$. *From the figure we see the Law of Large Numbers in action: For small sample sizes, the sample mean* $\overline{X}_n$ *can be far from the population mean* $\mu_0 = 3.5$, *but as the sample size n increases, the sample mean invariably* *approaches the population mean. This behavior is even more evident in the right panel of Figure 11.1, which plots the same data but where the x-axis, denoting the sample size n used to construct the sample mean $\overline{X}_n$, is now on a log-scale.*

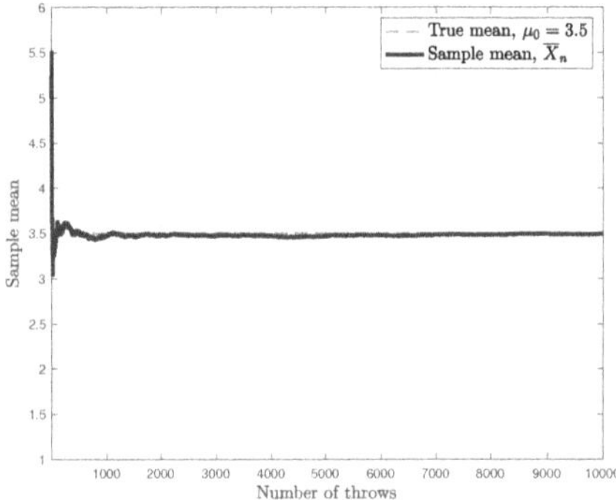
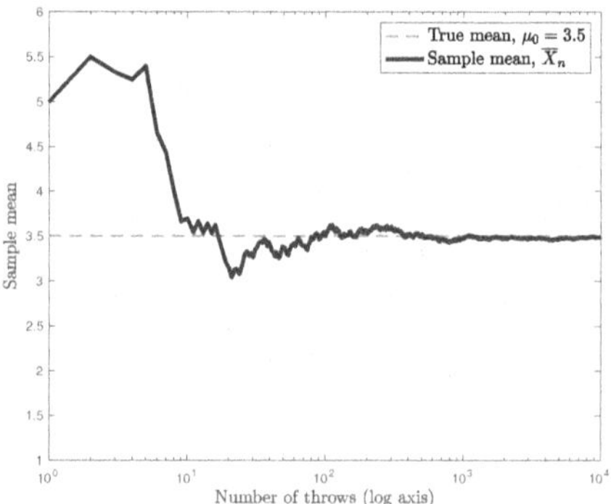

FIGURE 11.1

Sample mean of dice throws as a function of the number of throws, n, illustrating the Law of Large Numbers. See Example 11.7.

To see how we can apply the Law of Large Numbers to other estimators than the sample mean, we extend it to the situation where a function $g()$ is applied to each X_i. The reason that the Law of Large Numbers still holds in this case, is that the random variables $(g(X_1), \ldots, g(X_n))$ inherit the properties of mutual statistical independence and identical distribution from $(X_1, \ldots, X_n)$. This feature is stated as a separate theorem.

Theorem 11.5 (Transformation of a simple random sample) *Let $(X_1, \ldots, X_n)$ be mutual statistically independent and identically distributed random variables. For any function $g()$, the transformed random variables $(g(X_1), \ldots, g(X_n))$ are mutual statistically independent and identically distributed.*

Theorem 11.5 implies that if $(\widetilde{X}_1, \ldots, \widetilde{X}_n)$ is a simple random sample, then $(g(\widetilde{X}_1), \ldots, g(\widetilde{X}_n))$ can also be considered as a simple random sample. As shown in the proof of Theorem 11.5, statistical independence between, say, X_1 and X_2, also implies that $g(X_1)$ and $h(X_2)$ are statistically independent for any choice of functions $g()$ and $h()$. This holds no matter the distribution of X_1 and X_2.

With a simple random sample $(\widetilde{X}_1, \ldots, \widetilde{X}_n)$, Theorem 11.5, together with the Law of Large Numbers (Theorem 11.4), implies the following result.

Corollary 11.5.2 (Law of Large Numbers for a transformation of a simple random sample) *Assume $(X_1, \ldots, X_n)$ are mutual statistically independent and identically distributed random variables from a population represented by the random variable X. Let $g()$ be a function such that $\mu_g = E(g(X))$ and $\sigma_g^2 = Var(g(X))$ exist. Then*

$$\underset{n \to \infty}{plim} \, \frac{1}{n} \sum_{i=1}^{n} g(X_i) = \mu_g. \tag{11.7}$$

Example 11.8 (Example 11.7, continued) *Consider again the dice throws in Example 11.7, but now define the random variable $\widetilde{Y}_j$ to be equal to k^2 if the outcome of the j'th roll* is $k = 1, 2, \ldots, 6$. *In other words, we define* $\widetilde{Y}_j = g\left(\widetilde{X}_j\right)$, *where the function* $g()$ *is given by* $g(x) = x^2$. *Let* $Y = g(X) = X^2$ *and* $\mu_Y = E(Y)$ *be the population mean of* Y. *Using the rules for calculating the expected value of the function of a discrete random variable (Theorem 5.1), we can show that* $\mu_Y = E(Y) = E(X^2) = \frac{91}{6} \approx 15.167$. *Similar to Figure 11.1, Figure 11.2 shows one realization of the sample mean* $\overline{Y}_n$ *as a function of the sample size* n *(solid line), the horizontal dashed line denoting* $\mu_Y = \frac{91}{6}$. *The figure corroborates the discussion above, in particular Equation* (11.7).

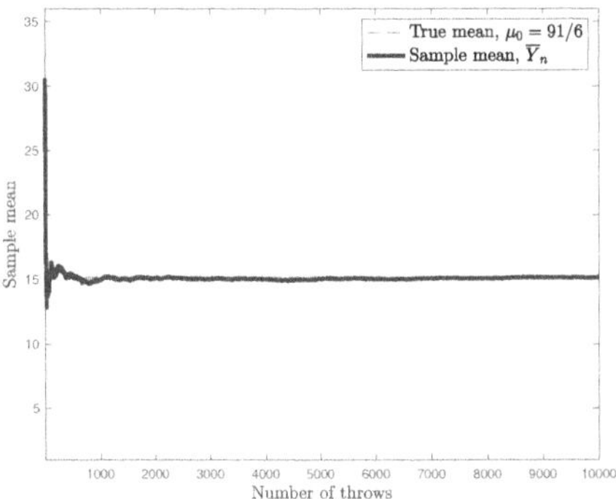

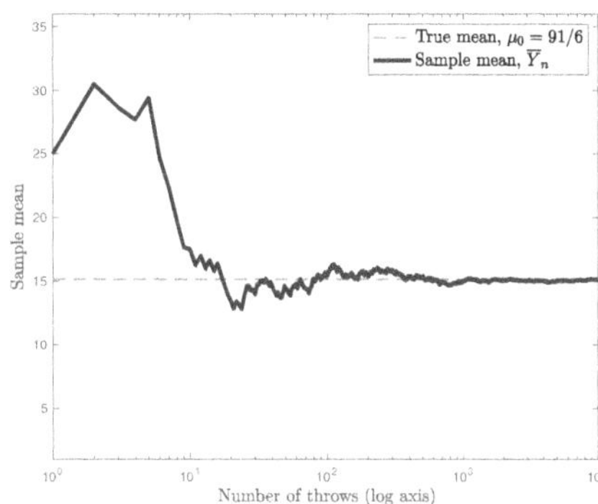

FIGURE 11.2
Sample mean of squared dice throws as a function of the number of throws, n, illustrating the Law of Large Numbers. See Example 11.8.

11.6 Comparison of estimators

We can compare, and choose among, estimators using a measure of precision of an estimator. If one estimator is more precise than another estimator, then we will say that the former estimator is more *efficient* than the latter. Here, we use the MSE (Definition 11.6) as a measure of precision to define the concept of efficiency. We make this explicit in the next definition.

Definition 11.10 (Efficiency) *Let* $\widehat{\theta}$ *and* $\widehat{\tau}$ *be two estimators of the estimand* θ_0, *where* θ_0 *is assumed to be in the parameter space* $\Theta \subseteq \mathbb{R}$. *Then* $\widehat{\theta}$ *is said to be at least as* ***efficient*** *as* $\widehat{\tau}$ *if and only if*

$$MSE\left(\widehat{\theta}, \theta_0\right) \leq MSE\left(\widehat{\tau}, \theta_0\right)$$

for all values of $\theta_0 \in \Theta$. *Further,* $\widehat{\theta}$ *is said to be more* ***efficient*** *than* $\widehat{\tau}$ *if, in addition,*

$$MSE\left(\widehat{\theta}, \theta_0\right) < MSE\left(\widehat{\tau}, \theta_0\right)$$

for at least one value of $\theta_0 \in \Theta$.

Example 11.9 *In this example, we compare two estimators of a population mean. Assume a simple random sample* $(\widetilde{X}_1, \ldots, \widetilde{X}_n)$ *from a population represented by* X. *Assume the* *population mean and variance are*

$$E(X) = \mu_0,$$
$$Var(X) = \sigma_0^2,$$

where $\mu_0 \in \mathbb{R}$ and $\sigma_0^2 \in (0, \infty)$. Suppose the quantity of interest, and therefore the estimand, is the population mean μ_0.

First consider the sample average $\overline{X}$ as an estimator of the mean μ_0

$$\overline{X} = \frac{1}{n} \sum_{i=1}^{n} \widetilde{X}_i.$$

In Theorem 11.2, we found the MSE of the sample average to be

$$MSE\left(\overline{X}, \mu_0\right) = \frac{1}{n}\sigma_0^2.$$

Second, consider the estimator $\widehat{\tau}$ which is the first observation only

$$\widehat{\tau} = \widetilde{X}_1.$$

To calculate the MSE of $\widehat{\tau}$, first calculate its bias

$$Bias\left(\widehat{\tau}, \mu_0\right) = E_{\widehat{\tau}}(\widehat{\tau}) - \mu_0 = E_{\widetilde{X}_1}(\widetilde{X}_1) - \mu_0 = \mu_0 - \mu_0 = 0.$$

Thus, the estimator $\widehat{\tau}$ is an unbiased estimator of μ_0. The variance of the estimator $\widehat{\tau}$ is

$$Var\left(\widehat{\tau}\right) = Var(\widetilde{X}_1) = \sigma_0^2.$$

Using Theorem 11.1, we deduce that the MSE of $\widehat{\tau}$ is

$$MSE\left(\widehat{\tau}, \mu_0\right) = \sigma_0^2.$$

Now it can be seen that for any sample size $n \geq 2$,

$$MSE\left(\overline{X}, \mu_0\right) = \frac{1}{n}\sigma_0^2 < MSE\left(\widehat{\tau}, \mu_0\right) = \sigma_0^2,$$

regardless of the value of μ_0, i.e. for all $\mu_0 \in \mathbb{R}$. We conclude that the sample average $\overline{X}$ is a more efficient estimator of μ_0 than the estimator $\widehat{\tau}$ (Definition 11.10). This result aligns with the intuition that, other things being equal, more information, that is more observations, should increase the precision of an estimator.

When using the concept of efficiency, we will typically restrict attention to some class of estimators, such as the class of all unbiased estimators. If an estimator is at least as efficient to any other estimator in the class of estimators under consideration, then we say that the estimator is efficient in that class of estimators. If we do not restrict the class of estimators, the concept of efficiency (Definition 11.10) becomes vacuous. The following example illustrates this.

Example 11.10 (Efficiency in an unrestricted class of estimators) *Suppose we are interested in estimating the estimand θ_0, where the parameter space of θ_0 is $\Theta = \mathbb{R}$. Suppose further, that we impose no restrictions on the class of estimators we consider. Then, no estimator can be efficient. To see this, consider the estimator $\widehat{\tau}$ that always outputs the number 42, i.e. $\widehat{\tau} = 42$ no matter the random sample. Now, for $\theta_0 = 42$, it is the case that $\widehat{\tau}$ is as precise as can be, i.e. $MSE(\widehat{\tau}, \theta_0) = MSE(\widehat{\tau}, 42) = MSE(42, 42) = 0$. Of course, in general, $\widehat{\tau}$ is terrible estimator. Indeed, it is biased for all values of the estimand $\theta_0 \neq 42$ and it is inconsistent. However, if $\widehat{\theta}$ is a continuous random variable which is also an estimator of θ_0, we will have $MSE(\widehat{\theta}, 42) > 0$. Hence, by Definition 11.10, $\widehat{\theta}$ cannot be efficient since $42 \in \Theta$.*

Example 11.10 shows that to make the concept of efficiency useful in practice, we need to consider a restricted class of estimators so that obviously "pathological" estimators, such as $\hat{\tau} = 42$, are excluded. For instance, as mentioned above, we can exclude such estimators by considering only the class of unbiased estimators.

Before we close our discussion of efficiency, we note that there are other approaches, besides efficiency, that can be used to choose between estimators. To choose an estimator based on efficiency requires that such an estimator has a MSE no larger than any other estimator for all possible values of the estimand θ_0 in the parameter space Θ. Such estimators may be hard to find. Alternatively, we may be concerned with how imprecise an estimator is when it is most imprecise. That is, we may consider how an estimator performs under the least favorable circumstances. Then we could choose the estimator that is least imprecise when circumstances are least favorable. Mathematically, we find the value of $\theta \in \Theta$ that leads to the maximum risk of an estimator $\hat{\tau}$. Denote this value $MaxRisk(\hat{\tau})$. Do this for all estimators considered, and then choose the estimator with the minimum $MaxRisk$. This approach to choosing an estimator is called a *minimax* (minimum of the maximum risk) approach.

11.7 Monte Carlo simulation

In Section 11.3, we considered two examples, which allowed us to derive a closed-form solution to the sampling distribution of the estimator in question. In many cases, however, doing the analytical derivation and finding a closed-form solution for the sampling distribution of an estimator is difficult, if not practically impossible. As an alternative approach, it is possible to use simulation to estimate a sampling distribution, by assuming a specific distribution for the population and a sampling mechanism. This very powerful approach is called *Monte Carlo simulation* and the result is called a *Monte Carlo approximation* of the sampling distribution. In effect, a Monte Carlo simulation seeks to implement the thought experiment used to interpret probabilities; namely, repeat the statistical experiment infinitely many times and interpret the fraction of times an event occurs as the probability of that event. The only difference to the thought experiment is that the number of repetitions is necessarily finite in the Monte Carlo simulation. By increasing the number of repetitions in the Monte Carlo simulation, the precision of the Monte Carlo approximation may be improved.

The following example describes the historical origins of the Monte Carlo method, and at the same time illustrates how the approach may be used to solve analytically intractable problems by simulation methods.

Example 11.11 (The origins of Monte Carlo simulation) *In 1946, the Polish mathematician and nuclear physicist Stanislaw Ulam was recovering from surgery. The doctors said that, to ease his recovery, Ulam needed to avoid strain, physically as well as intellectually. Ulam therefore decided to play (a lot) of Canfield (solitaire). While playing, he was trying to work out the probability that a given deal of solitaire would end up in a win. However, the calculations quickly got out of hand: the number of different ways the card can fall is enormous. Ulam realized that he could approximate the probability of winning by simply playing out the game a great many times, and use the empirical fraction of wins as an approximation to the true probability of a win. Playing out the games could be done physically, of course, but would be very time consuming. At that time, the modern computer was in its infancy. Ulam realized that a computer could be programmed with the rules of* *solitaire and, by simulating random draws of solitaire decks, the computer could play the games quicker than a human, allowing many more games to be played, and, thus, increase the precision of the approximation. The Monte Carlo method was born.*

11.7.1 Elements in Monte Carlo simulation

Monte Carlo simulation works by drawing a sample from the population distribution F_X in accordance with the sampling mechanism. To avoid confusion with a sample from a population, we will denote the sample drawn in a Monte Carlo simulation for a *Monte Carlo sample*. If we implement the Monte Carlo simulation on a computer, we can draw many Monte Carlo samples, and, for each of the Monte Carlo samples, we can calculate Monte Carlo sample statistics. The distribution of all the sample statistics calculated is the Monte Carlo distribution of the sample statistics and, thus, the Monte Carlo approximation of the sampling distribution of the sample statistics.

To be specific, consider a Monte Carlo approximation to a sample statistics T. Suppose we draw a Monte Carlo sample $\left(\tilde{x}_1^{\#}, \ldots, \tilde{x}_n^{\#}\right)$ M times, where M is a large number, e.g. $M = 10,000$. For every Monte Carlo sample, the sample statistics T is calculated. We denote the sample statistics $t_j^{\#}$ when it is based on the jth Monte Carlo sample. Similarly to the interpretation of a probability, we calculate the fraction of times the Monte Carlo sample statistics is no larger than some value, say t. This is the Monte Carlo approximation of the probability $\Pr\left(T \leq t\right)$,

$$F_T^{\#}(t) = \frac{1}{M} \sum_{j=1}^{M} I\left(t_j^{\#} \leq t\right),$$

where $I()$ is the indicator function. The difference between the Monte Carlo CDF $F_T^{\#}(t)$ of the sample statistics and the CDF $F_T(t)$ of the sampling distribution of the sample statistics can be made arbitrarily small by choosing a sufficiently high number of Monte Carlo repetitions, M. Below, the main steps in a Monte Carlo simulation are illustrated.

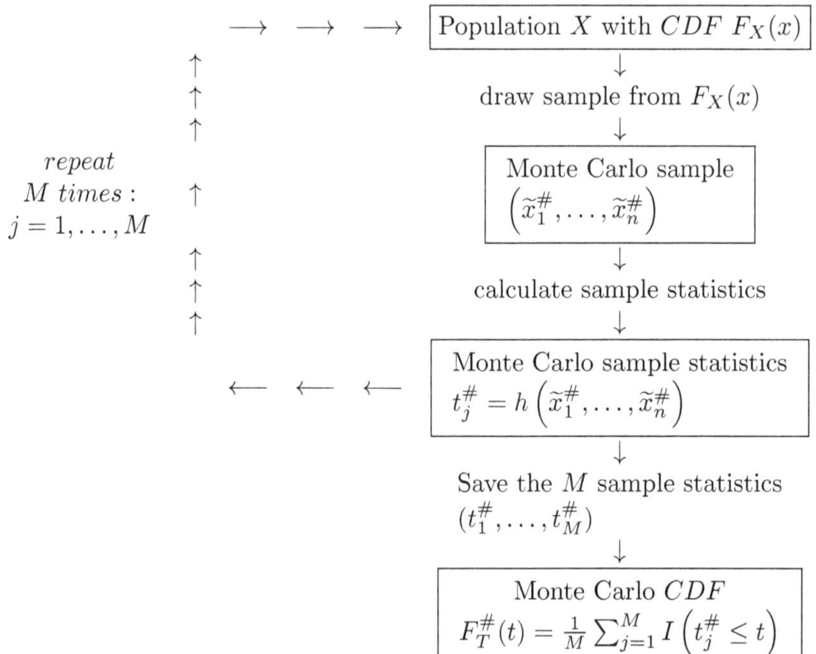

The following example illustrates how the Monte Carlo simulation approach can be used to approximate the sampling distribution of a sample statistics.

Example 11.12 (Monte Carlo approximation of sampling distribution of the sample mean) *Consider a population represented by the random variable $X \sim Exp(2)$, i.e. X has the Exponential distribution with parameter $\lambda = 2$. Assume that $(\widetilde{X}_1, \ldots, \widetilde{X}_{20})$ is a simple random sample of size $n = 20$ from X and consider the sample mean*

$$\overline{X} = \frac{1}{20} \sum_{i=1}^{20} \widetilde{X}_i.$$

Since $\overline{X}$ is a (scaled) sum of iid Exponentially distributed random variables, it has the Erlang distribution (Definition 8.4), $\overline{X} \sim Erlang(2 \cdot 20, 20) = Erlang(40, 20)$.

We are now going to use Monte Carlo simulation to approximate the sampling distribution of the sample mean $\overline{X}$ and compare the Monte Carlo approximation to the true distribution, i.e. $Erlang(40, 20)$. To this end, we follow the procedure outlined above. Let M denote the number of Monte Carlo samples and, for each sample $j = 1, \ldots, M$, we simulate a simple random sample from X, $(\widetilde{X}_1^{\#}, \ldots, \widetilde{X}_{20}^{\#})$ where $\widetilde{X}_i^{\#} \sim Exp(2)$ are iid, and construct the Monte Carlo sample average

$$\overline{X}_j^{\#} = \frac{1}{20} \sum_{i=1}^{20} \widetilde{X}_i^{\#}.$$

Note that most statistical software packages contain routines for simulation from the Exponential distribution, see also Example 11.17 below. The Monte Carlo approximation to the CDF $F_{\overline{X}}$ of $\overline{X}$ is

$$F_{\overline{X}}^{\#}(t) = \frac{1}{M} \sum_{j=1}^{M} I\left(\overline{X}_j^{\#} \le t\right).$$

The left panel of Figure 11.3 shows the histogram of the Monte Carlo sample means, $\overline{X}_1^{\#}, \ldots, \overline{X}_M^{\#}$, where $M = 100\ 000$, and the right panel of the figure shows the Monte Carlo approximation of the CDF of $\overline{X}$ (red line), i.e. $F_{\overline{X}}^{\#}(t)$. In the figure, solid lines denote the true distribution of $\overline{X}$, i.e. in the left panel it is the PDF of the $Erlang(40, 20)$ distribution and in the right panel it is the CDF of the $Erlang(40, 20)$ distribution. From the figure, we see that the Monte Carlo approximation is very close to the true distribution of $\overline{X}$.

In Example 11.12, the true sampling distribution of the sample statistics $\overline{X}$ is known and, hence, there would, in practice, not be any reason to approximate the distribution of $\overline{X}$ using Monte Carlo simulation. However, the example illustrates that when the population distribution and sample mechanism are known, Monte Carlo simulation can provide an arbitrarily accurate approximation to the sampling distribution of a sample statistics by choosing M large. The following example considers Monte Carlo approximation of the sampling distribution of a sample statistics in a case where the sampling distribution cannot be derived in closed-form.

Example 11.13 (Example 11.12, continued) *Consider the same setup as in Example 11.12, but suppose now that the sample statistics of interest is the sample mean of the squares of the observations*

$$\overline{X^2} = \frac{1}{20} \sum_{i=1}^{20} \widetilde{X}_i^2.$$

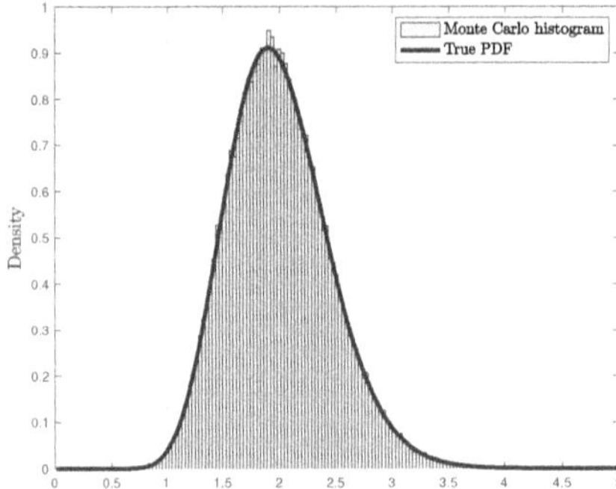
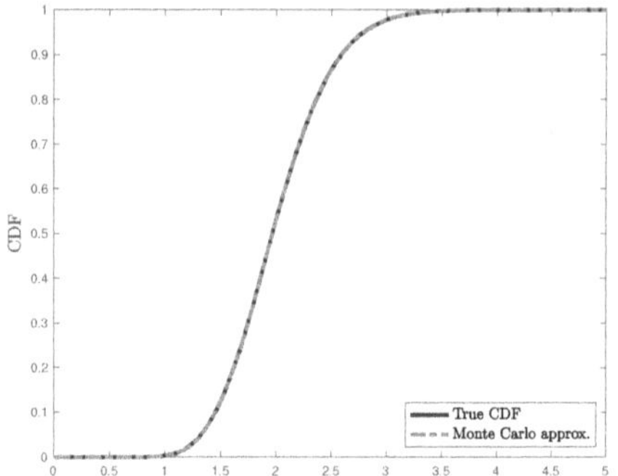

FIGURE 11.3
Monte Carlo approximation to the sampling distribution of the sample mean (Example 11.12). Left panel: Histogram of Monte Carlo sample means $\overline{X}_j^{\#}$, $j = 1, \ldots, M$ with $M = 100\,000$, along with true PDF of the sample mean $\overline{X}$ (solid curve). Right panel: True CDF of the sample mean $\overline{X}$ (solid curve) along with Monte Carlo approximation $F_{\overline{X}}^{\#}(t)$ (dashed curve).

In this case, an analytic expression of the sampling distribution of the statistics $\overline{X^2}$ is not readily available. However, following the Monte Carlo approach, we can approximate this distribution using simulation. To this end, we proceed as above. Specifically, we let M denote the number of Monte Carlo samples and, for each sample $j = 1, \ldots, M$, we simulate a simple random sample from X, $(\widetilde{X}_1^{\#}, \ldots, \widetilde{X}_{20}^{\#})$ where $\widetilde{X}_i^{\#} \sim Exp(2)$ are iid, and construct the Monte Carlo sample statistics

$$\overline{X^2}_j^{\#} = \frac{1}{20} \sum_{i=1}^{20} (\widetilde{X}_i^{\#})^2.$$

The Monte Carlo approximation to the CDF of $\overline{X^2}$ is then

$$F_{\overline{X^2}}^{\#}(t) = \frac{1}{M} \sum_{j=1}^{M} I\left(\overline{X^2}_j^{\#} \le t \right).$$

The left panel of Figure 11.4 shows the histogram of the Monte Carlo sample, $\overline{X^2}_1^{\#}, \ldots, \overline{X^2}_M^{\#}$, where $M = 100\,000$, and the right panel of the figure shows the Monte Carlo approximation of the CDF of $\overline{X^2}$ (dashed line), i.e. $F_{\overline{X^2}}^{\#}(t)$. As mentioned above, an analytic expression for the distribution of $\overline{X^2}$ is not readily available, and hence this cannot be plotted in the figure.

Note that, similarly to the assumption in Section 11.3, the approach described above relies on knowing the population distribution F_X as well as the sampling mechanism. This approach is infeasible in practice. In Chapter 13, we use the insights obtained in this section, to show how we can use simulation to estimate the sampling distribution of an estimator in a feasible way, i.e. without relying on knowing the population distribution F_X.

If the Monte Carlo samples are drawn independently from the underlying population, then, due to the Law of Large Numbers, the Monte Carlo approximation to the CDF of a sample statistics T is a consistent estimator for the CDF of T, as the number of Monte Carlo repetitions goes to infinity. This is shown in the next example.

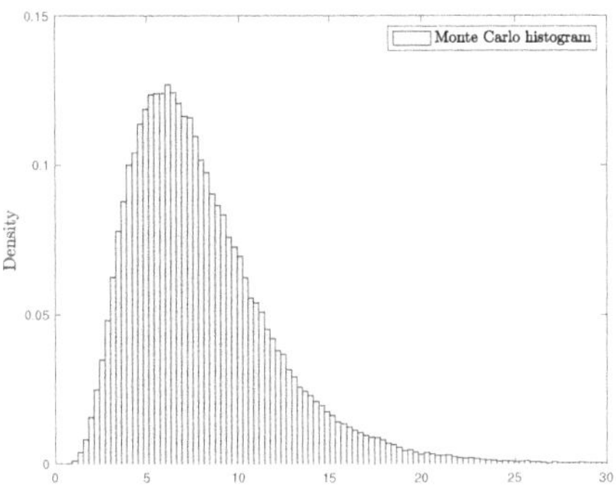
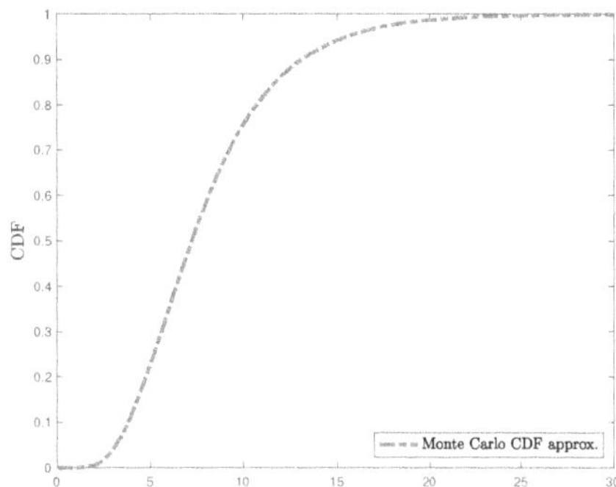

FIGURE 11.4
Monte Carlo approximation to the sampling distribution of the sample statistics $\overline{X^2}$ (Example 11.13). Left panel: Histogram of Monte Carlo sample statistics $\overline{X^2}_j^{\#}$, $j = 1, \ldots, M$ with $M = 100\ 000$. Right panel: Monte Carlo approximation to the CDF of the sample statistics $\overline{X^2}$, $F_{\overline{X^2}}^{\#}(t)$ (dashed curve).

Example 11.14 *Consider a sample statistics T, given by*

$$T = h\left(\widetilde{X}_1, \ldots, \widetilde{X}_n\right),$$

where $(\widetilde{X}_1, \ldots, \widetilde{X}_n)$ is a random sample. Our aim is to use Monte Carlo simulation to approximate the CDF F_T of T. By definition, F_T is given by

$$F_T(t) = \Pr\left(T \leq t\right) = \Pr\left(h\left(\widetilde{X}_1, \ldots, \widetilde{X}_n\right) \leq t\right).$$

Let $\left(\widetilde{X}_{1,j}^{\#}, \ldots, \widetilde{X}_{n,j}^{\#}\right)$ be the jth Monte Carlo sample and let

$$T_j^{\#} = h\left(\widetilde{X}_{1,j}^{\#}, \ldots, \widetilde{X}_{n,j}^{\#}\right)$$

be the jth Monte Carlo sample statistics. Suppose we do M Monte Carlo repetitions and, thereby, obtain $j = 1, \ldots, M$ Monte Carlo sample statistics $T_j^{\#}$. The Monte Carlo approximation $F_T^{\#}(t)$ to the CDF $F_T(t)$ is given by

$$F_T^{\#}(t) = \frac{1}{M} \sum_{j=1}^{M} I\left(T_j^{\#} \leq t\right).$$

Assume each Monte Carlo sample is drawn according to the same statistical experiment as the one leading to the random sample $(\widetilde{X}_1, \ldots, \widetilde{X}_n)$. This implies that $F_T = F_{T^{\#}}$. In addition, assume we draw each Monte Carlo sample such that it is independent of the other Monte Carlo samples. These two assumptions imply that $F_{T_j^{\#}} = F_{T^{\#}} = F_T$ for all $j = 1, \ldots, M$. In other words, $(T_1^{\#}, \ldots, T_M^{\#})$ constitutes a simple random sample from T. We apply the Law of Large Numbers to the transformation

$$g\left(T_j^{\#}\right) = I\left(T_j^{\#} \leq t\right),$$

where $I()$ is the indicator function. According to Corollary 11.5.2,

$$\operatorname*{plim}_{M \to \infty} \frac{1}{M} \sum_{j=1}^{M} I\left(T_j^{\#} \leq t\right) = E\left(I\left(T_j^{\#} \leq t\right)\right) = \Pr\left(T^{\#} \leq t\right) = F_{T^{\#}}(t) = F_T(t),$$

where we used that for an event A, then $E(I(A)) = \Pr(A)$. Hence, the Monte Carlo approximation consistently estimates the CDF of T.

*   Notice, this consistently result does not require anything about the random sample $(\widetilde{X}_1, \ldots, \widetilde{X}_n)$. The only requirements are that the Monte Carlo samples are drawn similarly to the random sample $(\widetilde{X}_1, \ldots, \widetilde{X}_n)$ and that the Monte Carlo samples are drawn independently of each other.*

From the discussion above, and as seen in Examples 11.12–11.13, to run a Monte Carlo simulation, it is necessary to have a method to draw observations from a population. In other words, we need a method to generate observations from a *CDF* $F_X(x)$. Most software packages will have routines to generate draws from some distributions using a so-called *pseudo-random number generator*. In the appendix (Chapter C.1), we give a brief overview of how to draw pseudorandom numbers from a uniform distribution and in the next section, we show how to use numbers drawn from a uniform distribution to generate draws of random variables from any distribution of interest.

11.7.2   Random draws from a distribution

In this section, we show how drawing observations from any CDF can be done by drawing observations from a uniform $U(0,1)$ distribution.

To understand how a uniform distribution is related to any other distribution, we first focus on continuous random variables. For a continuous random variable X with CDF F_X, the random variable $F_X(X)$ is distributed as a uniform random variable on the interval $[0,1]$. This result is known as the *probability integral transform*.

Theorem 11.6 (Probability integral transform) *Assume that the random variable X is continuous with a distribution according to the CDF F_X. Then the random variable*

$$Y = F_X(X)$$

has a uniform distribution on $[0,1]$, that is, $Y \sim U(0,1)$.

Theorem 11.6 suggests a way to obtain a random variable X with distribution F_X based on a uniform distribution. Suppose F_X has an inverse function F_X^{-1}. Then $X = F_X^{-1}(Y)$. Thus, the transformation of $Y \sim U(0,1)$ by the function $F_X^{-1}(Y)$ results in a random variable X with the distribution F_X.

In Theorem 11.7 below, we show how we can use draws from a uniform distribution to generate a draw from any distribution of random variables. To state this theorem, we make use of p-quantiles q_p, for $p \in [0,1]$, see Section 5.6. Recall that, if X is a continuous random variable, then a p-quantile satisfies that

$$F_X(q_p) = p.$$

In general, there may be more than one value of q_p that can be p-quantiles. For our purpose here, we need the smallest of them. That is, let $\underline{q}_p$ be the smallest of the p-quantiles. This can be compactly written as

$$\underline{q}_p = \min \ q_p. \tag{11.8}$$

Example 11.15 *In Example 5.13, we considered a discrete random variable X for which all values in the interval $[10, 12]$ are 0.20-quantiles, that is $q_{0.20} = [10, 12]$. By definition $\underline{q}_{0.20}$ is the smallest of these values, i.e. $\underline{q}_{0.20} = \min q_p = \min[10, 12] = 10$.*

Based on $\underline{q}_p$, the next theorem contains the result on how to generate observations from any distribution based on draws from the uniform $U(0, 1)$ distribution.

Theorem 11.7 (Generating observations from CDF using uniform random numbers) *Let $F_X(x)$ be a CDF and $\underline{q}_p$ the smallest of the p-quantiles for each value of p, see (11.8). Assume $U \sim U(0, 1)$. Then the random variable*

$$X = \underline{q}_U$$

is distributed according to the CDF $F_X(x)$.

Example 11.16 *Consider the Binomial distribution $X \sim Bin(n, p)$ with $n = 2$ and $p = 0.4$. This distribution has the PMF*

$$f_X(x) = \begin{cases} 0.36 & if \quad x = 0, \\ 0.48 & if \quad x = 1, \\ 0.16 & if \quad x = 2, \end{cases}$$

and the corresponding CDF

$$F_X(x) = \begin{cases} 0 & if \quad x < 0, \\ 0.36 & if \quad 0 \le x < 1, \\ 0.84 & if \quad 1 \le x < 2, \\ 1 & if \quad x \ge 2. \end{cases}$$

Then the p-quantiles are

$$\underline{q}_p = \begin{cases} 0 & if \quad p < 0.36, \\ 1 & if \quad 0.36 \le p < 0.84, \\ 2 & if \quad 0.84 \le p < 1. \end{cases}$$

For example, the median $q_{0.5}$ equals 1.
Let $U \sim U(0, 1)$, then

$$X = \underline{q}_U = \begin{cases} 0 & if \quad U \in [0, 0.36), \\ 1 & if \quad U \in [0.36, 0.84), \\ 2 & if \quad U \in [0.84, 1]. \end{cases}$$

With this construction, the distribution of X is in accordance with $X \sim Bin(2, 0.4)$, as ensured by Theorem 11.7. Indeed, by the properties of the uniform distribution, we have that $Pr(U \in [0, 0.36]) = 0.36$, $Pr(U \in (0.36, 0.84]) = 0.84 - 0.36 = 0.48$, and $Pr(U \in (0.84, 1]) = 1 - 0.84 = 0.16$, which matches the PMF values $f_X(0), f_X(1)$, and $f_X(2)$, given above.

Example 11.17 *Consider the Exponential distribution $X \sim Exp(2)$. The CDF for this distribution is*

$$F_X(x) = 1 - \exp(-2 \cdot x), \quad x \ge 0.$$

Since F_X is continuous and monotonically increasing for all $x \ge 0$, its inverse exists. Hence, as discussed above, we have $q_p = F_X^{-1}(p)$, where q_p is the p-quantile of the distribution of X. This p-quantile can be found by solving the equation $F_X(q_p) = p$, which the reader is invited *to verify yields $q_p = F_X^{-1}(p) = -\frac{1}{2}\ln(1 - p)$. To generate outcomes from this distribution using realizations of a uniformly distributed random variable $U \sim U(0,1)$, let*

$$Y = -\frac{1}{2}\ln(1 - U).$$

Theorem 11.7 now ensures that Y generated this way has CDF F_X, i.e. $Y \sim Exp(2)$.

To implement this method on a computer, the last remaining obstacle is to generate observations from the uniform distribution. This can be done using pseudo-random numbers, see, e.g. Chapter C in the appendix for details. This appendix also contains further methods for simulating from a distribution.

11.8   Proofs

Proof of Theorem 11.1 *By definition,*

$$MSE\left(\widehat{\theta}\right) = E\left(\left(\widehat{\theta} - \theta_0\right)^2\right).$$

"Add 0" in the form of $\left(E(\widehat{\theta}) - E(\widehat{\theta})\right)$ and rearrange

$$MSE\left(\widehat{\theta}\right) = E\left(\left(\widehat{\theta} - \theta_0 + \left(E(\widehat{\theta}) - E(\widehat{\theta})\right)\right)^2\right) = E\left(\left(\left(\widehat{\theta} - E(\widehat{\theta})\right) + \left(E(\widehat{\theta}) - \theta_0\right)\right)^2\right).$$

Square the term inside the expectation to get

$$
\begin{aligned}
MSE\left(\widehat{\theta}\right) &= E\left(\left(\widehat{\theta} - E(\widehat{\theta})\right)^2 + \left(E(\widehat{\theta}) - \theta_0\right)^2 + 2\left(E_{\widehat{\theta}}(\widehat{\theta}) - \theta_0\right)\left(\widehat{\theta} - E(\widehat{\theta})\right)\right) \\
&= \underbrace{E\left(\left(\widehat{\theta} - E(\widehat{\theta})\right)^2\right)}_{(I)} + \underbrace{E\left(\left(E(\widehat{\theta}) - \theta_0\right)^2\right)}_{(II)} + \underbrace{E\left(2\left(E(\widehat{\theta}) - \theta_0\right)\left(\widehat{\theta} - E(\widehat{\theta})\right)\right)}_{(III)}.
\end{aligned}
$$

Consider each of the three terms. By definition of variance, term (I) is

$$E\left(\widehat{\theta} - E(\widehat{\theta})\right)^2 = Var\left(\widehat{\theta}\right).$$

By definition of bias, the inside of term (II) is

$$E\left(E(\widehat{\theta}) - \theta_0\right)^2 = E\left(Bias(\widehat{\theta})\right)^2.$$

Since $Bias(\widehat{\theta})$ is a number, i.e. it is non-random, then

$$E\left(Bias(\widehat{\theta})\right)^2 = \left(Bias(\widehat{\theta})\right)^2.$$

In term (III), the first part is also number, i.e. non-random. It can therefore be taken outside $E()$ to get

$$E\left(2\left(E(\widehat{\theta}) - \theta_0\right)\left(\widehat{\theta} - E(\widehat{\theta})\right)\right) = 2\left(E(\widehat{\theta}) - \theta_0\right)E\left(\widehat{\theta} - E(\widehat{\theta})\right).$$

Since

$$E\left(\widehat{\theta} - E(\widehat{\theta})\right) = E\left(\widehat{\theta}\right) - E(\widehat{\theta}) = 0,$$

term (III) equals 0.

Insert the results for terms (I) to (III) to get

$$MSE\left(\widehat{\theta}\right) = Var\left(\widehat{\theta}\right) + \left(Bias(\widehat{\theta})\right)^2 + 0,$$

which completes the proof.

Proof of Theorem 11.2 *Consider the sample average $\overline{X}$ defined in Equation (11.1). The bias of the sample average is*

$$
\begin{aligned}
Bias\left(\overline{X}, \mu_0\right) &= E\left(\overline{X}\right) - \mu_0 \\
&= E\left(\frac{1}{n}\sum_{i=1}^{n}\widetilde{X}_i\right) - \mu_0 \\
&= \frac{1}{n}\sum_{i=1}^{n}E\left(\widetilde{X}_i\right) - \mu_0 \\
&= \frac{1}{n}\sum_{i=1}^{n}\mu_0 - \mu_0 \\
&= \frac{1}{n}n\mu_0 - \mu_0 \\
&= \mu_0 - \mu_0 \\
&= 0.
\end{aligned}
$$

Hence, the sample average $\overline{X}$ with a simple random sample is an unbiased estimator of the population mean $\mu_0 = E(X)$.

The variance of the sample average is

$$
\begin{aligned}
Var\left(\overline{X}\right) &= E\left(\left(\overline{X} - (\overline{X})\right)^2\right) \\
&= E\left(\left(\overline{X} - \mu_0\right)^2\right) \\
&= E\left(\left(\frac{1}{n}\sum_{i=1}^{n}\widetilde{X}_i - \mu_0\right)^2\right) \\
&= \left(\frac{1}{n}\right)^2 \cdot E\left(\left(\sum_{i=1}^{n}\left(\widetilde{X}_i - \mu_0\right)\right)^2\right) \\
&= \frac{1}{n^2} \cdot \sum_{i=1}^{n}\sum_{j=1}^{n}E\left(\left(\widetilde{X}_i - \mu_0\right)\left(\widetilde{X}_j - \mu_0\right)\right).
\end{aligned}
$$

Now, due to the statistical independence between $\widetilde{X}_i$ and $\widetilde{X}_j$ when $i \neq j$, we have (Theorem 7.4)

$$E\left(\left(\widetilde{X}_i - \mu_0\right)\left(\widetilde{X}_j - \mu_0\right)\right) = Cov(\widetilde{X}_i, \widetilde{X}_j) = 0.$$

When $i = j$, we find

$$E\left(\left(\widetilde{X}_i - \mu_0\right)\left(\widetilde{X}_j - \mu_0\right)\right) = E\left(\left(\widetilde{X}_i - \mu_0\right)^2\right) = Var(\widetilde{X}_i) = Var(X),$$

where $Var(X) = \sigma_0^2$ is the population variance. Insert this to get

$$
\begin{aligned}
Var\left(\overline{X}\right) &= \frac{1}{n^2} \cdot \sum_{i=1}^{n} E\left(\left(\tilde{X}_i - \mu_0\right)\left(\tilde{X}_i - \mu_0\right)\right) \\
&= \frac{1}{n^2} \cdot \sum_{i=1}^{n} \sigma_0^2 \\
&= \frac{1}{n^2} \cdot n\sigma_0^2 \\
&= \frac{1}{n} \cdot \sigma_0^2.
\end{aligned}
$$

Finally, use Theorem 11.1 to conclude that

$$
MSE\left(\overline{X}\right) = \left(Bias\left(\overline{X}\right)\right)^2 + Var\left(\overline{X}\right) = 0^2 + \frac{1}{n} \cdot \sigma_0^2 = \frac{1}{n} \cdot \sigma_0^2.
$$

Proof of Theorem 11.3 *The proof relies on a result called* Markov's Inequality, *which we state below in Lemma 11.8.1. Let $\varepsilon > 0$. By Markov's Inequality (Lemma 11.8.1) applied to the non-negative random variable $X = \left|\widehat{\theta}_n - \theta_0\right|^2$, we get*

$$
\begin{aligned}
Pr\left(\left|\widehat{\theta}_n - \theta_0\right| \geq \varepsilon\right) &= Pr\left(\left|\widehat{\theta}_n - \theta_0\right|^2 \geq \varepsilon^2\right) \\
&\leq \frac{E\left(\left|\widehat{\theta}_n - \theta_0\right|^2\right)}{\varepsilon^2} \\
&= \frac{MSE(\widehat{\theta}_n)}{\varepsilon^2}.
\end{aligned}
$$

Using the assumption that $MSE(\widehat{\theta}_n) \to 0$ as $n \to \infty$, we get that

$$
\begin{aligned}
Pr\left(\left|\widehat{\theta}_n - \theta_0\right| < \varepsilon\right) &= 1 - Pr\left(\left|\widehat{\theta}_n - \theta_0\right| \geq \varepsilon\right) \\
&\geq 1 - \frac{MSE(\widehat{\theta}_n)}{\varepsilon^2} \\
&\to 1 - 0 \\
&= 1,
\end{aligned}
$$

as $n \to \infty$. Thus, $Pr\left(\left|\widehat{\theta}_n - \theta_0\right| < \varepsilon\right) \geq 1$, which implies that $Pr\left(\left|\widehat{\theta}_n - \theta_0\right| < \varepsilon\right) = 1$. This concludes the proof that $\widehat{\theta}_n$ is a consistent estimator for θ_0 (Definition 11.9).

Lemma 11.8.1 (Markov's Inequality) *Let X be a non-negative random variable and $a > 0$ a number. It holds that*

$$
Pr(X \geq a) \leq \frac{E(X)}{a}.
$$

Proof of Lemma 11.8.1 *Let $I(A)$ be the indicator function of the event A, i.e. $I(A) = 1$ if A is true and $I(A) = 0$ otherwise. We have that*

$$a \cdot I(X \geq a) \leq X.$$

Taking expectations on both sides, and using that the expectation operator is a monotonically increasing function (i.e. $E(X) \leq E(Y)$ if $X \leq Y$), we get that

$$a \cdot E\left(I(X \geq a)\right) \leq E(X).$$

As shown in Theorem 12.1 in the next chapter, the expectation of an indicator function is equal to the probability of the event it is defined with respect to, i.e. $E\left(I(X \geq a)\right) = Pr\left(X \geq a\right)$. Using this, we get

$$a \cdot Pr(X \geq a) \leq E(X).$$

Dividing both sides by a, which is valid since $a > 0$, yields the result.

Proof of Theorem 11.4 *In Example 11.6, we used Theorem 11.3 to show that $\overline{X}_n$ is a consistent estimator of the population mean μ_0. As remarked in the end of Section 11.5, this is simply a different way of saying $\operatorname*{plim}_{n \to \infty} \overline{X}_n = \mu_0$, which is what we wanted to show.*

Proof of Theorem 11.5 *We prove the result for two random variables, the extension to more than two random variables is straight forward. Let X_1 and X_2 be two statistically independent and identically distributed random variables. Note first, for any function $g()$, it is clear that $g(X_1)$ and $g(X_2)$ have the same distribution. To prove independence, define $Y_1 = g(X_1)$ and $Y_2 = h(X_2)$, where $g()$ and $h()$ are functions. Let a_1 and a_2 be numbers and define the sets $B_1 = \{x : g(x) \leq a_1\}$ and $B_2 = \{x : h(x) \leq a_2\}$. Note that $Y_1 \leq a_1$ if and only if $X_1 \in B_1$, and, similarly, $Y_2 \leq a_2$ if and only if $X_2 \in B_2$. Now,*

$$
\begin{aligned}
F_{Y_1,Y_2}(y_1, y_2) &= Pr(Y_1 \leq y_1, Y_2 \leq y_2) \\
&= Pr(X_1 \in B_1, X_2 \in B_1) \\
&= Pr(X_1 \in B_1)Pr(X_2 \in B_2) \\
&= Pr(Y_1 \leq y_1)Pr(Y_2 \leq y_2) \\
&= F_{Y_1}(y_1)F_{Y_2}(y_2),
\end{aligned}
$$

which means that Y_1 and Y_2 are statistically independent (Definition 7.2), as we wanted to show. Note that, in the third equality above, we used the fact that for two statistically independent random variables X_1 and X_2, then $Pr(X_1 \in C, X_2 \in D) = Pr(X_1 \in C)Pr(X_2 \in D)$ for all Borel measurable sets C and D.

Proof of Theorem 11.6 *Define $Y = F_X(X)$. Suppose first that the CDF F_X has an inverse F_X^{-1}. Then, for $y \in [0, 1]$,*

$$
\begin{aligned}
F_Y(y) &= Pr(Y \leq y) \\
&= Pr(F_X(X) \leq y) \\
&= Pr(X \leq F_X^{-1}(y)) \\
&= F_X(F_X^{-1}(y)) \\
&= y.
\end{aligned}
$$

In other words, the CDF of Y, F_Y, is the CDF of a uniform distribution on the unit interval, i.e. $U(0,1)$. We conclude that $Y \sim U(0,1)$, as we wanted to show.

If F_X does not have an inverse, i.e. if F_X^{-1} does not exist, we can use the fact that $F_X(x)$ is right-continuous (Property 4 of Definition 2.1) and repeat the proof in the same manner, but with the (modified) generalized inverse function of F_X, $\tilde{F}_X^{-1}$, in place of F_X^{-1}, where $\tilde{F}_X^{-1}(y)$ is defined via $\tilde{F}_X^{-1}(y) = \min\{x \in \mathbb{R} : F_X(x) \geq y\}$ for $y \in (0,1)$. Notice, $\tilde{F}_X^{-1}(y)$ equals the minimum p-quantile $\underline{q}_p$, as defined in (11.8). At the end-points $y = 0$ and $y = 1$, we modify the function such that $\tilde{F}_X^{-1}(0) = -\infty$ and $\tilde{F}_X^{-1}(1) = \infty$.

Proof of Theorem 11.7 *Suppose first that the CDF F_X has an inverse F_X^{-1}. Let $U \sim U(0,1)$ and define $Y = F_X^{-1}(U)$. We have*

$$
\begin{aligned}
Pr(Y \leq y) &= Pr(F_X^{-1}(U) \leq y) \\
&= Pr(U \leq F_X(y)) \\
&= F_X(y).
\end{aligned}
$$

In other words, the random variable $Y = F_X^{-1}(U)$ has CDF F_X, as we wanted to show.

If F_X does not have an inverse, we repeat the proof above with the generalized inverse $\tilde{F}_X^{-1}(y)$ of F_X, given by $\tilde{F}_X^{-1}(y) = \min\{x \in \mathbb{R} : F_X(x) \geq y\}$ for $y \in [0,1]$, in place of F_X, using also the fact that F_X is right-continuous, by Property 4 of Definition 2.1, see also the proof of Theorem 11.6 above.

11.9   Exercises

Problem 11.9.1 *Let $\widehat{\theta}$ be an estimator of the estimand θ_0.*

1. *Suppose the outcome of the estimator results in the estimate $\widehat{\theta} = \theta_0 + 4$. Calculate the losses resulting from the squared error loss function (11.2) and the absolute value loss function in Equation (11.3). Re-do the calculations with $\widehat{\theta} = \theta_0 + \frac{1}{4}$.*

2. *Compare the results found in the previous question and reflect on how the two loss functions assigns different relative losses to a case where the estimate is an "outlier", i.e. far from the true value of the estimand (first case), and to a case where the estimate is relatively more close to the true value of the estimand (second case).*

Problem 11.9.2 *Suppose a simple random sample $\left(\tilde{X}_1, \ldots, \tilde{X}_n\right)$ is available from a population X with a Bernoulli distribution $X \sim Ber(p_0)$, where $p_0 \in [0,1]$. Consider the sample average of the Bernoulli distributed random variables*

$$
\overline{X} = \frac{1}{n} \sum_{i=1}^{n} \tilde{X}_i.
$$

1. *Show that $E(\overline{X}) = p_0$ and $Var(\overline{X}) = \frac{p_0(1-p_0)}{n}$.*

2. *Argue that $\overline{X}$ is an unbiased and consistent estimator for p_0.*

Problem 11.9.3 *Suppose a simple random sample $\left(\widetilde{X}_1, \ldots, \widetilde{X}_n\right)$ is available from a population X with population mean $\mu_0 = E(X)$ and variance $\sigma_0^2 = Var(X) < \infty$. Let $k_n = \lfloor \frac{n}{2} \rfloor$ be the integer part of $\frac{n}{2}$ and consider the two estimators of μ_0:*

$$
\begin{aligned}
\overline{X}_n &= \frac{1}{n} \sum_{i=1}^{n} \widetilde{X}_i, \\
\overline{X}_{k_n} &= \frac{1}{k_n} \sum_{i=1}^{k_n} \widetilde{X}_i.
\end{aligned}
$$

In effect, the estimator $\overline{X}_n$ uses the whole sample, while $\overline{X}_{k_n}$ uses only the first half of the observations of the sample.

1. *Argue that both $\overline{X}_n$ and $\overline{X}_{k_n}$ are unbiased estimators of μ_0. Argue also that both $\overline{X}_n$ and $\overline{X}_{k_n}$ are consistent estimators of μ_0 as $n \to \infty$.*

2. *Show that when $n \geq 2$, $\overline{X}_n$ is a more efficient estimator than $\overline{X}_{k_n}$, i.e. show that the variance of $\overline{X}_n$ is lower than the variance of $\overline{X}_{k_n}$.*

Problem 11.9.4 *Suppose a simple random sample $\left(\widetilde{X}_1, \ldots, \widetilde{X}_n\right)$ is available from a population X with population mean $\mu_0 = E(X)$ and variance $\sigma_0^2 = Var(X) < \infty$. Consider the following estimator of μ_0,*

$$
\widehat{\theta} = \overline{X} + \frac{1}{n}.
$$

1. *Calculate the MSE of $\widehat{\theta}$.*

2. *Show that $\widehat{\theta}$ is a biased, but consistent, estimator of μ_0.*

3. *Which estimator is more efficient, $\widehat{\theta}$ or $\overline{X}$?*

Problem 11.9.5 *Suppose a random sample $\left(\widetilde{X}_1, \ldots, \widetilde{X}_n\right)$ is available from a population X with population mean $\mu_0 = E(X)$ and variance $\sigma_0^2 = Var(X) < \infty$. In Theorem 11.2 and Example 11.6, we showed that when $(\widetilde{X}_1, \ldots, \widetilde{X}_n)$ is a simple random sample, then the sample average $\overline{X}$ is an unbiased and consistent estimator of the population mean $\mu_0 = E(X)$. In this problem, you will show that the sample average $\overline{X}$ may still be an unbiased and consistent estimator of the population mean μ_0, even when the sample used to construct $\overline{X}$ is not a simple random sample. In particular we show that, under certain conditions, $\overline{X}$ is a consistent estimator of the population mean when the elements in the sample are not independent, e.g. if X is a time series.*

Assume that the random sample $\left(\widetilde{X}_1, \ldots, \widetilde{X}_T\right)$ is collected such that $Cov(\widetilde{X}_t, \widetilde{X}_s) = 0$ if $|t - s| > 1$, while $Cov(\widetilde{X}_t, \widetilde{X}_s) = \frac{1}{2}\sigma_0^2$ if $|t - s| = 1$. That is, observations adjacent to each other are correlated, while observations more than one period apart are uncorrelated. Consider $\overline{X} = \frac{1}{T} \sum_{t=1}^{T} \widetilde{X}_t$ as an estimator of μ_0.

1. *Show that $\overline{X}$ is an unbiased estimator of μ_0.*

2. *Derive the MSE of $\overline{X}$.*

3. *Argue that $\overline{X}$ is a consistent estimator of μ_0.*

Problem 11.9.6 *Suppose a simple random sample $\left(\tilde{X}_1, \ldots, \tilde{X}_n\right)$ is available from a population X with an exponential distribution $Exp(\lambda_0)$. Let the estimand be $\theta_0 = 1/\lambda_0$.*

1. *Show that $\overline{X}$ is an unbiased estimator of θ_0.*

2. *Derive an expression for the distribution of the sample average $\overline{X}$. (Hint: use the Erlang distribution.)*

3. *Argue that $\overline{X}$ is a consistent estimator of θ_0.*

Problem 11.9.7 *Let a population be given by a random variable X with an exponential distribution $X \sim Exp(\lambda)$ and density function*

$$f_X(x) = \lambda \exp(-\lambda x), \quad for \ x \geq 0.$$

Suppose a simple random sample $(\tilde{X}_1, \tilde{X}_2)$ with $n = 2$ observations is available.

1. *Derive the sample distribution of $(\tilde{X}_1, \tilde{X}_2)$ (expressed as a joint density function).*

2. *Consider the sample average*

$$\overline{X} = \frac{1}{2} \sum_{i=1}^{2} \tilde{X}_i.$$

Assume $\lambda = 0.25$. Notice, this implies that the mean of the population is $E(X) = 1/0.25 = 4$.

Express the sampling distribution of $\overline{X}$ by the CDF $F_{\overline{X}}$. Then calculate the probability that $\overline{X}$ has an outcome less than or equal to 3. (Hint: the sum $\sum_{i=1}^{2} \tilde{X}_i$ follows the Erlang distribution $Erlang(\lambda = 0.25, n = 2)$. Use the CDF of the Erlang distribution.)

Problem 11.9.8 *Let a population be given by a random variable X with the normal distribution $X \sim N(\mu_0, 1)$, that is, a normal distribution with known variance equal to 1. Assume the mean μ_0 can be any value in the interval $[-100, 100]$. Also assume a simple random sample $(\tilde{X}_1, \ldots, \tilde{X}_n)$ is available.*

To estimate μ_0, consider two estimators. The first estimator is the sample average

$$\overline{X} = \frac{1}{n} \sum_{i=1}^{n} \tilde{X}_i,$$

and the other estimator is

$$\hat{\tau} = 3,$$

that is, the estimator $\hat{\tau}$ always gives the outcome 3.

1. *Using the Mean Squared Error MSE as risk function, derive the MSE of each of the estimators*

2. *Are $\overline{X}$ and $\hat{\tau}$ consistent estimators of μ_0?*

3. *Is one of the estimators at least as efficient as the other estimator?*

4. *Based on your answers above, which of the estimators do you prefer?*

Problem 11.9.9 *Consider a population of legal claims. Assume the population is represented by X, which can be either 0, 1, or 2. Assume the PMF of X is*

$$f_X(x) = \begin{cases} 0.5 & if \quad x = 0, \\ 0.3 & if \quad x = 1, \\ 0.2 & if \quad x = 2. \end{cases}$$

Assume a simple random sample $(\widetilde{X}_1, \widetilde{X}_2)$ with $n = 2$ observations is available. Consider the sample average

$$\overline{X} = \frac{1}{n} \sum_{i=1}^{n} \widetilde{X}_i.$$

1. *Calculate the mean number of legal claims, that is, calculate $E(X)$.*

2. *Calculate the sampling distribution $f_{\overline{X}}(x)$ of $\overline{X}$, where $f_{\overline{X}}(x)$ is a probability function. (Hint: first write up the possible outcomes of $\overline{X}$ for $n = 2$. Then calculate the probabilities for each of those possible outcomes.)*

3. *Based on the sampling distribution $f_{\overline{X}}(x)$ of $\overline{X}$ calculated in part 2, calculate the mean of $\overline{X}$. Is $\overline{X}$ an unbiased estimator of the mean of X?*

Problem 11.9.10 *Let X be a random variable such that $E(X) = 3$ and $Var(X) = 5$.*

1. *Calculate $E(X^2)$. (Hint: You may use the rules of calculation for the variance in Theorem 5.3.)*

2. *Suppose a simple random sample $(\widetilde{X}_1, \ldots, \widetilde{X}_n)$ from a population, represented by X, is available. Let $\widehat{\tau}$ be an estimator defined as*

$$\widehat{\tau} = \frac{1}{n} \sum_{i=1}^{n} \left(\widetilde{X}_i \right)^2.$$

Show that $\widehat{\tau}$ is an unbiased estimator of $E(X^2)$. Is $\widehat{\tau}$ a consistent estimator of $E(X^2)$?

Problem 11.9.11 *The purpose of this problem is to conduct a Monte Carlo simulation to demonstrate the working of the Law of Large Numbers.*

Assume the population is a t-distribution with 6 degrees of freedom, i.e. $X \sim t(6)$. (See, e.g. Section 8.6.2 for details on the t-distribution.)

Consider the sample average $\overline{X}$ as used in the Law of Large Numbers (Theorem 11.4).

1. *Use Monte Carlo simulation to calculate the Monte Carlo distribution of the sample average $\overline{X}$ for $n = 200$.*

   - *Make a plot of the Monte Carlo distribution, e.g. the CDF $F^*_{\overline{X}^*}(t)$.*
   - *Report the $0.1, 0.5$ and 0.9 quantiles of the Monte Carlo distribution.*
   - *Report the probability of an outcome in the neighborhood $(-0.05, 0.05)$ of 0.*

2. *Redo part 1 for $n = 800$. Compare your results for the two cases $n = 200$ and $n = 800$.*

3. *Redo parts 1 to 2 for a t-distribution with 1 degree of freedom. Compare your results for the two different t-distributions. Relate your comments to the fact that the t-distribution with 1 degree of freedom does not have a mean.*

Problem 11.9.12 *Examples 11.12 and 11.13 illustrated how to use Monte Carlo simulation to approximate the sampling distribution of two sample statistics when the population characteristics was Exponentially distributed. In this exercise, you are asked to perform a similar analysis with a different population distribution.*

Let $Y = \exp(X)$, where $X \sim N(0,1)$. The distribution of the population random variable Y is called the log-normal distribution.

1. *Simulate a simple random sample of size $n = 10$, $\widetilde{Y}_1, \ldots, \widetilde{Y}_{10}$, from Y. (Hint: You can simulate from the log-normal distribution by simulating $\widetilde{X}$ as a normal random variable and setting $\widetilde{Y} = \exp(\widetilde{X})$.) Calculate the sample mean $\overline{Y}$ and sample mean of the squares $\overline{Y^2}$.*

2. *Use Monte Carlo simulation to approximate the sampling distribution of the sample mean $\overline{Y}$. Plot a histogram of the Monte Carlo sample and the approximated CDF, as done in Example 11.12.*

3. *Use Monte Carlo simulation to approximate the sampling distribution of the sample statistics $\overline{Y^2}$. Plot a histogram of the Monte Carlo sample and the approximated CDF, as done in Example 11.13.*

12

Estimating aspects of a univariate distribution

12.1   Introduction

In this chapter, we consider estimation of various aspects of an unknown univariate distribution using a random sample. The aspects can be the distribution itself or other quantities of interest, such as the mean, the variance, other higher-order moments, the median, and other quantiles. We present and discuss a general approach, called the *analogy principle*, which is useful for constructing estimators of various quantities of interest. The analogy principle directly applies the definition of a population quantity to the construction of an estimator of that population quantity. This is done by replacing the distribution function in the definition of a population quantity by an estimator of the distribution function. The resulting estimators are denoted *analog estimators*.

We start by considering an estimator of the CDF of a univariate distribution called the *empirical distribution function*. This estimator is the basis for developing analog estimators for various aspects of a univariate distribution. Aspects of a univariate distribution can also be estimated by other alternative estimators than the analog estimators, such as the maximum likelihood estimator, the least squares estimator, and the methods of moments estimator. We discuss the maximum likelihood estimator in Chapters 17 and 18 of this book, and treat all three estimators thoroughly in Volume II. These estimators will prove useful when estimating more complex quantities of multivariate distributions. For many of the aspects of a univariate distribution, however, all these estimators are either equivalent or, at least, very similar. For this reason, we focus on only one of the approaches in this chapter, namely, the analog estimators.

12.2   Empirical distribution function

In this section, we consider how to estimate a CDF $F_X(x)$ based on a random sample from X. Though it may be interesting in itself to estimate the CDF, what makes it particularly interesting for our purposes is the fact that the estimator of the CDF can in turn be used to construct estimators of many different quantities of interest e.g. the mean and quantiles, as will be demonstrated in the following sections.

To construct an estimator of the CDF, let $I(X \leq x)$ denote the indicator function of the event $\{X \leq x\}$, i.e. the function taking the value 1 if $X \leq x$ and the value 0 if $X > x$. That is

$$I(X \leq x) = \left\{ \begin{array}{ll} 1 & \text{if} \quad X \leq x, \\ 0 & \text{if} \quad X > x. \end{array} \right.$$

DOI: 10.1201/9781003591191-12

We can get inspiration on how to construct an estimator of the population CDF $F_X(x) = Pr(X \leq x)$ by noting that $F_X(x)$ is equal to the expectation of the random variable $I(X \leq x)$. This is shown in the following theorem.

Theorem 12.1 (Expectation of indicator function) *Let X be a random variable with CDF $F_X(x)$. It holds that*

$$E(I(X \leq x)) = F_X(x).$$

Since $F_X(x) = E\left(I(X \leq x)\right)$ is an expected value, the Law of Large Numbers (Theorem 11.4) is suggestive that a sample average may be used as an estimator of $F_X(x)$. This estimator is denoted the empirical distribution function, or the empirical CDF (ECDF), and it is defined next.

Definition 12.1 (Empirical distribution function, ECDF) *Let $(\widetilde{X}_1, \ldots, \widetilde{X}_n)$ be a random sample. The **empirical distribution function** (ECDF) $\widehat{F}_X^{emp}(x)$ is given by*

$$\widehat{F}_X^{emp}(x) = \frac{1}{n} \sum_{i=1}^{n} I(\widetilde{X}_i \leq x),$$

for $x \in \mathbb{R}$.

The empirical distribution function $\widehat{F}_X^{emp}(x)$ simply counts the number of observations that are smaller than or equal to the number x, and divides by the sample size n. Equivalently, the empirical distribution function $\widehat{F}_X^{emp}(x)$ denotes the fraction of the observations that are smaller than or equal to x. Figure 12.1 illustrates the ECDF for two sample sizes, $n = 10$ and $n = 100$, when $(\widetilde{X}_1, \ldots, \widetilde{X}_n)$ is a simple random sample from a population represented by the random variable $X \sim N(0, 1)$.

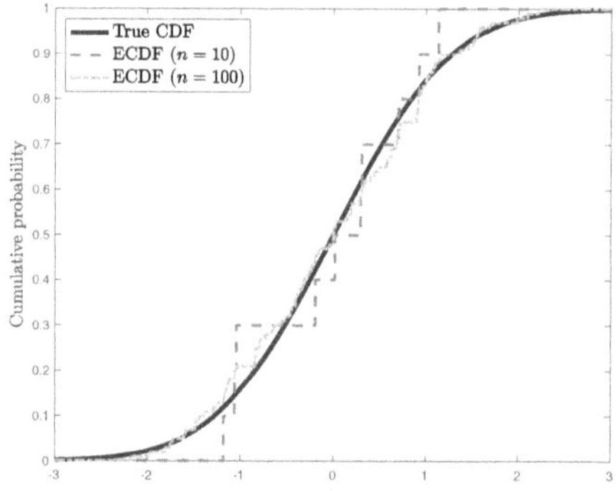

FIGURE 12.1
Empirical distribution function (Definition 12.1) obtained from two simple random samples $(\widetilde{X}_1, \ldots, \widetilde{X}_n)$ with $n = 10$ (dashed line) and $n = 100$ (dashed and dotted line), respectively. Here, the population random variable is $X \sim N(0, 1)$ and the solid line denotes the CDF F_X of the $N(0, 1)$ distribution.

With a simple random sample, the Law of Large Numbers can be invoked to prove consistency of the empirical distribution function. The result is stated in the next theorem.

Theorem 12.2 (Consistency of Empirical Distribution function) *Assume $(\widetilde{X}_1, \ldots, \widetilde{X}_n)$ is a simple random sample. For every $x \in \mathbb{R}$, the empirical distribution function $\widehat{F}_X^{emp}(x)$ is a consistent estimator of the CDF $F_X(x)$, i.e.*

$$\underset{n \to \infty}{plim} \; \widehat{F}_X^{emp}(x) = F_X(x).$$

Figure 12.1 illustrates Theorem 12.2: It is evident how the estimator $\hat{F}_X(x)$ of $F_X(x)$ becomes better as the sample size increases from $n = 10$ to $n = 100$.

Since the empirical distribution function is a consistent estimator of the CDF of a random variable, we may use it to estimate probabilities of outcomes of a random variable. The next example illustrates.

Example 12.1 (Estimating probability of positive stock returns) *Let X denote the return (measured in percent) of a financial stock, e.g. Apple Inc., in a randomly selected month. As an investor, you are interested in the probability that the stock will have a positive return in a given month, i.e. you would like to know $\Pr(X > 0)$. Let $F_X()$ denote the CDF of X and suppose $(\widetilde{X}_1, \ldots, \widetilde{X}_n)$ is a simple random sample from X. For any $x \in \mathbb{R}$, we may estimate $\Pr(X \leq x) = F_X(x)$ using the empirical CDF,*

$$\widehat{F}_X^{emp}(x) = \frac{1}{n} \sum_{i=1}^{n} I(\widetilde{X}_i \leq x).$$

Hence, the probability of a positive return, $\Pr(X > 0) = 1 - \Pr(X \leq 0) = 1 - F_X(0)$, may be estimated as

$$\widehat{\Pr}(X > 0) = 1 - \widehat{F}_X^{emp}(0) = 1 - \frac{1}{n} \sum_{i=1}^{n} I(\widetilde{X}_i \leq 0).$$

Using that $1 - I(\widetilde{X}_i \leq x) = I(\widetilde{X}_i > x)$ this can also be written as

$$\widehat{\Pr}(X > 0) = \frac{1}{n} \sum_{i=1}^{n} \left(1 - I(\widetilde{X}_i \leq 0)\right) = \frac{1}{n} \sum_{i=1}^{n} I(\widetilde{X}_i > 0).$$

That is, the estimate of $\Pr(X > 0)$ is simply the fraction of observations for which $\widetilde{X}_i > 0$.

A stronger version of Theorem 12.2, called the *Glivenko-Cantelli Theorem*, can be proven. This result is also known as the *Fundamental Theorem of Statistics*. The reason for this nomenclature is that the theorem connects observable quantities $(\widetilde{X}_1, \ldots, \widetilde{X}_n)$ with the basic building block of probability, namely the distribution function $F_X(x)$ of the random variable X.

Since $\widehat{F}_X^{emp}$ has the same structure as a *CDF* of a discrete random variable, namely a stair function, a PMF can be derived from it. The PMF $\widehat{f}_X^{emp}(x)$ is here called the *empirical probability mass function*, or the Empirical PMF (EPMF). It is defined next.

Definition 12.2 (Empirical probability mass function, EPMF) *Let $(\widetilde{X}_1, \ldots, \widetilde{X}_n)$ be a random sample. The **empirical probability mass function** (EPMF) is given by*

$$\widehat{f}_X^{emp}(x) = \frac{1}{n} \sum_{i=1}^{n} I(\widetilde{X}_i = x),$$

for $x \in \mathbb{R}$.

The empirical PMF only differs from 0 at the observations $\widetilde{X}_i$, and it assigns the probability $1/n$ to each of them. In case X is a discrete random variable with PMF $f_X(x)$, and the sample is a simple random sample, then $\widehat{f}_X^{emp}(x)$ is a consistent estimator of the PMF $f_X(x)$ (Problem 12.9.2).

For each x, the ECDF $\widehat{F}_X^{emp}(x)$ is a random variable since it is a function of a random sample $(\widetilde{X}_1, \ldots, \widetilde{X}_n)$. Thus, for each x, $\widehat{F}_X^{emp}(x)$ has a distribution. When $(\widetilde{X}_1, \ldots, \widetilde{X}_n)$ is a simple random sample, the sampling distribution of $\widehat{F}_X^{emp}(x)$ can be expressed using the Binomial distribution. The result is given in the next theorem.

Theorem 12.3 (Properties of ECDF) *Assume $(\widetilde{X}_1, \ldots, \widetilde{X}_n)$ is a simple random variable. Then, for each x and a, the CDF $F_{\widehat{F}_X^{emp}(x)}(a)$ of the ECDF $\widehat{F}_X^{emp}(x)$ is*

$$F_{\widehat{F}_X^{emp}(x)}(a) = \Pr\left(\widehat{F}_X^{emp}(x) \le a\right) = F_{S_n}(n \cdot a),$$

where $F_{S_n}()$ is the CDF of a Binomial random variable $S_n \sim Bin\,(n, F_X(x))$.
For each x, the bias, variance, and MSE of the ECDF are

$$
\begin{aligned}
Bias\left(\widehat{F}_X^{emp}(x), F_X(x)\right) &= 0, \\
Var\left(\widehat{F}_X^{emp}(x)\right) &= \frac{1}{n}F_X(x)\left(1 - F_X(x)\right), \\
MSE\left(\widehat{F}_X^{emp}(x), F_X(x)\right) &= \frac{1}{n}F_X(x)\left(1 - F_X(x)\right).
\end{aligned}
$$

It can be seen that the $MSE\left(\widehat{F}_X^{emp}(x), F_X(x)\right)$ of the empirical CDF is largest when $F_X(x) = 0.5$. This implies that for a given sample size, the precision is higher for estimation of $F(x)$ for $F(x)$ close to 0 or 1 than for $F(x)$ around 0.5.

Example 12.2 *Consider a simple random sample of $n = 10$ observations from a population X. Suppose we are interested in how precisely we can estimate $F(x) = 0.5$ using the ECDF. The MSE is*

$$MSE\left(\widehat{F}_X^{emp}(x), 0.5\right) = \frac{1}{10}0.5\left(1 - 0.5\right) = 0.025.$$

We can use the distribution of $\widehat{F}_X^{emp}(x)$ to calculate the probability that we obtain an estimate less than, say, 0.4. This probability is

$$F_{\widehat{F}_X^{emp}(x)}(0.4) = \Pr\left(\widehat{F}_X^{emp}(x) \le 0.4\right) = F_{S_n}(10 \cdot 0.4) = F_{S_n}(4),$$

where $S_n \sim Bin\,(n, F_X(x)) = Bin\,(10, 0.5)$. The CDF of the $Bin\,(10, 0.5)$ is

$$F_{S_{10}}(4) = f_{S_{10}}(0) + f_{S_{10}}(1) + f_{S_{10}}(2) + f_{S_{10}}(3) + f_{S_{10}}(4) = 0.38,$$

where $f_{S_{10}}$ is the PMF of the $Bin\,(10, 0.5)$, see Section 8.2. Thus, there is a probability of 38% that we obtain an estimate of $F(x) = 0.5$ less than 0.4.

12.3 The analogy principle

In this section, we consider the so-called analogy principle, which is a general approach to constructing estimators. We then illustrate how the analogy principle can be used to construct estimators of various aspects of a distribution, by using it to construct estimators of the mean, the variance, and a p-quantile.

Consider the estimation of a population parameter, e.g. the mean $E(X)$ of a population characterized by the random variable X. Suppose X is a discrete random variable with probability function $f_X(x)$. Then the mean of X is (Definition 5.1)

$$E(X) = \sum_x x \cdot f_X(x).$$

In practice, we cannot calculate the mean because we do not know the probability function $f_X(x)$. However, as we saw above, with a simple random sample, we can consistently estimate $f_X(x)$ by the empirical PMF $\widehat{f}_X^{emp}(x)$. The idea of the analogy principle is to replace $f_X(x)$ by $\widehat{f}_X^{emp}(x)$ in the formula for $E(X)$ and use this modified formula as an estimator $\widehat{E}(X)$ of $E(X)$:

$$\widehat{E}(X) = \sum_x x \cdot \widehat{f}_X^{emp}(x). \tag{12.1}$$

The analogy principle can be stated in the words of the econometrician Arthur Goldberger as follows:

> *A population parameter is a feature of the population. To estimate it, use the corresponding feature of the sample.*[1]

In other words, a feature is calculated by the appropriate formula imagining the sample is a population. When we imagine each observation is an element of a population, it has $1/n$ probability of being drawn. Thus, the distribution of these elements is the empirical CDF, or, equivalently, the empirical PMF. Thus, another way of stating the analogy principle is that an estimator can be constructed by replacing the population distribution in the formula for a population parameter with the empirical distribution. An estimator constructed from the analogy principle is called an analog estimator.

After having constructed an analog estimator, it still remains to be checked if the analog estimator has desirable properties, such as unbiasedness and consistency. In particular, consistency of an analog estimator often follows by invoking the Law of Large Numbers similar to how the empirical CDF was shown to be a consistent estimator of the population CDF in the first place (Theorem 12.2).

12.4   Analog estimators of PDF, mean, p-quantiles, and higher-order moments

12.4.1   Analog estimator of a PDF

A PDF $f_X(x)$ of a continuous random variable X can be estimated using the empirical CDF via the analogy principle. Suppose we want to estimate the value of the PDF at $x = x_0$. Recall from Theorem 2.12 that a PDF can be derived from a differentiable CDF. This implies that for $\varepsilon > 0$ small,

$$f_X(x_0) \approx \frac{F_X(x_0 + 0.5 \cdot \varepsilon) - F_X(x_0 - 0.5 \cdot \varepsilon)}{\varepsilon},$$

---

[1] Arthur S. Goldberger, *A Course in Econometrics* (1991), Harvard University Press, p. 117.

where $F_X()$ is the CDF of X. According to the analogy principle, we can replace F_X by the empirical distribution function $\widehat{F}_X^{emp}$ to obtain an estimator of the PDF, i.e.

$$\widehat{f}_X^{emp}(x_0) \approx \frac{\widehat{F}_X^{emp}(x_0 + 0.5 \cdot \varepsilon) - \widehat{F}_X^{emp}(x_0 - 0.5 \cdot \varepsilon)}{\varepsilon}. \tag{12.2}$$

This implies that an interval of width ε is chosen around x_0, and the average change of $\widehat{F}_X^{emp}$ is calculated over this interval. In practice, the value of ε is important. If ε is chosen too small, then either $\widehat{F}_X^{emp}$ does not change over the interval or, if $\widehat{F}_X^{emp}(x)$ jumps in the interval, the estimate of $\widehat{f}_X^{emp}(x)$ becomes too large. If ε is chosen too large, then the estimate of $f_X(x_0)$ is too imprecise. There exist methods to choose ε but a visual inspection of the estimated density function for different values of ε is typically suggestive for a good choice of ε.

12.4.2   Analog estimator of the mean

By Definition 5.1, the mean of a discrete random variable X with PMF $f_X(x)$ is

$$\mu_0 = E(X) = \sum_x x \cdot f_X(x).$$

As explained above, the analogy principle leads to the analog estimator

$$\widehat{\mu} = \sum_x x \cdot \widehat{f}_X^{emp}(x), \tag{12.3}$$

where $\widehat{f}_X^{emp}(x)$ is the empirical PMF (Definition 12.2). As discussed above, the empirical PMF only differs from 0 at the observations $\widetilde{X}_i$, and it assigns the probability $1/n$ to each of them. Hence, we can rewrite (12.3) as

$$\widehat{\mu} = \sum_{i=1}^{n} \widetilde{X}_i \cdot \frac{1}{n} = \frac{1}{n} \sum_{i=1}^{n} \widetilde{X}_i. \tag{12.4}$$

We see that the analog estimator of the mean is the sample average (Definition 11.3), i.e. $\widehat{\mu} = \overline{X}$. Thus, with a simple random sample, $\widehat{\mu}$ is unbiased, has $MSE(\widehat{\mu}, \mu_0) = \sigma_0^2/n$ and in case $X \sim N(\mu_0, \sigma_0^2)$, $\widehat{\mu} \sim N(\mu_0, \sigma_0^2/n)$. Further, the consistency of this estimator follows directly from the Law of Large Numbers (Theorem 11.4).

In case X is a continuous random variable with density function $f_X(x)$, the mean is defined as

$$\mu_0 = E(X) = \int_{-\infty}^{\infty} x \cdot f_X(x)dx.$$

One could be tempted to define the analog estimator of μ_0 via this equation by replacing the PDF with the analog estimator of the PDF (12.2), derived in the previous section. However, this estimator will be sensitive to the choice of ε, which can lead to bad performance of the estimator in practice. Instead, we use the fact mentioned above, namely that the analogy principle can be viewed as treating the sample as a population and use the appropriate formula for a discrete random variable with a distribution according to the empirical distribution. For the mean of a continuous random variable, this, in effect, implies that the density function $f_X(x)$ is replaced by the empirical PMF $\widehat{f}_X^{emp}(x)$ and the integral $\int_{-\infty}^{\infty} \cdot \, dx$ is replaced by the summation $\sum_{i=1}^{n}$. This leads to the sample average (12.4), i.e. the analog estimator is $\widehat{\mu}_0 = \overline{X}$, as was also the case when X is discrete. In general, whether X is discrete or continuous, the analogy principle typically leads to the same analog estimator of the quantity of interest, similarly to what happened here with the mean.

12.4.3 Analog estimator of a p-quantile

The analogy principle can be used to construct an estimator of a p-quantile. This is done by replacing the CDF in the definition of a p-quantile (Definition 5.7) with the empirical CDF. Since each observation is assigned the probability $1/n$ in the empirical distribution, estimating e.g. the median by the empirical distribution function is equivalent to sorting all the observations in increasing values, and letting the median estimator be the middle observation among the sorted observations.

The idea of sorting the observations to find the median can be extended to all p-quantiles. The sorted observations are called *order statistics*. These are formally defined next.

Definition 12.3 (Order statistics) *Let $(\widetilde{X}_1, \ldots, \widetilde{X}_n)$ be a random sample and $k \in \{1, \ldots, n\}$. Then the kth-**order statistics** is the kth smallest observation in the sample. It is denoted by $\widetilde{X}_{(k)}$.*

Example 12.3 *Let a realized sample be $(\widetilde{x}_1, \ldots, \widetilde{x}_5) = (4, 1, 10, 7, -2)$. Then the five order statistics are $\widetilde{x}_{(1)} = -2$, $\widetilde{x}_{(2)} = 1$, $\widetilde{x}_{(3)} = 4$, $\widetilde{x}_{(4)} = 7$, and $\widetilde{x}_{(5)} = 10$.*

In case of two or more observations are the same, then two or more of the order statistics are the same.

Example 12.4 *Let a realized sample be $(\widetilde{x}_1, \ldots, \widetilde{x}_5) = (4, 1, 10, -2, -2)$. Then the five order statistics are $\widetilde{x}_{(1)} = -2$, $\widetilde{x}_{(2)} = -2$, $\widetilde{x}_{(3)} = 1$, $\widetilde{x}_{(4)} = 4$, and $\widetilde{x}_{(5)} = 10$.*

An analog estimator of a p-quantile can now be expressed using order statistics. It is given in the next definition.

Definition 12.4 (p-quantile estimator) *Let $(\widetilde{X}_1, \ldots, \widetilde{X}_n)$ be a random sample and $\widetilde{X}_{(k)}$, $k = 1, \ldots, n$ the corresponding order statistics. Then the p-**quantile estimator** $\widehat{q}_p$ is defined by*

$$\widehat{q}_p = \begin{cases} X_{(Int(n \cdot p + 1))} & if \quad Int(n \cdot p) \neq n \cdot p, \\ \frac{1}{2}\left(X_{(n \cdot p)} + X_{(n \cdot p + 1)}\right) & if \quad Int(n \cdot p) = n \cdot p, \end{cases} \tag{12.5}$$

where $Int(a)$ is the integer part of the number a.

The p-quantile estimator is written in two pieces depending on whether $n \cdot p$ is an integer, because there exists an interval of solutions for a p-quantile if $n \cdot p$ is an integer. In this situation, the second part of (12.5) defines the midpoint of this interval as the estimator of the p-quantile.

Example 12.5 *Let a realized sample be $(\widetilde{x}_1, \ldots, \widetilde{x}_5) = (4, 1, 10, 7, -2)$. To calculate the 0.3-quantile estimate, first notice that*

$$0.3 \cdot 5 = 1.5 \neq Int(1.5) = 1.$$

Therefore use the first row in (12.5) and the result on the realized order statistics from Example 12.3 to get

$$\widehat{q}_{0.3} = \widetilde{x}_{(Int(5 \cdot 0.3 + 1))} = \widetilde{x}_{(2)} = 1.$$

With a simple random sample, the sampling distribution CDF $F_{X_{(k)}}$ of the kth-order statistics can be shown to be

$$F_{X_{(k)}}(x) = \sum_{j=k}^{n} \binom{n}{j} F_X(x)^j \left(1 - F_X(x)\right)^{n-j},$$

where $F_X()$ is the CDF of X and

$$\binom{n}{j} = \frac{n!}{j!(n-j)!}$$

is the binomial coefficient. In case X is a continuous random variable with PDF $f_X(x)$, the PDF of the kth-order statistics can be shown to be

$$f_{X_{(k)}}(x) = \frac{n!}{k!(n-k)!} F_X(x)^{k-1} \left(1 - F_X(x)\right)^{n-k} f_X(x).$$

Since $\widehat{q}_p$ equals an order statistics when $Int(n \cdot p) \neq n \cdot p$, then the above provides the distribution for those $\widehat{q}_p$.

Analytical expressions for the mean and variance of the kth-order statistics are difficult to compute. Next is an example, where it is possible and which shows that $\widehat{q}_k$ is a biased estimator of q_k.

Example 12.6 *Let $X \sim U(0,1)$. Then it can be shown that*

$$E\left(X_{(k)}\right) = \frac{k}{n+1}.$$

In case we estimate the median, that is, $p = 0.5$, then $k = n \cdot p + 1 = 0.5n + 1$ for n odd and

$$E\left(\widehat{q}_{0.5}\right) = \frac{k}{n+1} = \frac{0.5n+1}{n+1} = \frac{0.5(n+1)+0.5}{n+1} = 0.5 + \frac{0.5}{n+1}.$$

The median in the $Unif(0,1)$ distribution is 0.5. Hence, the estimator $\widehat{q}_{0.5}$ is upward biased by $0.5/(n+1)$.

12.4.4 Analog estimators of the variance and higher order moments

Let the variance of a population be $\sigma_0^2 = Var(X) < \infty$. By Definition 5.3, the variance of the random variable X is

$$Var(X) = E\left((X - E(X))^2\right).$$

It is seen that the variance is not only defined as an expected value $E()$, but inside $E()$, there is another expected value, namely, $E(X)$.

The analogy principle proposes to replace the PMF/PDF $f_X(x)$ used to calculate the two expectations with the empirical PMF. This leads to the analog estimator of the variance given by

$$\widehat{Var}(X) = \widehat{E}\left((X - \widehat{E}(X))^2\right),$$

where $\widehat{E}()$ denotes the analog estimator of the mean given in (12.1). Writing this out using the empirical PMF, we have

$$\widehat{Var}(X) = \sum_x \left(x - \left(\sum_x x \cdot \widehat{f}_X^{emp}(x)\right)\right)^2 \cdot \widehat{f}_X^{emp}(x) = \frac{1}{n}\sum_{i=1}^{n}\left(\tilde{X}_i - \left(\frac{1}{n}\sum_{i=1}^{n}\tilde{X}_i\right)\right)^2.$$

Since the inner parenthesis equals the sample average $\overline{X}$, the analog estimator of the variance, denoted $\widehat{\sigma}^2$, can be written as

$$\widehat{\sigma}^2 = \widehat{Var}(X) = \frac{1}{n}\sum_{i=1}^{n}(\tilde{X}_i - \overline{X})^2. \tag{12.6}$$

The analog estimator of the variance $\widehat{\sigma}^2$ in Equation (12.6) has the following properties

$$E\left(\widehat{\sigma}^2\right) = \frac{n-1}{n}\sigma_0^2,$$

$$Var\left(\widehat{\sigma}^2\right) = \left(\frac{n-1}{n}\right)^2 \frac{1}{n}\left(m_4^* - \frac{n-3}{n-1}\sigma_0^4\right),$$

where $m_4^* = E\left((X - E(X))^4\right)$ is the 4th central moment of X. The estimator $\widehat{\sigma}^2$ is downward biased because

$$Bias\left(\widehat{\sigma}^2\right) = E\left(\widehat{\sigma}^2\right) - \sigma_0^2 = \frac{n-1}{n}\sigma_0^2 - \sigma_0^2 = -\frac{1}{n}\sigma_0^2 < 0.$$

To derive the sampling distribution of the estimator $\widehat{\sigma}^2$, it is necessary to assume a specific distribution of the population. For instance, in case the population is normal $X \sim N\left(\mu_0, \sigma_0^2\right)$, it can be shown that

$$\widehat{\sigma}^2 \cdot \frac{n}{\sigma_0^2} \sim \chi_{n-1}^2,$$

where χ_{n-1}^2 is the χ^2-distribution with $n-1$ degrees of freedom.

We may be interested in higher order moments of a distribution. Estimators of moments can be straightforwardly constructed using the analogy principle. The analog estimator of the kth moment $m_k = E(X^k)$ is

$$\widehat{m}_k = \widehat{E}(X^k) = \sum_x x^k \cdot \widehat{f}_X^{emp}(x) = \frac{1}{n}\sum_{i=1}^n \widetilde{X}_i^k.$$

By Definition 5.5, the kth central moment is the kth moment around the mean of the distribution e.g. the 2nd central moment is the variance. The central moments are functions of the first moment, namely, the mean $E(X)$. The analog estimator of the kth central moment $m_k^* = E((X - E(X))^k)$ is

$$\widehat{m}_k^* = \widehat{E}((X - \widehat{E}(X))^k) = \frac{1}{n}\sum_{i=1}^n \left(\widetilde{X}_i - \overline{X}\right)^k.$$

12.5   Analog estimators of parameters in distributions

Consider a random variable X with distribution given by the CDF $F_X(x;\theta)$, where θ is a parameter. In cases where θ is representable as an aspect of the distribution of X, such as the mean, the variance, a quantile, etc., then we may use the analog estimator of that aspect to formulate an analog estimator of the parameter θ. The following two examples illustrate this in cases where X is a discrete random variable and the parameter of interest can be expressed as the mean of X.

Example 12.7 (Estimating a proportion in a population) *Let $p_0 \in [0,1]$ be a number and let the population be represented by the random variable $X \sim Ber(p_0)$. For instance, p_0 may denote the fraction, or proportion, of individuals in a population with a given characteristic. Suppose we have a simple random sample $\widetilde{X}_1, \ldots, \widetilde{X}_n$ from X. We know that (Theorem 8.1)*

$$E(X) = p_0.$$

We know from the discussion above that the sample mean $\overline{X}$ is the analog estimator of the mean of X, $E(X)$. Hence, we may use this as an estimator of the proportion p_0, i.e.

$$\widehat{p} = \overline{X}.$$

The Law of Large Numbers imply that $\widehat{p}$ is a consistent estimator of p_0. Further, we can show that (Problem 12.9.4)

$$E(\widehat{p}) = p_0,$$

i.e. $\widehat{p}$ is an unbiased estimator of p_0, and

$$Var(\widehat{p}) = \frac{p_0(1 - p_0)}{n}.$$

Example 12.8 (Estimating the intensity parameter in the Poisson distribution)
Let $\lambda_0 > 0$ be a number and let the population be represented by the random variable $X \sim Poi(\lambda_0)$. For instance, X may denote the number of events happening in a given period of time and λ is the intensity of arrivals of the events. Suppose we have a simple random sample $\widetilde{X}_1, \ldots, \widetilde{X}_n$ from X. We know that (Theorem 8.5)

$$E(X) = \lambda_0.$$

We know from the discussion above that the sample mean $\overline{X}$ is the analog estimator of the mean of X, $E(X)$. Hence, we may use this as an estimator of the intensity parameter λ_0, i.e.

$$\widehat{\lambda} = \overline{X}.$$

The Law of Large Numbers imply that $\widehat{\lambda}$ is a consistent estimator of λ_0. Further, we can show that (Problem 12.9.5)

$$E(\widehat{\lambda}) = \lambda_0,$$

i.e. $\widehat{\lambda}$ is an unbiased estimator of λ_0, and

$$Var(\widehat{\lambda}) = \frac{\lambda_0}{n}.$$

If the parameter of interest can be expressed as a function of an aspect of the distribution of X, the analog estimator of the parameter can be formulated accordingly.

Example 12.9 (Estimating the parameter in the exponential distribution)
Let $\lambda_0 > 0$ be a number and let the population be represented by the random variable $X \sim Exp(\lambda_0)$. For instance, X may denote the waiting time until an event happens, given that the mean number of arrivals of the events in a given time unit is λ_0. Suppose we have a simple random sample $\widetilde{X}_1, \ldots, \widetilde{X}_n$ from X. We know that (Theorem 8.3)

$$E(X) = \frac{1}{\lambda_0},$$

or, equivalently,

$$\lambda_0 = \frac{1}{E(X)} \tag{12.7}$$

We know from the discussion above that the sample mean $\overline{X}$ is the analog estimator of the mean of X, $E(X)$. By plugging this into (12.7), we arrive at the analog estimator of the parameter λ_0, i.e.

$$\widehat{\lambda} = \frac{1}{\overline{X}}.$$

It can be shown that $\widehat{\lambda}$ is a consistent, but biased, estimator of λ_0. You are asked to provide Monte Carlo evidence of this in Problem 12.9.7 at the end of this chapter.

The last example of this section shows how to formulate analog estimators of the parameters in a normal distribution.

Example 12.10 (Estimating the parameters in the normal distribution)
Let $\mu_0 \in \mathbb{R}$ and $\sigma_0^2 > 0$ be numbers and let the population be represented by the random variable $X \sim N(\mu_0, \sigma_0^2)$. Suppose we have a simple random sample $\widetilde{X}_1, \ldots, \widetilde{X}_n$ from X. We know that (e.g. Theorem 11.2)

$$E(X) = \mu_0$$

and

$$Var(X) = \sigma_0^2$$

We know from the discussion above that the sample mean $\overline{X}$ is the analog estimator of the mean of X, $E(X)$, and $\widehat{Var}(X) = \frac{1}{n}\sum_{i=1}^{n}(\widetilde{X}_i - \overline{X})^2$ is the analog estimator of the variance of X, $Var(X)$. Hence, we may use these as analog estimators of the parameters μ_0 and σ_0^2, i.e.

$$\widehat{\mu} = \overline{X},$$

$$\widehat{\sigma^2} = \frac{1}{n}\sum_{i=1}^{n}(\widetilde{X}_i - \overline{X})^2.$$

The Law of Large Numbers imply that $\widehat{\mu}$ and $\widehat{\sigma^2}$ are consistent estimators of μ_0 and σ_0^2, respectively.

12.6 Adjustments to analog estimators

Although the analogy principle often results in good estimators, we may sometimes make adjustments to an analog estimator to improve its properties. For example, if an estimator is biased, we may try to correct the estimator to lower, or remove, the bias, thus leading to a new estimator. In this section, we illustrate how this can be done in the case of constructing an estimator of the variance.

As we saw above, the analog estimator of the variance σ_0^2 given in (12.6) has mean

$$E\left(\widehat{\sigma}^2\right) = \frac{n-1}{n}\sigma_0^2,$$

and, as a consequence, it is a biased estimator of σ_0^2. Since the form of the bias is known, we can make a correction to the estimator such that the bias disappears. The correction is made by multiplying $\widehat{\sigma}^2$ by $n/(n-1)$. This leads to an unbiased estimator, since

$$E\left(\frac{n}{n-1}\widehat{\sigma}^2\right) = \frac{n}{n-1}E\left(\widehat{\sigma}^2\right) = \frac{n}{n-1}\frac{n-1}{n}\sigma_0^2 = \sigma_0^2.$$

Hence, a new unbiased estimator of the population variance can be obtained by adjusting $\widehat{\sigma}^2$ appropriately. This new estimator is called the *sample variance* and it is typically denoted S^2. It is defined next.

Definition 12.5 (Sample variance) *Let $(\widetilde{X}_1, \ldots, \widetilde{X}_n)$ be a random sample. The **sample variance** S^2 is defined as*

$$S^2 = \frac{1}{n-1}\sum_{i=1}^{n}(\widetilde{X}_i - \overline{X})^2, \tag{12.8}$$

where $\overline{X}$ is the sample average (Definition 11.3).

The sample variance is probably the most commonly used estimator of a population variance. By default, many statistical software packages use the sample variance to estimate the population variance because it is an unbiased estimator, which the analog estimator in (12.6) is not.

The sample variance has the following properties when calculated based on a simple random sample:

$$E\left(S^2\right) = \sigma_0^2,$$
$$Var\left(S^2\right) = \frac{1}{n}\left(m_4^* - \frac{n-3}{n-1}\sigma_0^4\right),$$

where m_4^* is the 4th central moment of X. In case the population is normal $X \sim N\left(\mu_0, \sigma_0^2\right)$, then

$$S^2 \cdot \frac{n-1}{\sigma_0^2} \sim \chi_{n-1}^2.$$

12.7   Choice of variance estimator

In Section 12.6, we considered two estimators of the population variance, namely, the analog estimator $\widehat{\sigma}^2$ given in (12.6) and the sample variance S^2 given in (12.8). As discussed in Section 11.6, to choose between two estimators, we can compare them based on a measure of risk. As measure of risk, or precision, we here choose the mean squared error, MSE, see Section 11.4.2. Recall that, for this choice of risk measure, we say that an estimator is more efficient than another estimator if, for all possible values of the parameter of interest, the former estimator has a MSE no larger than the latter estimator, and smaller for at least one value of σ_0^2.

It turns out that the efficiency of the two estimators compared to each other depends on the distribution of the population. The next example derives an efficiency result in case the population is normal.

Example 12.11 *Assume that the population represented by X is normal, $X \sim N(\mu_0, \sigma_0^2)$. We use that the MSE can be decomposed into a squared bias part and a variance part, see Theorem 11.1. We also use that the 4th central moment of a normal distribution is $m_4^* = 3 \cdot \sigma_0^4$. Based on the mean and variance of the analog estimator $\widehat{\sigma}^2$ and the sample variance S^2, respectively, their MSEs are*

$$MSE\left(S^2, \sigma_0^2\right) = \frac{2\sigma_0^4}{n-1},$$

and

$$MSE\left(\widehat{\sigma}^2, \sigma_0^2\right) = \frac{2\sigma_0^4}{n-1}\left(\frac{n-1}{n}\right)^2 + \frac{\sigma_0^4}{n^2}.$$

Subtract the two and rearrange to get

$$MSE\left(S^2, \sigma_0^2\right) - MSE\left(\widehat{\sigma}^2, \sigma_0^2\right) = \frac{\sigma_0^4\left(3n-1\right)}{n^2\left(n-1\right)} > 0.$$

Thus, in terms of risk measured by MSE, the analog estimator $\widehat{\sigma}^2$ is a more efficient estimator of the population variance σ_0^2 than the sample variance S^2. Recall, this result is derived under the assumption that the population is normal. The result demonstrates that a biased estimator may have higher precision than an unbiased estimator because the lower variance of the biased estimator, in this case, makes up for its bias.

It is worth stressing that, without restricting what the possible distributions of the population can be, none of the two estimators is at least as efficient as the other. That is, the result found above, i.e. that the analog estimator is more efficient than the sample variance, relies on the assumption that the population X is normal, $X \sim N(\mu_0, \sigma_0^2)$.

Finally, there are other estimators of the population variance that may be more efficient than both the analog estimator and the sample variance. The next example shows such an estimator.

Example 12.12 *Assume the population is normal* $X \sim N\left(\mu_0, \sigma_0^2\right)$ *with known population mean* μ_0, *and let k be an integer. We will now consider estimators of the variance, where we use $n - k$ in the denominator instead of n (the analog estimator) or $n - 1$ (the sample variance). That is, we consider estimators of the form*

$$\frac{1}{n-k} \sum_{i=1}^{n} (\widetilde{X}_i - \mu_0)^2. \tag{12.9}$$

It turns out that the most efficient estimator of σ_0^2 is obtained by setting $k = -2$. In other words, the most efficient estimator of the population variance, among the estimators given by (12.9), is

$$\frac{1}{n+2} \sum_{i=1}^{n} (\widetilde{X}_i - \mu_0)^2. \tag{12.10}$$

In this example, where the mean μ_0 is assumed known, the analog estimator is unbiased since it is not necessary to replace μ_0 by an estimator. This can be seen as follows,

$$E\left(\frac{1}{n} \sum_{i=1}^{n} (\widetilde{X}_i - \mu_0)^2\right) = \frac{1}{n} \sum_{i=1}^{n} E\left((\widetilde{X}_i - \mu_0)^2\right) = \frac{1}{n} \sum_{i=1}^{n} \sigma_0^2 = \frac{1}{n} n \sigma_0^2 = \sigma_0^2.$$

The efficient estimator (12.10), however, is biased. It has the mean

$$E\left(\frac{1}{n+2} \sum_{i=1}^{n} (\widetilde{X}_i - \mu_0)^2\right) = \frac{n}{n+2} E\left(\frac{1}{n} \sum_{i=1}^{n} (\widetilde{X}_i - \mu_0)^2\right) = \frac{n}{n+2} \sigma_0^2.$$

Hence, the efficient estimator is biased toward 0.

Biasing an estimator toward 0, in exchange for a lower variance, sometimes turns out to be useful for obtaining higher precision. In Volume II of this book, we will study a technique called regularization, *which exploits this possibility. Regularization can be particularly useful for estimation in so-called high-dimensional problems, i.e. when there are many unknown parameters to be estimated.*

12.8 Proofs

Proof of Theorem 12.1 *If X is a discrete random variable with PMF $f_X(x)$, then*

$$
\begin{aligned}
E\left(I(X \leq x)\right) &= \sum_{x_i} I(x_i \leq x) \cdot f_X(x_i) \\
&= \sum_{x_i \leq x} 1 \cdot f_X(x_i) + \sum_{x_i > x} 0 \cdot f_X(x_i) \\
&= \sum_{x_i \leq x} 1 \cdot f_X(x_i) \\
&= F_X(x),
\end{aligned}
$$

where the second equality sign follows from splitting the summation over all x_i into the summation for all $x_i \leq x$ plus the summation for all $x_i > x$. In case X is a continuous random variable with PDF $f_X(x)$, then

$$
\begin{aligned}
E\left(I(X \leq x)\right) &= \int_{-\infty}^{\infty} I(y \leq x) \cdot f_X(y)dy \\
&= \int_{-\infty}^{x} 1 \cdot f_X(y)dy + \int_{x}^{\infty} 0 \cdot f_X(y)dy \\
&= \int_{-\infty}^{x} 1 \cdot f_X(y)dy \\
&= F_X(x).
\end{aligned}
$$

This concludes the proof.

Proof of Theorem 12.2　*For a given value of x, $I(\widetilde{X}_i \leq x)$ is a random variable with possible outcomes 0 and 1, i.e. it is a Bernoulli distributed random variable (Definition 8.1). In particular, $I(\widetilde{X}_i \leq x) \sim Ber(p)$ with parameter $p = E\left(I(\widetilde{X}_i \leq x)\right)$, where, by Theorem 12.1,*

$$
E\left(I(\widetilde{X}_i \leq x)\right) = F_X(x).
$$

From the variance of the Bernoulli distribution (Theorem 8.1), we have

$$
Var\left(I(\widetilde{X}_i \leq x)\right) = F_X(x)\left(1 - F_X(x)\right).
$$

Using these results, we find that

$$
Bias\left(\widehat{F}_X^{emp}(x), F_X(x)\right) = E\left(\frac{1}{n}\sum_{i=1}^{n} I(\widetilde{X}_i \leq x)\right) - F_X(x) = F_X(x) - F_X(x) = 0,
$$

and

$$
Var\left(\widehat{F}_X^{emp}(x)\right) = Var\left(\frac{1}{n}\sum_{i=1}^{n} I(\widetilde{X}_i \leq x)\right) = \frac{1}{n}Var\left(I(\widetilde{X}_i \leq x)\right) = \frac{1}{n}F_X(x)\left(1 - F_X(x)\right).
$$

For $n \to \infty$, $Bias\left(\widehat{F}_X^{emp}(x)\right) = 0$ and $V\left(\widehat{F}_X^{emp}(x)\right) \to 0$. By Theorem 11.1, we conclude that $\widehat{F}_X^{emp}(x)$ is a consistent estimator of $F_X(x)$, as we wanted to show.

Proof of Theorem 12.3　*The ECDF consists of a sum of random variables $I(\widetilde{X}_i \leq x)$, which are either 0 or 1. With a simple random sample, $(I(\widetilde{X}_1 \leq x), I(\widetilde{X}_2 \leq x), \ldots, I(\widetilde{X}_n \leq x))$ comprises a simple random sample from a Bernoulli distribution with parameter $p = F_X(x)$, that is $I(\widetilde{X}_i \leq x) \sim Ber\left(F_X(x)\right)$ for $i = 1, 2, \ldots, n$, see also the proof of Theorem 12.2. Recall that a sum of identically and independently distributed Bernoulli random variables is binomially distributed, that is $S_n = \sum_{i=1}^{n} I(\widetilde{X}_i \leq x) \sim Bin\left(n, F_X(x)\right)$, see Section 8.2. Hence, we can express the CDF $F_{\widehat{F}_X^{emp}(x)}$ of the random variable $\widehat{F}_X^{emp}(x)$ as* *follows*

$$F_{\widehat{F}_X^{emp}(x)}(a) \;=\; \Pr\left(\widehat{F}_X^{emp}(x) \le a\right)$$

$$=\; \Pr\left(\frac{1}{n}\sum_{i=1}^{n} I(\widetilde{X}_i \le x) \le a\right)$$

$$=\; \Pr\left(\sum_{i=1}^{n} I(\widetilde{X}_i \le x) \le n\cdot a\right)$$

$$=\; \Pr\left(S_n \le n\cdot a\right)$$

$$=\; F_{S_n}(n\cdot a),$$

where $a \in \mathbb{R}$ and F_{S_n} is the CDF of the $Bin\left(n, F_X(x)\right)$ distribution. The mean and variance of the Empirical Distribution function can be shown to be (Problem 12.9.3)

$$E\left(\widehat{F}_X^{emp}(x)\right) \;=\; F_X(x),$$

$$Var\left(\widehat{F}_X^{emp}(x)\right) \;=\; \frac{1}{n}F_X(x)\left(1 - F_X(x)\right).$$

12.9 Exercises

Problem 12.9.1 *Consider the population represented by the random variable $X \sim Exp(2)$.*

1. *Write down the CDF of X, $F_X(x)$.*

2. *Consider a simple random sample from the population X, $(\widetilde{X}_1,\ldots,\widetilde{X}_n)$. Use a software programme to simulate such a simple random sample when $n = 10$. Use this simple random sample to estimate the CDF $F_X(x)$ using the empirical CDF (Definition 12.1). Plot your estimated CDF along with the true CDF from part 1, similarly to what was done in Figure 12.1.*

3. *Redo part 2, but now with three additional simple random samples with $n = 100$, $n = 1000$, and $n = 10000$, respectively. Plot all four estimates of F_X, together with the true CDF from part 1. Comment on how your results relate to the consistency of the empirical CDF as an estimator of F_X (Theorem 12.2).*

Problem 12.9.2 *Suppose that $(\widetilde{X}_1, \widetilde{X}_2,\ldots,\widetilde{X}_n)$ is a simple random sample drawn from a population represented by the discrete random variable random variable X with PMF $f_X(x)$. Consider the empirical PDF (Definition 12.2),*

$$\widehat{f}_X^{emp}(x) = \frac{1}{n}\sum_{i=1}^{n} I(\widetilde{X}_i = x).$$

Fix $x \in \mathbb{R}$. Prove that the empirical PMF $\widehat{f}_X^{emp}(x)$ is a consistent estimator of the population PMF $f_X(x)$. (Hint: You may use the fact that $E(I(X = x)) = Pr(X = x)$.)

Problem 12.9.3 *Suppose that $(\widetilde{X}_1, \widetilde{X}_2,\ldots,\widetilde{X}_n)$ is a simple random sample drawn from a population represented by the random variable X. Consider the empirical CDF (Definition*

12.1),

$$\widehat{F}_X^{emp}(x) = \frac{1}{n} \sum_{i=1}^{n} I(\widetilde{X}_i \leq x).$$

Prove that

$$E\left(\widehat{F}_X^{emp}(x)\right) = F_X(x),$$

$$Var\left(\widehat{F}_X^{emp}(x)\right) = \frac{1}{n} F_X(x)\left(1 - F_X(x)\right).$$

Problem 12.9.4 *Consider the setup in Example 12.7. Prove that*

$$E(\widehat{p}) = p_0,$$

$$Var(\widehat{p}) = \frac{p_0(1 - p_0)}{n}.$$

Problem 12.9.5 *Consider the setup in Example 12.8. Prove that*

$$E(\widehat{\lambda}) = \lambda_0,$$

$$Var(\widehat{\lambda}) = \frac{\lambda_0}{n}.$$

Problem 12.9.6 *Consider the setup in Example 12.8. In that example, we used the fact that $E(X) = \lambda_0$ to justify the sample mean $\overline{X}$ as an estimator $\widehat{\lambda}$ of the intensity parameter λ_0, based on the analogy principle. But since we also have that (Theorem 8.5)*

$$Var(X) = \lambda_0,$$

we may equally well have used the analogy principle to suggest the alternative estimator

$$\check{\lambda} = \widehat{\sigma}^2,$$

where $\widehat{\sigma}^2$ is the analog estimator of $Var(X)$ given in (12.6).

Let $\widetilde{X}_1, \ldots, \widetilde{X}_n$ be a simple random sample from $X \sim Poi(\lambda_0)$.

1. *Let $n = 10$ and $\lambda_0 = 2$. Setup and conduct a Monte Carlo study that compares the two estimators $\widehat{\lambda}$ and $\check{\lambda}$ in terms of bias, variance, and mean squared error.*

2. *Repeat the Monte Carlo study for the sample sizes $n = 20, 40, 80, 160, 320, 640, 1280$.*

3. *Repeat 1. and 2. but now with $\lambda_0 = 10$. Compare your results with those from 1. and 2.*

Problem 12.9.7 *Consider the setup in Example 12.9. In that example, we used the fact that $E(X) = 1/\lambda_0$ to justify the inverse of the sample mean $1/\overline{X}$ as an estimator $\widehat{\lambda}$ of the parameter λ_0, based on the analogy principle. In this exercise, you are asked to use Monte Carlo simulation to illustrate that $\widehat{\lambda} = 1/\overline{X}$ is a consistent, but biased, estimator of λ_0.*

Let $\widetilde{X}_1, \ldots, \widetilde{X}_n$ be a simple random sample from $X \sim Exp(\lambda_0)$.

1. *Let $n = 5$ and $\lambda_0 = 0.5$. Setup and conduct a Monte Carlo study that estimates the bias of the analog estimator $\widehat{\lambda} = 1/\overline{X}$.*

2. *Repeat the Monte Carlo study for the sample sizes $n = 10, 20, 40, 80, 160, 320, 640, 1280$. What happens to the bias as n increases? Comment on whether the Monte Carlo study provides evidence for $\widehat{\lambda}$ being a consistent estimator of λ_0.*

Problem 12.9.8 *Let $X \sim N(0,1)$ represent a population. In this problem, we will use Monte Carlo simulation to illustrate the fact that the sample variance, S^2, is an unbiased estimator of the population variance $\sigma_0^2 = 1$, while the analog estimator of the variance, $\widehat{\sigma}^2$ is a biased estimator of $\sigma_0^2 = 1$. Although in this case of the variance of a simple random sample, we were able to derive these results analytically above, it might be difficult or impossible to derive similar results analytically in other cases and for other estimators. In such instances, a Monte Carlo analysis can help in assessing the bias of an estimator. In particular, consider the following steps:*

1. *Simulate the simple random samples $(\widetilde{X}_1^*, \ldots, \widetilde{X}_n^*)$ from the population $X \sim N(0,1)$.*

2. *Use the simple random sample to estimate σ_0^2 via both the sample variance, S^2, and the analog estimator of the variance, $\widehat{\sigma}^2$.*

3. *Calculate the bias resulting from the two estimates.*

   Perform the steps 1.–3. a large number of times (e.g. $M = 100,000$) for various sample sizes, e.g. $n = 2, 3, \ldots, 25$. Calculate the average Monte Carlo bias for the two estimators and compare with their theoretical biases. (The comparison can, for instance, be made in a plot. I.e. show the theoretical and Monte Carlo biases as a function of n in the same plot.)

Problem 12.9.9 *The purpose of this problem is to conduct a Monte Carlo simulation to investigate the distribution of a median estimator. This exercise is similar to the Monte Carlo simulation in Problem 11.9.11, where the sample average is investigated.*

   Assume the population is a t-distribution with 6 degrees of freedom and consider an estimator of the median of the population. The median of any t-distribution is 0. Use as estimator the analog estimator of the 0.5-quantile given in (12.5).

1. *Calculate the Monte Carlo distribution of the median estimator.*

   - *Report a plot of the Monte Carlo distribution.*
   - *Report 0.1, 0.5 and 0.9 quantiles of the Monte Carlo distribution, e.g. the CDF $F_{\overline{X}^*}^*(t)$.*
   - *Report the probability of an outcome in the neighborhood $(-0.05, 0.05)$ of 0.*

2. *Redo part 1 for $n = 800$. Compare your results for the two cases $n = 200$ and $n = 800$.*

3. *Redo parts 1 and 2 for a t-distribution with 1 degree of freedom. Compare your results over the two different t-distributions.*

Problem 12.9.10 *Let X be a random variable such that $\sigma_0^2 = Var(X) < \infty$. Assume a simple random sample of X is available. Prove that the sample variance S^2 as defined in Definition 12.5 is a consistent estimator of the population variance σ_0^2. (Hint: You might use Theorem 11.3.)*

Problem 12.9.11 *Suppose a realized sample is $(4, 9, 3, 4, 1, 7, 6, 9)$. Calculate an estimate of the 0.3-quantile and of the median.*

Problem 12.9.12 *Suppose a realized sample is $(5, 3, 5, 8, 2)$. Calculate the estimate of the empirical distribution function and of the empirical probability function.*

Problem 12.9.13 *Consider the time series population $(S_1, S_2, \ldots, S_T)$, where S_t represents the price of a financial stock at time t. We specify the time series structure using the process formulation, see Section 10.4.2. In particular, let*

$$S_1 = 100,$$

and define, for $t = 2, \ldots, T$,

$$S_t = S_{t-1} + \epsilon_t,$$

where $\epsilon_2, \ldots, \epsilon_T$ are mutual statistically independent and identically distributed. Assume $\epsilon_t \sim t(4)$, where $t(4)$ denotes the t-distribution with 4 degrees of freedom (see, e.g. Section 8.6.2 for details on the t-distribution). We are interested in the distribution of a transformation of the stock price at the final time point, i.e. in a transformation of the random variable S_T. More precisely, we will study the distribution of the random variable

$$X = (S_T - 100)^+,$$

where $(x)^+ = \max(x, 0)$. We note that the random variable X denotes the pay-off of a so-called call option on the stock, a popular financial asset. We also note that, in the case studied here where $\epsilon_t \sim t(4)$, the distribution of X does not have a closed-form solution. Instead, you are asked to use Monte Carlo simulation to approximate the distribution of X. In the following, let $T = 100$ which could, e.g. be interpreted as the stock being observed over 100 days.

1. *Use Monte Carlo simulation to simulate a Monte Carlo sample from S_T, i.e. $S_{T,j}^{\#}$ for $j = 1, \ldots, M$, where $M = 10\,000$. Plot a histogram of this Monte Carlo sample and plot the Monte Carlo approximation to the CDF of S_T, $F_{S_T}^{\#}(t)$.*

2. *Use the Monte Carlo sample of S_T to construct a Monte Carlo sample from X, i.e. $X_j^{\#}$ for $j = 1, \ldots, M$, where $M = 10\,000$. Plot a histogram of this Monte Carlo sample and plot the Monte Carlo approximation to the CDF $F_X^{\#}(t)$ of X.*

3. *Use the Monte Carlo sample of S_T to estimate the mean $E(X)$ of X.*

4. *Use the Monte Carlo sample of S_T to estimate the variance $Var(X)$ of X.*

13

Estimation of a sampling distribution

13.1 Introduction

As discussed in the preceding two chapters, the statistical quality of an estimator can be evaluated using the sampling distribution of the estimator. The sampling distribution of an estimator describes the uncertainty of the estimator due to the distribution of the population and the sampling mechanism. In Chapters 11 and 12, we considered instances where we had sufficient knowledge of the population and the sampling mechanism to derive the distribution of the estimator in question. In practice, however, the sampling distribution is unknown. In this chapter, we consider two approaches to estimate the sampling distribution of an estimator. One is the *bootstrap distribution* of the estimator and the other is the *asymptotic distribution* of the estimator.

The bootstrap distribution of an estimator is obtained through estimation of the distribution of the population. Using this estimate of the population distribution, together with knowledge of the sampling mechanism, the sampling distribution of the estimator can be estimated.

The asymptotic distribution of an estimator is obtained through asymptotic analysis. Asymptotic analysis is performed by studying what happens with the distribution of an estimator for sample sizes increasing toward infinity. The reason this may lead to useful results is that many estimators exhibit certain regularities which dominate their distributions as the sample size grows toward infinity. An important consequence of such regularities is prominently captured by so-called *central limit theorems* (e.g. Theorem 13.1 below), which turn out to be essential for deriving asymptotic distributions of estimators.

13.2 Bootstrap distribution

A bootstrap distribution is based on an estimator of the population distribution, e.g. represented by the CDF $F_X(x)$. The estimator of the population distribution is used, together with the sampling mechanism, to calculate the sampling distribution $F_{\widehat{\theta}}(x)$ of an estimator $\widehat{\theta}$. We will consider two methods of estimating the population distribution that lead to two types of bootstrapping, denoted, respectively, the *non-parametric bootstrap* and the *parametric bootstrap*. The non-parametric bootstrap is based on the empirical CDF $\widehat{F}_X^{emp}(x)$ as an estimator of the population distribution, whereas the parametric bootstrap is based on assuming a class of population distributions $F_X(x;\theta)$ and replacing unknown parameters θ with estimators $\widehat{\theta}$ to arrive at an estimator of the population distribution $F_X(x;\widehat{\theta})$.

DOI: 10.1201/9781003591191-13

13.2.1   The non-parametric bootstrap

With a simple random sample from a population, the non-parametric bootstrap is based on the empirical CDF $\widehat{F}_X(x)$ (Definition 12.1) or, similarly, the empirical PMF $\widehat{f}_X^{emp}(x)$ (Definition 12.2). In the following, we present the non-parametric bootstrap using the empirical PMF $\widehat{f}_X^{emp}(x)$, i.e.

$$\widehat{f}_X^{emp}(x) = \frac{1}{n} \sum_{i=1}^{n} I\left(x = \widetilde{X}_i\right),$$

where $\left(\widetilde{X}_1, \ldots, \widetilde{X}_n\right)$ is a random sample and $I()$ is the indicator function.

As discussed in Chapter 12, we can think of the empirical probability mass function as representing a population with n elements $\widetilde{X}_1, \ldots, \widetilde{X}_n$, where each element $\widetilde{X}_i$ is assigned the probability $1/n$. In the context of bootstrapping, we denote such a population a *bootstrap population*. In other words, with the non-parametric bootstrap, we consider the sample $\left(\widetilde{X}_1, \ldots, \widetilde{X}_n\right)$ as the bootstrap population. Since the empirical probability mass function $\widehat{f}_X(x)$ can be calculated from the empirical CDF $\widehat{F}_X^{emp}(x)$, and with a simple random sample, $\widehat{F}_X^{emp}(x)$ is a consistent estimator of the CDF $F_X(x)$ of the population (Theorem 12.2), the bootstrap population is a consistent estimator of the population distribution.

Consider a parameter θ, which is a parameter from, or an aspect of, a population distribution $F_X(x)$, and assume that an estimator $\widehat{\theta}$ of θ is available using the random sample $\left(\widetilde{X}_1, \ldots, \widetilde{X}_n\right)$. We are interested in the sampling distribution of the estimator $\widehat{\theta}$, which we denote by $f_{\widehat{\theta}}$. To obtain an estimator $\widehat{f}_{\widehat{\theta}}$ of this quantity, we imagine drawing observations from the bootstrap population by a similar sampling mechanism as is used to obtain the original random sample $\left(\widetilde{X}_1, \ldots, \widetilde{X}_n\right)$. The result of sampling from the bootstrap population is a sample denoted a *bootstrap sample* $\left(\widetilde{X}_1^*, \ldots, \widetilde{X}_n^*\right)$, where $\widetilde{X}_1^*$ denotes the first draw from the bootstrap population, $\widetilde{X}_2^*$ denotes the second draw, and so forth. Since the bootstrap population is known, and assuming the sampling mechanism is also known, the distribution of the bootstrap sample can be calculated, either analytically or using simulations. Finally, knowing how the estimator is constructed as a function of a sample, a distribution of the estimator can be calculated. Since this distribution is based on the bootstrap population, it is called a bootstrap distribution of the estimator. The next example illustrates the process of calculating a bootstrap distribution by the non-parametric bootstrap. The example is sufficiently simple to make it possible to write up the bootstrap distribution analytically, i.e. without resorting to simulation.

Example 13.1 *In this example, we calculate the bootstrap distribution of the sampling distribution of the sample average $\overline{X}$. Let $\left(\widetilde{X}_1, \ldots, \widetilde{X}_n\right)$ denote a sample of size n. The sample average is*

$$\overline{X} = \frac{1}{n} \sum_{i=1}^{n} \widetilde{X}_i.$$

Assume a simple random sample with $n = 3$ observations. Suppose the realized sample is $(\widetilde{x}_1, \widetilde{x}_2, \widetilde{x}_3) = (1, 2, 3)$. Then $(1, 2, 3)$ are the elements of the bootstrap population and the empirical probability mass function, which assigns the probability $1/3$ to each of these three observations, is the distribution of the bootstrap population. That is, the bootstrap distribution can be represented by the PMF $f^*(x)$, where $f^*(\widetilde{x}_1) = f^*(\widetilde{x}_2) = f^*(\widetilde{x}_3) = 1/3$.

Based on the bootstrap population, the bootstrap distribution of the sample average can be calculated. This can be done by considering all possible samples from the bootstrap population, their probabilities, and the corresponding values of the sample average. Conditional *on the realized sample with $n = 3$, there are $3^3 = 27$ different bootstrap samples since the elements are drawn with replacement in order to obtain a simple random sample from the bootstrap population. Each different bootstrap sample has a probability $1/27$ of being drawn. For instance, the bootstrap sample $(\widetilde{x}_1^*, \widetilde{x}_2^*, \widetilde{x}_3^*) = (2, 2, 1)$ obtains if the first two draws are $\widetilde{x}_2 = 2$ and the last draw is $\widetilde{x}_1 = 1$. Each draw has a probability of $1/n = 1/3$, and since the sampling mechanism implies that the observations are statistically independent, the probability of obtaining $(\widetilde{x}_1^*, \widetilde{x}_2^*, \widetilde{x}_3^*) = (2, 2, 1)$ is $1/3 \times 1/3 \times 1/3 = 1/27$. In this way, we can calculate the probabilities of all possible bootstrap samples. In the table below, we list the bootstrap samples, and their associated probabilities, as well as the resultant bootstrap sample average $\overline{x}^* = \frac{1}{3}\left(\widetilde{x}_1^* + \widetilde{x}_2^* + \widetilde{x}_3^*\right)$.*

Bootstrap sample $(\widetilde{x}_1^*, \widetilde{x}_2^*, \widetilde{x}_3^*)$	Probability bootstrap sample	$\overline{x}^*$
$(1, 1, 1)$	$1/27$	1
$(1, 1, 2)$	$1/27$	$4/3$
$(1, 1, 3)$	$1/27$	$5/3$
$(1, 2, 1)$	$1/27$	$4/3$
$\vdots$	$\vdots$	$\vdots$
$(3, 3, 2)$	$1/27$	$8/3$
$(3, 3, 3)$	$1/27$	3

Given all the possible outcomes $\overline{x}^*$ of the bootstrap sample average, and the probability of each of those outcomes, we can write the non-parametric bootstrap distribution $\widehat{f}_{\overline{X}}^*(x)$ of $\overline{X}$ as

$$
\widehat{f}_{\overline{X}}^{\#}(x) = \begin{cases}
1/27 & \text{if} & x = 1, \\
3/27 & \text{if} & x = \frac{4}{3}, \\
6/27 & \text{if} & x = \frac{5}{3}, \\
7/27 & \text{if} & x = 2, \\
6/27 & \text{if} & x = \frac{7}{3}, \\
3/27 & \text{if} & x = \frac{8}{3}, \\
1/27 & \text{if} & x = 3, \\
0 & & \text{otherwise.}
\end{cases}
$$

For example, there is a probability $3/27$ that the sample average x^* equals $\frac{4}{3}$, because exactly three bootstrap samples result in $\overline{x}^* = \frac{4}{3}$, namely $(\widetilde{x}_1^*, \widetilde{x}_2^*, \widetilde{x}_3^*) = (1, 1, 2)$, $(1, 2, 1)$, and $(2, 1, 1)$.

Notice that the bootstrap distribution $\widehat{f}_{\overline{X}}^*(x)$ of $\overline{X}$ is derived given a realized sample, here $(\widetilde{x}_1, \widetilde{x}_2, \widetilde{x}_3) = (1, 2, 3)$. A different realized sample from the population typically leads to another bootstrap distribution. In applications, the bootstrap distribution is calculated for the specific realized sample at hand.

In practice, it will usually be prohibitively time consuming to evaluate the estimator at all the possible bootstrap samples. For instance, if the n observations of the realized sample are different, then there are n^n different bootstrap samples. For example, for $n = 10$, there are $10^{10} = 10$ billion different bootstrap samples. To make bootstrapping practical, we can simulate the bootstrap distribution using Monte Carlo simulation. This will be discussed in Section 13.2.3.

13.2.2 The parametric bootstrap

The parametric bootstrap can be performed if we make an assumption regarding the possible distributions that the population may have. For example, we may have reason to believe that the population distribution is one of the exponential distributions $Exp(\lambda)$. Under this assumption, the only unknown about the population distribution is the value of λ. Therefore, the problem of obtaining an estimate of the population distribution is reduced to obtaining an estimate of a finite number of unknown parameters, in this case λ. This is the reason, that this type of bootstrap is called a parametric bootstrap.

Assume the population distribution is represented by the CDF $F_X(x; \theta_0)$, where $F_X(x; \theta)$ is known, but the true value θ_0 of θ is unknown. Also assume that an estimator $\widehat{\theta}$ of θ_0 is available e.g. using an analog estimator. Then $F_X\left(x; \widehat{\theta}\right)$ is an estimate of the population distribution. Notice the difference to the non-parametric bootstrap, where the estimate of the population distribution is the empirical distribution function $\widehat{F}_X(x)$.

Example 13.2 *Consider a situation where a population mean μ_0 is estimated by the sample average $\overline{X}$ of a simple random sample $(\widetilde{X}_1, \widetilde{X}_2, \ldots, \widetilde{X}_n)$ from a population represented by the random variable X. Our goal is to estimate the sampling distribution of the sample average $\overline{X}$, i.e. $F_{\overline{X}}(x)$.*

Suppose we have reason to believe that the population distribution is a normal distribution, $X \sim N\left(\mu_0, \sigma_0^2\right)$. As explained above, the parametric bootstrap approach estimates the population distribution by substituting estimators for the parameters into the population distribution. In other words, the population distribution is approximated by the $N(\widehat{\mu}, \widehat{\sigma}^2)$ distribution, where $\widehat{\mu}$ and $\widehat{\sigma}^2$ are estimators of μ_0 and σ_0^2, respectively. For instance, we may use the sample average to estimate μ_0 and the sample variance to estimate σ_0^2, i.e. $\widehat{\mu} = \overline{X}$ and $\widehat{\sigma}^2 = S^2$.

Since the original sample is a simple random sample, the sampling distribution of $\overline{X}$ is derived in Example 11.4, where it is shown that

$$\overline{X} \sim N\left(\mu_0, \sigma_0^2 \frac{1}{n}\right). \tag{13.1}$$

The bootstrap population is $N(\widehat{\mu}, \widehat{\sigma}^2) = N\left(\overline{X}, S^2\right)$. This implies that the sampling distribution of the bootstrap sample average $\overline{X}^*$ is

$$\overline{X}^* \sim N\left(\overline{X}, S^2 \frac{1}{n}\right). \tag{13.2}$$

This is the parametric bootstrap distribution for the sample average. It is seen that the parametric bootstrap amounts to replacing μ_0 and σ_0^2 in the sampling distribution for the sample average (13.1) with estimates of these parameters.

In Example 13.2, we used our knowledge of the sampling distribution (13.1) of the sample average to derive the sampling distribution of the bootstrap sample average (13.2) by plugging in estimates of the parameters into (13.1). If an estimator is a more complicated function of the random sample than a sample average, then it may not be easy to calculate the sampling distribution, at least not in a form into which the estimated parameters can be inserted. Instead, we can use Monte Carlo simulation by drawing bootstrap samples from the bootstrap distribution $F_X\left(x; \widehat{\theta}\right)$ to calculate the bootstrap distribution $\widehat{F}^*_{\widehat{\theta}}(x)$ of an estimator $\widehat{\theta}$. This is discussed in the next section.

13.2.3   Implementation of the bootstrap using Monte Carlo simulation

Monte Carlo simulation, as discussed in Section 11.7, can be used to approximate the bootstrap distribution $\widehat{F}^*_T(x)$ of a sample statistics T, where T, for instance, can be an estimator $\widehat{\theta}$.

Calculation of a bootstrap distribution can be implemented using Monte Carlo simulation in exactly the same way as done in Section 11.7, with the exception that the population distribution $F_X(x)$ in the Monte Carlo simulation is replaced by an estimate $\widehat{F}_X(x)$ of the population distribution. Here, $\widehat{F}_X(x)$ may be either the empirical CDF $\widehat{F}_X(x)$ (Definition 12.1), in the case of the non-parametric bootstrap, or $F(x;\widehat{\theta})$, in the case of the parametric bootstrap. To distinguish the notation used for bootstrapping, where the bootstrap samples are drawn from the bootstrap population, from Monte Carlo simulation, where the Monte Carlo samples are drawn from the population distribution $F_X(x)$, we will denote samples from the former with an asterisk superscript, $*$, whereas we used a hashtag superscript, $\#$, in Section 11.7 for Monte Carlo samples from $F_X(x)$. The bootstrapping procedure is illustrated in the diagram below.

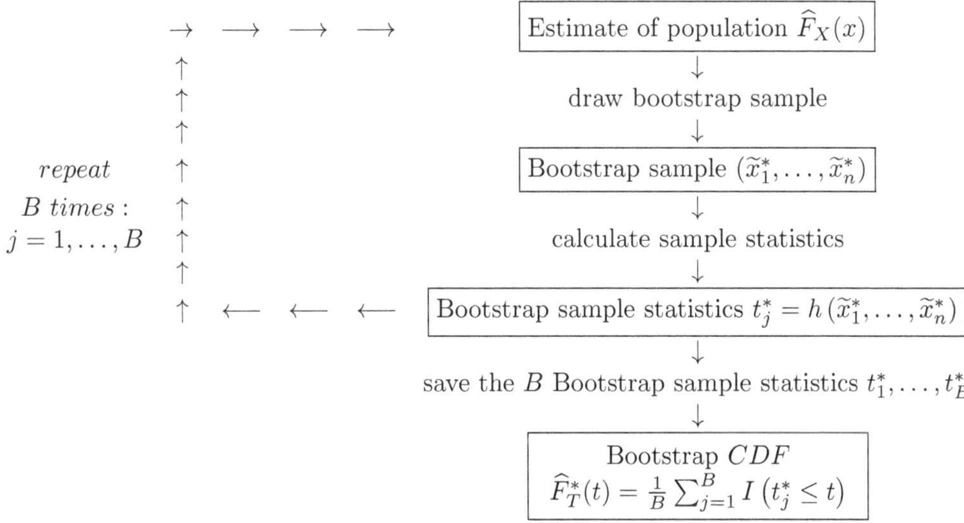

For the non-parametric bootstrap, drawing a simple random sample from a bootstrap population is equivalent to drawing n observations from the realized sample with replacement, where each observation has a $1/n$ probability of being drawn each time. Sampling with replacement implies that in almost all bootstrap samples, some observations from the realized sample are drawn more than once. It can be shown that on average 63% of the observations from the realized sample is in a bootstrap sample.

Similarly to a Monte Carlo simulation, drawing about $B = 10,000$ bootstrap samples is sufficient for most purposes, although if the computational burden is high for drawing a bootstrap sample, a smaller B could be entertained as well.

Example 13.3 below illustrates how the Monte Carlo simulation approach can be used to estimate the non-parametric bootstrap distribution of an estimator. It also illustrates the effect of the choice of the number of bootstrap replications, B. In Problem 13.7.3 you are asked to re-do this analysis in a situation where the sample size n is larger. In Problem 13.7.4 you are asked to use a parametric bootstrap to revisit Example 13.2.

Example 13.3 (Simulation-based estimate of the bootstrap distribution of Example 13.1) *In Example 13.1, a simple random sample of $n = 3$ observations was considered. In this case, because the number of observations is very small, we were able to calculate the exact non-parametric bootstrap distribution $F_{\overline{X}}^*(x)$ of the sample average $\overline{X}$. We now demonstrate how Monte Carlo simulation alternatively can be used to approximate the exact bootstrap distribution $F_{\overline{X}}^*(x)$.*

We follow the process set out in the diagram above. That is, for $b = 1, 2, \ldots, B$, do the following:

1. *Simulate a bootstrap sample $(\widetilde{x}_1^*, \widetilde{x}_2^*, \widetilde{x}_3^*)$ using the bootstrap PMF $f^*(\widetilde{x}_1) = f^*(\widetilde{x}_2) = f^*(\widetilde{x}_3) = 1/3$.*

2. *Calculate the bootstrap sample average, i.e. $\overline{x}_b^* = \frac{1}{3} \sum_{i=1}^{3} \widetilde{x}_i^*$, where $\widetilde{x}_i^*$, $i = 1, 2, 3$, are the bootstrap observations from step 1.*

The two steps result in B bootstrap sample averages $\overline{x}_b^*$, $b = 1, 2, \ldots, B$. The empirical distribution of these constitute the bootstrap approximation to the sampling distribution of $\overline{X}$, i.e. to $F_{\overline{X}}^*(x)$. In particular, since there are only 7 different values that $\overline{x}_b^*$ may take (see the discussion in Example 13.1), we can consider the bootstrap empirical mass function of these 7 values. To illustrate the effect of the number of bootstrap samples, B, we do this analysis for $B = 100, 1000, 10000$. In Figure 13.1 we compare the exact non-parametric bootstrap PMF $\widehat{f}_{\overline{X}}^*$ (vertical lines) to the simulation-based approximation (circles and squares), as considered in this example. In particular, we see how the simulation-based approach becomes increasingly accurate as we increase B. Indeed, for $B = 10,000$ there is hardly any difference between the exact non-parametric bootstrap PMF $\widehat{f}_{\overline{X}}^*$, as derived analytically in Example 13.1, and the simulation-based non-parametric bootstrap PMF.

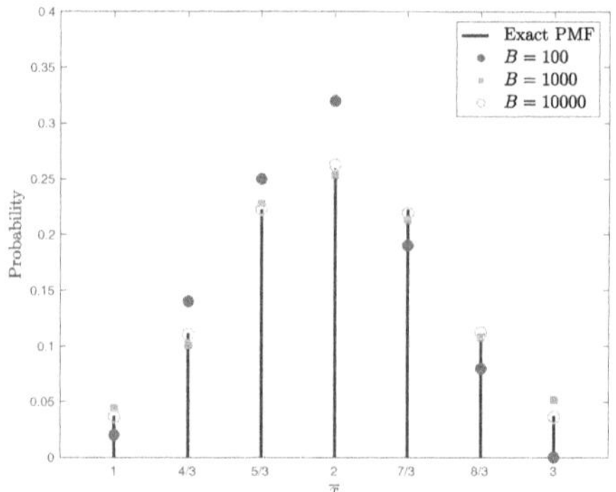

FIGURE 13.1

Bootstrap approximation to the sampling distribution $F_{\overline{X}}(x)$ in the setting from Example 13.1. The vertical lines denote the exact bootstrap distribution $F_{\overline{X}}^*(x)$, as calculated in Example 13.1. The filled circles, squares, and open circles denote the simulation-based approximation to $F_{\overline{X}}^*(x)$ for $B = 100$, $B = 1,000$, and $B = 10,000$, respectively, as described in Example 13.3.

13.2.4   Other types of bootstrapping

The bootstrapping should mimic, as closely as possible, how the random sample is drawn from the population. In the section above, the bootstrap is discussed in the context of a population where the distribution is either completely unknown, leading to the non-parametric bootstrap, or known except for a finite number of parameters, leading to the parametric bootstrap. It is also discussed in a context where the sampling mechanism implies a simple random sample. In other contexts, the implementation of the bootstrap should reflect those contexts. For many of these contexts, the specific implementation of the bootstrap has given rise to a specific name of the bootstrap.

In a context where the observations are not statistically independent, e.g. with time series, the bootstrap must take this into account. This has led to bootstrapping known as the *Block Bootstrap*, where the statistical dependence is taken into account when sampling from the bootstrap population. Another context is one where the variance is different over subpopulations. This has led to bootstrapping known as the *Wild Bootstrap*.

There are contexts where we may know some aspects of the population, for instance that the conditional expectation function is linear. This is a context in between assuming the population distribution is completely unknown and assuming it is known except for a finite number of parameters. These types of bootstraps are often called *semiparametric bootstraps*. It is typically an advantage to use known aspects of the population when drawing bootstrap samples. In Volume II of this book, we will encounter situations where context-specific knowledge may be used in the construction of the bootstrap procedure, resulting in a bootstrap distribution that is a more precise estimator of a sampling distribution.

13.3 Asymptotic distribution

It is a remarkable fact that, for many estimators, their sampling distributions have shapes similar to normal distributions when the sample size is sufficiently large. This happens regardless of the population distribution. The reason is that, under mild assumptions, there are some regularities of the estimators which dominate their distributional behavior when the sample size is large, and these regularities "pull" the distributions of the estimators toward normal distributions. The regularities stem from the distribution of sample averages, which are a component of many estimators. The fundamental result about the distribution of a sample average, as the number of observations increases toward infinity, is called the *Central Limit Theorem* (Theorem 13.1). Since many estimators are functions of some kind of sample averages, the Central Limit Theorem is applicable to many estimators. An estimate of the sampling distribution of an estimator, based on large sample arguments, is called an asymptotic distribution.

To derive an asymptotic distribution, we need to introduce the concept of *convergence in distribution*. This concept formalizes what it means that distributions become similar for increasing sample sizes. Following this introduction of convergence in distribution, we discuss the Central Limit Theorem.

13.3.1 Convergence in distribution

A key concept in asymptotic distribution theory is convergence in distribution. Convergence in distribution means that the distribution of a random variable indexed by the sample size n, say, X_n, is about the same as the distribution of another random variable, say, X, when the sample size n is large. This is formalized as the CDFs $F_{X_n}(x)$ of the X_n's gets closer to the CDF $F_X(x)$ of X when n goes to infinity. The concept is formally defined next.

Definition 13.1 (Convergence in distribution) *Let $X_1, \ldots, X_n$ be a sequence of random variables each with marginal CDF $F_{X_n}(x)$. Then X_n **converges in distribution** to X if and only if*

$$F_{X_n}(x) \to F_X(x) \text{ as } n \to \infty$$

at every x for which the function $F_X(x)$ is continuous.

We will often write

$$\text{``} X_n \xrightarrow{d} X \text{ as } n \to \infty \text{''}$$

to denote that X_n converges to X in distribution.

The distribution $F_X(x)$ of X is called the *limiting distribution* for the sequence of X_n. Sometimes for shorthand, we simply write the limiting distribution in place of X. That is, if the limiting distribution is, say, the standard normal distribution, we may write "$X_n \xrightarrow{d} N(0,1)$" instead of "$X_n \xrightarrow{d} X$, where $X \sim N(0,1)$".

Example 13.4 *Let*

$$X_n \sim Exp\left(\lambda + \frac{1}{n}\right).$$

Then the CDF for X_n is

$$F_{X_n}(x) = 1 - \exp\left(-\left(\lambda + \frac{1}{n}\right)x\right).$$

Consider a value of x and keep it fixed. If $n \to \infty$, then

$$1 - \exp\left(-\left(\lambda + \frac{1}{n}\right)x\right) \to 1 - \exp\left(-\lambda x\right).$$

The right hand side is the CDF $F_X(x)$ of an exponentially distributed random variable $X \sim Exp(\lambda)$. Since this is true for any value of x, we conclude that

$$X_n \xrightarrow{d} X \ \text{as } n \to \infty.$$

In other words, the $Exp(\lambda)$ distribution is the limiting distribution of the sequence of random variables $X_1, X_2, \ldots$.

The usefulness of convergence in distribution is that we may not know the distribution of X_n but we may be able to show that the sequence converges in distribution to that of a random variable X for which we do know the distribution. Then we can use the distribution of X as an estimate, or approximation, of the distribution of X_n. When the sample size n is large, we are often justified in assuming this approximation is good.

Example 13.5 *Assume*

$$X_n = Z + \frac{1}{n}W.$$

where $Z \sim N(0,1)$ but the distribution of the random variable W is unknown. Then

$$X_n \xrightarrow{d} Z \ \text{as } n \to \infty.$$

Therefore, when n is large, we can use the distribution of Z, which here is $N(0,1)$, to estimate the distribution of X_n.

13.3.2   The Central Limit Theorem

The concept of convergence in distribution is widely applicable because the distribution of a standardized sample average will, under certain conditions, converge to the normal distribution. Such results go under the name "central limit theorem" (CLTs), and different assumptions or conditions may lead to different central limit theorems. For instance, the simplest central limit theorem, called the *Lindeberg-Lévy Central Limit Theorem*, is stated under the assumption that the sample behind the sample average is a simple random sample (Theorem 13.1).

A standardized sample average is constructed as follows. Let $\overline{X}_n$ be the sample average and assume the variance $Var\left(\overline{X}_n\right)$ exists. Define the standardized sample average Z_n as the random variable

$$Z_n = \frac{\overline{X}_n - E\left(\overline{X}_n\right)}{\sqrt{Var\left(\overline{X}_n\right)}}.$$

By this construction, the standardized sample average has the properties that (Problem 13.7.1)

$$E\left(Z_n\right) = 0,$$

and

$$Var\left(Z_n\right) = 1,$$

i.e. it has zero mean and unit variance, which is the reason for the "standardized" nomenclature.

Example 13.6 (Standardized sample average of simple random sample)

Let $(\tilde{X}_1, \ldots, \tilde{X}_n)$ be a simple random sample from a population represented by the random variable X, and let $\mu_0 = E(X)$ and $\sigma_0^2 = Var(X) < \infty$. In this case, we have (e.g. Theorem 11.2)

$$\begin{aligned} E\left(\overline{X}_n\right) &= \mu_0, \\ Var\left(\overline{X}_n\right) &= \frac{\sigma_0^2}{n}, \end{aligned}$$

and, hence,

$$Z_n = \frac{\overline{X}_n - \mu_0}{\sqrt{\sigma_0^2/n}}$$

is the standardized sample average.

Next we state a central limit theorem appropriate for the case of a simple random sample.

Theorem 13.1 (Central Limit Theorem, Lindeberg-Lévy CLT)

Assume $X_1, \ldots, X_n$ are mutual statistically independent and identically distributed random variables with $E(X_i) = \mu_0$ and $Var(X_i) = \sigma_0^2 < \infty$. Let the sample average be

$$\overline{X}_n = \frac{1}{n}\sum_{i=1}^{n} X_i.$$

Then, as $n \to \infty$,

$$Z_n = \frac{\overline{X}_n - \mu_0}{\sqrt{\sigma_0^2/n}} \xrightarrow{d} N\left(0,1\right), \tag{13.3}$$

where $N(0,1)$ is the standard normal distribution.

Due to properties of the normal distributions, the result (13.3) in Theorem 13.1 can also be written as

$$\sqrt{n}\left(\overline{X}_n - \mu_0\right) = \frac{1}{\sqrt{n}}\sum_{i=1}^{n}(X_i - \mu_0) \xrightarrow{d} N\left(0,\sigma_0^2\right). \tag{13.4}$$

Equation (13.4) provides some intuition for the magic of the Central Limit Theorem. We know from the Law of Large Numbers (Theorem 11.4) that the sample average $\overline{X}_n$ will approach μ_0 as the sample size n increases. In other words, $\overline{X}_n - \mu_0$ will approach zero, as

n increases. One may now ask *how fast*, expressed by the sample size n, the sample average $\overline{X}_n$ approaches μ_0. It turns out that for all $\epsilon > 0$ then $n^{1/2-\epsilon}\left(\overline{X}_n - \mu_0\right)$ will converge to zero, while $n^{1/2+\epsilon}\left(\overline{X}_n - \mu_0\right)$ will diverge. However, as stated in the Central Limit Theorem, $n^{1/2}\left(\overline{X}_n - \mu_0\right)$ neither converges to zero nor diverges. Instead, it converges (in distribution) to *a random variable* (in contrast to convergence to a number). This shows that the "speed" at which $\overline{X}_n$ approaches μ_0 is exactly $n^{1/2} = \sqrt{n}$. Related, another remarkable fact is that the random variable, to which the standardized sample average converges, is normally distributed, regardless of the distribution of the random variables X_i. In fact, this is the main reason why the normal distribution is so special: Under very mild conditions, such as those stated in Theorem 13.1, standardized sample averages will *always* converge in distribution to the normal distribution. An important consequence of this is, as we will illustrate presently, that we can often use the normal distribution as an approximation for the distribution of estimators and other sample statistics built from sample averages.

The limiting distribution of the standardized sample average Z_n can be used to construct an estimate of the sampling distribution of the sample average $\overline{X}_n$ as follows. First, an estimate of the sampling distribution of Z_n is $N(0, 1)$ since Z_n converges in distribution to $N(0, 1)$. We indicate an estimate of a sampling distribution based on asymptotic behavior as

$$Z_n \stackrel{A}{\sim} N\left(0, 1\right), \tag{13.5}$$

where "$\stackrel{A}{\sim}$" means asymptotically, or approximately, distributed. In other words, by "$Z_n \stackrel{A}{\sim} N\left(0, 1\right)$", we mean that the $N(0, 1)$-distribution is a good approximation to the distribution of the random variable Z_n, when n is large.

Secondly, re-write Equation (13.3) for the standardized sample average Z_n to arrive at

$$\overline{X}_n = \mu_0 + Z_n \cdot \sqrt{\sigma_0^2 \frac{1}{n}}.$$

Now, use the approximate distribution of Z_n, given in (13.5), and the properties of the normal distribution (see Section 8.5), to conclude that

$$\overline{X}_n = \mu_0 + Z_n \cdot \sqrt{\sigma_0^2 \frac{1}{n}} \stackrel{A}{\sim} N\left(\mu_0, \frac{1}{n} \cdot \sigma_0^2\right). \tag{13.6}$$

Thus, the asymptotic distribution of the sample average $\overline{X}_n$ of a simple random sample is $N\left(\mu_0, \frac{1}{n} \cdot \sigma_0^2\right)$. For finite sample sizes n, we expect this distribution to be a reasonable approximation to the distribution of the sample average $\overline{X}_n$, provided that n is not too small.

Example 13.7 (Illustration of the Law of Large Numbers and Central Limit Theorem) *Suppose the population has an exponential distribution given by $X \sim Exp(0.25)$. By Theorem 8.3, the mean and variance of this distribution are $E(X) = 1/0.25 = 4$ and $Var(X) = 1/0.25^2 = 16$. Assume a simple random sample of size n is available.*

Consider the sample average $\overline{X}_n$. In Problem 13.7.6 at the end of this chapter, you are asked to show that the sample average has the Erlang distribution, $\overline{X}_n \sim Erlang(\lambda n, n)$. The left panel of Figure 13.2 shows the PDF of the sample average $\overline{X}_n$, i.e. the $Erlang(\lambda n, n)$-distribution, for sample sizes n equal to 2, 10, and 100. From the figure, it is seen that the distributions concentrate around the mean $E(X) = 4$, indicated by the vertical line, for larger n. This is a consequence of the Law of Large Numbers (Theorem 11.4). It can also be seen that in the process of the distributions concentrating around $E(X) = 4$, the distributions become more "bell" shaped like normal distributions.

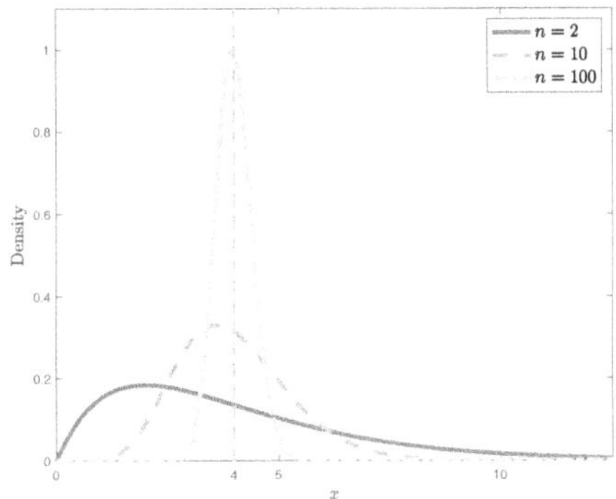 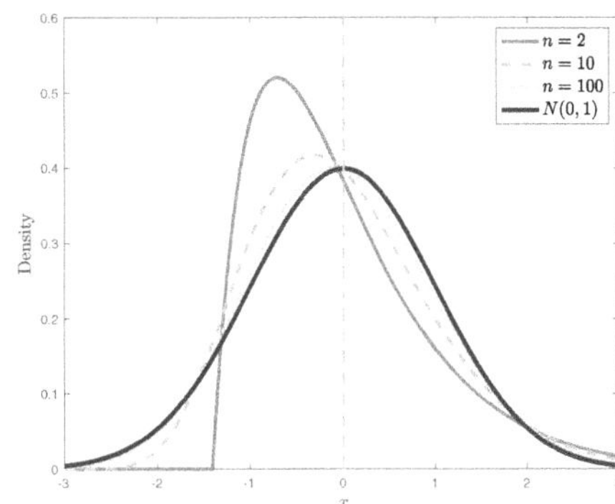

FIGURE 13.2
Sampling distributions of two sample statistics from a simple random sample from an exponentially distributed population, see Example 13.7. Left: Sampling distribution of sample average $\overline{X}_n$. Right: Sampling distribution of standardized sample average Z_n.

Consider now the standardized sample average

$$Z_n = \frac{\overline{X}_n - E(X)}{\sqrt{Var(X)/n}}.$$

In Problem 13.7.6 you are asked to show that the PDF of Z_n can be written as $f(x + \sqrt{n})$, where $f(x)$ is the PDF of the Erlang($\sqrt{n}, n$)-distribution. The PDF of this distribution is shown in the right panel of Figure 13.2. It is seen that the distribution gets a shape closer to that of a standard normal distribution $N(0,1)$, shown by the thick black curve, as the sample size n grows. This is a consequence of the Central Limit Theorem (Theorem 13.1).

In this example, it was possible to derive the PDFs of the sample average $\overline{X}_n$ and standardized sample average Z_n analytically using the Erlang distribution. In general, it may not be possible to derive these PDFs in closed form, even if the population distribution is known. In such a case, we may approximate the distributions of $\overline{X}_n$ and Z_n using Monte Carlo simulation. You are asked to do this in Problem 13.7.6.

To apply the asymptotic distribution in practice, the unknown parameters in the asymptotic distribution need to be estimated. In particular, for the asymptotic distribution of the sample average, estimators of μ_0 and σ_0^2 are needed, cf. Equation (13.6).

13.3.3 Extending asymptotic results beyond a sample average

By using results on the distribution of functions of random variables, the results on the asymptotic distributional properties of the sample average $\overline{X}_n$, discussed above, can be extended to estimators which are not directly formulated as sample averages. In the following, we state a number of results which are particularly useful and frequently encountered in asymptotic theory.

The first result states that if the function $g()$ is continuous and a sequence of random variables Y_n converges, either in distribution or in probability, to some limit Y, then the sequence $Z_n = g(Y_n)$ also converges, either in distribution or in probability, to the limit $g(Y)$. This result is known as the *continuous mapping theorem*.

Theorem 13.2 (Continuous mapping theorem) *Let $g()$ be a continuous function.*

1. *If $Y_n \overset{p}{\to} Y$ as $n \to \infty$, then*

$$g(Y_n) \overset{p}{\to} g(Y), \quad as \ n \to \infty.$$

2. *If $Y_n \overset{d}{\to} Y$ as $n \to \infty$, then*

$$g(Y_n) \overset{d}{\to} g(Y), \quad as \ n \to \infty.$$

Note that, if $g()$ is continuous and $\widehat{\theta}$ is a consistent estimator of the parameter θ, then the first part of the continuous mapping theorem implies that $g(\widehat{\theta})$ is a consistent estimator of $g(\theta)$.

The next result is useful when we have two sequences of random variables, say Y_n and W_n, where one sequence converges in probability and the other in distribution.

Theorem 13.3 (Convergence in probability and in distribution) *Assume $Y_n \overset{d}{\to} Y$ and $W_n \overset{p}{\to} b$ as $n \to \infty$, where Y is a random variable and $b \in \mathbb{R}$ a constant. Let $g(x, w)$ be a continuous function. Then*

$$g(Y_n, W_n) \overset{d}{\to} g(Y, b) \quad as \ n \to \infty.$$

The following three results are useful implications of Theorem 13.3. They are known under the name *Slutsky's theorem*.

Theorem 13.4 (Slutsky's theorem) *Assume $Y_n \overset{d}{\to} Y$ and $W_n \overset{p}{\to} b$ as $n \to \infty$, where Y is a random variable and $b \in \mathbb{R}$ a constant. Then, as $n \to \infty$,*

$$\begin{aligned}
W_n + Y_n &\overset{d}{\to} b + Y, \\
W_n \cdot Y_n &\overset{d}{\to} bY, \quad and \\
Y_n / W_n &\overset{d}{\to} Y/b, \quad for \ b \neq 0.
\end{aligned}$$

The next example illustrates how Slutsky's theorem can be applied. The application is relevant to the much used *ordinary least squares estimator*, studied in Volume II of this book.

Example 13.8 *Suppose $X_n \overset{d}{\to} X$, where $X \sim N\left(\mu, \sigma^2\right)$ and assume $\underset{n \to \infty}{plim}\,(W_n) = b$. Then, according to the second part of Slutsky's theorem (Theorem 13.4), we have*

$$W_n X_n \overset{d}{\to} bX.$$

Further, by properties of the normal distribution,

$$bX \sim N\left(b\mu, b^2\sigma^2\right).$$

Consequently, we conclude that the asymptotic distribution of the product $W_n X_n$ is

$$W_n X_n \overset{A}{\sim} N\left(b\mu, b^2\sigma^2\right).$$

The Central Limit Theorem (Theorem 13.1) applies to the standardized sample average (13.3),

$$Z_n = \frac{\overline{X}_n - \mu_0}{\sqrt{\sigma_0^2/n}}.$$

In practice, however, the variance σ_0^2 will often be unknown and thus has to be estimated, e.g. by the sample variance S^2. In this case, the more practically relevant sequence of random variables to study may be

$$W_n = \frac{\overline{X}_n - \mu_0}{\sqrt{S^2/n}}.$$

We can apply Slutsky's theorem to show that W_n does also satisfy a Central Limit Theorem, i.e. that

$$W_n \overset{d}{\to} N(0,1),$$

as $n \to \infty$. We call this a *feasible* Central Limit Theorem, because it does not rely on the unknown variance. The following theorem presents this result.

Theorem 13.5 (Feasible Central Limit Theorem) *Assume $X_1, \ldots, X_n$ are mutual statistically independent and identically distributed random variables with $E(X_i) = \mu_0$ and $Var(X_i) = \sigma_0^2 < \infty$. Let the sample average be*

$$\overline{X}_n = \frac{1}{n} \sum_{i=1}^{n} X_i$$

and suppose that $\widehat{Var}(X)$ is a consistent estimator of σ_0^2. Then, as $n \to \infty$,

$$W_n = \frac{\overline{X}_n - \mu_0}{\sqrt{\widehat{Var}(X)/n}} \overset{d}{\to} N(0,1),$$

where $N(0,1)$ is the standard normal distribution.

We can extend the Central Limit Theorem to hold for a known function $g()$ of a sample average. In fact, we can formulate the result more generally: If Y_n is a sequence of random variables satisfying a central limit theorem result, then the sequence of random variables $Z_n = g(Y_n)$, where $g()$ is a function, satisfies a similar central limit theorem. This is known as the *delta method*.

Theorem 13.6 (Delta method) *Let Y_n be a sequence of random variables such that,*

$$\sqrt{n}(Y_n - \theta) \overset{d}{\to} N(0, \sigma^2) \quad as\ n \to \infty,$$

where $\theta \in \mathbb{R}$ and $\sigma^2 > 0$ are constants. Let $g()$ be a function such that $g'(\theta)$ exists and is different from zero. Then, as $n \to \infty$,

$$\sqrt{n}(g(Y_n) - g(\theta)) \overset{d}{\to} N(0, (g'(\theta))^2 \sigma^2).$$

The delta method can be very useful in cases where we can estimate the parameter θ, but the quantity of interest is not θ itself, but some function of θ, i.e. $g(\theta)$. Recall that, by the continuous mapping theorem (Theorem 13.2), if $\widehat{\theta}$ is a consistent estimator of θ, then $g(\widehat{\theta})$ is a consistent estimator of $g(\theta)$. If $\widehat{\theta}$ also satisfies a central limit theorem-type result, we can then use the (asymptotic) sampling distribution of $\widehat{\theta}$, along with the delta method, to ascertain the sampling distribution of $g(\widehat{\theta})$. Note that, although we will not state it here, a straightforward multivariate extension of the delta method, which is relevant in the case where θ is a vector, exists.

In the setup of the Central Limit Theorem (Theorem 13.1), where Y_n is the sample average $\overline{X}_n$ and θ is the population mean μ, Theorem 13.6 can be stated as follows.

Corollary 13.3.1 (Delta method for sample average) *Assume* $X_1, \ldots, X_n$ *are mutual statistically independent and identically distributed random variables with* $E(X_i) = \mu_0$ *and* $Var(X_i) = \sigma_0^2 < \infty$. *Let the sample average be*

$$\overline{X}_n = \frac{1}{n} \sum_{i=1}^{n} X_i.$$

Then, as $n \to \infty$,

$$\sqrt{n}(g(\overline{X}_n) - g(\mu_0)) \xrightarrow{d} N\left(0, (g'(\mu_0))^2 \sigma_0^2\right),$$

for any function $g()$ *such that* $g'(\theta)$ *exists and is different from zero.*

Example 13.9 *Suppose* $(\widetilde{X}_1, \widetilde{X}_2, \ldots, \widetilde{X}_n)$ *is a simple random sample from a population* X *such that* $\mu_0 = E(X)$ *and* $\sigma_0^2 = Var(X) < \infty$. *Let* $\overline{X}$ *be the sample average. By the Central Limit Theorem (Theorem 13.1), it holds that*

$$\sqrt{n}(\overline{X}_n - \mu_0) \xrightarrow{d} N\left(0, \sigma_0^2\right), \quad n \to \infty.$$

By Corollary 13.3.1 we have, say,

$$\sqrt{n}(\exp(\overline{X}_n) - \exp(\mu_0)) \xrightarrow{d} N\left(0, (\exp(\mu_0))^2 \sigma_0^2\right), \quad n \to \infty,$$

and, if $\mu_0 \neq 0$,

$$\sqrt{n}((\overline{X}_n)^2 - (\mu_0)^2)) \xrightarrow{d} N\left(0, (2\mu_0)^2 \sigma_0^2\right), \quad n \to \infty.$$

In the first case, we applied the delta method using the function $g(x) = \exp(x)$, *while we in the second case applied the delta method using the function* $g(x) = x^2$.

13.4 The standard error of an estimator

In this chapter, we have studied how to estimate the sampling distribution of an estimator $\widehat{\theta}$. We may, in turn, apply the methods discussed in Chapter 12 to estimate various aspects of the sampling distribution. To get a single measure for the dispersion of the sampling distribution, the variance $Var(\widehat{\theta})$ is useful. If the estimator has a high variance, it indicates that we should be inclined to put less trust in a particular estimate, since it is a draw from a distribution with high variance, i.e. where outcomes can plausibly be far away from the mean. Conversely, if the variance of the estimator is small, we would be inclined to put more trust in a particular estimate.

To report a measure of dispersion using the same units of measurement as the estimator $\widehat{\theta}$, we focus on the standard deviation of the sampling distribution, that is, $\sqrt{Var(\widehat{\theta})}$. Traditionally, an estimator of a standard deviation of a sample statistics, e.g. an estimator, is called the *standard error*. The formal definition is given next.

Definition 13.2 (Standard error of estimator) *Let θ be the estimand and $\widehat{\theta}$ be an estimator of θ such that $Var(\widehat{\theta})$ exists. Then, the **standard error** of θ is defined by*

$$SE(\widehat{\theta}) = \sqrt{\widehat{Var(\widehat{\theta})}},$$

where $\widehat{Var}(\widehat{\theta})$ is an estimator of $Var(\widehat{\theta})$.

For an unbiased estimator $\widehat{\theta}$, the standard error of $\widehat{\theta}$ can be interpreted as a measure of precision. That is, if $E(\widehat{\theta}) = \theta$, then

$$SE(\widehat{\theta}) = \sqrt{\widehat{MSE}(\widehat{\theta})}.$$

The next example considers the standard error of the sample mean $\overline{X}$ when this is used as an estimator of the mean. This is the so-called *standard error of the mean* (SEM).

Example 13.10 (Standard error of the mean) *Let $(\widetilde{X}_1, \ldots, \widetilde{X}_n)$ be a simple random sample from a population represented by the random variable X, and let $\mu_0 = E(X)$ and $\sigma_0^2 = Var(X) < \infty$. Let $\overline{X}$ be the sample average, which we know is an unbiased estimator of the mean μ_0. As also noted in Example 13.6, we have (e.g. Theorem 11.2)*

$$Var\left(\overline{X}\right) = \frac{\sigma_0^2}{n}.$$

We may estimate the variance of $\overline{X}$ by substituting an estimator for σ_0^2 into this expression. We may, for instance, use the sample variance S^2 (Definition 12.5) as an estimator of σ_0^2, leading to the following estimator of $Var\left(\overline{X}\right)$

$$\widehat{Var}\left(\overline{X}\right) = \frac{S^2}{n}.$$

Using this, the standard error of the mean becomes

$$SE(\overline{X}) = \sqrt{\frac{S^2}{n}}.$$

Note that, because $\overline{X}$ is an unbiased estimator of μ_0 (Theorem 11.2), $SE(\overline{X})$ is an estimator of the root mean squared error of $\overline{X}$.

When the sampling distribution of the estimator $\widehat{\theta}$ is the normal distribution $N\left(\theta_0, Var\left(\widehat{\theta}\right)\right)$, then the estimated sampling distribution of $\widehat{\theta}$ is $N\left(\widehat{\theta}, Var\left(\widehat{\theta}\right)\right)$, or, expressed by the standard error, $N\left(\widehat{\theta}, \left(SE(\widehat{\theta})\right)^2\right)$. Hence, the estimator together with the standard error completely specify the estimated sample distribution of $\widehat{\theta}$. Since many estimators are asymptotically normally distributed, then having the estimators $\widehat{\theta}$ and $SE(\widehat{\theta})$ are sufficient to specify the asymptotic distribution of $\widehat{\theta}$. This is the reason that estimates and their accompanying standard errors are typically reported in empirical studies.

13.5 Comparisons of estimators of a sampling distribution

When compared to each other, the bootstrap and the asymptotic distribution have advantages and disadvantages. We briefly comment on some of them in the following.

The asymptotic distribution is commonly used in practice because it turns out that many estimators have an asymptotic distribution that can be derived using a central limit theorem and, thus, the asymptotic distribution is normally distributed, or a derived distribution from the normal distribution. For theoretical considerations, the asymptotic distribution is useful because it often can be expressed by relatively simple functions. On the other hand, the asymptotic distribution may be imprecise if the sample size is small, and, for complicated estimators, it may be difficult to derive the asymptotic distribution.

The bootstrap distribution is useful in practice because it can be implemented using simulations which are relatively easy to do on a computer. Often, the bootstrap distribution is a good approximation to the sampling distribution, also for small sample sizes. On the other hand, if the sampling distribution is needed to derive theoretical properties of an estimator, the bootstrap distribution may be difficult to use, because it may be difficult to write the bootstrap distribution in a simple closed-form expression.

Compared to the non-parametric bootstrap, the parametric bootstrap typically provides a better estimate of the sampling distribution. The intuitive reason is that more information is included in the estimation of the population distribution under the parametric bootstrap. This implies higher precision of the estimator of the population distribution, and this higher precision spills over to a higher precision of the estimator of the sampling distribution. All of this, however, only hold if the assumption on the population distributions $F_X(x; \theta)$, needed to perform the parametric bootstrap, is a reasonable description of the population.

13.6 Proofs

Proof of Theorem 13.1 *The proof of the Central Limit Theorem is beyond the scope of this book. The interested reader may find a proof of this fact in most textbooks on advanced probability theory.*

Proof of Theorem 13.2 *The proof of the continuous mapping theorem is beyond the scope of this book. The interested reader may find a proof of this fact in most textbooks on advanced probability theory.*

Proof of Theorem 13.3 *The proof of this result is beyond the scope of this book. The interested reader may find a proof of this fact in most textbooks on advanced probability theory.*

Proof of Theorem 13.4 *Slutsky's theorem may be deduced from Theorem 13.3. In particular, the first part follows by considering the function $g(Y_n, W_n) = W_n + Y_n$ and applying Theorem 13.3; the second part follows similarly using the function $g(Y_n, W_n) = Y_n \cdot W_n$; the third part by using the function $g(Y_n, W_n) = Y_n / W_n$.*

Proof of Theorem 13.5 *Let $\widehat{Var}(X)$ be a consistent estimator of σ_0^2 and write*

$$W_n = \frac{\overline{X}_n - \mu_0}{\sqrt{\widehat{Var}(X)/n}} = \frac{\overline{X}_n - \mu_0}{\sqrt{\sigma_0^2/n}} \cdot \sqrt{\frac{\sigma_0^2}{\widehat{Var}(X)}}.$$

From the Central Limit Theorem (Theorem 13.1), we know that

$$\frac{\overline{X}_n - \mu_0}{\sqrt{\sigma_0^2/n}} \xrightarrow{d} N(0, 1),$$

as $n \to \infty$. Further, since $\widehat{Var}(X)$ is a consistent estimator of σ_0^2, we have, by the continuous mapping theorem (Theorem 13.2), that

$$\sqrt{\frac{\sigma_0^2}{\widehat{Var}(X)}} \xrightarrow{p} \sqrt{\frac{\sigma_0^2}{\sigma_0^2}} = 1.$$

Combining these results using Slutsky's theorem (Theorem 13.4), we get

$$W_n = \frac{\overline{X}_n - \mu_0}{\sqrt{\sigma_0^2/n}} \cdot \sqrt{\frac{\sigma_0^2}{\widehat{Var}(X)}} \xrightarrow{d} N(0,1),$$

as $n \to \infty$, as we wanted to show.

Proof of Theorem 13.6 *The proof of the Delta method relies on a result known as Taylor's theorem, which states that we may write*

$$g(Y_n) = g(\theta) + g'(\theta)(Y_n - \theta) + R(Y_n, \theta), \tag{13.7}$$

where $g'()$ is the derivative of the function $g()$ and $R()$ is a remainder term, which satisfies

$$\sqrt{n}R(Y_n, \theta) \xrightarrow{p} 0, \tag{13.8}$$

as $n \to \infty$. Using (13.7), we have

$$\sqrt{n}(g(Y_n) - g(\theta)) = \sqrt{n}g'(\theta)(Y_n - \theta) + \sqrt{n}R(Y_n, \theta).$$

By the Central Limit Theorem, we have

$$\sqrt{n}g'(\theta)(Y_n - \theta) \xrightarrow{d} N(0, g'^2\sigma^2),$$

as $n \to \infty$. This, together with (13.8), implies, by way of Slutsky's theorem (Theorem 13.4), that

$$\sqrt{n}(g(Y_n) - g(\theta)) \xrightarrow{d} N(0, g'^2\sigma^2),$$

as $n \to \infty$. This completes the proof.

Proof of Corollary 13.3.1 *This follows from the Central Limit Theorem (Theorem 13.1) and the Delta method (Theorem 13.6).*

13.7 Exercises

Problem 13.7.1 *Let Y be a random variable such that $Var(Y)$ exists. Define the standardized version of Y via*

$$Z = \frac{Y - E(Y)}{Var(Y)}.$$

Show that $E(Z) = 0$ and $Var(Z) = 1$. (Note: By setting $Y = \overline{X}$, this problem shows that the standardized sample average has mean zero and unit variance, as used in Section 13.3.2.)

Problem 13.7.2 *Consider a population X with a mean μ_0, variance σ_0^2, and fourth central moment $m_4^* < \infty$. Assume a simple random sample $\left(\tilde{X}_1, \ldots, \tilde{X}_n\right)$ is available. Finally, assume it is known that the mean is $\mu_0 = 5$.*

An estimator of the population variance σ_0^2 is

$$\hat{\sigma}^2 = \frac{1}{n} \sum_{i=1}^{n} \left(\tilde{X}_i - \mu_0\right)^2$$

Derive the asymptotic distribution of $\hat{\sigma}^2$.

$\Big($ Hint: You may define a new random variable by $Y_i = \left(\tilde{X}_i - \mu_0\right)^2$. Hence, the estimator $\hat{\sigma}^2$ is the sample average taken over $Y_1, \ldots, Y_n$. Notice that

$$E(Y_i) = E\left(\left(\tilde{X}_i - \mu_0\right)^2\right) = Var(\tilde{X}_i),$$

and

$$Var(Y_i) = E(Y_i^2) - (E(Y_i))^2 = E\left(\tilde{X}_i - \mu_0\right)^4 - \left(Var(\tilde{X}_i)\right)^2.\Big)$$

Problem 13.7.3 *Consider a setup similar to Examples 13.1 and 13.3, but assume now that the realized (simple random) sample has $n = 10$ elements given by*

$$(\tilde{x}_1, \tilde{x}_2, \ldots, \tilde{x}_{10}) = (1, 2, 3, 3, 1, 4, 5, 4, 3, 2).$$

1. *Calculate the realized sample average $\bar{x}$.*

2. *Use a non-parametric bootstrap to approximate the sampling distribution of $\bar{X}$, $F_{\bar{X}}$. (Hint, it is not advisable to try to derive the closed-form solution to the bootstrap distribution $F_{\bar{X}}^*$, so you might want to use a simulation-based approach to approximate $F_{\bar{X}}^*$, see Example 13.3.)*

Problem 13.7.4 *Consider the setup from Example 13.2, and assume that you observe a simple random sample with $n = 10$ elements,*

$$(\tilde{x}_1, \tilde{x}_2, \ldots, \tilde{x}_{10}) = (0.74, 1.36, -1.01, -1.22, 0.22, -0.58, -0.21, -0.66, -0.20, 0.52).$$

1. *Use the sample to calculate the realized sample average $\bar{x}$ and the realized sample variance s^2.*

2. *Use $\bar{x}$ and s^2 to specify the sampling distribution of the bootstrap sample average $\overline{X}^*$. (Hint, recall that the population distribution is assumed to be normal in Example 13.2.)*

3. *Use a non-parametric bootstrap approach, implemented via Monte Carlo simulation, to estimate the sampling distribution of the sample average $\overline{X}^*$. Run three instances of the Monte Carlo simulations, one with $B = 100$, one with $B = 1000$, and one with $B = 10000$. Compare your simulation-based approximation of the bootstrap sampling distribution with the theoretically derived bootstrap sampling distribution found in 2 in the following ways:*

   (a) *Compare the PDF $f_{\overline{X}^*}$, implied by the distribution found in 2, with a histogram of the bootstrap sample averages $\bar{x}_b^*$, $b = 1, 2, \ldots, B$, found in 3.*

    (b) Compare the quantiles q_p for $p = 0.01, 0.05, 0.10, 0.50, 0.90, 0.95, 0.99$ from the distribution found in 2, with the empirical quantiles of the bootstrap sample averages $\overline{x}_b^*$, $b = 1, 2, \ldots, B$, found in 3.

4. *Reflect on how the choice of the number of bootstrap replications B affects the precision of the simulation-based approximation to the bootstrap sampling distribution.*

Problem 13.7.5 *In this exercise, you are asked to verify parts of Example 13.7. Consider a simple random sample $\widetilde{X}_1, \ldots, \widetilde{X}_n$ from a population represented by $X \sim Exp(\lambda)$, where $\lambda > 0$.*

1. *Show that $\overline{X}_n \sim Erlang(\lambda n, n)$. (Hint: You may use that if $Y \sim Exp(\lambda)$ when $aY \sim Exp(\lambda/a)$ for any $a > 0$.)*

2. *Let Z_n be the standardized sample average and show that you can write $Z_n = Y_n - \sqrt{n}$ where $Y_n \sim Erlang(\sqrt{n}, n)$. Conclude that the PDF of Z_n is given by $f(x + \sqrt{(n)})$ where $f(x)$ is the PDF of Y_n. (Hint: For this latter part, you may use Theorem 2.14.)*

3. *Let $\lambda = 1$. Use a parametric bootstrap, implemented using Monte Carlo simulation, to approximate the distributions of $\overline{X}_n$ and Z_n, and compare your results with the analytical PDFs derived above. Consider $n = 2, 10, 100$ as in Example 13.7.*

Problem 13.7.6 *Consider the two random variables $X_1 \sim Exp(\lambda_1)$ and $X_2 \sim Exp(\lambda_2)$, where $\lambda_1 \neq \lambda_2$. Define the population random variable $X = X_1 + X_2$. Since $\lambda_1 \neq \lambda_2$, the distribution of X is not exponential. It is possible to derive the distribution of X analytically, but in this exercise, you are asked to approximate it using Monte Carlo simulation.*

1. *Let $\lambda = 1$, $\lambda_2 = 0.5$. Use a parametric bootstrap, implemented using Monte Carlo simulation, to approximate the distribution of X.*

    Consider a simple random sample $\widetilde{X}_1, \ldots, \widetilde{X}_n$ from X.

2. *Let $\lambda = 1$, $\lambda_2 = 0.5$, $n = 20$. Use a parametric bootstrap, implemented using Monte Carlo simulation, to approximate the distribution of the sample average $\overline{X}_n$.*

3. *Let $\lambda = 1$, $\lambda_2 = 0.5$, $n = 20$. Use a parametric bootstrap, implemented using Monte Carlo simulation, to approximate the distribution of the standardized sample average Z_n. Compare with the $N(0,1)$ distribution.*

14

Confidence intervals

14.1 Introduction

One major challenge in empirical work is to decide how much confidence to have in results based on sampling from a population. In this chapter, we discuss how uncertainty can be included as an element of an estimator by formulating the estimator as an interval. Such an estimator is called an *interval estimator*. This should be contrasted with a *point estimator*, i.e. an estimator outputting a single value as the estimate, which is the type of estimator considered thus far in the book. Specifically, we consider interval estimators which are called "confidence intervals" due to their specific focus on the probability of covering the estimand. Typically, a confidence interval is constructed based on a point estimator together with an estimator of the uncertainty of the point estimator, such as the standard error.

14.2 The concept of a confidence interval

Our aim is to find an interval estimator of an estimand θ_0. Let $\left(\tilde{X}_1, \ldots, \tilde{X}_n\right)$ be a random sample and $L = g_L\left(\tilde{X}_1, \ldots, \tilde{X}_n\right)$ and $U = g_U\left(\tilde{X}_1, \ldots, \tilde{X}_n\right)$ be two sample statistics such that $L \leq U$. Define the *random interval* $\widehat{CI} = [L, U]$. For each realized sample, the interval consists of all the real numbers between the realized values of L and U. The aim will be to choose the two sample statistics L and U such that the interval $\widehat{CI} = [L, U]$ contains the estimand θ_0 with a certain pre-specified probability called the *confidence level*. In this case, the interval $\widehat{CI}$ is said to be a *confidence interval* for θ_0 corresponding to the chosen confidence level. In terms of the statistical experiment, it means that if we were to repeat the statistical experiment infinitely many times, then the fraction of experiments, where the estimand is included in the confidence interval, equals the confidence level. The formal definition of a confidence interval is given next.

Definition 14.1 (Confidence interval) *Let θ_0 be the estimand in a parameter space Θ and $\alpha \in (0, 0.5]$ a number. Then the random interval $\widehat{CI}_{1-\alpha}$ is a $(1 - \alpha)$-**confidence interval** for θ if and only if α is the smallest number such that*

$$\Pr\left(\theta_0 \in \widehat{CI}_{1-\alpha}\right) \geq 1 - \alpha \tag{14.1}$$

for all values of $\theta_0 \in \Theta$.

The number $(1 - \alpha)$ is denoted the confidence level of the confidence interval. The confidence level can also be stated as the percentage $100(1 - \alpha)\%$.

DOI: 10.1201/9781003591191-14

256

In Definition 14.1, the confidence interval is defined by the smallest α that satisfies (14.1). The reason for this formulation is that if, say, $\Pr\left(\theta \in \widehat{CI}_{1-\alpha}\right) \geq 1 - 0.1 = 0.9$, then it is also true that, e.g. $\Pr\left(\theta \in \widehat{CI}_{1-\alpha}\right) \geq 1 - 0.2 = 0.8$. By using this formulation, $1 - \alpha$ is the highest possible probability not larger than $\Pr\left(\theta \in \widehat{CI}_{1-\alpha}\right)$, and it is this probability we denote the confidence level.

In many cases, in particular when the random variables under consideration are continuous, it will be possible to find an $(1 - \alpha)$-confidence interval $\widehat{CI}_{1-\alpha}$ such that (14.1) is satisfied with an equality, i.e. such that

$$\Pr\left(\theta \in \widehat{CI}_{1-\alpha}\right) = 1 - \alpha,$$

for any value of $\theta \in \Theta$. If this is not possible, then we call the confidence interval $\widehat{CI}_{1-\alpha}$ *conservative*. That is, there are values of θ for which the probability that θ is contained in $\widehat{CI}_{1-\alpha}$ is larger than the confidence level.

14.3 Construction of a confidence interval

In this section, we consider how to construct a confidence interval based on a point estimator $\hat{\theta}$ of the estimand θ_0. Assuming all the random variables are continuous, the task is to determine L and U such that

$$\Pr\left(\theta_0 \in [L, U]\right) = 1 - \alpha. \tag{14.2}$$

The left hand side of this expression can be rewritten as

$$\begin{aligned}
\Pr\left(\theta_0 \in [L, U]\right) &= \Pr\left(L \leq \theta_0 \leq U\right) \\
&= \Pr\left(\theta_0 \leq U\right) - \Pr\left(\theta_0 \leq L\right) \\
&= 1 - \Pr\left(\theta_0 \geq U\right) - \Pr\left(\theta_0 \leq L\right) \\
&= 1 - \left(\Pr\left(\theta_0 \leq L\right) + \Pr\left(\theta_0 \geq U\right)\right).
\end{aligned}$$

This, together with Equation (14.2), implies that L and U must solve

$$\Pr\left(\theta_0 \leq L\right) + \Pr\left(\theta_0 \geq U\right) = \alpha. \tag{14.3}$$

In words, the probability that the true parameter θ_0 is below the lower boundary L plus the probability that it is above the upper bounder U must equal α.

By varying L and U such that (14.3) holds, many confidence intervals with the same confidence level can be constructed. Similarly to the consideration for point estimators in Section 11.4, the "optimal" choice of L and U depends on the loss incurred when $\theta_0 \notin [L, U]$. In what follows, we will assume that the loss incurred when $\theta_0 < L$ is the same as the loss incurred when $\theta_0 > U$. Therefore, we divide the probability α equally between $(\theta_0 \leq L)$ and $(\theta_0 \geq U)$. That is, we want to choose L and U such that the following two equations hold simultaneously,

$$\begin{aligned}
\Pr\left(\theta_0 \leq L\right) &= \alpha/2, \tag{14.4} \\
\Pr\left(\theta_0 \geq U\right) &= \alpha/2. \tag{14.5}
\end{aligned}$$

To solve this problem, we seek a sample statistics, formulated as a transformation, i.e. a function, $T()$, of θ_0, such that the distribution of $T(\theta_0)$ is known. For instance, if the distribution of the sample statistics $T(\theta_0) = \theta_0 - \widehat{\theta}$ is known, a solution to the problem is to set (Problem 14.9.1)

$$
\begin{aligned}
L &= \widehat{\theta} + q_{\alpha/2}, \\
U &= \widehat{\theta} + q_{1-\alpha/2},
\end{aligned}
$$

where q_p is the p-quantile of the distribution of the sample statistics $T(\theta_0) = \theta_0 - \widehat{\theta}$. In practice, however, the distribution of $T(\theta_0) = \theta_0 - \widehat{\theta}$ is seldom known, e.g. because it depends on unknown parameters. To possibly circumvent this problem, a different transformation $T()$ is needed. The following section gives the details under the assumptions that the sample is a simple random sample and the population is normally distributed. These assumptions allow us to derive the distribution of a particular transformation $T(\theta_0)$. Section 14.4.1 considers the case for constructing a confidence interval for the mean, while Section 14.4.2 considers the case for constructing a confidence interval for the variance. As we will see, the appropriate transformation $T()$ in these two cases differ.

14.4 Confidence intervals for mean and variance in the case of a normally distributed population

In this section, we consider a normally distributed population for which we want to estimate the mean and the variance. We show how confidence intervals can be constructed for the mean and the variance.

14.4.1 Confidence intervals for the mean

Let the estimand be the mean μ_0 in the normally distributed population $N\left(\mu_0, \sigma_0^2\right)$. The variance σ_0^2 is unknown. Assume a simple random sample $\left(\tilde{X}_1, \ldots, \tilde{X}_n\right)$ is available. As point estimators of μ_0 and σ_0^2, we use the sample average $\overline{X}$ and the sample variance S^2, respectively. The sampling distribution of $\overline{X}$ is

$$
\overline{X} \sim N\left(\mu_0, \sigma_0^2/n\right),
$$

implying the following distribution for the transformation

$$
T\left(\mu_0\right) = \frac{\overline{X} - \mu_0}{\sqrt{\sigma_0^2/n}} \sim N(0, 1).
$$

While this distribution is known, i.e. it does not depend on unknown parameters, the transformation cannot be used in practice unless σ_0^2 is known. If σ_0^2 is unknown, we can insert our estimator S^2 of σ_0^2, and apply the transformation

$$
T\left(\mu_0\right) = \frac{\overline{X} - \mu_0}{\sqrt{S^2/n}} \sim t(n-1),
$$

where $t(n-1)$ is the t-distribution with $(n-1)$ degrees of freedom. This transformation is a known distribution because the sample size n is known in practice.

To see how these transformations may be used to choose L and U such that (14.4)–(14.5) hold, consider the case where σ_0^2 is known. Apply the transformation $T(\mu_0)$ to both sides of the inequality in (14.4). Since the transformation is a strictly decreasing function of the argument μ_0, this reverses the inequality with the result that

$$\Pr(\mu_0 \leq L) = \Pr(T(\mu_0) \geq T(L)) = \alpha/2.$$

Insert for the transformation $T()$

$$\Pr\left(\frac{\overline{X} - \mu_0}{\sqrt{\sigma_0^2/n}} \geq \frac{\overline{X} - L}{\sqrt{\sigma_0^2/n}}\right) = \alpha/2.$$

Let $Z = \frac{\overline{X} - \mu_0}{\sqrt{\sigma_0^2/n}}$ and note that $Z \sim N(0,1)$ due to Theorem 8.9. Then

$$\Pr\left(Z \geq \frac{\overline{X} - L}{\sqrt{\sigma_0^2/n}}\right) = 1 - \Pr\left(Z \leq \frac{\overline{X} - L}{\sqrt{\sigma_0^2/n}}\right) = \alpha/2$$

which implies

$$\Pr\left(Z \leq \frac{\overline{X} - L}{\sqrt{\sigma_0^2/n}}\right) = 1 - \alpha/2.$$

Since $Z \sim N(0,1)$, a solution for L is such that

$$\frac{\overline{X} - L}{\sqrt{\sigma_0^2/n}} = q_{1-\alpha/2},$$

where $q_{1-\alpha/2}$ is the $(1 - \alpha/2)$-quantile of the $N(0,1)$ distribution. We can then solve for L to get

$$L = \overline{X} - q_{1-\alpha/2}\sqrt{\sigma_0^2/n}.$$

Using similar arguments, we can use the transformation $T(\mu_0)$ on both sides of the inequality in (14.5) to find U. This leads to

$$\Pr\left(\frac{\overline{X} - \mu_0}{\sqrt{\sigma_0^2/n}} \leq \frac{\overline{X} - U}{\sqrt{\sigma_0^2/n}}\right) = \alpha/2,$$

which implies

$$\frac{\overline{X} - U}{\sqrt{\sigma_0^2/n}} = q_{\alpha/2}.$$

Solving for U yields

$$U = \overline{X} - q_{\alpha/2}\sqrt{\sigma_0^2/n}.$$

Collecting the expressions for the lower and upper limits L and U, the confidence interval becomes

$$\widehat{CI}_{1-\alpha} = \left[\overline{X} - q_{1-\alpha/2}\sqrt{\sigma_0^2/n}, \overline{X} - q_{\alpha/2}\sqrt{\sigma_0^2/n}\right].$$

As you are asked to show in Problem 14.9.3 at the end of this chapter, then, the symmetry of the standard normal distribution around 0 implies that $q_{\alpha/2} = -q_{1-\alpha/2}$. Hence, the confidence interval can also be written as

$$\widehat{CI}_{1-\alpha} = \left[\overline{X} - q_{1-\alpha/2}\sqrt{\sigma_0^2/n}, \overline{X} + q_{1-\alpha/2}\sqrt{\sigma_0^2/n}\right].$$

In the case where σ_0^2 is unknown and the transformation is $T(\mu_0) = \frac{\overline{X}-\mu_0}{\sqrt{S^2/n}} \sim t(n-1)$, similar lines of argument can be used. The resulting confidence interval takes the same form, although S^2 is inserted in place of σ_0^2 and $q_{\alpha/2}, q_{1-\alpha/2}$ represent the quantiles in the t-distribution with $n-1$ degrees of freedom, instead of the quantiles in the normal distribution. These insights are collected in the following theorem.

Theorem 14.1 (Confidence interval for population mean, normal case) *Assume the population X is normally distributed, $X \sim N(\mu_0, \sigma_0^2)$, with $\mu_0 \in \mathbb{R}$ and $\sigma_0^2 < \infty$. Let $\widetilde{X}_1, \ldots, \widetilde{X}_n$ be a simple random sample from X, $\overline{X}$ the sample average, and S^2 the sample variance. Define an interval estimator $\widehat{CI}_{1-\alpha}$ by*

$$\widehat{CI}_{1-\alpha} = \left[\overline{X} - q_{1-\alpha/2} \cdot \sqrt{S^2/n} \ , \ \overline{X} + q_{1-\alpha/2} \cdot \sqrt{S^2/n}\right],$$

where $q_{1-\alpha/2}$ is the $(1-\alpha/2)$-quantile of the t-distribution with $(n-1)$ degrees of freedom. Then

$$\Pr\left(\mu_0 \in \widehat{CI}_{1-\alpha}\right) = 1 - \alpha,$$

i.e. $\widehat{CI}_{1-\alpha}$ is a $(1-\alpha)$-confidence interval for μ_0.

In case the population variance σ_0^2 is known, define an interval estimator $\widehat{CI}_{1-\alpha}$ by

$$\widehat{CI}_{1-\alpha} = \left[\overline{X} - q_{1-\alpha/2} \cdot \sqrt{\sigma_0^2/n} \ , \ \overline{X} + q_{1-\alpha/2} \cdot \sqrt{\sigma_0^2/n}\right],$$

where $q_{1-\alpha/2}$ is the $(1-\alpha/2)$-quantile of the standard normal distribution. Then

$$\Pr\left(\mu_0 \in \widehat{CI}_{1-\alpha}\right) = 1 - \alpha,$$

i.e. $\widehat{CI}_{1-\alpha}$ is a $(1-a)$-confidence interval for μ_0.

Note that the confidence intervals presented in Theorem 14.1 are symmetric around the estimator $\widehat{\mu} = \overline{X}$, due to the symmetry of the $N(0,1)$ and $t(n-1)$ distributions. Indeed, the confidence intervals have $\widehat{\theta}$ in the center and the half-width of the confidence is equal to $q_{1-\alpha/2} \cdot SE(\widehat{\theta})$, where $SE(\widehat{\theta}) = \sqrt{\widehat{Var}(\widehat{\mu})}$ is the standard error of $\widehat{\mu}$ (Definition 13.2). Recall that the standard error is a measure of the uncertainty of the estimator $\widehat{\mu}$, which means that the width of a confidence interval is increasing in the uncertainty of the estimator in question.

The following example illustrates how Theorem 14.1 can be used to construct confidence intervals when the population is normal and the variance is known.

Example 14.1 *Let $\mu_0 \in \mathbb{R}$ and $\sigma_0^2 > 0$ be numbers and consider a population represented by the random variable $X \sim N(\mu_0, \sigma_0^2)$. Based on a simple random sample $\widetilde{X}_1, \ldots, \widetilde{X}_n$ we may estimate μ_0 using the sample mean $\overline{X}$. Recall that $Var(\overline{X}) = \sigma_0^2/n$. Suppose $n = 50$ and that $\sigma_0^2 = 10$ is known. Let $\alpha = 5\%$. Then, by Theorem 14.1, a 95%-confidence interval for μ_0 is*

$$\begin{aligned}
\widehat{CI}_{95\%} &= \left[\overline{X} - q_{1-\alpha/2} \cdot \sqrt{\sigma_0^2/n} \ , \ \overline{X} + q_{1-\alpha/2} \cdot \sqrt{\sigma_0^2/n}\right] \\
&= \left[\overline{X} - 1.96 \cdot \sqrt{10/50} \ , \ \overline{X} + 1.96 \cdot \sqrt{10/50}\right] \\
&= \left[\overline{X} - 0.8765 \ , \ \overline{X} + 0.8765\right],
\end{aligned}$$

where $q_{1-\alpha/2} = q_{0.975} = 1.96$ is the 97.5%-quantile of the standard normal distribution. A 90% confidence interval is obtained by letting $\alpha = 10\%$,

$$\widehat{CI}_{90\%} = \left[\overline{X} - q_{1-\alpha/2} \cdot \sqrt{\sigma_0^2/n}\ ,\ \overline{X} + q_{1-\alpha/2} \cdot \sqrt{\sigma_0^2/n}\right]$$
$$= \left[\overline{X} - 1.6449 \cdot \sqrt{10/50}\ ,\ \overline{X} + 1.6449 \cdot \sqrt{10/50}\right]$$
$$= \left[\overline{X} - 0.7356\ ,\ \overline{X} + 0.7356\right],$$

where $q_{1-\alpha/2} = q_{0.95} = 1.6449$ is the 95%-quantile of the standard normal distribution.

If, say, the realized random sample is such that $\overline{X} = 1.42$, then the realized confidence intervals become

$$\widehat{CI}_{95\%} = \left[\overline{X} - 0.8765\ ,\ \overline{X} + 0.8765\right] = [0.5435\ ,\ 2.2965],$$

and

$$\widehat{CI}_{90\%} = \left[\overline{X} - 0.7356\ ,\ \overline{X} + 0.7356\right] = [0.6844\ ,\ 2.1556].$$

Notice how the 95%-confidence interval is wider than the 90%-confidence interval.

In Example 14.1, we assume that the variance σ_0^2 is known. This is seldom the case in practice. The next example is similar to Example 14.1, but now under the more realistic assumption that the variance σ_0^2 is unknown.

Example 14.2 *Let $\mu_0 \in \mathbb{R}$ and $\sigma_0^2 > 0$ be numbers and consider a population represented by the random variable $X \sim N(\mu_0, \sigma_0^2)$. Based on a simple random sample $\widetilde{X}_1, \ldots, \widetilde{X}_n$ we may estimate μ_0 using the sample mean $\overline{X}$. Recall that $Var(\overline{X}) = \sigma_0^2/n$. Suppose $n = 50$ and consider the sample variance S^2 as an estimator of σ_0^2. Let $\alpha = 5\%$. Then, by Theorem 14.1, a 95%-confidence interval for μ_0 is*

$$\widehat{CI}_{95\%} = \left[\overline{X} - q_{1-\alpha/2} \cdot \sqrt{\sigma_0^2/n}\ ,\ \overline{X} + q_{1-\alpha/2} \cdot \sqrt{S^2/n}\right]$$
$$= \left[\overline{X} - 2.0096 \cdot \sqrt{S^2/50}\ ,\ \overline{X} + 2.0096 \cdot \sqrt{S^2/50}\right],$$

where $q_{1-\alpha/2} = q_{0.975} = 2.0096$ is the 97.5%-quantile of the t distribution with $n - 1 = 49$ degrees of freedom. A 90% confidence interval is obtained by letting $\alpha = 10\%$,

$$\widehat{CI}_{90\%} = \left[\overline{X} - q_{1-\alpha/2} \cdot \sqrt{S^2/n}\ ,\ \overline{X} + q_{1-\alpha/2} \cdot \sqrt{S^2/n}\right]$$
$$= \left[\overline{X} - 1.6766 \cdot \sqrt{S^2/50}\ ,\ \overline{X} + 1.6766 \cdot \sqrt{S^2/50}\right],$$

where $q_{1-\alpha/2} = q_{0.95} = 1.6766$ is the 95%-quantile of the t distribution with $n - 1 = 49$ degrees of freedom.

If, say, the realized random sample is such that $\overline{X} = 1.42$ and $S^2 = 10$, then the realized confidence intervals become

$$\widehat{CI}_{95\%} = \left[\overline{X} - 2.0096 \cdot \sqrt{S^2/50}\ ,\ \overline{X} + 2.0096 \cdot \sqrt{S^2/50}\right] = [0.5213\ ,\ 2.3187],$$

and

$$\widehat{CI}_{90\%} = \left[\overline{X} - 1.6766 \cdot \sqrt{S^2/10}\ ,\ \overline{X} + 1.6766 \cdot \sqrt{S^2/10}\right] = [0.6702\ ,\ 2.1698].$$

Notice when $S^2 = \sigma^2 (= 10)$, the confidence intervals calculated here are wider than the corresponding ones in Example 14.1. This happens because the quantiles of the t-distribution are larger (in absolute value) than the corresponding quantiles in the normal distribution. The intuition behind this is that the uncertainty introduced by having to estimate σ_0^2 translates into wider confidence intervals to ensure that the true parameter is covered with probability equal to the chosen significance level.

14.4.2   Confidence interval for the variance

Consider the same setup as above with a $N\left(\mu_0, \sigma_0^2\right)$ population and a simple random sample, but now let the estimand be the variance σ_0^2 and the estimator the sample variance S^2. In this case, it turns out that using the transformation considered in the previous section will not result in a random variable with known distribution. Instead, consider the transformation

$$T\left(\sigma_0^2\right) = \frac{S^2(n-1)}{\sigma_0^2}.$$

We saw in Section 12.6 that, for this transformation,

$$T\left(\sigma_0^2\right) \sim \chi^2(n-1),$$

where $\chi^2(n-1)$ denotes the χ^2-distribution with $n-1$ degrees of freedom. This is a known distribution, because the sample size n will be known in practice. Using arguments similar to those leading to Theorem 14.1 above, we may use the knowledge of the distribution of $T\left(\sigma_0^2\right)$ to determine L and U such that Equations (14.4)–(14.5) hold. The following theorem contains the details.

Theorem 14.2 (Confidence interval for population variance, normal case)
Assume the population X is normally distributed, $X \sim N(\mu_0, \sigma_0^2)$, with $\mu_0 \in \mathbb{R}$ and $\sigma_0^2 < \infty$. Let $\widetilde{X}_1, \ldots, \widetilde{X}_n$ be a simple random sample from X and S^2 the sample variance. Define an interval estimator $\widehat{CI}_{1-\alpha}$ by

$$\widehat{CI}_{1-\alpha} = \left[\frac{(n-1) \cdot S^2}{q_{1-\alpha/2}} \, , \, \frac{(n-1) \cdot S^2}{q_{\alpha/2}} \right], \tag{14.6}$$

where q_p is the p-quantile of the χ^2-distribution with $(n-1)$ degrees of freedom. Then

$$\Pr\left(\sigma_0^2 \in \widehat{CI}_{1-\alpha}\right) = 1 - \alpha,$$

i.e. $\widehat{CI}_{1-\alpha}$ is a $(1-\alpha)$-confidence interval for σ_0^2.

In contrast to the confidence interval for μ_0, the confidence interval for σ_0^2 given in Theorem 14.2 is not symmetric around S^2 despite S^2 being an unbiased estimator of σ_0^2. This typically happens when the sampling distribution of the point estimator is asymmetric. The sampling distribution of S^2 is not symmetric and, in particular, there are no negative outcomes possible.

Example 14.3 *Let $\mu_0 \in \mathbb{R}$ and $\sigma_0^2 > 0$ be numbers and consider a population represented by the random variable $X \sim N(\mu_0, \sigma_0^2)$. Based on a simple random sample $\widetilde{X}_1, \ldots, \widetilde{X}_n$ we may estimate σ_0^2 using the sample variance S^2. Suppose $n = 50$. Let $\alpha = 5\%$. Then, by Theorem 14.2, a 95%-confidence interval for σ_0^2 is*

$$\widehat{CI}_{95\%} = \left[\frac{(n-1) \cdot S^2}{q_{1-\alpha/2}} \, , \, \frac{(n-1) \cdot S^2}{q_{\alpha/2}} \right] = \left[\frac{(50-1) \cdot S^2}{q_{0.975}} \, , \, \frac{(50-1) \cdot S^2}{q_{0.025}} \right]$$

$$= \left[\frac{49 \cdot S^2}{70.22} \, , \, \frac{49 \cdot S^2}{31.55} \right],$$

where $q_{1-\alpha/2} = q_{0.975} = 70.22$ and $q_{\alpha/2} = q_{0.025} = 31.55$ are the 97.5%- and 2.5%-quantiles of the χ^2-distribution with $n - 1 = 49$ degrees of freedom, respectively. A 90% confidence interval is obtained by letting $\alpha = 10\%$,

$$\widehat{CI}_{90\%} = \left[\frac{(n-1)\cdot S^2}{q_{1-\alpha/2}} , \frac{(n-1)\cdot S^2}{q_{\alpha/2}}\right] = \left[\frac{(50-1)\cdot S^2}{q_{0.95}} , \frac{(50-1)\cdot S^2}{q_{0.05}}\right]$$

$$= \left[\frac{49\cdot S^2}{66.34} , \frac{49\cdot S^2}{33.93}\right],$$

where $q_{1-\alpha/2} = q_{0.95} = 66.34$ and $q_{\alpha/2} = q_{0.05} = 33.93$ are the 95%- and 5%-quantiles of the χ^2-distribution with $n - 1 = 49$ degrees of freedom, respectively.

If, say, the realized random sample is such that $S^2 = 10$, then the realized confidence intervals become

$$\widehat{CI}_{95\%} = \left[\frac{49\cdot S^2}{70.22} , \frac{49\cdot S^2}{31.55}\right] = [6.98 , 15.53],$$

and

$$\widehat{CI}_{90\%} = \left[\frac{49\cdot S^2}{66.34} , \frac{49\cdot S^2}{33.93}\right] = [7.39 , 14.44].$$

Notice how the 95%-confidence interval is wider than the 90%-confidence interval. This is in order to ensure that σ_0^2 is included in the latter confidence interval $95\% - 90\% = 5\%$ more often than in the former confidence interval in repeated statistical experiments.

The discussion above shows that we may construct confidence intervals when we can find a transformation of θ_0 with known distribution. In cases where we cannot find such a transformation with known distribution, we may nonetheless be able to construct an *approximate* confidence interval using bootstrap or asymptotic methods, similar to how we estimated a sampling distribution in Chapter 13. This possibility is explored in the following sections.

14.5 Bootstrap confidence intervals

The bootstrap can be applied in several manners to construct approximate confidence intervals. In Section 14.5.1, we show how to construct approximate confidence intervals by bootstrapping the sampling distribution of the point estimator $\widehat{\theta}$. Then, in Section 14.5.2 we show how to construct approximate confidence intervals by bootstrapping the sampling distribution of the standardized point estimator.

14.5.1 Bootstrap confidence interval based on point estimator

Let the estimand be θ_0 and consider an estimator $\widehat{\theta}$ of θ_0. We first propose a bootstrap approach that results in a confidence interval that is symmetric around $\widehat{\theta}$, which is achieved by setting $L = \widehat{\theta} - h$ and $U = \widehat{\theta} + h$ for some $h > 0$. That is, our goal is to select a positive h such that

$$\Pr\left(\theta_0 \in [\widehat{\theta} - h, \widehat{\theta} + h]\right) = 1 - \alpha$$

or, equivalently (Problem 14.9.2),

$$\Pr\left(\widehat{\theta} - \theta_0 \leq h\right) - \Pr\left(\widehat{\theta} - \theta_0 < -h\right) = 1 - \alpha.$$

We can estimate the distribution of $\widehat{\theta}$ by the non-parametric bootstrap (Section 13.2.1) or by the parametric bootstrap (Section 13.2.2). Denote the CDF of the bootstrap distribution by $\widehat{F}^*_{\widehat{\theta}}$. To solve for h in the above equation, we need an estimator of the distribution of $\left(\widehat{\theta} - \theta_0\right)$. This can be accomplished by defining

$$\widehat{F}^*_{(\widehat{\theta}-\theta_0)}(x) = \widehat{F}^*_{\widehat{\theta}}(x + \hat{t}),$$

where $\hat{t}$ is the point estimate of θ_0 obtained from the original random sample. Now, using this equation along with the bootstrapped distribution for $\widehat{\theta}$, it is possible to solve for h in

$$\widehat{F}^*_{(\widehat{\theta}-\theta_0)}(h) - \widehat{F}^*_{(\widehat{\theta}-\theta_0)}(-h) = 1 - \alpha. \tag{14.7}$$

These considerations lead to the following definition of a symmetric bootstrap confidence interval.

Definition 14.2 (Symmetric bootstrap confidence interval) *Let $\theta_0 \in \mathbb{R}$ be the estimand and $\widehat{\theta}$ an estimator of θ_0. Let $\alpha \in (0, 0.5]$. A **symmetric bootstrap** $(1 - \alpha)$-**confidence interval** $\widehat{CI}^*_{1-\alpha,symmetric}$ is given by*

$$\widehat{CI}^*_{1-\alpha,symmetric} = \left[\widehat{\theta} - h \, , \, \widehat{\theta} + h\right],$$

where h solves (14.7).

Note that the solution of h in (14.7) typically does not result in the probability of the left tail, $\widehat{F}^*_{\widehat{\theta}}(\hat{t} - h)$, being the same as the probability of the right tail, $\left(1 - \widehat{F}^*_{\widehat{\theta}}(\hat{t} + h)\right)$, since the bootstrapped distribution $\widehat{F}^*_{\widehat{\theta}}$ may not be symmetric.

A bootstrap-based confidence interval with equal probability in the left and right tails can be calculated by picking the $\alpha/2$-quantile of the bootstrap distribution of $\widehat{\theta}$ as the lower boundary of the confidence interval, and by picking the $(1 - \alpha/2)$-quantile of the bootstrap distribution of $\widehat{\theta}$ as the upper boundary of the confidence interval. The next definition contains the details.

Definition 14.3 (Equal-tailed bootstrap confidence interval) *Let $\theta_0 \in \mathbb{R}$ be the estimand and $\widehat{\theta}$ an estimator of θ_0. Let $\alpha \in (0, 0.5]$. An **equal-tailed bootstrap** $(1 - \alpha)$-**confidence interval** $\widehat{CI}^*_{1-\alpha,equal\ tail}$ is given by*

$$\widehat{CI}^*_{1-\alpha,equal\ tail} = \left[q^*_{\alpha/2} \, , \, q^*_{1-\alpha/2}\right],$$

where $q^*_{\alpha/2}$ and $q^*_{1-\alpha/2}$ are the $\alpha/2$ and $1 - \alpha/2$ quantiles of $\widehat{F}^*_{\widehat{\theta}}$, respectively.

The equal-tailed bootstrap confidence interval is easier to calculate than the symmetric bootstrap confidence interval because the quantiles can be calculated easier than solving for h in (14.7). Indeed, the equal-tailed bootstrap confidence interval only requires a bootstrap estimate of the distribution of $\widehat{\theta}$, i.e. $F^*_{\widehat{\theta}}$, and then inferring the quantiles $q_{\alpha/2}$ and $q_{1-\alpha/2}$ from this distribution. Note that since the bootstrap distribution $F^*_{\widehat{\theta}}$ is unlikely to be symmetric, the equal-tailed bootstrap confidence interval will likely not be symmetric around $\widehat{\theta}$ either.

14.5.2 Bootstrap confidence interval based on studentized point estimator

The bootstrap approaches presented in the previous section relied on bootstrapping the distribution of the estimator $\widehat{\theta}$. In this section, we propose to construct a confidence interval by bootstrapping the sampling distribution of the so-called *studentized* point estimator, namely

$$T = \frac{\widehat{\theta} - \theta_0}{\sqrt{\widehat{Var}(\widehat{\theta})}}, \tag{14.8}$$

where $\widehat{Var}(\widehat{\theta})$ is an estimator of $Var(\widehat{\theta})$. In many situations, it turns out that constructing an estimator of a confidence interval based on bootstrapping the distribution of the studentized point estimator provides a more precise interval estimator compared to bootstrapping the distribution of the point estimator $\widehat{\theta}$. Let B be the number of bootstrap replications and $\widehat{F}(x)$ a bootstrap estimator of the population distribution. The bootstrap distribution of (14.8) can be obtained by the following algorithm.

Algorithm 14.5.1 (Boostrap confidence interval using studentization)

1. Use the original sample $(\widetilde{X}_1, \ldots, \widetilde{X}_n)$ to calculate the estimate $\widehat{\theta}$.

2. For $b = 1, \ldots, B$, simulate a bootstrap sample $(\widetilde{X}_1^{(b)*}, \ldots, \widetilde{X}_n^{(b)*})$ from $\widehat{F}(x)$ and calculate

$$t_b^* = \frac{\widehat{\theta}_b^* - \widehat{\theta}}{\sqrt{\widehat{Var}(\widehat{\theta}^*)}},$$

where $\widehat{\theta}_b^*$ is the realized bootstrap estimate and $\widehat{Var}(\widehat{\theta}^*)$ is the realized bootstrap estimate of the variance of $\widehat{\theta}^*$ in the b'th bootstrap sample $(\widetilde{X}_1^{(b)*}, \ldots, \widetilde{X}_n^{(b)*})$, respectively.

3. The distribution of $\{t_b^*\}_{b=1}^B$ is the bootstrap distribution $\widehat{F}_T^*$ of T in (14.8).

As in the section above, the estimated bootstrap population distribution $\widehat{F}(x)$ may be estimated by the non-parametric bootstrap (Section 13.2.1) or by the parametric bootstrap (Section 13.2.2). The next definition states how the bootstrap distribution of the studentized statistics of the point estimator can be used to construct a confidence interval.

Definition 14.4 (Studentized bootstrap confidence interval) *Let $\theta_0 \in \mathbb{R}$ be the estimand, $\widehat{\theta}$ an estimator of θ_0, and $\widehat{Var}(\widehat{\theta})$ an estimator of $Var(\widehat{\theta})$. Let $\alpha \in (0, 0.5]$. A* **studentized bootstrap** $(1-\alpha)$**-confidence interval** $\widehat{CI}_{1-\alpha,student}^*$ *is given by*

$$\widehat{CI}_{1-\alpha,student}^* = \left[\widehat{\theta} - t_{1-\alpha/2}^* \cdot \sqrt{\widehat{Var}(\widehat{\theta})} \; , \; \widehat{\theta} - t_{\alpha/2}^* \cdot \sqrt{\widehat{Var}(\widehat{\theta})} \right],$$

where $t_{\alpha/2}^*$ and $t_{1-\alpha/2}^*$ are the $\alpha/2$- and $(1-\alpha/2)$-quantiles of the bootstrap distribution $\widehat{F}_T^*$ of T in (14.8).

Similar for all the bootstrap-based confidence intervals, applying the parametric bootstrap typically leads to higher precision than the non-parametric bootstrap provided the assumption on the population distribution $F_X(x; \theta)$ is appropriate.

14.6   Asymptotic confidence intervals

We first define an *asymptotically valid confidence interval*, i.e. an interval fulfilling Definition 14.1 in the limit of an infinite sample size. Analogous to the estimation of a sampling distribution in Chapter 13 via asymptotic arguments, we will in the following define an approximate confidence interval as an asymptotically valid confidence interval constructed using a finite sample size. First, we give the formal definition of an asymptotically valid confidence interval.

Definition 14.5 (Asymptotically valid confidence interval) *Let θ_0 be the estimand in a parameter space Θ and $\alpha \in (0, 0.5]$ be a number. Then a random interval $\widehat{CI}_{1-\alpha}$ is an* **asymptotically valid $(1 - \alpha)$-confidence interval** *for θ_0 if and only if α is the smallest number such that*

$$\lim_{n \to \infty} \Pr\left(\theta_0 \in \widehat{CI}_{1-\alpha}\right) \geq 1 - \alpha$$

for all values of $\theta_0 \in \Theta$.

An asymptotically valid $(1 - \alpha)$-confidence interval (Definition 14.5) is used much more often in practice than a $(1 - \alpha)$-confidence interval (Definition 14.1). The reason is that the former can, in general, be constructed under milder assumptions than the latter, see, e.g. the discussion following Theorem 14.3 below.

14.6.1   Constructing asymptotic confidence interval

In this section, we describe how to construct an asymptotically valid confidence interval based on an asymptotic distribution of a point estimator. We consider the case of a $(1 - \alpha)$-confidence interval for an estimator $\widehat{\theta}$ of the estimand θ_0. The key assumption we will require is that $\widehat{\theta}$ satisfies a central limit type-theorem, i.e. that $\widehat{\theta} \overset{A}{\sim} N(\theta_0, Var(\widehat{\theta}))$. The approximate normality of $\widehat{\theta}$ will allow us to use the normal quantiles, similar to that in Theorem 14.1, for formulating an approximate confidence interval. The following theorem contains the details.

Theorem 14.3 (Asymptotic confidence interval) *Let $\theta_0 \in \mathbb{R}$ be the estimand and $\widehat{\theta}$ an estimator of θ_0 such that*

$$\frac{\widehat{\theta} - \theta_0}{\sqrt{\widehat{Var}(\widehat{\theta})}} \overset{d}{\to} N(0, 1),$$

as $n \to \infty$. Let $\alpha \in (0, 0.5]$ and define an interval estimator $\widehat{CI}_{1-\alpha,asymp}$ by

$$\widehat{CI}_{1-\alpha,asymp} = \left[\widehat{\theta} - q_{1-\alpha/2} \cdot \sqrt{\widehat{Var}(\widehat{\theta})} \ , \ \widehat{\theta} + q_{1-\alpha/2} \cdot \sqrt{\widehat{Var}(\widehat{\theta})}\right],$$

where $q_{1-\alpha/2}$ is the $(1 - \alpha/2)$-quantile of the standard normal distribution $N(0, 1)$. Then

$$\lim_{n \to \infty} \Pr\left(\theta_0 \in \widehat{CI}_{1-\alpha,asymp}\right) = 1 - \alpha,$$

i.e. $\widehat{CI}_{1-\alpha,asymp}$ is an asymptotically valid $(1 - \alpha)$-confidence interval for θ_0.

In the case of a simple random sample, Theorem 14.3 may be combined with the Central Limit Theorem (Theorem 13.1) to construct a confidence interval for the mean. The following theorem contains the details.

Theorem 14.4 (Asymptotic confidence interval for the mean) *Let $\tilde{X}_1, \ldots, \tilde{X}_n$ be a simple random sample of the random variable X with $\mu_0 = E(X)$ and $\sigma_0^2 = Var(X) < \infty$. Consider the sample mean $\overline{X}$ as an estimator of μ_0 and the sample variance S^2 as an estimator of σ_0^2. Let $\alpha \in (0, 0.5]$ and define an interval estimator $\widehat{CI}_{1-\alpha,asymp}$ by*

$$\widehat{CI}_{1-\alpha,asymp} = \left[\overline{X} - q_{1-\alpha/2} \cdot \sqrt{S^2/n} \,,\, \overline{X} + q_{1-\alpha/2} \cdot \sqrt{S^2/n} \right], \tag{14.9}$$

where $q_{1-\alpha/2}$ is the $(1-\alpha/2)$-quantile of the standard normal distribution $N(0,1)$. Then

$$\lim_{n \to \infty} \Pr\left(\mu_0 \in \widehat{CI}_{1-\alpha,asymp} \right) = 1 - \alpha,$$

i.e. $\widehat{CI}_{1-\alpha,asymp}$ is an asymptotically valid $(1-\alpha)$-confidence interval for μ_0.

Theorem 14.4 can be compared to Theorem 14.1 above. In both cases a confidence interval for the mean is constructed from a simple random sample. To get a $(1-\alpha)$-confidence interval in the sense of Definition 14.1, we have to assume that the underlying population is normally distributed (Theorem 14.1), which might be a dubious assumption in many practical applications. In contrast, if we are content with an asymptotically valid $(1-\alpha)$-confidence interval in the sense of Definition 14.5, the only necessary assumption is that the population distribution has finite variance, ensuring that the Central Limit Theorem holds, a much milder assumption and one that is likely to be fulfilled in many practical applications.

Example 14.4 *Consider a population represented by the random variable X with $\mu_0 = E(X)$ and $\sigma_0^2 = Var(X) < \infty$. Based on a simple random sample $\tilde{X}_1, \ldots, \tilde{X}_n$ we may estimate μ_0 using the sample mean $\overline{X}$ and the variance σ_0^2 with the sample variance S^2. Suppose $n = 50$. Let $\alpha = 5\%$. Then, by Theorem 14.3, an approximate 95%-confidence interval for μ_0 is*

$$\begin{aligned}
\widehat{CI}_{95\%,asymp} &= \left[\overline{X} - q_{1-\alpha/2}\sqrt{S^2/n} \,,\, \overline{X} + q_{1-\alpha/2}\sqrt{S^2/n} \right] \\
&= \left[\overline{X} - 1.96 \cdot \sqrt{S^2/50} \,,\, \overline{X} + 1.96 \cdot \sqrt{S^2/50} \right],
\end{aligned}$$

where $q_{1-\alpha/2} = q_{0.975} = 1.96$ is the 97.5%-quantile of the standard normal distribution. An approximate 90% confidence interval is obtained by letting $\alpha = 10\%$,

$$\begin{aligned}
\widehat{CI}_{90\%,asymp} &= \left[\overline{X} - q_{1-\alpha/2}\sqrt{S^2/n} \,,\, \overline{X} + q_{1-\alpha/2}\sqrt{S^2/n} \right] \\
&= \left[\overline{X} - 1.6449 \cdot \sqrt{S^2/50} \,,\, \overline{X} + 1.6449 \cdot \sqrt{S^2/50} \right],
\end{aligned}$$

where $q_{1-\alpha/2} = q_{0.95} = 1.6449$ is the 95%-quantile of the standard normal distribution.

If, say, the realized random sample is such that $\overline{X} = 1.42$ and $S^2 = 10$, then the realized confidence intervals become

$$\widehat{CI}_{95\%,asymp} = \left[\overline{X} - 2.2622 \cdot \sqrt{S^2/50} \,,\, \overline{X} + 2.2622 \cdot \sqrt{S^2/50} \right] = [0.5435 \,,\, 2.2965],$$

and

$$\widehat{CI}_{90\%,asymp} = \left[\overline{X} - 1.8331 \cdot \sqrt{S^2/50} \,,\, \overline{X} + 1.8331 \cdot \sqrt{S^2/50} \right] = [0.6844 \,,\, 2.1556].$$

Notice how the confidence intervals calculated here exactly the same as those in Example 14.1, where the population was assumed normal and the variance known. Here, in contrast, no assumption on the population distribution, except finite variance, is required. The price to pay is that the confidence intervals are approximate, in the sense that they are only asymptotically valid.

We note that another difference between Theorems 14.1 and 14.4 is that in the case of an unknown variance, the former uses the quantiles of the t-distribution to construct the confidence interval, while the latter suggests to use the quantiles of the standard normal distribution to construct the asymptotically valid confidence interval. Since the t-distribution with $n-1$ degrees of freedom approaches the standard normal distribution as $n \to \infty$, the confidence interval (14.9) is also asymptotically valid if $q_{1-\alpha/2}$ is taken to be the $(1-\alpha/2)$-quantile from the t-distribution instead of the $(1-\alpha/2)$-quantile from the standard normal distribution. Hence, to adjust for the fact that the variance has to be estimated, the asymptotically valid confidence interval (14.9) is sometimes constructed using the quantiles from the t-distribution with $n-1$ degrees of freedom in place of the quantiles from the normal distribution.

An asymptotic confidence interval can also be constructed for the variance σ_0^2 of a population. This is done in the next example.

Example 14.5 *Assume a population with variance σ_0^2. Furthermore, assume a simple random sample is available. As point estimator of the estimand σ_0^2, use the sample variance S^2. If the fourth moment of the population exists, then*

$$\frac{S^2 - \sigma_0^2}{\sqrt{\widehat{Var}(S^2)}} \xrightarrow{d} N(0,1),$$

where $\widehat{Var}(S^2)$ is an estimator of the variance of the sample variance. The variance of S^2 can be shown to be

$$Var\left(S^2\right) = \frac{1}{n}\left(\mu_4 - \frac{n-3}{n-1}\sigma_0^4\right),$$

where $\mu_4 = E\left(X - \mu_0\right)^4$ is the fourth central moment. A typical used estimator $\widehat{Var}(S^2)$ of $Var(S^2)$ is

$$\widehat{Var}(S^2) = \frac{1}{n}\left(\frac{1}{n}\sum_{i=1}^{n}\left(\tilde{X}_i - \overline{X}\right)^4 - S^4\right).$$

Then the asymptotic $(1-\alpha)$-confidence interval for σ_0^2 is

$$\widehat{CI}_{1-\alpha,asymp} = \left[S^2 - q_{1-\alpha/2}\cdot\sqrt{\widehat{Var}(S^2)}\,,\ S^2 + q_{1-\alpha/2}\cdot\sqrt{\widehat{Var}(S^2)}\right],$$

where $q_{1-\alpha/2}$ is the $(1-\alpha/2)$-quantile of the standard normal $N(0,1)$ distribution.

Notice how the asymptotic confidence interval has quite a different form than the confidence interval (14.6) in case the population is normal. The difference is caused by the reliance on the central limit theorem in the former case.

14.7 Comparisons of different confidence intervals

Various results suggest the following relative precision of the different confidence interval approximations. The $\widehat{CI}^*_{1-\alpha,student}$ has a higher precision than the asymptotic confidence interval, and the symmetric and the equal tail bootstrap confidence interval partly due to the nature of the statistics being bootstrapped. For instance, when $\hat{\theta}$ satisfies a central limit-type theorem, the statistics T being bootstrapped has an asymptotic distribution which does not depend on unknown parameters. That is, the asymptotic distribution of the statistics is $T \overset{A}{\sim} N(0,1)$, where $N(0,1)$ does not depend on unknown parameters. In contrast, the symmetric and equal-tailed bootstraps rely on the distribution of $\widehat{\theta}$, which asymptotically is $\widehat{\theta} \overset{A}{\sim} N(\theta, Var(\widehat{\theta}))$, and thus depends on the unknown quantities θ and $Var(\widehat{\theta})$. If a statistic does not depend on unknown parameters, we call it *pivotal*. In this case, T is called *asymptotically pivotal* since its asymptotic distribution does not rely on unknown parameters. Asymptotic pivotality has been found to be important for explaining when bootstrapping gives a higher precision compared to an asymptotic distribution or bootstrapping a statistics which is not asymptotically pivotal. Hence, rather than estimating the distribution of a statistics of interest directly, it is typically better, if possible, to estimate the distribution of the statistics of interest by use of an asymptotically pivotal statistics.

14.8 Proofs

Proof of Theorem 14.1 *Let $\alpha \in (0, 0.5]$. Suppose first that σ_0^2 is known. Let q_p be the p-quantile in the standard normal distribution. According to Definition 14.1, to prove the result, it is enough to show that*

$$\Pr\left(\mu_0 \in \left[\widehat{\mu} - q_{1-\alpha/2} \cdot \sqrt{\sigma_0^2/n} \,,\, \widehat{\mu} + q_{1-\alpha/2} \cdot \sqrt{\sigma_0^2/n}\right]\right) = 1 - \alpha.$$

Note that

$$\mu_0 \in \left[\widehat{\mu} - q_{1-\alpha/2} \cdot \sqrt{\sigma_0^2/n} \,,\, \widehat{\mu} + q_{1-\alpha/2} \cdot \sqrt{\widehat{Var(\mu)}}\right]$$

$$\Longleftrightarrow \quad \widehat{\mu} - q_{1-\alpha/2} \cdot \sqrt{\sigma_0^2/n} \leq \mu_0 \leq \widehat{\mu} + q_{1-\alpha/2} \cdot \sqrt{\sigma_0^2/n}$$

$$\Longleftrightarrow \quad -q_{1-\alpha/2} \cdot \sqrt{\sigma_0^2/n} \leq \mu_0 - \widehat{\mu} \leq q_{1-\alpha/2} \cdot \sqrt{\sigma_0^2/n}.$$

Dividing by $\sqrt{\sigma_0^2/n}$ and multiplying with minus one, the above is equivalent to

$$q_{1-\alpha/2} \geq \frac{\widehat{\mu} - \mu_0}{\sqrt{\sigma_0^2/n}} \geq -q_{1-\alpha/2}$$

$$\Longleftrightarrow \quad -q_{1-\alpha/2} \leq T \leq q_{1-\alpha/2},$$

where $T = T(\widehat{\mu}) = \frac{\widehat{\mu} - \mu_0}{\sqrt{\sigma_0^2/n}} \sim N(0,1)$. This shows that

$$\Pr\left(\mu_0 \in \left[\widehat{\mu} - q_{1-\alpha/2} \cdot \sqrt{\sigma_0^2/n} \,,\, \widehat{\mu} - q_{\alpha/2} \cdot \sqrt{\sigma_0^2/n}\right]\right) = \Pr\left(-q_{1-\alpha/2} \leq T \leq q_{1-\alpha/2}\right).$$

Since the $N(0,1)$ distribution is symmetric around zero, it holds that $-q_{1-\alpha/2} = q_{\alpha/2}$ (Problem 14.9.3). Using this, together with the usual rules for calculation of probabilities for continuous random variables and the fact that $q_{\alpha/2}$ and $q_{1-\alpha/2}$ are quantiles in the distribution of T, we get

$$\Pr\left(-q_{1-\alpha/2} \leq T \leq q_{1-\alpha/2}\right) = \Pr\left(q_{\alpha/2} \leq T \leq q_{1-\alpha/2}\right) \begin{aligned}[t] &= \Pr(T \leq q_{1-\alpha/2}) - \Pr(T \leq q_{\alpha/2}) \\ &= 1 - \alpha/2 - \alpha/2 \\ &= 1 - \alpha, \end{aligned}$$

which completes the proof.

   Suppose now that σ_0^2 is unknown. Recall that an estimator of $Var(\overline{X}) = \sigma_0^2/n$ is $\widehat{Var}(\overline{X}) = S^2/n$. In this case, it can be shown that

$$T = \frac{\overline{X} - \mu_0}{\sqrt{\widehat{Var}(\overline{X})}} = \frac{\overline{X} - \mu_0}{\sqrt{S^2/n}} \sim t(n-1),$$

where $t(n-1)$ denotes the t-distribution with $n-1$ degrees of freedom. Now, the result follows from arguments similar to those when σ_0^2, by noting that $q_{\alpha/2} = -q_{1-\alpha/2}$ because the t-distribution is symmetric around zero.

Proof of Theorem 14.2   *The proof follows similar arguments as those in the proof of Theorem 14.1. Let $\alpha \in (0, 0.5]$. According to Definition 14.1, to prove the result, it is enough to show that*

$$\Pr\left(\sigma_0^2 \in \left[\frac{(n-1)\cdot S^2}{q_{1-\alpha/2}} \ , \ \frac{(n-1)\cdot S^2}{q_{\alpha/2}}\right]\right) = 1 - \alpha.$$

Note that

$$\sigma_0^2 \in \left[\frac{(n-1)\cdot S^2}{q_{1-\alpha/2}} \ , \ \frac{(n-1)\cdot S^2}{q_{\alpha/2}}\right]$$

$$\Longleftrightarrow \quad \frac{(n-1)\cdot S^2}{q_{1-\alpha/2}} \leq \sigma_0^2 \leq \frac{(n-1)\cdot S^2}{q_{\alpha/2}}$$

$$\Longleftrightarrow \quad \frac{1}{q_{1-\alpha/2}} \leq \frac{\sigma_0^2}{(n-1)\cdot S^2} \leq \frac{1}{q_{\alpha/2}}$$

$$\Longleftrightarrow \quad q_{1-\alpha/2} \geq \frac{(n-1)\cdot S^2}{\sigma_0^2} \geq q_{\alpha/2}$$

$$\Longleftrightarrow \quad q_{\alpha/2} \leq W \leq q_{1-\alpha/2},$$

where the random variable $W = T(S^2)$ is given by

$$W = \frac{(n-1)\cdot S^2}{\sigma_0^2}.$$

This shows that

$$\Pr\left(\sigma_0^2 \in \left[\frac{(n-1)\cdot S^2}{q_{1-\alpha/2}} \ , \ \frac{(n-1)\cdot S^2}{q_{\alpha/2}}\right]\right) = \Pr\left(q_{\alpha/2} \leq W \leq q_{1-\alpha/2}\right).$$

It can be shown that $W \sim \chi^2(n-1)$, where $\chi^2(n-1)$ denotes the χ^2-distribution with $n-1$ degrees of freedom. Using the usual rules for calculation of probabilities for continuous random variables and the fact that $q_{\alpha/2}$ and $q_{1-\alpha/2}$ are quantiles in the χ^2-distribution with $n-1$ degrees of freedom, we get

$$\Pr\left(q_{\alpha/2} \leq W \leq q_{1-\alpha/2}\right) = \Pr(W \leq q_{1-\alpha/2}) - \Pr(W \leq q_{\alpha/2}) = 1 - \alpha/2 - \alpha/2 = 1 - \alpha,$$

which completes the proof.

Proof of Theorem 14.3   *Using similar arguments as in the proof of Theorem 14.1, we may conclude that*

$$\Pr\left(\theta_0 \in \widehat{CI}_{1-\alpha,asymp}\right) = \Pr\left(q_{\alpha/2} \le Z_n \le q_{1-\alpha/2}\right) = \Pr(Z_n \le q_{1-\alpha/2}) - \Pr(Z_n \le q_{\alpha/2}),$$

where Z_n is given by

$$Z_n = \frac{\widehat{\theta} - \theta_0}{\sqrt{\widehat{Var}(\widehat{\theta})}}.$$

By assumption, $Z_n \overset{d}{\to} Z \sim N(0,1)$ as $n \to \infty$. Therefore, by the definition of convergence in distribution (Definition 13.1), we get

$$
\begin{aligned}
\lim_{n\to\infty} \Pr\left(\theta_0 \in \widehat{CI}_{1-\alpha,asymp}\right) &= \lim_{n\to\infty} \left(\Pr(Z_n \le q_{1-\alpha/2}) - \Pr(Z_n \le q_{\alpha/2})\right) \\
&= \lim_{n\to\infty} \Pr(Z_n \le q_{1-\alpha/2}) - \lim_{n\to\infty} \Pr(Z_n \le q_{\alpha/2}) \\
&= \Pr(Z \le q_{1-\alpha/2}) + \Pr(Z \le q_{\alpha/2}) \\
&= 1 - \alpha/2 - \alpha/2 \\
&= 1 - \alpha,
\end{aligned}
$$

where we in the second-to-last equality used that $Z \sim N(0,1)$ and that $q_{\alpha/2}$ and $q_{1-\alpha/2}$ are quantiles in the standard normal distribution $N(0,1)$. This concludes the proof.

Proof of Theorem 14.4   *The result follows directly from Theorem 14.3 by noting that the Central Limit Theorem implies that*

$$\frac{\overline{X} - \mu_0}{\sqrt{S^2/n}} \overset{d}{\to} N(0,1),$$

as $n \to \infty$, see, e.g. Theorem 13.5.

14.9   Exercises

Problem 14.9.1 *Let q_p be the p-quantile of the random variable $T = \theta_0 - \widehat{\theta}$. Set*

$$
\begin{aligned}
L &= \widehat{\theta} + q_{\alpha/2}, \\
U &= \widehat{\theta} + q_{1-\alpha/2}.
\end{aligned}
$$

Show that, for these choices of L and U, Equations (14.4)–(14.5) hold. Use this to conclude that Equation (14.2) holds.

Problem 14.9.2 *Let θ_0 be the estimand, $\widehat{\theta}$ an estimator, and $h > 0$ a number. Prove that*

$$\Pr\left(\theta_0 \in [\widehat{\theta} - h, \widehat{\theta} + h],\right) = \Pr(\widehat{\theta} \le \theta_0 + h) - \Pr(\widehat{\theta} < \theta_0 - h).$$

Problem 14.9.3 *Let X be a continuous random variable with PDF $f_X(x)$. Suppose that the function $f_X(x)$ is symmetric around zero, i.e. $f_X(-x) = f_X(x)$ for all $x \in \mathbb{R}$. Let $p \in (0,1)$ and define q_p as the p-quantile of X. Prove that*

$$q_p = -q_{1-p}.$$

Problem 14.9.4 *Consider estimation of the mean μ of a population. Suppose a simple random sample with $n = 10$ is available and the realized values of the sample average and the sample variance are $\bar{x} = 34$ and $s^2 = 12$.*

1. *Calculate an asymptotic 0.9-confidence interval for μ using Theorem 14.3.*

2. *Assume the population is normal distributed. Calculate an 0.9-confidence interval for μ using Theorem 14.1. Compare your answers in 1 and 2 and discuss the difference.*

Problem 14.9.5 *In this exercise, you are asked to use Monte Carlo simulation to corroborate the results of Theorem 14.3, i.e. the fact that, when the sample size is not too small, we can use the confidence interval based on the normal distribution, even when the underlying population X is non-normal.*

1. *Choose a non-normal distribution for the random variable X representing the population. For example, $X \sim Ber(p)$ for some $p \in (0,1)$, $X \sim Exp(2)$, or $X \sim U(a,b)$ for some $a < b$.*

2. *Calculate the mean of X, $\mu_0 = E(X)$.*

3. *Let $\alpha = 0.05$ and $n = 1000$. Choose M large. For $m = 1, 2, \ldots, M$ do:*

   (a) *Simulate a simple random Monte Carlo sample of size n, $(\widetilde{X}^*_{1,m}, \widetilde{X}^*_{2,m}, \ldots, \widetilde{X}^*_{n,m})$ from the distribution of X.*

   (b) *Use the Monte Carlo sample from (a) to calculate the Monte Carlo sample average $\bar{x}^{\#}_m$ and construct a $(1 - \alpha)$ confidence interval $\widehat{CI}_{1-\alpha,m}$ for μ_0 using Theorem 14.3.*

   (c) *Record whether or not the true mean of X is in the confidence interval constructed in (b), by saving the variable $d_m = I(\mu_0 \in \widehat{CI}_{1-\alpha,m})$, where $I()$ is the indicator function.*

4. *Use the variables d_m to assess whether the confidence interval $\widehat{CI}_{1-\alpha}$, constructed using Theorem 14.3 is accurate. (Hint: Note that the Monte Carlo sample average of $d_m, m = 1, 2, \ldots, M$ is an estimate of $Pr(\mu_0 \in \widehat{CI}_{1-\alpha})$ which should equal $1 - \alpha$ if Theorem 14.3 is accurate.)*

5. *Re-do the above analysis for other values of the sample size n. For what sample sizes are the results of Theorem 14.3 (in)accurate?*

Problem 14.9.6 *A polling institute is conducting a poll for whether there is majority support for a particular political issue. Let the true fraction of people in the population who are "for" the issue be denoted by the number $p_0 \in [0,1]$. The polling institute collects a simple random sample of size $n = 1000$. In this sample, 580 people are "for" the issue and 420 people are "against".*

1. *Use the sample to estimate p_0.*

2. *Use Theorem 14.3 to construct a 90% confidence interval for p_0. Is $p = 50\%$ inside the confidence interval? What can the polling institute conclude from this?*

Problem 14.9.7 *In this exercise, you are asked to use Monte Carlo simulation to investigate the performance of the various approaches to constructing confidence intervals.*

1. *Choose a distribution for the random variable X representing the population. (For instance, $X \sim N(0,1)$, $X \sim Ber(p)$ for some $p \in (0,1)$, $X \sim Exp(2)$, or $X \sim U(a,b)$ for some $a < b$.)*

2. *Calculate the mean of X, $\mu_0 = E(X)$.*

3. *Let $\alpha = 0.05$ and $n = 100$. Choose M large. For $m = 1, 2, \ldots, M$ do:*

   (a) *Simulate a simple random Monte Carlo sample of size n, $(\tilde{X}^*_{1,m}, \tilde{X}^*_{2,m}, \ldots, \tilde{X}^*_{n,m})$ from the distribution of X.*

   (b) *Use the Monte Carlo sample from (a) to calculate the Monte Carlo sample average $\bar{x}^{\#}_m$ and construct four different $(1 - \alpha)$ confidence interval $\widehat{CI}_{1-\alpha,m}$ for μ_0 using, respectively, (i) Definition 14.2, (ii) Definition 14.3, (iii) Definition 14.4, and (iv) Theorem 14.3. Use the non-parametric bootstrap for (i)–(iii).*

   (c) *Record whether or not the true mean of X is in the confidence intervals constructed in (b), by saving the variable $d_m^{(i)} = I(\mu_0 \in \widehat{CI}^{(i)}_{1-\alpha,m})$, where $I()$ is the indicator function, and $i = 1, 2, 3, 4$ denotes the four different ways of constructing the confidence interval.*

4. *Use the variables $d_m^{(i)}$ to assess compare the accuracy of the four different confidence intervals $\widehat{CI}^{(i)}_{1-\alpha}$. (Hint: Note that the Monte Carlo sample average of $d_m^{(i)}$, $m = 1, 2, \ldots, M$ is an estimate of $Pr(\mu_0 \in \widehat{CI}^{(i)}_{1-\alpha})$ which should equal $1 - \alpha$ if the confidence interval is accurate.)*

5. *Re-do the above analysis for other choices of population distributions X. Do the conclusions for instance change if, say X is normal versus non-normal, or if the distribution of X is symmetric vs. non-symmetric?*

Problem 14.9.8 *Let $X \sim N(\mu, \sigma^2)$ represent the population distribution. Set $n = 1000$.*

1. *Set $\mu = 0$ and $\sigma^2 = 1$ and simulate n iid random variables $(\tilde{X}_1, \tilde{X}_2, \ldots, \tilde{X}_n)$ from the distribution represented by X, i.e. from the $N(0,1)$ distribution. These random variables constitute your simple random sample, i.e. they are to be thought of as the "data".*

2. *Use the sample to estimate μ and σ^2. Estimate also the fourth central moment of X, $m_4^* = E((X - E(X))^4)$.*

3. *Use a parametric bootstrap to construct 95% construct confidence intervals for μ, σ^2, and m_4^*. (Hint: Run B bootstrap simulations, where the bootstrap samples are drawn from $N(\hat{\mu}, \widehat{\sigma^2})$ and use these samples to construct bootstrap estimates of μ, σ^2, and, m_4^*. The quantiles of the B estimates of these quantities can deliver the confidence intervals via, e.g. Definition 14.3.)*

4. *Run a Monte Carlo simulation study to assess the accuracy of the three confidence intervals from 3. (Hint: See Problem 14.9.5 for inspiration for how such a Monte Carlo simulation can be set up.)*

15

Analysis of hypotheses

15.1 Introduction

In the previous chapters, we saw how to represent characteristics in a population by distributions. In Chapter 11, we then discussed how to use a random sample to estimate various aspects of these distributions, most notably we studied estimating one or more parameters θ, describing parts or all of a given distribution. In this chapter, we will study how hypotheses concerning the underlying population can be formulated and analyzed. Then, in the next chapter, we will show how to use a random sample to choose between different hypotheses in practice.

Scientific theories may lead to the formulation of hypotheses. For example, a scientific theory may claim that there is a positive effect of implementing some policy, such as instituting a specific job training program for people who are unemployed. Such a claim can be represented by two hypotheses, one being that there is a positive effect, and the other one being that there is a no or a negative effect. Both hypotheses cannot be true simultaneously. The first hypothesis supports the scientific theory whereas the other rejects the scientific theory. In the following, we consider how to analyze hypotheses such as these.

In general, we are interested in analyzing hypotheses because knowledge of which hypothesis is true will have some influence over the acts which we might take now or in the future. When we act, we may incur losses, or gains. For example, if we choose a hypothesis that claims a positive effect of a policy, and therefore enact that policy, then we may get a gain from that policy, if the hypothesis is true, but incur a loss if the hypothesis is false. We will discuss how the sizes of losses and gains may influence the choice between hypotheses.

The general framework to formulate, analyze, and test hypotheses, presented in this chapter and the next, is powerful, because it will allow us to define a statistically rigorous guide to choosing between different available hypotheses of interest. However, as we shall see, care must be taken. Because the procedure relies on a random sample, its inherently statistical nature means there is always a risk of choosing a hypothesis that is, in fact, false. Such errors may ultimately result in acts that are costly when applied in practice. In what follows, we will design our framework for choosing hypotheses in a way that takes account of such costs whenever possible.

15.2 Hypotheses

Let Y be a random variable representing the population. We consider hypotheses which can be formulated by certain restrictions of the values of a parameter θ, where θ is a parameter in the distribution $f_Y(y; \theta)$ of Y. The unknown parameter θ describes all or parts of the distribution, and we assume that θ belongs to some fixed parameter set Θ, i.e. $\theta \in \Theta$. A

DOI: 10.1201/9781003591191-15

hypothesis is a restriction that θ belongs to some specific subset of Θ. To be precise, we consider $J + 1$ hypotheses, denoted H_0, H_1, ..., H_J, where $J \geq 1$, and hypothesis H_j, $j = 0, 1, \ldots, J$, states that θ belongs to the set $\Theta_j \subseteq \Theta$, which we write as "$H_j : \theta \in \Theta_j$". We say that the hypothesis H_j is *true* if the true value of the parameter θ is in the set Θ_j, i.e. if $\theta \in \Theta_j$, while H_j is said to be *false* if the true value of the parameter θ is not in Θ_j, i.e. if $\theta \notin \Theta_j$. The aim of the analysis of hypotheses is to *select* true hypotheses and *reject* false hypotheses.

We assume that all the hypotheses are mutually exclusive and that exactly one of the hypotheses is true, the remaining being false. The former means that θ can be in at most one of the subsets Θ_j, i.e. $\Theta_j \cap \Theta_i = \emptyset$ for $j \neq i$. The latter means that, no matter the value of $\theta \in \Theta$, it must be in exactly one of the sets, i.e. $\Theta_0 \cup \Theta_1 \ldots \cup \Theta_J = \Theta$. Most often, we only consider two hypotheses, that is, $J = 1$. Therefore, unless otherwise mentioned, we assume there are two hypotheses only. In this case, one hypothesis is called the *null hypothesis*, denoted H_0, and the other hypothesis is called the *alternative hypothesis*, denoted H_1.

Example 15.1 *Let $\theta = E(Y)$ describe the mean of a population represented by the random variable Y. Assume that the parameter space is $\Theta = \mathbb{R}$. We may be interested in the hypothesis that the mean of Y is zero, i.e. $\theta = 0$. This can be stated as the null hypothesis:*

$$H_0 : \theta = 0.$$

We state the alternative hypothesis H_1 as the hypothesis that θ is not zero,

$$H_1 : \theta \neq 0.$$

In the notation introduced above, we would thus have $\Theta_0 = \{0\}$ and $\Theta_1 = \mathbb{R} \setminus \{0\} = (-\infty, 0) \cup (0, \infty)$. As required, $\Theta_0 \cap \Theta_1 = \emptyset$ and $\Theta_0 \cup \Theta_1 = \Theta$.

We classify hypotheses into two types, *simple hypotheses* and *composite hypotheses*, depending on whether a hypothesis completely describes the population distribution. The definition is next.

Definition 15.1 (Simple and composite hypotheses) *A hypothesis H_j is called a **simple hypothesis** if, and only if, it completely specifies the distribution of the population. If the hypothesis is not simple, it is called a **composite hypothesis**.*

For a hypothesis H_j to be simple, the set Θ_j must only have one element. It also requires that the distribution function is assumed known, apart from the value of the parameters θ.

Example 15.2 *Assume $Y \sim N\left(\mu, \sigma^2\right)$. Suppose we know that $\sigma^2 = 16$. Then the hypothesis $H_0 : \mu = 0$ is a simple hypothesis since the implication of the hypothesis is that $Y \sim N(0, 16)$, which is a complete specification of the distribution of Y.*

Example 15.3 *Assume $Y \sim N\left(\mu, \sigma^2\right)$ with the value of σ^2 being unknown. Then the hypothesis $H_0 : \mu = 0$ is a composite hypothesis because $Y \sim N\left(0, \sigma^2\right)$ can have infinitely many distributions, namely, one for each value of σ^2. In the notation introduced above, we have*

$$
\begin{aligned}
\Theta &= \left\{\left(\mu, \sigma^2\right) : \mu \in (-\infty, \infty), \sigma^2 \in (0, \infty)\right\}, \\
\Theta_0 &= \left\{\left(\mu, \sigma^2\right) : \mu = 0, \sigma^2 \in (0, \infty)\right\}, \\
\Theta_1 &= \left\{\left(\mu, \sigma^2\right) : \mu \neq 0, \sigma^2 \in (0, \infty)\right\}.
\end{aligned}
$$

In case the distribution function of the population is not specified, a hypothesis is composite.

Example 15.4 *Consider a hypothesis about the mean $\mu = E(Y)$ of a population. Suppose the hypothesis is $H_0 : \mu = 0$. This is a composite hypothesis because there exists many distributions of Y, which have a mean equal to 0.*

A hypothesis is also composite if the parameter of interest θ is allowed to take different values under the hypothesis.

Example 15.5 *Consider again the setup in Example 15.2 with $Y \sim N(\mu, 16)$, but now assume that we are interested in the hypothesis that the mean of Y equals 0 or is negative versus the hypothesis that the mean is greater than zero. This can be stated as the null hypothesis*

$$H_0 : \mu \leq 0$$

and the alternative hypothesis

$$H_1 : \mu > 0.$$

In the notation introduced above, we have

$$
\begin{aligned}
\Theta &= \left\{ (\mu, \sigma^2) : \mu \in (-\infty, \infty), \sigma^2 \in (0, \infty) \right\}, \\
\Theta_0 &= \left\{ (\mu, \sigma^2) : \mu \leq 0, \sigma^2 \in (0, \infty) \right\}, \\
\Theta_1 &= \left\{ (\mu, \sigma^2) : \mu > 0, \sigma^2 \in (0, \infty) \right\}.
\end{aligned}
$$

Both hypotheses are composite since many normal distributions of Y satisfy the hypotheses.

In Example 15.3, the parameter σ^2 is not restricted under H_0 nor H_1 compared to the permitted values as specified by the parameter space. We call a parameter not restricted by a hypothesis a *nuisance parameter*. This is formally stated next.

Definition 15.2 (Nuisance parameter) *Let Θ be the parameter space and Θ_0 and Θ_1 be the parameter values satisfying H_0 and H_1, respectively. Then a parameter λ is a **nuisance parameter** if and only if all permitted values of λ by Θ are also permitted by both Θ_0 and Θ_1.*

Example 15.6 *As mentioned above, in Example 15.3 σ^2 is a nuisance parameter because it is unknown but not restricted by either the null nor the alternative hypothesis. The parameter μ is not a nuisance parameter since different values of μ are permitted under H_0 and H_1.*

The term nuisance parameter is also used outside the context of analysis of hypotheses. In general, a nuisance parameter refers to a parameter, which is not of interest for the specific problem studied. This is consistent with our definition of a nuisance parameter, since the formulation of the hypotheses directly state which parameters are of interest, and which parameters are not of interest.

15.3   Decision rules and test rules

In practice, we will use data, in the form of a random sample, to decide between the available hypotheses. A *decision rule* is a function that connects a random sample to the choice of a particular action. In analysis of hypotheses, the choice of action will often be equivalent to a choice of hypothesis, unless several actions may be undertaken when choosing a hypothesis. This can happen if the hypothesis is composite. The definition is next.

Definition 15.3 (Decision rule) *Let $\widetilde{Y} = \left(\widetilde{Y}_1, \ldots, \widetilde{Y}_n\right)$ be a random sample. A **decision rule**, $\delta\left(\widetilde{Y}\right)$, is a function with the sample space of $\widetilde{Y}$ as domain, and a set of actions $\mathcal{A}$ as co-domain.*

Example 15.7 *Suppose the purpose of considering the two hypotheses H_0 and H_1 is descriptive, that is, we use the choice of hypothesis to gain insight into a problem without having any action in mind due to the choice of hypothesis. For this purpose, the co-domain of a decision function may be represented by the hypotheses, that is $\mathcal{A} = \{H_0, H_1\}$.*

Example 15.8 *Consider a situation with two hypotheses, where the purpose is to determine what type of action should be taken. Such a situation may arise with, say, the hypotheses $H_0 : \theta = \theta_0$ against $H_1 : \theta \neq \theta_0$, where θ_0 is some number. This was the setup considered in Example 15.2, where $\theta_0 = 0$. Since H_0 is simple, only one action is undertaken if H_0 is chosen. With slight abuse of notation, we will refer to this action as θ_0. Under the alternative hypothesis H_1, since H_1 is composite, suppose many actions are available. Assume the action undertaken, when H_1 is chosen, depends on the estimate $\widehat{\theta}$ of θ. Therefore, we refer to the action taken, when H_1 is chosen, as $\widehat{\theta}$.*

Suppose the choice of hypothesis is determined by whether the estimate $\widehat{\theta}$ of θ belongs to an interval I_0 or to an interval I_1, disjoint from I_0. Later, we shall discuss how to determine such intervals but for now we take them as given. This leads to the decision function

$$\delta\left(\widetilde{y}\right) = \begin{cases} \theta_0 & \text{if } \widehat{\theta}\left(\widetilde{y}\right) \in I_0 \quad (\text{alias } H_0), \\ \widehat{\theta}\left(\widetilde{y}\right) & \text{if } \widehat{\theta}\left(\widetilde{y}\right) \in I_1 \quad (\text{alias } H_1), \end{cases}$$

where we have made explicit that $\widehat{\theta}$ is a function of the random sample $\widetilde{y}$. In case we choose H_0, then the same action is taken no matter the value of $\widehat{\theta}$ as long as it belongs to I_0. Contrary to this, if H_1 is chosen, then the action depends on $\widehat{\theta}$. For example, if H_1 means that a change should be made to a policy, then this decision function guides the size of the change to be made through the evidence given by the random sample and, thus, $\widehat{\theta}$.

In the context of analysis of hypotheses, when there is only one action for each hypothesis, we call the decision rule a *test rule*. The definition is next.

Definition 15.4 (Test rule) *A decision rule is called a **test rule** when the set of actions is the set of hypotheses, $\mathcal{A} = \{H_0, H_1, \ldots, H_J\}$.*

To simplify the discussion, we will, unless otherwise mentioned, assume from here on that there is only one action taken for each hypothesis, i.e. $\mathcal{A} = \{H_0, H_1, \ldots, H_J\}$. Hence, when we consider two hypotheses only, there is an action for H_0 and an action for H_1 and, thus, we focus on test rules.

15.4 Properties of a test rule

15.4.1 Power function

Let $\widetilde{Y} = (\widetilde{Y}_1, \ldots, \widetilde{Y}_n)$ denote a random sample. Due to $\widetilde{Y}$ being random, the test rule $\delta(\widetilde{Y})$ is also random. The probability of a test rule resulting in a particular choice of hypothesis can be calculated using the sample distribution $f_{\widetilde{Y}}(\widetilde{y}; \theta)$. Since the test rule is a function, we can find the probability of all the random samples which lead the test rule to choose a particular hypothesis. For example, if Y is a continuous random variable, then

$$\Pr\left(\delta\left(\widetilde{Y}\right) = H_j\right) = \int I\left(\delta\left(\widetilde{y}\right) = H_j\right) \cdot f_{\widetilde{Y}}(\widetilde{y}; \theta)\, d\widetilde{y},$$

for $j = 0, 1$, where $I()$ is the indicator function and the integral is over all possible values of $\widetilde{Y}$.

The probability of making a particular decision plays a key role in characterizing the properties of a test rule. This probability depends on the true value of θ. For that purpose, we next define the *power function* as the probability of the test rule to choose H_1, as a function of θ.

Definition 15.5 (Power function) *Let $\delta\left(\widetilde{Y}\right)$ be a test rule with co-domain $\{H_0, H_1\}$, and let $\theta \in \Theta$ be the true value of the parameter. Then the **power function** $\pi\left(\theta\right)$ is defined as*

$$\pi\left(\theta\right) = \Pr\left(\delta\left(\widetilde{Y}\right) = H_1 \; ; \; \theta\right), \quad \theta \in \Theta,$$

where we use the notation "$\Pr\left(\cdot \; ; \; \theta\right)$" to specify that the probability is determined using $f_{\widetilde{Y}}(\widetilde{y}; \theta)$ as the distribution of the random sample $\widetilde{Y}$.

Note that, formally, the power function depends on the choice of test rule $\delta()$, but we often suppress this for notational convenience.

The power function $\pi(\theta)$ gives the probability of the test rule to choose H_1. The probability of the test rule to choose H_0 is $1 - \pi(\theta)$.

Example 15.9 *Assume the population is $Y \sim N(\theta, 25)$, where the mean θ is unknown. Consider the hypotheses*

$$\begin{aligned} H_0 &: \quad \theta \leq 1, \\ H_1 &: \quad \theta > 1. \end{aligned}$$

With a simple random sample $\widetilde{Y}$ of size n, we can estimate θ using the sample average $\widehat{\theta} = \overline{Y} = \frac{1}{n}\sum_{i=1}^{n}\widetilde{Y}_i$. We will consider the test rule given by

$$\delta(\widetilde{Y}) = \begin{cases} H_0 & \text{if } \dfrac{\overline{Y}-1}{\sqrt{\frac{25}{n}}} \leq 1.28, \\[2ex] H_1 & \text{if } \dfrac{\overline{Y}-1}{\sqrt{\frac{25}{n}}} > 1.28. \end{cases} \tag{15.1}$$

The reason for this particular choice of test rule, called a "t-test rule", is given in Chapter 16. Since $\overline{Y} \sim N\left(\theta, \frac{25}{n}\right)$, the expression $(\overline{Y} - 1)/\sqrt{\frac{25}{n}}$ is a standardization of $\overline{Y}$ when $\theta = 1$ is the true value of the mean of Y. An intuition for why this can be a sensible test rule for the two hypotheses is that a small or negative value of $(\overline{Y} - 1)/\sqrt{\frac{25}{n}}$ is suggestive that the truth is a value of θ less than 1 and, thus, H_0 being true, whereas a large value of $(\overline{Y} - 1)/\sqrt{\frac{25}{n}}$ is suggestive of a value of θ larger than 1 and, thus, H_1 being true. The cut-off value 1.28, which delineates "small" values of $(\overline{Y} - 1)/\sqrt{\frac{25}{n}}$ from "large" values, is decided by technical considerations, which we will study further in Chapter 16.

We can derive the power function because the distribution of $\overline{Y}$ is known for given values of θ and n. Using the rules for probability calculations,

$$
\begin{aligned}
\pi(\theta) \;&=\; \Pr\left(\delta\left(\widetilde{Y}\right) = H_1 \;;\; \theta\right) \\[2mm]
&=\; \Pr\left(\frac{\overline{Y}-1}{\sqrt{\frac{25}{n}}} > 1.28\right) \\[2mm]
&=\; 1 - \Pr\left(\frac{\overline{Y}-1}{\sqrt{\frac{25}{n}}} \le 1.28\right).
\end{aligned}
$$

Now, add and subtract θ in the numerator to get

$$
\begin{aligned}
\pi(\theta) \;&=\; 1 - \Pr\left(\frac{\overline{Y}-1+\theta-\theta}{\sqrt{\frac{25}{n}}} \le 1.28\right) \\[2mm]
&=\; 1 - \Pr\left(\frac{\overline{Y}-\theta}{\sqrt{\frac{25}{n}}} + \frac{\theta-1}{\sqrt{\frac{25}{n}}} \le 1.28\right) \\[2mm]
&=\; 1 - \Pr\left(\frac{\overline{Y}-\theta}{\sqrt{\frac{25}{n}}} \le 1.28 - \frac{\theta-1}{\sqrt{\frac{25}{n}}}\right) \\[2mm]
&=\; 1 - \Pr\left(Z \le 1.28 - \frac{\theta-1}{\sqrt{\frac{25}{n}}}\right),
\end{aligned}
$$

where $Z = (\overline{Y}-\theta)/\sqrt{\frac{25}{n}}$ is a random variable with distribution $N(0,1)$. Denoting the CDF of a standard normal random variable by $\Phi(x)$, the power function becomes

$$
\pi(\theta) = 1 - \Phi\left(1.28 - \frac{\theta-1}{\sqrt{\frac{25}{n}}}\right). \tag{15.2}
$$

If the hypothesis H_0 is true such that $\theta = 1$, we have

$$
\pi(1) = 1 - \Phi(1.28) = 1 - 0.8 = 0.1.
$$

When $\theta \neq 1$, the power function $\pi(\theta)$ will depend on the sample size n. In Figure 15.1, the graphs of the power function (15.2) for $n = 25$ and $n = 100$, respectively, are shown.

It is seen that the power function is increasing in θ. This means that the probability of choosing H_1 increases with higher values of θ. The influence of n can also be seen. When the sample size n increases, the probability of choosing H_1 when in fact H_1 is true, that is, when $\theta > 1$, also increases. Similarly, and recalling $\Pr(\delta\left(\overline{Y}\right) = H_0; \theta) = 1 - \pi(\theta)$, when the sample size increases, the probability of choosing H_0 when $\theta < 1$, i.e. when in fact H_0 is true, also increases.

15.4.2   Type I error and Type II error

With two hypotheses, we assume that exactly one of the hypotheses is true, while the other hypothesis is false. Since the test rule returns a hypothesis, the test rule either chooses correctly (if it outputs the true hypothesis) or makes an error (if it outputs a false hypothesis).

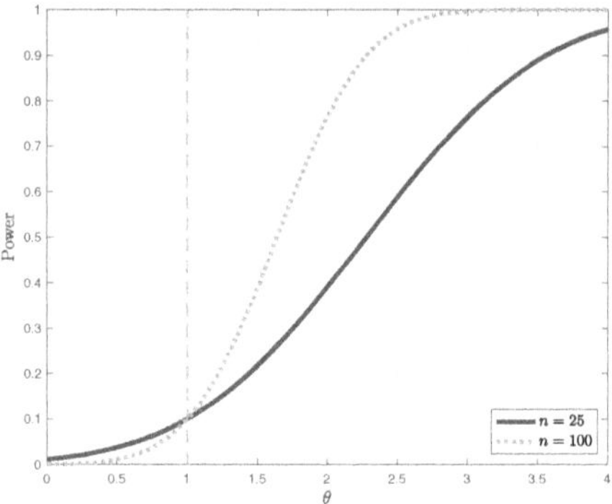

FIGURE 15.1
Power function for testing $H_0 : \theta \leq 1$ against $H_1 : \theta > 1$ using test rule (15.1). The vertical line is at $\theta = 1$. See Example 15.9 for details.

Since choosing between two hypotheses, H_0 and H_1, is common in empirical studies, the erroneous decisions have been given special names. A *Type I error* occurs when the test rule outputs H_1 but H_0 is true, while a *Type II error* occurs when the test rule outputs H_0 but H_1 is true. This is summarized in the following table.

Error types		Truth	
		H_0	H_1
Decision	H_0	Correct decision	Type II error
	H_1	Type I error	Correct decision

Example 15.10 *Let $H_0 : \theta \leq 0$ and $H_1 : \theta > 0$. Suppose the true value of the parameter is $\theta = -2$. Then the null hypothesis is true and the correct decision is to choose H_0. If, however, the realized sample $\widetilde{y}$ results in the test rule choosing H_1, then a Type I error is committed.*

Conversely, if the true value of the parameter is $\theta = 1$, say, then H_0 is false and H_1 is true. Then the correct decision is to choose H_1. If $\widetilde{y}$ results in the test rule choosing H_0, then a Type II error is committed.

The probability of making a correct choice and of making an error can be calculated based on the probability of the test rule choosing a particular hypothesis. Conveniently, we can use the power function to calculate the probabilities of committing Type I and Type II errors. We find the probability of a Type I error to be the power function evaluated at a θ under H_0, i.e. for $\theta \in \Theta_0$,

$$\Pr\left(Type\ I\ error\ ;\ \theta\right) = \Pr\left(\delta\left(\widetilde{Y}\right) = H_1\ ;\ \theta\right) = \pi\left(\theta\right), \quad \theta \in \Theta_0. \tag{15.3}$$

Conversely, we can find the probability of a Type II error as one minus the power function, evaluated at a θ under H_1, i.e. for $\theta \in \Theta_1$,

$$\Pr\left(Type\ II\ error\ ;\ \theta\right) = \Pr\left(\delta\left(\widetilde{Y}\right) = H_0\ ;\ \theta\right) = 1 - \Pr\left(\delta\left(\widetilde{Y}\right) = H_1\ ;\ \theta\right) = 1 - \pi\left(\theta\right).$$

Figure 15.2 illustrates where the probabilities of Type I errors and Type II errors can be found on a graph of the power function. For example, if θ_a is the true value and $\theta_a \in \Theta_0$, then the probability of choosing H_1 is the probability of a Type I error. In contrast, if θ_b is the true value and $\theta_b \in \Theta_1$, then the probability of choosing H_1 is $\pi(\theta)$, whereas the probability of a Type II error is $1 - \pi(\theta)$. The figure also illustrates that there can be many different probabilities of Type I errors and Type II errors, depending on the true value of θ.

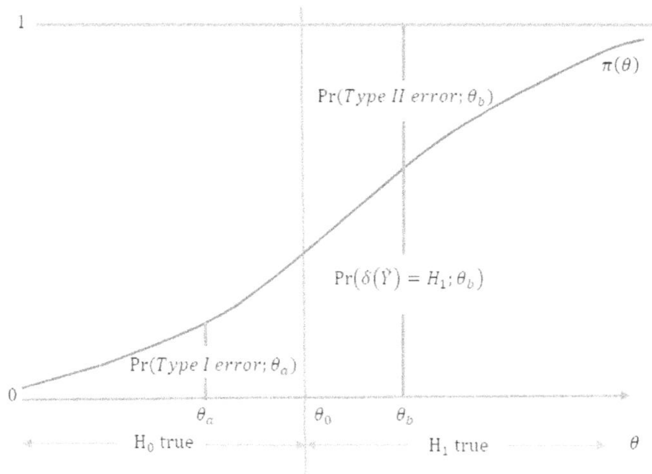

FIGURE 15.2
Power function of a test rule with $H_0 : \theta \leq \theta_0$ against $H_1 : \theta > \theta_0$.

15.4.3   Size of a test rule

When the null hypothesis is composite, that is, it is satisfied by many values of θ, there may be different probabilities of a Type I error depending on the value of θ. The maximum probability of making a Type I error is called the *size* of a test rule and it is defined next.

Definition 15.6 (Size of decision rule) *Let δ be a test rule. The **size** of the test rule δ is the number $s \in [0, 1]$, defined as*

$$s = \sup_{\theta \in \Theta_0} \Pr\left(\textit{Type I error} \; ; \; \theta\right).$$

In the definition, we use the supremum, sup, instead of maximum, for the technical reason that a maximum may not exists. For thinking about the condition, interpreting the sup as a max is adequate in most cases.

Using the relationship in (15.3), we may express the size of a test rule via the power function,

$$s = \sup_{\theta \in \Theta_0} \pi(\theta).$$

Example 15.11 (Example 15.9, continued) *In Example 15.9, we found the power function*

$$\pi(\theta) = 1 - \Phi\left(1.28 - \frac{\theta - 1}{\sqrt{\frac{25}{n}}}\right),$$

see Equation (15.2) and the calculations preceding it. Since $\Phi(x)$ is increasing in x, we have that $\pi(\theta)$ is increasing in θ. Assuming that $\theta \in \Theta_0$, i.e. $\theta \leq 1$, we therefore conclude that

$$s = \sup_{\theta \in \Theta_0} \Pr\left(Type\ I\ error\ ;\ \theta\right) = \sup_{\theta \leq 1} \pi(\theta) = \pi(1) = 0.1.$$

Thus, in this example, the size of the test rule equals 0.1 or 10%.

15.4.4 Consistency of a test rule

We usually require that decision rules become better as the evidence grow, e.g. if the sample size increases. Therefore, we next discuss properties of a test rule for large sample sizes. To do so, consider a sequence of test rules $\delta_n()$ indexed by the sample size n of the random sample $\widetilde{Y}$ used as input in the decision procedure. That a test rule becomes better with larger sample sizes is conceptualized by defining the *consistency* of a test rule. We will say that a decision rule is consistent if the probability of committing a Type II error goes to zero, as the sample size increases to infinity. In other words, a decision rule is consistent if it always rejects false null hypotheses in the limit of an infinite sample size. Since it is trivial to construct a decision rule that never commits a Type II error (e.g. the decision rule that always rejects H_0), we will simultaneously require that the size of the test is bounded. The definition is next.

Definition 15.7 (Consistent decision rule) *Let $\delta_n()$ be a sequence of test rules indexed by the sample size n and let s_{δ_n} be the size of δ_n. Let $\alpha \in (0,1)$ and assume that*

$$s_{\delta_n} \leq \alpha$$

*for all n. Then $\delta_n(\widetilde{y})$ is a **consistent test rule** for that α if, and only if, for all $\theta \in \Theta_1$,*

$$\lim_{n \to \infty} \Pr\left(\delta_n\left(\widetilde{Y}\right) = H_1\ ;\ \theta\right) = 1. \tag{15.4}$$

In words, Definition 15.7 says that a sequence of decision rules δ_n is consistent if we can "control" the probability of making a Type I error, i.e. ensure that this probability is less than or equal to the number α, while still making sure that the probability of rejecting H_0, when it is false, increases to 1 as the sample size n increases. Typically, we do not use a decision rule in practice unless it is consistent.

We may express the condition (15.4) in Definition 15.7 using the power function. That is, the sequence of test rules $\delta_n()$ with size bounded by $\alpha \in (0,1)$ is consistent if, for all $\theta \in \Theta_1$,

$$\lim_{n \to \infty} \pi(\theta) = 1.$$

Example 15.12 (Example 15.11, continued) *Consider the test rule (15.1) in Example 15.9 as a function of n resulting in a sequence of test rules. As discussed in Example 15.11, the size of the test equals 0.1. This is independent of the sample size n. Thus, for any $\alpha \geq 0.1$, the first requirement of Definition 15.7 is satisfied.*

Next we check the second requirement of Definition 15.7. Suppose H_1 is true, i.e. $\theta > 1$. Then, for $n \to \infty$,

$$\pi(\theta) = 1 - \Phi\left(1.28 - \frac{\theta - 1}{\sqrt{\frac{25}{n}}}\right) \to 1 - \Phi\left(1.28 - \infty\right) = 1 - \Phi\left(-\infty\right) = 1 - 0 = 1.$$

Therefore, the second requirement of Definition 15.7 is satisfied. In conclusion, the sequence of test rules is consistent for any $\alpha \geq 0.1$.

15.5   Loss of a test rule

In the following sections, we further develop the analysis of hypotheses based on the *loss* of making incorrect decisions.

15.5.1   Two approaches to measure loss

We initially consider two different approaches to define a loss relevant for decision making based on hypothesis testing. One approach is to measure a loss for each decision and true hypothesis. In many cases, however, we are interested in *avoidable loss*. Avoidable loss is the additional loss we incur from a decision, compared to the best decision available had we known the true hypothesis. This is because we cannot control which hypothesis is true, and hence the magnitudes of the losses may be out of our control, but we can strive to minimize the loss given the true hypothesis. Avoidable loss is also called *regret*. The following example illustrates.

Example 15.13 *Consider the case where we have to choose between two simple hypotheses $H_0 : \theta = \theta_{H_0}$ and $H_1 : \theta = \theta_{H_1}$ for two numbers $\theta_{H_0}, \theta_{H_1} \in \mathbb{R}$. Write C_{ij} for the loss incurred when the true hypothesis is H_i and the decision made is H_j. The relevant losses can be stated in tabular form as follows.*

		Truth	
		H_0	H_1
Decision	H_0	C_{00}	C_{10}
	H_1	C_{01}	C_{11}

Suppose, for the sake of illustration, that we make a decision of either H_0 or H_1 independent of any data. Then one approach would be to consider the maximum possible loss for each decision and then make the decision with the lowest maximum possible loss. For example, suppose the losses are as follows.

		Truth	
		H_0	H_1
Decision	H_0	1	1002
	H_1	50	1000

Here, the maximum loss of decision H_0 is 1002 and the maximum loss of decision H_1 is 1000. Then, according to the approach suggested above, we should make the decision H_1, because H_1 has the smallest maximum loss.

The regrets (avoidable losses) can be calculated by adjusting the loss of each decision with the best decision that could have been made for a given true hypothesis. In the example, if H_0 is true, then we could avoid losing $50 - 1 = 49$ by choosing the optimal H_0 rather than H_1. Similarly, if H_1 is true, then we could avoid losing $1002 - 1000 = 2$ by choosing the optimal H_1 rather than H_0. For the numerical example, the regrets are given in the following table.

		Truth		
		H_0		H_1
Decision	H_0	0 $(= 1 - 1)$		2 $(= 1002 - 1000)$
	H_1	49 $(= 50 - 1)$		0 $(= 1000 - 1000)$

The optimal decision based on regrets may be derived by finding the decision which has the lowest maximum regret. The maximum regret taking decision H_0 is 2, and the maximum regret taking decision H_1 is 49. The lowest maximum regret is 2 and, thus, the optimal decision, in this setting, where a hypothesis is chosen without any reference to a data sample, is H_0.

As the example illustrates, depending on the approach to defining losses, different decisions may be optimal. The context should determine which approach to apply. In the remainder, we shall only consider losses, which can be interpreted as regrets.

Example 15.14 *Consider a company that has to decide whether to produce a batch of commodities. Suppose the company is unsure whether or not they will be able to sell the batch in the market. Let the null hypothesis H_0 be that they cannot sell the batch and let the alternative hypothesis H_1 be that they can sell it. Suppose the cost of producing the batch is 70 and, if it is sold, there is an income of 90, resulting in an earnings of 20. If the company deems H_0 to be true, they will take the action "not produce", which has a cost of zero. Conversely, if the company deems H_1 to be true, they will take the action "produce". The loss (regret) function is given in the following table.*

		Truth	
		H_0	H_1
Decision	H_0	0	20
	H_1	70	0

For example, if the company decides to produce, that is H_1 is chosen, but cannot sell, that is, H_0 is true, then they have committed a Type I error. The loss from their choice is 70, i.e. what they could have saved, had they not produced. If it turns out that they could have sold the batch, that is H_1 is true, but they did not produce, that is H_0 is chosen, then the loss of the Type II error is the foregone earning of 20.

15.5.2 Loss function

In practice, we will base our test rule on evidence in the form of a random sample. The loss of a test rule depends on the action implicit in the choice of hypothesis. The definition is next.

Definition 15.8 (Loss of a decision rule) *Let $L\left(\delta\left(\widetilde{y}\right),\theta\right)$ be a function of the decision rule $\delta\left(\widetilde{y}\right)$, evaluated at a random sample $\widetilde{y}$ and the true value θ. Then $L\left(\delta\left(\widetilde{y}\right),\theta\right)$ is a **loss function for the decision rule** $\delta\left(\widetilde{y}\right)$ if, and only if,*

$$L\left(\delta\left(\widetilde{y}\right),\theta\right) \geq 0$$

for all $\theta \in \Theta$.

For ease of notation, we continue focusing on decision rules which are test rules, that is, there is only one action attached to each hypothesis.

Example 15.15 *Let $H_0 : \theta = \theta_{H_0}$ and $H_1 : \theta = \theta_{H_1}$ for two numbers $\theta_{H_0}, \theta_{H_1} \in \mathbb{R}$. Let L_{01} be the loss when H_0 is true but decision H_1 is taken (a Type I error), and L_{10} the loss when H_1 is true but decision H_0 is taken (a Type II error). Then the loss function can be represented in the following table format.*

		Truth	
		H_0	H_1
Decision	H_0	0	L_{10}
	H_1	L_{01}	0

Example 15.15 considered the case of two simple hypotheses. In case of one or both of the hypotheses being composite, the loss may vary over the true parameter values θ. Assuming $\delta(\widetilde{y})$ is a test rule, the loss function can be written as

$$L(\delta(\widetilde{y}),\theta) = L_{01}(\theta) \cdot \underbrace{I(\theta \in \Theta_0) \cdot I(\delta(\widetilde{y}) = H_1)}_{indicator\ of\ Type\ I\ error} + L_{10}(\theta) \cdot \underbrace{I(\theta \in \Theta_1) \cdot I(\delta(\widetilde{y}) = H_0)}_{indicator\ of\ Type\ II\ error},$$

where $I()$ is the indicator function, and $L_{01}()$ and $L_{10}()$ are functions of θ, describing the losses in case of Type I and Type II errors, respectively. The next example illustrates such a situation.

Example 15.16 *Consider the hypotheses $H_0 : \theta \leq 1$ against $H_1 : \theta > 1$. Suppose the context is such that the loss function for the test rule $\delta(\widetilde{y})$ is*

$$L(\delta(\widetilde{y}),\theta) = \begin{cases} 0 & if\ \theta \leq 1\ and\ \delta(\widetilde{y}) = H_0 & (Correct\ decision), \\ 0 & if\ \theta > 1\ and\ \delta(\widetilde{y}) = H_1 & (Correct\ decision), \\ 5 \cdot |\theta - 1| & if\ \theta \leq 1\ and\ \delta(\widetilde{y}) = H_1 & (Type\ I\ error), \\ 2 \cdot |\theta - 1| & if\ \theta > 1\ and\ \delta(\widetilde{y}) = H_0 & (Type\ II\ error). \end{cases} \tag{15.5}$$

The loss function is shown in Figure 15.3. The solid and dashed lines show the loss from choosing H_0 and H_1, respectively, as a function of the true value of θ. It is seen that the loss of a faulty decision depends on the distance of the true value θ to 1, which is the value separating H_0 and H_1. In addition, the loss is larger when making a Type I error compared to a Type II error for the true value being the same absolute distance to 1. For example, if we choose H_1 (i.e. $\delta(\widetilde{y}) = H_1$) but the true value of the parameter is $\theta = -1$ (i.e. H_0 is true), then the loss is $5 \cdot |-1-1| = 5 \cdot 2 = 10$. In contrast, if we choose H_0 (i.e. $\delta(\widetilde{y}) = H_0$) but the true value is $\theta = 3$ (i.e. H_1 is true), then we get a loss of $2 \cdot |3-1| = 2 \cdot 2 = 4$. That is, the loss is 2.5 times larger for making the Type I error compared to making a Type II error for these two values of θ, even though they are the same distance from 1.

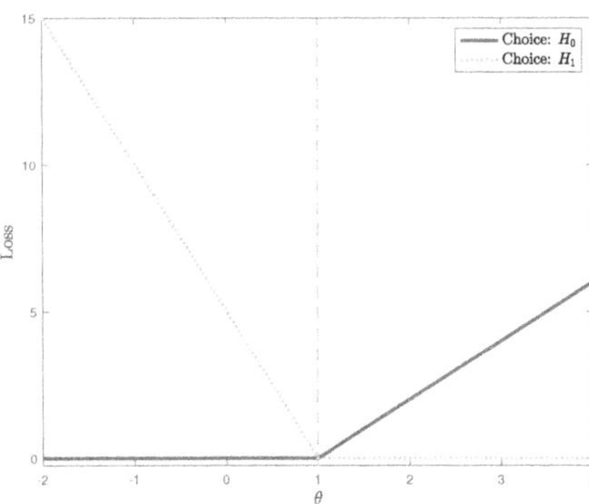

FIGURE 15.3

Asymmetric loss function with hypotheses $H_0 : \theta \leq 1$ against $H_1 : \theta > 1$. See Example 15.16 for details.

15.5.3 Risk of a decision rule

To evaluate the quality of a decision rule that depends on a random sample, we may consider the average loss of the decision rule taken over all possible samples. This leads to the *risk of a decision rule*. The definition is next.

Definition 15.9 (Risk function of decision rule) *Let $L(\delta(\widetilde{y}),\theta)$ be the loss of the decision rule $\delta(\widetilde{y})$. Assuming the expectation exists, the **risk of the decision rule** as a function of the true value θ is defined as*

$$Risk_\delta(\theta) = E\left(L\left(\delta\left(\widetilde{Y}\right),\theta\right)\right),$$

where the expectation is taken over all possible random samples $\widetilde{Y}$, i.e. it is calculated using the distribution $f_{\widetilde{Y}}(\widetilde{y};\theta)$.

As above, our focus is on decision rules which are also test rules. With a test rule and two hypotheses, the risk function can be formulated using the power function as shown in the next theorem.

Theorem 15.1 (Risk as function of power function) *Let $\delta\left(\widetilde{Y}\right)$ be a test rule with set of actions (co-domain) $\mathcal{A} = \{H_0, H_1\}$. Then the risk function of the test rule is*

$$Risk_\delta(\theta) = \begin{cases} L_{01}(\theta) \cdot \pi(\theta) & \text{if } \theta \in \Theta_0, \\ L_{10}(\theta) \cdot (1 - \pi(\theta)) & \text{if } \theta \in \Theta_1, \end{cases}$$

where $\pi(\theta)$ is the power function of the test rule, and $L_{01}()$ and $L_{10}()$ denote the losses of Type I errors and Type II errors, respectively.

When the two hypotheses are simple, or if the loss function does not depend on θ, then the risk can be written as

$$Risk_\delta(\theta) = \begin{cases} L_{01} \cdot \Pr\left(Type\ I\ error\ ;\ \theta\right) & \text{if } \theta \in \Theta_0 \\ L_{10} \cdot \Pr\left(Type\ II\ error\ ;\ \theta\right) & \text{if } \theta \in \Theta_1 \end{cases}.$$

The next example illustrates how the risk varies with the true value of θ for two reasons. One reason is that the probability of choosing an untrue hypothesis is a function of θ, and the other reason is that the magnitude of the loss incurred may also depend on θ.

Example 15.17 (Example 15.9, continued) *Let $H_0 : \theta \leq 1$ against $H_1 : \theta > 1$ and consider again the test rule $\delta()$ in Equation (15.1). The power function $\pi(\theta)$ of the test rule, derived in Example 15.9, may be used to calculate the probabilities of Type I and Type II errors. These are shown in the top two panels of Figure 15.4. Consider also the loss function*

$$L(\delta(\widetilde{y}),\theta) = \begin{cases} 0 & \text{if } \theta \leq 1 \text{ and } \delta(\widetilde{y}) = H_0 \ \ (\textit{Correct decision}), \\ 0 & \text{if } \theta > 1 \text{ and } \delta(\widetilde{y}) = H_1 \ \ (\textit{Correct decision}), \\ 3 \cdot |\theta - 1| & \text{if } \ \ \theta \leq 1 \text{ and } \delta(\widetilde{y}) = H_1 \ \ (\textit{Type I error}), \\ 3 \cdot |\theta - 1| & \text{if } \ \ \theta > 1 \text{ and } \delta(\widetilde{y}) = H_0 \ \ (\textit{Type II error}). \end{cases}$$

The graph of this loss function is shown in the bottom left panel of Figure 15.4. The losses of the errors is symmetric around $\theta = 1$. The risk of using this test rule with the loss function $L()$ given above is,

$$Risk_\delta(\theta) = \begin{cases} 3 \cdot |\theta - 1| \cdot \pi(\theta) & \text{if } \theta \leq 1, \\ 3 \cdot |\theta - 1| \cdot (1 - \pi(\theta)) & \text{if } \theta > 1, \end{cases}$$

where $\pi()$ is the power function associated with the test rule δ. The bottom right panel in Figure 15.4 shows the graphs of the risk functions for two different sample sizes. It is seen that the risk function is not symmetric around $\theta_0 = 1$. This is due to the probabilities of Type I errors and of Type II errors are not the same around $\theta_0 = 1$, see the top left and top right panels in Figure 15.4. It can also be seen that the risk is lower for a higher sample size. The reason for this is that the probabilities of making Type I and Type II errors are lower for the higher sample size.

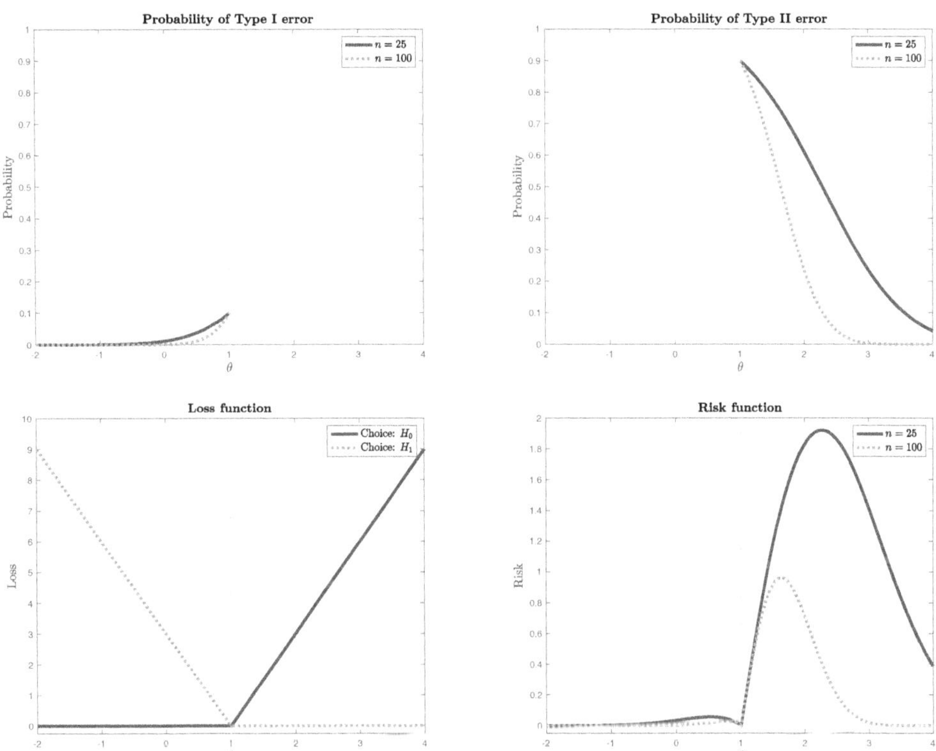

FIGURE 15.4
Deciding between hypotheses $H_0 : \theta \leq 1$ against $H_1 : \theta > 1$ with a symmetric loss function. Top left panel: Probability of Type I error as a function of the true value of θ. Top right panel: Probability of Type II error as a function of the true value of θ. Bottom left panel: Loss function as a function of the true value of θ. Bottom right panel: Risk function as a function of the true value of θ. See Example 15.17 for details.

The previous example considered a case where the losses of the errors were symmetric, but the probabilities of committing Type I and Type II errors were very asymmetric. The following example illustrates the converse case, where the loss function is asymmetric and the probabilities of Type I and Type II errors are symmetric.

Example 15.18 *Let $H_0 : \theta \leq 1$ against $H_1 : \theta > 1$. Let the test rule be*

$$\delta(\widetilde{Y}) = \begin{cases} H_0 & \textit{if} \quad \frac{\overline{Y}-1}{\sqrt{\frac{25}{n}}} \leq 0, \\[2ex] H_1 & \textit{if} \quad \frac{\overline{Y}-1}{\sqrt{\frac{25}{n}}} > 0. \end{cases}$$

This test rule has the power function

$$\pi(\theta) = 1 - \Phi\left(0 - \frac{\theta - 1}{\sqrt{\frac{25}{n}}}\right) = 1 - \Phi\left(\frac{\theta - 1}{\sqrt{\frac{25}{n}}}\right).$$

At $\theta = 1$, we have $\pi(1) = 1 - \Phi(0) = 1 - 0.5 = 0.5$. The probability of a Type I error, which can happen for $\theta \leq 1$, is $1 - \Phi\left(\frac{\theta-1}{\sqrt{\frac{25}{n}}}\right)$, whereas the probability of a Type II error is $1 - \left(1 - \Phi\left(\frac{\theta-1}{\sqrt{\frac{25}{n}}}\right)\right) = \Phi\left(\frac{\theta-1}{\sqrt{\frac{25}{n}}}\right)$. These probabilities of Type I and Type II errors are symmetric around $\theta = 1$, see top left and top right panels in Figure 15.5.

Consider the same asymmetric loss function as in Example 15.16, shown in the bottom left panel of Figure 15.5. This loss function leads to the risk function being

$$Risk_\delta(\theta) = \begin{cases} 5 \cdot |\theta - 1| \cdot \pi(\theta) & \text{if } \theta \leq 1, \\ 2 \cdot |\theta - 1| \cdot (1 - \pi(\theta)) & \text{if } \theta > 1, \end{cases}$$

where $\pi()$ is the power function associated with the test rule $\delta()$. The risk function for the test rule in this example is shown bottom right panel in Figure 15.5. It is seen that the risk function is not symmetric around $\theta = 1$. Since the probabilities of Type I and Type II errors are symmetric around $\theta = 1$, the asymmetric risk function is due to the asymmetric loss function, that is, there is a relatively higher loss of making a Type I error than a Type II error.

Deciding between hypotheses $H_0 : \theta \leq 1$ against $H_1 : \theta > 1$ with a symmetric loss function. Top left panel: Probability of Type I error as a function of the true value of θ. Top right panel: Probability of Type II error as a function of the true value of θ. Bottom left panel: Loss function as a function of the true value of θ. Bottom right panel: Risk function as a function of the true value of θ. See Example 15.17 for details.

15.6 Selecting a test rule when the risk function is known

In this section and the next, we consider how to select a test rule if several test rules are available. In this section, we present two concepts that may be used when the risk function $Risk_\delta(\theta)$ is known. In Section 15.7, we consider an approach to selecting a test rule when the risk function is not known.

15.6.1 Admissibility

Since the risk $Risk_\delta(\theta)$ of a test rule δ typically varies with the true value of θ, we must take this into account when evaluating the quality of a test rule. Suppose we are given two test rules δ_A and δ_B, and we have to select which test rule to apply in practice. If we compare the two test rules, then we would certainly select test rule δ_A if

$$Risk_{\delta_A}(\theta) \leq Risk_{\delta_B}(\theta), \quad \text{for all } \theta \in \Theta,$$

and

$$Risk_{\delta_A}(\theta) < Risk_{\delta_B}(\theta), \quad \text{for at least one } \theta \in \Theta.$$

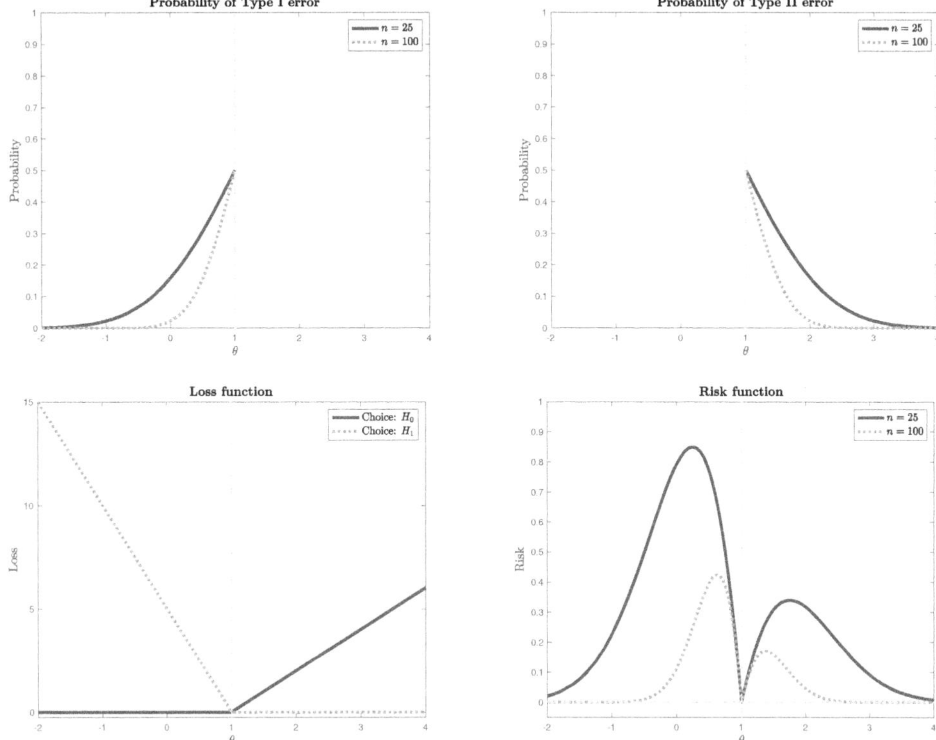

FIGURE 15.5
Deciding between the hypotheses $H_0 : \theta \leq 1$ against $H_1 : \theta > 1$ with an asymmetric loss function. Top left panel: Probability of Type I error as a function of the true value of θ. Top right panel: Probability of Type II error as a function of the true value of θ. Bottom left panel: Loss function as a function of the true value of θ. Bottom right panel: Risk function as a function of the true value of θ. See Example 15.18 for details.

That is, no matter the value of θ, test rule δ_A does not have higher risk than test rule δ_B, and there is at least one value of θ where test rule δ_A has lower risk than test rule δ_B. If these conditions hold, we say that the test rule δ_B is *inadmissible*, or we say that test rule δ_A is *uniformly better*, in terms of risk, than test rule δ_B, where the "uniformly" refers to "over all values of θ". This leads us to a formal definition of admissibility of a test rule within a collection, or *class*, of test rules.

Definition 15.10 (Admissible test rule) *Let $\mathcal{C}$ be a class of test rules. Then a test rule $\delta \in \mathcal{C}$ is **admissible** in the class $\mathcal{C}$ of test rules if and only if there is no test rule $\widetilde{\delta} \in \mathcal{C}$ such that*

$$Risk_{\widetilde{\delta}}(\theta) \leq Risk_{\delta}(\theta) \quad \text{for all } \theta \in \Theta, \tag{15.6}$$

and

$$Risk_{\widetilde{\delta}}(\theta) < Risk_{\delta}(\theta) \quad \text{for at least one } \theta \in \Theta. \tag{15.7}$$

In words, a test rule δ is admissible in the class $\mathcal{C}$ if there is no other test rule in $\mathcal{C}$ that is uniformly better than δ. The concept of admissibility of a test rule is similar in spirit to the concept of efficiency of an estimator in a class of estimators (Definition 11.10). In most cases, we will not use test rules that are inadmissible in the same way that we would not use an estimator if another more efficient estimator is available.

Admissibility of a test rule can be investigated by solely considering the power function at θ's where there is a loss of making a faulty decision. The next theorem describes this result.

Theorem 15.2 (Admissibility by power function) *Let C be a class of test rules. Then a test rule $\delta \in C$ is admissible in the class C of test rules if there is no test rule $\widetilde{\delta} \in C$ such that*

$$\pi_{\widetilde{\delta}}(\theta) \;\leq\; \pi_\delta(\theta) \quad \text{for all } \theta \in \Theta_0 \tag{15.8}$$

$$\text{and}$$

$$\pi_{\widetilde{\delta}}(\theta) \;\geq\; \pi_\delta(\theta) \quad \text{for all } \theta \in \Theta_1, \tag{15.9}$$

and

$$\pi_{\widetilde{\delta}}(\theta) \;<\; \pi_\delta(\theta) \quad \text{for at least one } \theta \in \Theta_0$$

$$\text{or}$$

$$\pi_{\widetilde{\delta}}(\theta) \;>\; \pi_\delta(\theta) \quad \text{for at least one } \theta \in \Theta_1.$$

Example 15.19 *Consider again the setting of Example 15.9 with the hypotheses $H_0 : \theta \leq 1$ against $H_1 : \theta > 1$. We extend the example by considering a class of test rules defined as*

$$\delta_c(\widetilde{Y}) = \begin{cases} H_0 & \text{if } \dfrac{\overline{Y}-1}{\sqrt{\frac{25}{n}}} \leq c, \\[3ex] H_1 & \text{if } \dfrac{\overline{Y}-1}{\sqrt{\frac{25}{n}}} > c, \end{cases}$$

where c can be any real number. For example, the test rules in Examples 15.9 and 15.18 are obtained by setting $c = 1.28$ and $c = 0$, respectively. Call this class of test rules C, i.e. $C = \{\delta_c() : c \in \mathbb{R}\}$. By replacing 1.28 with c in the derivations of the power function in Example 15.9, the power function can be shown to be

$$\pi_{\delta_c}(\theta) = 1 - \Phi\left(c - \frac{\theta - 1}{\sqrt{\frac{25}{n}}}\right).$$

To show that all test rules in C are admissible, we can use Theorem 15.2. Let $c \in \mathbb{R}$ be any number. We want to show that δ_c is admissible, i.e. that there is no $\widetilde{c}$ such that $\delta_{\widetilde{c}}$ satisfies the conditions of Theorem 15.2.

First suppose $\widetilde{c} < c$. Then

$$\pi_{\delta_{\widetilde{c}}}(\theta) = 1 - \Phi\left(\widetilde{c} - \frac{\theta - 1}{\sqrt{\frac{25}{n}}}\right) > 1 - \Phi\left(c - \frac{\theta - 1}{\sqrt{\frac{25}{n}}}\right) = \pi_{\delta_c}(\theta),$$

because $\widetilde{c} - \frac{\theta-1}{\sqrt{\frac{25}{n}}} < c - \frac{\theta-1}{\sqrt{\frac{25}{n}}}$ and $\Phi()$ is an increasing function. Hence, for a $\widetilde{c}$ such that $\widetilde{c} < c$, then $\delta_{\widetilde{c}}$ does not satisfy (15.8).

Now suppose instead that $\widetilde{c} > c$. Then, using a similar argument as above,

$$\pi_{\delta_{\widetilde{c}}}(\theta) = 1 - \Phi\left(\widetilde{c} - \frac{\theta - 1}{\sqrt{\frac{25}{n}}}\right) < 1 - \Phi\left(c - \frac{\theta - 1}{\sqrt{\frac{25}{n}}}\right) = \pi_{\delta_c}(\theta).$$

Hence, for a $\widetilde{c}$ such that $\widetilde{c} > c$, then $\delta_{\widetilde{c}}$ does not satisfy (15.9).

We have now checked all test rules in $\mathcal{C}$ to see if one of them makes δ_c inadmissible. None of them do, so we conclude that δ_c is admissible. Since this holds for any $c \in \mathbb{R}$, all the test rules in $\mathcal{C}$ are admissible.

15.6.2 Minimax test rule

To select a test rule among admissible test rules, we impose further criteria for the selection. Inspired by the minimax criteria for selection of estimators, the *minimax test rule δ^** is the test rule with lowest maximum risk over $\theta \in \Theta$. The definition is next.

Definition 15.11 (Minimax test rule) *Let $\mathcal{C}$ be a class of test rules. Then δ^* is a **minimax test rule** in $\mathcal{C}$ if, and only if, it is a solution to the problem*

$$\arg\min_{\delta \in \mathcal{C}} \left(\max_{\theta \in \Theta} Risk_\delta(\theta) \right).$$

The problem in the definition can be read as follows. Pick a test rule $\delta()$ in $\mathcal{C}$. For that test rule, find the θ for which the test rule has the highest risk. Do the same for all other test rules in the class of test rules $\mathcal{C}$. The smallest of those risks, call it r^*, is the smallest obtainable maximum risk among the test rules. The test rules δ^* in $\mathcal{C}$ for which their maximum risk equals the smallest in the set of test rules, i.e. for which $\max_{\theta \in \Theta} Risk_{\delta^*}(\theta) = r^*$, is referred to as minimax test rules.

For a given class of test rules, the minimax test rule is always admissible, at least when there is only one minimax test rule in the class. This fact is stated in the next result.

Theorem 15.3 (The minimax test rule is admissible) *Let $\mathcal{C}$ be a set of test rules, and suppose that the test rule $\delta^* \in \mathcal{C}$ is the only minimax test rule in $\mathcal{C}$. Then δ^* is admissible in $\mathcal{C}$.*

Theorem 15.3 shows that if there is only one admissible test rule in $\mathcal{C}$, then this test rule is necessarily admissible. Note, however, that there can be more than one test rule which is minimax and that not all of them may be admissible. In particular, two test rules may have the same maximum risk (and thus both be minimax), but one of the test rules may dominate the other away from the maximum. If a class of test rules contains several minimax test rules, then we assume that one of them is chosen by some means, in such a way that the chosen test rule is admissible.

Example 15.20 *Consider the hypotheses $H_0 : \theta = 1$ against $H_1 : \theta = 3$ in a setting similar to Example 15.19, with $Y \sim N(\theta, 25)$. Assume the loss function is as given in the following table.*

<table>
<tr><td></td><td></td><td colspan="2">*Truth*</td></tr>
<tr><td></td><td></td><td>H_0</td><td>H_1</td></tr>
<tr><td>*Decision*</td><td>H_0</td><td>0</td><td>2</td></tr>
<tr><td></td><td>H_1</td><td>5</td><td>0</td></tr>
</table>

Let the class of test rules be $\mathcal{C} = \{\delta_c : c \in \mathbb{R}\}$, where

$$\delta_c(\widetilde{Y}) = \begin{cases} H_0 & \text{if } \dfrac{\overline{Y}-1}{\sqrt{\frac{25}{n}}} \le c, \\[2mm] H_1 & \text{if } \dfrac{\overline{Y}-1}{\sqrt{\frac{25}{n}}} > c. \end{cases}$$

We shall select the test rule δ_c according to the minimax criterion. That is, we choose the value of c in such a way as to minimize the maximum risk of δ_c.

The risk at $\theta = 1$ is

$$Risk_{\delta_c}(1) = 5 \cdot \pi(1) = 5 \cdot \left(1 - \Phi\left(c - \frac{1-1}{\sqrt{\frac{25}{n}}}\right)\right) = 5 \cdot (1 - \Phi(c)),$$

and the risk at $\theta = 3$ is

$$Risk_{\delta_c}(3) = 2 \cdot (1 - \pi(3)) = 2 \cdot \Phi\left(c - \frac{3-1}{\sqrt{\frac{25}{n}}}\right).$$

Note that $Risk_{\delta_c}(1)$ is a decreasing function of c, and that $Risk_{\delta_{-\infty}}(1) = 5$ and $Risk_{\delta_\infty}(1) = 0$. Similarly $Risk_{\delta_c}(3)$ is an increasing function of c, and $Risk_{\delta_{-\infty}}(3) = 0$ and $Risk_{\delta_\infty}(3) = 2$. Since both functions are continuous in c, there must exist a $c^* \in \mathbb{R}$ such that $Risk_{\delta_{c^*}}(1) = Risk_{\delta_{c^*}}(3)$.

To minimize the maximum risk, the best solution is where $Risk_{\delta_c}(1) = Risk_{\delta_c}(3)$ because at any other c, either $Risk_{\delta_c}(1)$ or $Risk_{\delta_c}(3)$ must be higher, and we must guard against the maximum risk over θ. Hence, we need to solve for c in

$$Risk_{\delta_c}(1) = Risk_{\delta_c}(3),$$

that is,

$$5 \cdot (1 - \Phi(c)) = 2 \cdot \Phi\left(c - \frac{3-1}{\sqrt{\frac{25}{n}}}\right).$$

The value of c that solves this equation will depend on the sample size n. Using a numerical search over c, we find that the solution is $c = 1.30$ when $n = 25$ and $c = 2.19$ when $n = 100$. For instance, for $n = 25$, the minimax test rule is $\delta^* = \delta_{1.3}$ and the minimum maximum risk is

$$Risk_{\delta_{1.3}}(3) = Risk_{\delta_{1.3}}(1) = 0.484.$$

In the next example, we consider the same loss function as in the previous example, but now consider two composite hypotheses instead of two simple hypotheses.

Example 15.21 *Consider the hypotheses $H_0 : \theta \leq 1$ against $H_1 : \theta > 1$ in a setting similar to Example 15.19, with $Y \sim N(\theta, 25)$. Assume the loss function is given as in the following table.*

		Truth	
		H_0	H_1
Decision	H_0	0	2
	H_1	5	0

Assume the class of test rules is $\mathcal{C} = \{\delta_c : c \in \mathbb{R}\}$, where

$$\delta_c(\widetilde{Y}) = \begin{cases} H_0 & \text{if } \frac{\overline{Y}-1}{\sqrt{\frac{25}{n}}} \leq c, \\ H_1 & \text{if } \frac{\overline{Y}-1}{\sqrt{\frac{25}{n}}} > c. \end{cases}$$

We shall use the minimax criterion to choose a particular test rule in this class of test rules.

In this setting, the maximum risk happens just around $\theta = 1$ because at $\theta = 1$, the probabilities of Type I errors and Type II errors are largest and the loss is the same regardless of θ, see the first three panels of Figure 15.6. This implies that we need to solve

$$5 \cdot (1 - \Phi(c)) = 2 \cdot \Phi(c)$$

for $c \in \mathbb{R}$. The solution is

$$c = \Phi^{-1}\left(\frac{5}{7}\right) = 0.57.$$

Note that while this minimax solution for test rule $\delta^* = \delta_{0.57}$ does not depend on the sample size, the risk does. Indeed, the risk goes to 0 as n goes to infinity. The fourth panel of Figure 15.6 shows the risk function for $n = 25$ and $n = 100$.

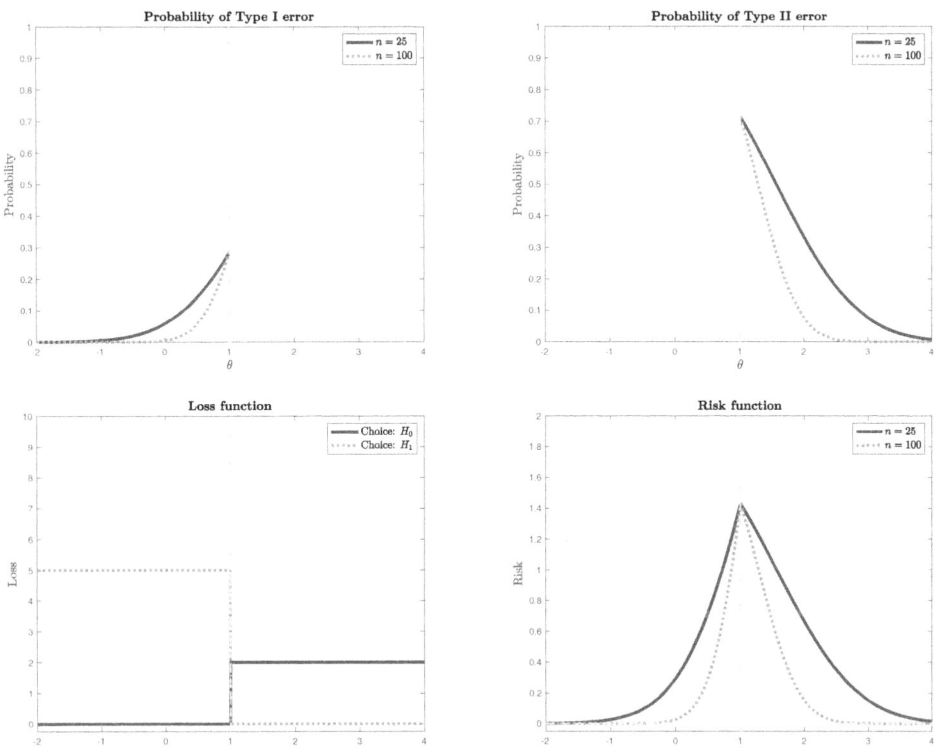

FIGURE 15.6
Deriving the minimax decision rule with hypotheses $H_0 : \theta = 1$ against $H_1 : \theta = 3$ and a constant loss of Type I and Type II errors. Top left panel: Probability of Type I error as a function of the true value of θ. Top right panel: Probability of Type II error as a function of the true value of θ. Bottom left panel: Loss function as a function of the true value of θ. Bottom right panel: Risk function as a function of the true value of θ. See Example 15.21 for details.

In Examples 15.20 and 15.21, the losses of Type I and Type II errors did not depend on the specific value of θ. This simplified the derivation of the minimax decision rule considerably. In the last example, we illustrate how the minimax decision rule can be derived in the case of two composite hypotheses and a loss function that depends on the true value of θ.

Example 15.22 *Consider the hypotheses $H_0 : \theta \leq 1$ against $H_1 : \theta > 1$ in a setting similar to Example 15.19, with $Y \sim N(\theta, 25)$. Consider again the class of test rules be $\mathcal{C} = \{\delta_c : c \in \mathbb{R}\}$, where*

$$\delta_c(\widetilde{Y}) = \begin{cases} H_0 & \text{if} \quad \dfrac{\overline{Y}-1}{\sqrt{\frac{25}{n}}} \leq c, \\[2ex] H_1 & \text{if} \quad \dfrac{\overline{Y}-1}{\sqrt{\frac{25}{n}}} > c. \end{cases}$$

The power function $\pi(\theta)$ implied by these test rules is derived in Example 15.19. Assume the loss function is

$$
L\left(\delta\left(\widetilde{y}\right),\theta\right) = \begin{cases} 0 & \text{if } \theta \leq 1 \text{ and } \delta\left(\widetilde{y}\right) = H_0 \quad (\text{Correct decision}), \\ 0 & \text{if } \theta > 1 \text{ and } \delta\left(\widetilde{y}\right) = H_1 \quad (\text{Correct decision}), \\ 5 \cdot |\theta - 1| & \text{if } \theta \leq 1 \text{ and } \delta\left(\widetilde{y}\right) = H_1 \quad (\text{Type I error}), \\ 2 \cdot |\theta - 1| & \text{if } \theta > 1 \text{ and } \delta\left(\widetilde{y}\right) = H_0 \quad (\text{Type II error}). \end{cases}
$$

The risk function is then given by

$$
Risk_\delta(\theta) = \begin{cases} 5 \cdot |\theta - 1| \cdot \pi(\theta) & \text{if } \theta \leq 1, \\ 2 \cdot |\theta - 1| \cdot (1 - \pi(\theta)) & \text{if } \theta > 1, \end{cases}
$$

where $\pi(\theta)$ is the power function.

A numerical approximation to find the minimax test rule can be done by making a grid of c values and then searching over θ to find the maximum risk for each of them, followed by choosing the δ_c with the lowest of the maximum risks. It can be found that for $n = 25$, the minimax test rule is $\delta^* = \delta_{0.34}$ and for $n = 100$, the minimax test rule is also $\delta^* = \delta_{0.34}$. That they are the same for different sample size is due to properties of the normal distribution, and not a general result.

Figure 15.7 shows the probabilities of Type I and Type II errors, the loss function, and the risk function for the minimax test rule $\delta^* = \delta_{0.34}$. Due to the nature of the class of test rules considered here, the minimax selection of c is such that the highest risk for $\theta \in \Theta_0$ is equal to the highest risk for $\theta \in \Theta_1$. This is so because lowering the probability of a Type I error (i.e. increasing c) leads to a higher probability of a Type II error and vice versa. That is, starting from the minimax solution $c = 0.34$, increasing c results in the maximum risk becoming higher, because the probability of a Type II error increases. Conversely, lowering c results in the maximum risk becoming higher, because the probability of a Type I error increases.

At this point, it is worth mentioning that the approach described in this section of how to select a test rule is called a *frequentist approach*. In the frequentist approach, it is not meaningful to say, e.g. that we should select the test rule with the lowest expected loss *for the most likely values of θ*. The reason is that θ is not considered random in the frequentist approach and, therefore, the statement "most likely values of θ" is not meaningful. In an alternative to the frequentist approach, called the *Bayesian approach*, θ is considered a random variable, implying that such a statement is, on the contrary, meaningful. It should be the actual empirical context being investigated that should determine whether such a statement is meaningful and, thus, determine whether to use a frequentist or a Bayesian approach to analyze hypotheses. We will study the Bayesian approach to the analysis of hypotheses in Chapter 19.

15.7　Selecting a test rule when the risk function is unknown

In practice, assessing the loss of a particular decision may be difficult. For example, even with two simple hypotheses, it may be difficult to assess the values of the losses L_{01} and L_{10}. This implies that the risk function is unknown and, thus, we cannot use it for evaluating the quality of a test rule nor use it for selecting one test rule among many test rules. In the following, we discuss how to alleviate the problem of an unknown risk function by prioritizing

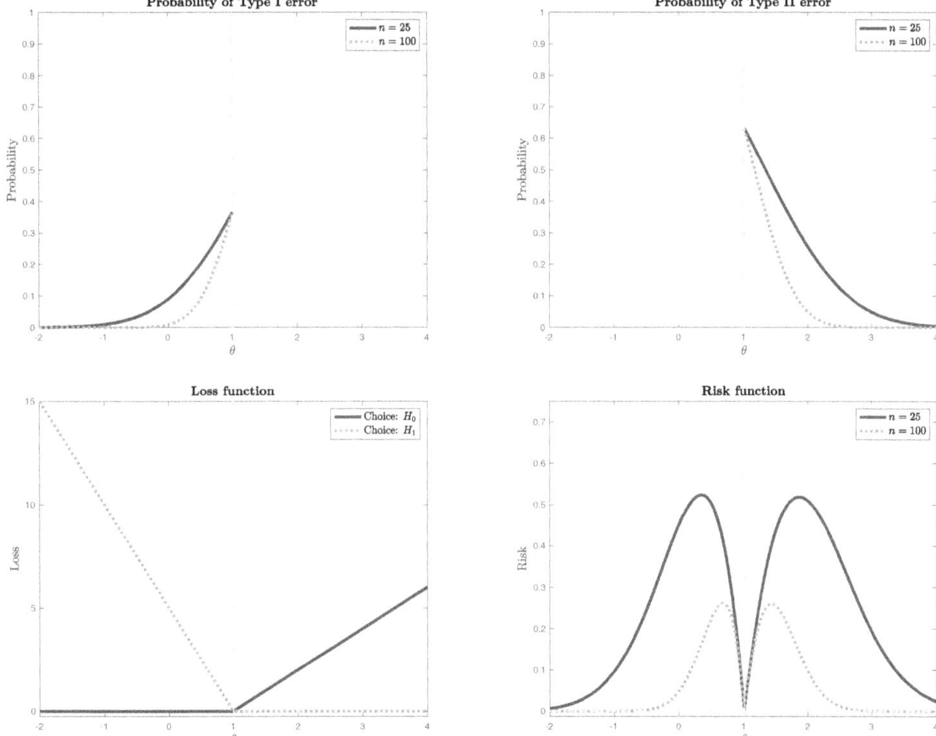

FIGURE 15.7
Deriving the minimax decision rule with hypotheses $H_0 : \theta \leq 1$ against $H_1 : \theta > 1$ when the losses depend on the value of θ. Top left panel: Probability of Type I error as a function of the true value of θ. Top right panel: Probability of Type II error as a function of the true value of θ. Bottom left panel: Loss function as a function of the true value of θ. Bottom right panel: Risk function as a function of the true value of θ. See Example 15.22 for details.

that the probability of making a Type I error is small, and then only secondarily considering the probability of making a Type II error. Such prioritization may be appropriate when the loss from committing a Type I error is believed to be (much) higher than the corresponding loss from committing a Type II error.

15.7.1 Type I error focus

To partly resolve the problem of an unknown loss function, we assume that the hypotheses can be formulated such that it is much more problematic to commit a Type I error than to commit a Type II error. In other words, we implicitly assume that it is more costly to wrongly select H_1 than to wrongly select H_0. For example, the null hypothesis H_0 could represent a status quo of a scientific belief about a relationship. The alternative hypothesis H_1 may represent a new theory of that relationship. We may argue that discarding the status quo comes with a high cost, unless we are really sure that the new, perhaps controversial, theory is right. A different way of saying this is that we only discard a current understanding of a topic if we get compelling new evidence. We give the current consensus on a topic more weight than a new theory, but it is hard to measure in terms of losses. In general, arguments to support that the null hypothesis deserves special attention need to be made beyond statistics. Once we have argued that the null hypothesis deserves special attention, it follows that the main focus should be on the risk associated with Type I errors.

The focus on the Type I error is implemented by setting a highest acceptable probability of making a Type I error. For example, we may find it acceptable that the highest probability of a Type I error is 5%. In words, we may find it acceptable to wrongly reject a true null hypothesis, on average, 1 in 20 times, were we to repeat the statistical experiment providing the random sample. The choice of the highest acceptable probability of a Type I error is called a *significance level*, and it is defined next.

Definition 15.12 (Significance level) *Let* $\alpha \in (0,1)$. *The number* α *is a **significance level** for a test rule* δ *if*

$$\Pr\left(Type\ I\ error\ ;\ \theta\right) = \Pr\left(\delta\left(\widetilde{Y}\right) = H_1\ ;\ \theta\right) \leq \alpha,$$

for all $\theta \in \Theta_0$.

The significance level α is chosen by the researcher, and, as we will see later on, this selection plays a key role in the construction of a test rule.

Example 15.23 *The test rule* (15.1) *studied in Examples 15.9 is constructed to have* $\pi(\theta_0) = 0.10$, *which means that it satisfies having significance level* $\alpha = 10\%$.

By Definition 15.6 of the size, s, of a test rule, all values of α such that $\alpha \geq s$ are also significance levels of that test rule. For example, in Example 15.23, where the size of the test rule was found to be $s = 10\%$, all values α such that $\alpha \geq 10\%$ are significance levels for the test rule. A test rule with a size s strictly smaller than the significance level, i.e. for which $s < \alpha$, is called *under-sized* or *conservative*. Conversely, a test rule for which the size is strictly larger than the significance level, i.e. for which $s > \alpha$, is called *over-sized*. In that case, the test rule does not satisfy having the significance level α. In practice, we may end up using a test rule with size larger than the significance level because the test rule has been chosen based on insufficient information about the population. For example, we often have to use a test rule chosen from asymptotic arguments, which do not necessarily hold true for a given finite sample size, see e.g. Section 16.7. A conservative test rule satisfies the significance level but a size much less than the significance level typically implies that the power of the test rule is much lower than it needs to be. Hence, we shall aim at using test rules with sizes equal to, or at least close to, the chosen significance level.

15.7.2   Type II error balance

Solely focusing on the Type I error and completely ignoring the Type II error implies that the analysis of hypotheses becomes uninteresting. Indeed, if we do not care about the Type II error, then we might as well choose a significance level α equal to 0, and thus avoid any Type I error altogether. This can be done with a test rule that always selects H_0. Such a test rule, however, implies that H_0 is never rejected and, thus, the probability of a Type II error equals 1 for any $\theta \in \Theta_1$. We would have reached a similar result had we known the loss function and that there is zero loss from committing a Type II error, $L_{10} = 0$. We implicitly assume that there is a loss associated with committing a Type II error and, thus, the Type II error should not be ignored.

After having selected a significance level $\alpha \in (0,1)$, e.g. $\alpha = 10\%$, we may design a test rule δ_α to have significance level α (Definition 15.12). See, e.g. Example 15.9. Given the test rule δ_α thus obtained, we may then examine its properties when H_1 is true, i.e.

when $\theta \in \Theta_1$. In particular, we are interested in the probability of making a Type II error, denoted by

$$\beta(\theta) = \Pr\left(Type\ II\ error\ ;\ \theta\right) = 1 - \pi(\theta), \quad \theta \in \Theta_1,$$

where $\pi(\theta)$ is the power function. We may compare the probability of making a Type II error, given by $\beta(\theta)$, to the probability of making a Type I error, given by α, and make a judgement about whether we find the test rule δ_α appropriate.

Example 15.24 *Consider the setup from Example 15.9 with $H_0 : \theta = 1$ against $H_1 : \theta = 3$. Let the class of test rules be $\mathcal{C} = \{\delta_\alpha : \alpha \in (0,1)\}$, where*

$$\delta_\alpha(\widetilde{Y}) = \begin{cases} H_0 & \text{if } \dfrac{\overline{Y}-1}{\sqrt{\frac{25}{n}}} \leq z_{1-\alpha}, \\[2ex] H_1 & \text{if } \dfrac{\overline{Y}-1}{\sqrt{\frac{25}{n}}} > z_{1-\alpha}, \end{cases}$$

and $z_{1-\alpha}$ is the $(1-\alpha)$-quantile of the standard normal distribution. When $\theta = 1$ and, thus, H_0 is true, the probability of a Type I error is

$$\pi_{\delta_\alpha}(1) = 1 - \Phi\left(z_{1-\alpha} - \frac{1-1}{\sqrt{\frac{25}{n}}}\right) = 1 - \Phi\left(z_{1-\alpha}\right) = 1 - (1-\alpha) = \alpha.$$

Hence, α is the size of the test rule.

When $\theta = 3$ and, thus, H_1 is true, the probability of a Type II error is

$$\beta(3) = 1 - \pi_{\delta_\alpha}(3) = \Phi\left(z_{1-\alpha} - \frac{3-1}{\sqrt{\frac{25}{n}}}\right).$$

For example, let $\alpha = 0.05$. Then, for $n = 25$,

$$\beta(3) = \Phi\left(z_{1-0.05} - \frac{3-1}{\sqrt{\frac{25}{25}}}\right) = \Phi\left(1.64 - 2\right) = 0.36,$$

and, for $n = 100$,

$$\beta(3) = \Phi\left(z_{1-0.05} - \frac{3-1}{\sqrt{\frac{25}{100}}}\right) = \Phi\left(1.64 - 4\right) = 0.0093.$$

Comparison of the probability of the Type I error relative to the Type II error can be done by considering their ratios. These ratios are

$$n = 25 \Rightarrow \frac{\beta(3)}{\alpha} = \frac{0.36}{0.05} = 7.2,$$

$$n = 100 \Rightarrow \frac{\beta(3)}{\alpha} = \frac{0.0093}{0.05} = 0.19.$$

It is seen that they are quite different. For $n = 25$, a Type II error occurs 7.2 times more often when $\theta = 3$ than a Type I error occurs if $\theta = 1$. For $n = 100$, it is the Type I error that occurs more than 5 times as often when $\theta = 1$ than a Type II error occurs if $\theta = 3$.

Example 15.24 illustrates how the behavior of a test rule may vary over n and, in particular, how that can influence the occurrences of Type I and Type II errors. Without a loss function, a possible criterion to select test rules can be that the ratio of the probability of a Type II error to the probability of a Type I error should remain fixed over n. The next example illustrates.

Example 15.25 (Example 15.24, continued) *The ratio of the probability of the Type II error to the probability of the Type I error is*

$$\frac{\beta(3)}{\alpha} = \frac{\Phi\left(z_{1-\alpha} - \frac{3-1}{\sqrt{\frac{25}{n}}}\right)}{\alpha} = \frac{\Phi\left(z_{1-\alpha} - \frac{2}{\sqrt{25}}\sqrt{n}\right)}{\alpha}.$$

To keep this ratio equal to a constant $d_{II/I}$ over n, we need to let α be a function of n. Hence, we need to solve

$$d_{II/I} = \frac{\Phi\left(z_{1-\alpha} - \frac{2}{\sqrt{25}}\sqrt{n}\right)}{\alpha}.$$

This can be done numerically. Figure 15.8 shows α as a function of n for two different values of the ratio $d_{II/I}$.

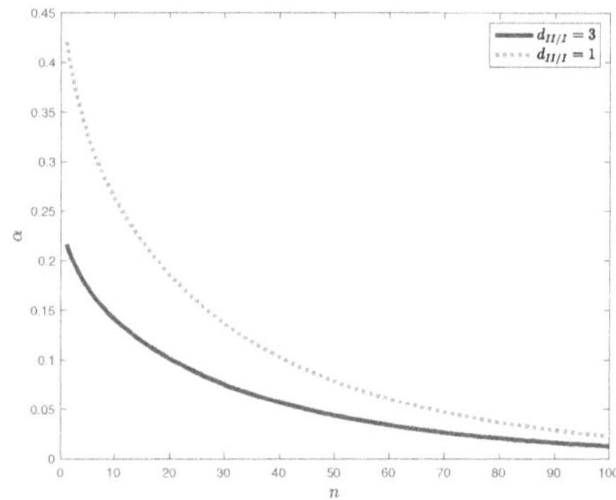

FIGURE 15.8
Significance level α as a function of sample size n for two values of the ratio $d_{II/I}$, where $d_{II/I}$ is the probability of a Type II error divided by the probability of a Type I error. The hypotheses are $H_0 : \theta = 1$ against $H_1 : \theta = 3$. See Example 15.25 for details.

Example 15.24 illustrates how a test rule may be chosen when the risk function is unknown. In particular, we may set an initial value of the significance level, say $\alpha = 5\%$, and then calculate the probability of making a Type II error, $\beta_\alpha(\theta)$. Using our judgement, we may then decide whether it is beneficial to adjust α so as to trade a higher (lower) probability of making a Type I error when H_0 is true, for a lower (higher) probability of making a Type II error when H_1 is true. The example illustrates that these trade-offs might be different depending on the sample size n. In practice, quantitative or qualitative assessment of the losses involved should of course also play a role, if such assessments are at all possible.

15.8 Significance of hypotheses and parameters

When considering values of θ and determining which values should be in which hypothesis, it is important to focus on the context. If $H_0 : \theta = 0$ is rejected, that is, if we find evidence that $\theta \neq 0$, then we say that θ is *statistically significant*, or *statistically significantly different from zero*. In contrast, we shall say that a parameter value is *context significant* if it has practical importance in the context considered. Statistical significance and context significance are not equivalent. For example, a parameter θ may measure the influence of a variable on some outcome we care about, $\theta = 0$ corresponding to no influence on the outcome. A value of, say, $\theta = 3$, may have real influence on some people's lives. In that case, we say that θ is context significant. In contrast, a value of, say, $\theta = 0.01$, may, although it is not exactly equal to zero, nonetheless have no real influence in practice. Then, we say that θ is not context significance or that θ is context insignificant. In this case, we may be able to reject $H_0 : \theta = 0$, and therefore conclude that θ is statistically significant, but, if $\widehat{\theta} = 0.01$, we may at the same time deem θ to be context insignificant.

The concepts of context significance and of statistical significance can be more aligned by formulating the hypotheses appropriately. For example, we may be interested in no influence, corresponding to $\theta = 0$, versus a positive influence, corresponding to $\theta > 0$. The influence implied by $\theta = 0.01$, however, may not be of any practical importance. That is, $\theta = 0.01$ is not context significant. If we assess that $\theta > 2.5$ is of practical importance, then we may formulate the hypotheses as

$$
\begin{aligned}
H_0 &: \quad \theta \leq 2.5 \\
H_1 &: \quad \theta > 2.5.
\end{aligned}
$$

Thus, H_0 is context insignificant and H_1 is context significant. Then statistical significance is aligned with context significance.

15.9 Proofs

Proof of Theorem 15.1 *The risk of the test rule depends on whether θ belongs to Θ_0 or Θ_1, and is given by*

$$
\begin{aligned}
Risk_\delta\left(\theta\right) &=
\begin{cases}
0 \cdot \Pr\left(\delta\left(\widetilde{Y}\right) = H_0 \; ; \; \theta\right) + L_{01}(\theta) \cdot \Pr\left(\delta\left(\widetilde{Y}\right) = H_1 \; ; \; \theta\right) & \text{if } \; \theta \in \Theta_0 \\
L_{10}(\theta) \cdot \Pr\left(\delta\left(\widetilde{Y}\right) = H_0 \; ; \; \theta\right) + 0 \cdot \Pr\left(\delta\left(\widetilde{Y}\right) = H_1 \; ; \; \theta\right) & \text{if } \; \theta \in \Theta_1 \\
\end{cases} \\
&=
\begin{cases}
L_{01}(\theta) \cdot \Pr\left(\delta\left(\widetilde{Y}\right) = H_1 \; ; \; \theta\right) & \text{if } \; \theta \in \Theta_0, \\
L_{10}(\theta) \cdot \Pr\left(\delta\left(\widetilde{Y}\right) = H_0 \; ; \; \theta\right) & \text{if } \; \theta \in \Theta_1.
\end{cases}
\end{aligned}
$$

We have that $\Pr\left(\delta\left(\widetilde{Y}\right) = H_1 \; ; \; \theta\right) = \pi(\theta)$ *and* $\Pr\left(\delta\left(\widetilde{Y}\right) = H_0 \; ; \; \theta\right) = 1 - \pi(\theta)$, *and, hence, then the risk can be written as*

$$
Risk_\delta\left(\theta\right) =
\begin{cases}
L_{01}(\theta) \cdot \pi\left(\theta\right) & \text{if } \; \theta \in \Theta_0, \\
L_{10}(\theta) \cdot \left(1 - \pi\left(\theta\right)\right) & \text{if } \; \theta \in \Theta_1.
\end{cases}
$$

Proof of Theorem 15.2   *Recall from Theorem 15.1 that we may represent the risk of a test rule as*

$$Risk_\delta\left(\theta\right) = \begin{cases} L_{01}(\theta) \cdot \pi\left(\theta\right) & \text{if} \quad \theta \in \Theta_0, \\ L_{10}(\theta) \cdot \left(1 - \pi\left(\theta\right)\right) & \text{if} \quad \theta \in \Theta_1, \end{cases}$$

where $\pi(\theta)$ is the power function of the test rule, and $L_{01}()$ and $L_{10}()$ denote the losses of Type I errors and Type II errors, respectively. Let $\delta \in \mathcal{C}$ be a test rule fulfilling the conditions of Theorem 15.2. Since the losses $L_{01}()$ and $L_{10}()$ do not depend on the chosen test rule δ, these conditions, together with the representation of the risk given above, imply that (15.6)–(15.7) hold, meaning that δ is admissible (Definition 15.10).

Proof of Theorem 15.3   *We prove this result using the technique "proof by contradiction". That is, we assume that δ^* is not admissible and show that this leads to a contradiction.*

To this end, suppose that δ^* is minimax in $\mathcal{C}$ but that it is not admissible. According to Definition 15.10, the latter means that there is a decision rule in $\widetilde{\delta} \in \mathcal{C}$ such that

$$Risk_{\widetilde{\delta}}\left(\theta\right) \le Risk_\delta\left(\theta\right) \quad \text{for all } \theta \in \Theta, \tag{15.10}$$

and

$$Risk_{\widetilde{\delta}}\left(\theta\right) < Risk_\delta\left(\theta\right) \quad \text{for at least one } \theta \in \Theta.$$

But by assumption, δ^* is the minimax test rule in $\mathcal{C}$, which implies that

$$\max_{\theta \in \Theta} Risk_{\widetilde{\delta}}(\theta) > \max_{\theta \in \Theta} Risk_{\delta^*}(\theta).$$

Clearly, this is a violation of (15.10). Hence, we conclude that δ^* is admissible.

15.10   Exercises

Problem 15.10.1 *Let $H_0 : \theta \le 0$ against $H_1 : \theta > 0$. Consider the test rule "always choose H_0". Argue that this is not a consistent test rule.*

Problem 15.10.2 *Consider the hypotheses $H_0 : \theta \le 0$ against $H_1 : \theta > 0$, and $Y \sim N(\theta, 25)$. Let the test rule be*

$$\delta(\widetilde{Y}) = \begin{cases} H_0 & \text{if} \quad \dfrac{\overline{Y}}{\sqrt{\frac{25}{n}}} \le \left(1.28 + n^{1/4}\right) \\ H_1 & \text{if} \quad \dfrac{\overline{Y}}{\sqrt{\frac{25}{n}}} > \left(1.28 + n^{1/4}\right) \end{cases} .$$

The power function for this test rule is

$$\pi(\theta) = 1 - \Phi\left(\left(1.28 + n^{1/4}\right) - \frac{\theta}{\sqrt{\frac{25}{n}}}\right).$$

1. *Show that the size of the test rule depends on n, and discuss that happens with the size of the test rule for n going to infinity.*

   (Hint, rewrite the last term in the power function as

   $$\frac{\theta}{\sqrt{\frac{25}{n}}} = \frac{\theta}{\sqrt{25}} \cdot n^{1/2},$$

   and notice that $n^{1/2}$ converges faster to infinity than $n^{1/4}$ for $n \to \infty$.)

2. *Show that the sequence of test rules, defined by the test rule as a function of n, is consistent.*

Problem 15.10.3 *Re-write the loss function in Equation (15.5) using indicator functions, i.e. to be of the form $L\left(\delta\left(\widetilde{y}\right),\theta\right) = \sum_i L_i(\theta)I_i(\theta)$, where $L_i()$ are functions and $I_i()$ are indicator functions.*

Problem 15.10.4 *Prove that there is at least one minimax test rule in a class of test rules with a finite number of test rules.*

Problem 15.10.5 *Let $\mathcal{C}$ be a class of test rules, and assume that there are exactly $N \in \mathbb{N}$ minimax test rules $\delta_1^*,\ldots,\delta_N^*$ in $\mathcal{C}$. Prove that at least one of these is admissible.*

16

Test of hypotheses

16.1 Introduction

In the previous chapter, we examined how to formulate hypotheses and ways to form decision rules used to choose between them. In this chapter, we consider how to use a random sample to propose so-called test statistics, which are sample statistics used for constructing decision rules for choosing between hypotheses. A test statistics and an associated decision rule is called a *hypothesis test*.

We begin by outlining a generic approach to constructing a hypothesis test. Then we delve into two specific examples of often-used hypothesis tests, namely the *t-test* and the *likelihood ratio test*. We discuss how to implement the t-test for combinations of simple and composite hypotheses, and show how the implementation depends on whether, or not, a risk function is known. We also discuss how to construct an approximate t-test, either using bootstrap or asymptotic methods, in cases where the distribution of the test statistics is unknown. Lastly, we study the likelihood ratio test for simple hypotheses, which turns out to possess certain optimal properties, in the sense that it is the test with the highest possible power for a given significance level.

16.2 Generic approach to hypothesis testing

In this section, we outline the steps in the construction of a *test* of two hypotheses, H_0 and H_1, given by

$$H_0 : \theta \in \Theta_{H_0}, \tag{16.1}$$

$$H_1 : \theta \in \Theta_{H_1}, \tag{16.2}$$

where $\Theta_{H_0}, \Theta_{H_1}$ are subsets of $\mathbb{R}$ such that $\Theta_{H_0} \cap \Theta_{H_1} = \emptyset$ and $\Theta_{H_0} \cup \Theta_1 = \Theta$, Θ being the parameter set under consideration for the parameter θ.

We consider hypothesis tests, which are constructed from so-called *test statistics*. A test statistics is a sample statistics (Definition 11.1) used in the context of choosing between hypotheses. The definition is next.

Definition 16.1 (Test statistics) *A random variable T is a **test statistics** if it is a sample statistics, $T : \Omega \to \mathbb{R}$, where Ω is the sample space, and it is used for construction of a decision rule for hypotheses.*

Assume a random sample $\widetilde{Y}$ is available. The construction of a test of the two hypotheses (16.1)–(16.2) may proceed via the following general steps.

DOI: 10.1201/9781003591191-16

1. Formulate the hypotheses H_0 and H_1 by specifying Θ_{H_0} and Θ_{H_1}.

2. From the random sample $\widetilde{Y}$, propose a test statistics T with a distribution that depends on the true value of $\theta \in \Theta$. In particular, the distribution of T should be different *under* H_0, i.e. when $\theta \in \Theta_{H_0}$, compared to the distribution of T *under* H_1, i.e. when $\theta \in \Theta_{H_1}$.

3. Derive the distribution of T as a function of the true value of $\theta \in \Theta$. If the distribution of T is unknown, it might be possible to estimate the distribution, e.g. using bootstrap or asymptotic methods.

4. Use the test statistics in step 2 to propose a class of decision rules δ_α, indexed by their significance level $\alpha \in (0, 1)$. Ideally, all the decision rules in the class should be admissible (Definition 15.10) and consistent (Definition 15.7).

5. The method of selecting a particular decision rule depends on whether or not the risk function is known.

    (a) Known risk function: Pick a decision rule δ_α with desirable risk properties, e.g. using a minimax decision rule.

    (b) Unknown risk function: Choose a significance level $\alpha \in (0, 1)$ and consider the corresponding decision rule δ_α. Derive the power function $\pi(\theta)$ and the associated probability of making a Type II error $\beta(\theta) = \Pr\left(Type\ II\ error\ ;\ \theta\right) = 1 - \pi(\theta)$ for $\theta \in \Theta_{H_1}$. Try to assess whether the relative probabilities of making a Type I and Type II error are reasonable, see, e.g. Example 15.24. If they are not reasonable, choose a new value of the significance level α and redo Step 5b.

6. Conduct the test by using the realized sample $\widetilde{y}$ to calculate the realized test statistic t and apply the decision rule chosen in Step 5.

In case of an unknown risk function, the test is designed such that the probability of making a Type I error is bounded by the number $\alpha \in (0, 1)$, chosen by the researcher. Lowering α will, in general, result in a higher probability of making a Type II error, $\beta(\theta)$. Hence, the optimal choice of significance level α should be guided by the associated probability of making a Type II error, $\beta(\theta)$, and of the possible losses incurred from making these errors.

In the following sections, we give specific examples of how to construct hypothesis tests, beginning with the so-called t-test. We first consider the case of two simple hypotheses.

16.3   t-test of simple hypotheses

The so-called t-tests are an approach to the construction of a decision rule based on an estimator of the parameter involved in the hypotheses. In this section, we consider the approach in the case where both the null and the alternative hypotheses are simple. Assume Y is known to be distributed according to $f(y; \theta)$, where θ is an unknown parameter. Let the null and the alternative hypotheses be

$$
\begin{aligned}
H_0 &: \quad \theta = \theta_{H_0}, \\
H_1 &: \quad \theta = \theta_{H_1},
\end{aligned}
$$

with $\theta_{H_0}, \theta_{H_1} \in \mathbb{R}$. Given f is known and only depends on θ, the two hypotheses are simple.

To test the hypotheses, we will define a decision rule using a so-called *t-test statistics*. A t-test statistics has a form similar to a standardized random variable. To do standardization, the mean and the variance of the random variable is needed. The t-test statistics is defined as a standardized version of the estimator $\widehat{\theta}$ of θ, where the standardization is made as if H_0 is true.

Definition 16.2 (t-test statistics) *Let $\widehat{\theta}$ be an estimator of θ and $\widehat{Var}\left(\widehat{\theta}\right)$ be an esti-mator of $Var(\widehat{\theta})$. Assume $\theta_{H_0} \in \Theta_{H_0}$. Then the **t-test statistics** t_{test} is defined as*

$$t_{test} = \frac{\widehat{\theta} - \theta_{H_0}}{\sqrt{\widehat{Var}\left(\widehat{\theta}\right)}}. \tag{16.3}$$

For any test statistics to distinguish between two hypotheses, it must have different properties depending on which hypothesis is true. For the t-test statistics, if the null hypothesis is true, i.e. $\theta = \theta_{H_0}$, then $\left(\widehat{\theta} - \theta_{H_0}\right)$ will approach 0 for large n, as long as $\widehat{\theta}$ is a consistent estimator of θ. This is in contrast to the situation where the alternative hypothesis is true, i.e. $\theta = \theta_{H_1}$. Then $\left(\widehat{\theta} - \theta_{H_0}\right)$ will approach $(\theta_{H_1} - \theta_{H_0}) \neq 0$ for large n, as long as $\widehat{\theta}$ is a consistent estimator of θ. Thus, the numerator in (16.3) reveals whether H_0 or H_1 is true, at least for large n.

To construct a decision rule based on the t-test statistics, and to assess the abilities of the t-test statistics to distinguish between hypotheses, the distribution of the t-test statistics is needed. The following example derives the distribution of the t-test statistics in the case of a normally distributed population with known variance.

Example 16.1 *Let the population be $Y \sim N\left(\mu, \sigma^2\right)$ with σ^2 known. Suppose the hypotheses under consideration are*

$$\begin{aligned} H_0 &: \quad \mu = 0, \\ H_1 &: \quad \mu = 1. \end{aligned}$$

Assume a simple random sample $\widetilde{Y}$ of size n is available and let the estimator of μ be the sample average $\overline{Y}$. In this case, we know that $\mu = E(\overline{Y})$ and $Var(\overline{Y}) = \sigma^2/n$, which implies

$$\frac{\overline{Y} - \mu}{\sqrt{\frac{\sigma^2}{n}}} \sim N\left(0, 1\right).$$

Therefore, by the properties of the normal distribution, we have that

$$t_{test} = \frac{\overline{Y} - 0}{\sqrt{\frac{\sigma^2}{n}}} = \frac{\overline{Y} - \mu + \mu}{\sqrt{\frac{\sigma^2}{n}}} = \frac{\overline{Y} - \mu}{\sqrt{\frac{\sigma^2}{n}}} + \frac{\mu}{\sqrt{\frac{\sigma^2}{n}}} \sim N\left(\frac{\mu}{\sqrt{\frac{\sigma^2}{n}}}, 1\right).$$

Under the null hypothesis where $\mu = 0$, then $t_{test} \sim N\left(0, 1\right)$. Under the alternative hypothesis where $\mu = 1$, then

$$t_{test} \sim N\left(\frac{1}{\sqrt{\frac{\sigma^2}{n}}}, 1\right) = N\left(\sqrt{n}\frac{1}{\sqrt{\sigma^2}}, 1\right). \tag{16.4}$$

Compared to the normal distribution of t_{test} under H_0, the distribution of the t_{test} under H_1 is also normal, with the same variance but a different mean.

The sample size influences the distribution of the t_{test} when H_1 is true. The higher the sample size, the further away the mean of the t_{test} is from 0, which is the mean of the t_{test} under H_0. Figure 16.1 illustrates this by showing the PDF of the t_{test} when H_0 is true (solid line), when H_1 is true and the sample size is $n = 100$ and population variance $\sigma^2 = 50$ (dashed and dotted line), and when H_1 is true and the sample size is $n = 1000$ and population variance $\sigma^2 = 50$ (dashed line). We see how the test statistics t_{test} has a distribution centered around zero when H_0 is true, while its distribution is shifted to the right when H_1 is true. The shift becomes more pronounced as the sample size increases. In fact, as seen in (16.4), when H_1 is true, then $plim_{n\to\infty}\, t_{test} = \infty$.

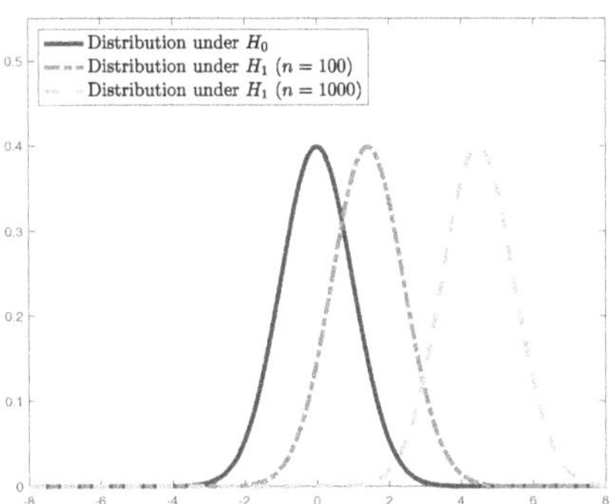

FIGURE 16.1
PDFs of the test statistics t_{test} from Example 16.1 when H_0 is true (solid line), when H_1 is true and the sample size is $n = 100$ and population variance $\sigma^2 = 50$ (dashed and dotted line), and when H_1 is true and the sample size is $n = 1000$ and population variance $\sigma^2 = 50$ (dashed line).

The observations in Example 16.1 suggest a way to design a decision rule to choose between H_0 and H_1 based on a realization $\widetilde{y}$ of the random sample $\widetilde{Y}$. First, use the realized sample $\widetilde{y}$ to calculate $\widehat{\theta}$ and $\widehat{Var}\left(\widehat{\theta}\right)$, and then use these and the value θ_{H_0} to calculate the realized test statistics t_{test} in (16.3). Intuitively, if the realized outcome of the test statistics t_{test} is very large, this suggests that it is the outcome of a distribution with a large mean, i.e. the distribution under H_1, confer Figure 16.1. Conversely, if the realized outcome of t_{test} is close to zero, this suggests that it is the outcome of a distribution with a mean close to zero, i.e. the distribution under the null hypothesis H_0. Thus, we will select H_1 in the former case and H_0 in the latter. The next definition presents a decision rule, formalizing these considerations.

Definition 16.3 (t-test of simple hypotheses) *Assume the two simple hypotheses $H_0 : \theta = \theta_{H_0}$ and $H_1 : \theta = \theta_{H_1}$. Let t_{test} be the t-test statistics (16.3) and $\tau \in \mathbb{R}$ a number. Then the **t-test of simple hypotheses** is defined by the decision rule $\delta_\tau()$ as follows.*

1. If $\theta_{H_1} < \theta_{H_0}$, then

$$\delta_\tau\left(\widetilde{Y}\right) = \begin{cases} H_0 & \text{if } t_{test} \geq \tau, \\ H_1 & \text{if } t_{test} < \tau. \end{cases}$$

2. If $\theta_{H_1} > \theta_{H_0}$, then

$$\delta_\tau\left(\widetilde{Y}\right) = \begin{cases} H_0 & \text{if } t_{test} \leq \tau, \\ H_1 & \text{if } t_{test} > \tau. \end{cases}$$

Thus, when $\theta_{H_1} > \theta_{H_0}$, the t-test chooses H_1 over H_0 when the value of the test statistics t_{test} is "large", and chooses H_0 over H_1 when this value is "small". The opposite decision rule is applied in the case where $\theta_{H_1} < \theta_{H_0}$. As can be seen from Definition 16.3, each value of τ describes a different decision rule, such that $\mathcal{C} = \{\delta_\tau : \tau \in \mathbb{R}\}$ defines a class of decision rules.

The number τ in Definition 16.3 is set by the researcher and denotes the threshold at which H_0 is chosen over H_1. A threshold that divides the values of a test statistics into regions leading to different decisions, is also called a *critical value* of a test. The set of values where H_0 is rejected, i.e. where H_1 is selected, is similarly called the *critical region* (or *rejection region*). For instance, in the case of $\theta_{H_1} > \theta_{H_0}$, the critical region for the test is the set (τ, ∞). Conversely, the set of values for which H_0 is not rejected is often called the *acceptance region* (or *non-rejection region*). Thus, the acceptance region is the complement of the critical region. For instance, in the case of $\theta_{H_1} > \theta_{H_0}$, the acceptance region for the test is the set $(\tau, \infty)^c = (-\infty, \tau]$.

The following two subsections discuss how the threshold τ might be chosen. If the loss function of the problem is known, we may set τ to a value such as to obtain a test that is optimal in some sense, confer the discussion in Section 15.6. This approach is illustrated in Section 16.3.1 where we show how to construct a minimax optimal test. Conversely, if the loss function of the problem is not known, τ needs to be chosen in a different manner. In Section 16.3.2, we discuss how this may be done by focusing on the Type I error of the test, confer the discussion in Section 15.7.

16.3.1 Known loss function

When the loss function is known, we may choose τ by considering the risk function $Risk_{\delta_\tau}(\theta)$, see Section 15.6. There are a continuum of t-tests, namely, one for each value of $\tau \in \mathbb{R}$, see Definition 16.3. The next theorem shows which one of the t-tests to select in case of a minimax risk function.

Theorem 16.1 (Minimax optimal t-test with known loss function) *Consider the simple hypotheses $H_0 : \theta = \theta_{H_0}$ and $H_1 : \theta = \theta_{H_1}$ and the t-test decision rule δ_τ of Definition 16.3. Assume L_{01} is the loss of making a Type I error and L_{10} is the loss of making a Type II error. Let $\pi_{\delta_\tau}(\theta)$ be the power function of the t-test δ_τ. Using the minimax criterion, the optimal t-test δ_{τ^*} is obtained using the threshold τ^* solving*

$$\frac{1 - \pi_{\delta_{\tau^*}}(\theta_{H_1})}{\pi_{\delta_{\tau^*}}(\theta_{H_0})} = \frac{L_{01}}{L_{10}}. \tag{16.5}$$

Example 16.2 *Let the population be $Y \sim N(\mu, \sigma^2)$ where σ^2 is known and consider the hypotheses*

$$
\begin{aligned}
H_0 &: \quad \mu = \mu_{H_0}, \\
H_1 &: \quad \mu = \mu_{H_1},
\end{aligned}
$$

where $\mu_{H_1} > \mu_{H_0}$. Suppose we have a simple random sample $\widetilde{Y}$ of size n and consider the t-test with decision rule as given in Definition 16.3. Using similar arguments as those in Example 16.1, we have that

$$t_{test} = \frac{\overline{Y} - \mu_{H_0}}{\sqrt{\frac{\sigma^2}{n}}} \sim N\left(\frac{\mu - \mu_{H_0}}{\sqrt{\frac{\sigma^2}{n}}}, 1\right).$$

Using the properties of the normal distribution, the power function is thus

$$\pi_{\delta_\tau}(\mu) = \Pr\left(t_{test} > \tau \ ; \ \mu\right) = 1 - \Pr\left(t_{test} \le \tau \ ; \ \mu\right) = 1 - \Phi\left(\tau - \frac{\mu - \mu_{H_0}}{\sqrt{\frac{\sigma^2}{n}}}\right).$$

Insert this into (16.5) to get

$$\frac{L_{01}}{L_{10}} = \frac{\Phi\left(\tau^* - \frac{\mu_{H_1} - \mu_{H_0}}{\sqrt{\frac{\sigma^2}{n}}}\right)}{1 - \Phi(\tau^*)} = \frac{\Phi(\tau^* - \lambda)}{1 - \Phi(\tau^*)}, \tag{16.6}$$

where

$$\lambda = \frac{\mu_{H_1} - \mu_{H_0}}{\sqrt{\frac{\sigma^2}{n}}}.$$

To solve (16.6) for τ^*, a numerical search is needed. There is however, a unique solution. The reason is that the right hand side of (16.6) is monotonically increasing in τ^* from 0 (for $\tau^* = -\infty$) to ∞ (for $\tau^* = \infty$).

As an example, suppose the sample size is $n = 100$ and the variance is $\sigma^2 = 50$, while $\mu_{H_0} = 0$ and $\mu_{H_1} = 1$. Then $\lambda = 1.41$. The next table shows different values of ratio of the Type I error loss relative to the Type II error loss, and the corresponding solution for τ^*, found by a numerical search.

$\frac{L_{01}}{L_{10}}$	τ^*	$\frac{L_{01}}{L_{10}}$	τ^*
0.01	−0.98	2.00	0.97
0.05	−0.42	5.00	1.32
0.10	−0.17	10.00	1.58
0.20	0.09	20.00	1.84
0.50	0.44	100.00	2.39
1.00	0.71	1000.00	3.10

In the table, it can be seen that if the two losses are equal ($L_{01}/L_{10} = 1.00$), then the optimal value of τ, in the minimax sense, is $\tau^* = 0.71$. It can also been seen that if we were to use a $\tau = 1.84$, then that τ would be optimal only if the loss of a Type I error is 20 times larger than the loss of a Type II error.

Example 16.2 shows how the relative losses and the optimal threshold τ vary together. The optimal τ would be different if λ was different, e.g. if the sample size or the variance were different. Likewise, if the population distribution was different, i.e. non-normal, then the optimal τ will, in general, be different as well.

16.3.2 Unknown loss function

When the loss function is unknown, we follow the approach in Section 15.7 and put initial focus on the Type I error. To be precise, we choose an $\alpha \in (0, 1)$, and seek a decision rule δ_τ, as given in Definition 16.3, such that α is the significance level for the decision rule (Definition 15.12). Since Θ_{H_0} consists of only one element, namely θ_{H_0}, this can be obtained by choosing τ such that

$$\alpha \ge \Pr\left(Type\ I\ error \ ; \ \theta_{H_0}\right) = \Pr\left(\delta_\tau(\widetilde{Y}) = H_1 \ ; \ \theta_{H_0}\right).$$

The calculation of this probability depends on whether $\theta_{H_1} < \theta_{H_0}$ or $\theta_{H_1} > \theta_{H_0}$, confer Definition 16.3. In the former case, $\theta_{H_1} < \theta_{H_0}$, we have

$$\alpha \geq \Pr\left(Type\ I\ error\ ;\ \theta_{H_0}\right) = \Pr\left(t_{test} \leq \tau\ ;\ \theta_{H_0}\right).$$

If the distribution of the test statistics t_{test} is continuous, as will often be the case, we can choose τ such that this relation is satisfied with an equality instead of an inequality, i.e. such that

$$\alpha = \Pr\left(Type\ I\ error\ ;\ \theta_{H_0}\right) = \Pr\left(t_{test} \leq \tau\ ;\ \theta_{H_0}\right).$$

In this case, the solution is to set τ equal to the α-quantile of the distribution of t_{test} under H_0, i.e. the quantile is to be calculated assuming that $\theta = \theta_{H_0}$. This quantile will, in general, depend on the distribution of the random sample $\widetilde{Y}$. In the latter case, $\theta_{H_1} > \theta_{H_0}$, we have

$$\alpha \geq \Pr\left(Type\ I\ error\ ;\ \theta_{H_0}\right) = \Pr\left(t_{test} > \tau\ ;\ \theta_{H_0}\right) = 1 - \Pr\left(t_{test} \leq \tau\ ;\ \theta_{H_0}\right).$$

Analogous to above, when the distribution of t_{test} is continuous, the solution is to set τ equal to the $(1 - \alpha)$-quantile of the distribution of t_{test} under H_0, i.e. the quantile is to be calculated assuming that $\theta = \theta_{H_0}$. This discussion is summarized in the following theorem.

Theorem 16.2 (Constructing a t-test with specific significance level)   *Let* $\alpha \in (0,1)$ *be a number and* Θ *a parameter space. Consider the hypotheses*

$$\begin{aligned} H_0 &:\quad \theta = \theta_{H_0}, \\ H_1 &:\quad \theta = \theta_{H_1}, \end{aligned}$$

for some $\theta_{H_0}, \theta_{H_1} \in \Theta$ *such that* $\theta_{H_0} \neq \theta_{H_1}$. *Suppose that the t-test statistics* t_{test} *in* (16.2) *is a continuous random variable and let* $\delta_\tau()$ *be as in Definition 16.3, where* τ *is given as follows.*

1. If $\theta_{H_1} < \theta_{H_0}$, *then*

$$\tau = q_\alpha,$$

   where q_α *denotes the* α-quantile of t_{test}, *calculated under the assumption that* $\theta = \theta_{H_0}$.

2. If $\theta_{H_1} > \theta_{H_0}$, *then*

$$\tau = q_{1-\alpha},$$

   where $q_{1-\alpha}$ *denotes the* $(1 - \alpha)$-quantile of t_{test}, *calculated under the assumption that* $\theta = \theta_{H_0}$.

Then,

$$\Pr\left(Type\ I\ error\ ;\ \theta_{H_0}\right) = \Pr\left(\delta_\tau\left(\widetilde{Y}\right) = H_1\ ;\ \theta_{H_0}\right) = \alpha,$$

i.e. α *is the smallest significance level for the decision rule* $\delta_\tau()$.

   Further, if $\widehat{\theta}$ *is a consistent estimator of* θ *and* $\widehat{Var}(\widehat{\theta})$ *is such that* $\lim_{n\to} \frac{\widehat{Var}(\widehat{\theta})}{Var(\widehat{\theta})} = 1$, *then*

$$\lim_{n\to\infty} \Pr\left(\delta_\tau\left(\widetilde{Y}\right) = H_1\ ;\ \theta_{H_1}\right) = 1,$$

i.e. $\delta_\tau()$ *is a consistent decision rule.*

We note that, if the distribution of t_{test} is symmetric around zero under H_0, then the α-quantile is equal to minus the $(1-\alpha)$-quantile, i.e. $q_\alpha = -q_{1-\alpha}$ in Theorem 16.2. Further, when the distribution of t_{test} is continuous, then each value of α will correspond to exactly one value of τ. Therefore, we can either think of the decision rules in Theorem 16.2 as being indexed by τ or, equivalently, as being indexed by the significance level α.

On a technical note, the reason for considering the limit of the ratio $\frac{\widehat{Var}(\theta)}{Var(\theta)}$ in the latter part of Theorem 16.2, and not just $\widehat{Var}(\theta)$, is that for most consistent estimators $\widehat{\theta}$ of θ, $Var(\widehat{\theta})$ converges to 0 as the sample size increases. The condition of the statement ensures that the estimator $\widehat{Var}(\theta)$ of $Var(\widehat{\theta})$ is consistent in the sense that if $nVar(\widehat{\theta})$ converges to a non-zero constant V, which is the case if the estimator $\widehat{\theta}$ obeys a Central Limit Theorem (e.g. Theorem 13.1), then $n \cdot \widehat{Var}(\widehat{\theta})$ is a consistent estimator of V. We note that V is called the *asymptotic variance* of the estimator $\widehat{\theta}$. In Chapter 18, we will study the asymptotic variance of estimators in more depth, in the context of maximum likelihood estimation.

The following example illustrates how to use Theorem 16.2 to determine the threshold τ, such that the t-test has significance level α in the case where the random sample $\widetilde{Y}$ leads to a normally distributed t-test statistics.

Example 16.3 (Example 16.1, continued) *Let $\alpha \in (0,1)$. We wish to select τ such that the decision rule of Definition 16.3 has significance level α. In Example 16.1, we saw that $t_{test} \sim N(0,1)$ under H_0. Since $\theta_{H_1} = 1 > 0 = \theta_{H_0}$, Theorem 16.2 implies that we should set*

$$\tau = z_{1-\alpha},$$

where $z_{1-\alpha} = \Phi^{-1}(1-\alpha)$ is the $(1-\alpha)$-quantile of the standard normal distribution. That is, the decision rule is to select H_1 if the value of the test statistics t_{test} is sufficiently large, where by "sufficiently large" we mean that $t_{test} > z_{1-\alpha}$. Conversely, we select H_0 if t_{test} is sufficiently small, where by "sufficiently small" we mean that $t_{test} \leq z_{1-\alpha}$. This will ensure that the resulting test has significance level α.

For example, with a significance level $\alpha = 0.10$, the threshold value $\tau = z_{0.90} = \Phi^{-1}(0.90) = 1.28$ is the 0.90-quantile of the standard normal distribution. Thus, with $\alpha = 0.10$, the decision rule is

$$\delta_\tau\left(\widetilde{Y}\right) = \begin{cases} H_0 & \text{if } t_{test} \leq 1.28, \\ H_1 & \text{if } t_{test} > 1.28. \end{cases}$$

This decision rule has a probability of a Type I error equal to the significance level $\alpha = 0.10$. It was the decision rule used throughout the previous chapter, see, e.g. Example 15.9.

Notice that τ does not depend on θ_{H_0}, σ^2, nor n. This is a consequence of the "standardization" used in the definition of the t-test statistics (16.3) and it is the main reason that we define the t-test statistics in this way.

Once the threshold value τ has been set to ensure that α is the significance level of the test, we may calculate the probability of committing a Type II error when $\theta = \theta_{H_1}$:

$$\beta(\theta_{H_1}) = \Pr\left(\textit{Type II error} \; ; \; \theta_{H_1}\right) = \Pr\left(\delta_\tau\left(\widetilde{Y}\right) = H_0 \; ; \; \theta_{H_1}\right).$$

Example 16.4 (Example 16.3, continued) *The distribution of the t-test statistics under H_1, where the true mean is $\mu = 1$, is*

$$t_{test} = \frac{\overline{Y} - 0}{\sqrt{\frac{\sigma^2}{n}}} \sim N\left(\sqrt{n}\frac{1}{\sqrt{\sigma^2}} \; , \; 1\right).$$

Using the properties of the normal distribution, we find the probability of committing a Type II error to be

$$\beta(\mu_{H_1}) = \Pr\left(t_{test} \le \tau \ ; \ \mu_{H_1}\right) = \Phi\left(\tau - \frac{\sqrt{n}}{\sqrt{\sigma^2}}\right).$$

Note that, as also shown in Theorem 16.2, this implies that the decision rule is consistent. Indeed,

$$\begin{aligned}
\lim_{n\to\infty} \Pr\left(\delta_\tau\left(\tilde{Y}\right) = H_1 \ ; \ \mu_{H_1}\right) = \lim_{n\to\infty} \Pr\left(t_{test} > \tau \ ; \ \mu_{H_1}\right) &= \lim_{n\to\infty}\left(1 - \beta(\mu_{H_1})\right) \\
&= 1 - \Phi(-\infty) \\
&= 1.
\end{aligned}$$

As a numerical example, suppose the sample size is $n = 100$ and the population variance is $\sigma^2 = 50$. This implies that the distribution of the t_{test} is

$$t_{test} \sim N\left(\frac{1}{\sqrt{50}}\sqrt{100}, 1\right) = N\left(1.41 \ , \ 1\right).$$

The PDF of this distribution is shown as the dotted curve in Figure 16.1 above. For a significance level of $\alpha = 0.05$, the threshold in the t-test is $\tau = z_{0.95} = 1.64$. This implies that the probability of a Type II error when $\mu = \mu_{H_1}$ is

$$\begin{aligned}
\Pr\left(Type\ II\ error\ ; \ \mu_{H_1}\right) &= \Phi\left(\tau - \frac{\sqrt{n}}{\sqrt{\sigma^2}}\right) \\
&= \Phi\left(1.64 - 1.41\right) \\
&= \Phi(0.23) \\
&= 0.59.
\end{aligned}$$

In terms of the power function, $\pi(\mu_{H_1}) = \pi(1) = 1 - \Pr\left(Type\ II\ error\ ; \ \mu_{H_1}\right) = 0.41$.

After having set the significance level α, and thus the probability of making a Type I error, and having calculated the power, and thus the probability of making a Type II error, we can take a step back and compare these probabilities. Based on our risk attitude and the perceived (relative) losses of making a Type I and a Type II error, we can re-evaluate whether the decision rule implied by α is adequate for our purpose. If the decision rule appears adequate in this light, we may go ahead and apply the decision rule in practice. If, however, the decision rule appears inadequate, we may examine a different decision rule by selecting a different significance level α and then comparing the resulting probabilities of making Type I and Type II errors. This process may be continued until the decision rule is deemed adequate for the problem at hand.

Example 16.5 (Example 16.4, continued) *In Example 16.4, a 5% significance level is used and that resulted in a critical value $\tau = 1.64$ and power $\pi(\mu_{H_1}) = 0.41$. That is, the probability of a Type I error is 5%, while the probability of a Type II error is much larger, namely 59%. Consider a minimax risk function. For the optimal τ^* to equal 1.64, the loss of a Type I error, relative to a the loss of a Type II error, would have to be (Theorem 16.1)*

$$\frac{1 - \pi_{\delta_{\tau^*}}(\mu_{H_1})}{\pi_{\delta_{\tau^*}}(\mu_{H_0})} = \frac{1 - 0.41}{0.05} \approx 12,$$

confer Equation (16.6). That is, the loss of the Type I error will have to be 12 times larger than the loss of a Type II error for $\tau = 1.64$ to be optimal.

Suppose that we judge that the relative loss of a Type I error is smaller than 12, compared to the loss of a Type II error. Then we may re-evaluate and specify a different decision rule, i.e. specify a different significance level. To this end, let now $\alpha = 10\%$. Then $\tau = z_{1-\alpha} = z_{0.90} = \Phi^{-1}(0.90) = 1.28$. Using calculations similar to those in Example 16.4, we find that

$$
\begin{aligned}
\Pr\left(\textit{Type II error} \; ; \; \mu_{H_1}\right) &= \Phi\left(\tau - \frac{\sqrt{n}}{\sqrt{\sigma^2}}\right) \\
&= \Phi\left(1.28 - 1.41\right) \\
&= \Phi(-0.13) \\
&= 0.45.
\end{aligned}
$$

In terms of the power function, $\pi(\mu_{H_1}) = \pi(1) = 1 - \Pr\left(\textit{Type II error} \; ; \; \mu_{H_1}\right) = 0.55$. For the optimal τ^* to equal 1.28, the loss of a Type I error, relative to a the loss of a Type II error, would have to be (Theorem 16.1)

$$
\frac{1 - \pi_{\delta_{\tau^*}}(\mu_{H_1})}{\pi_{\delta_{\tau^*}}(\mu_{H_0})} = \frac{1 - 0.55}{0.05} = 9.
$$

If a relative loss between making a Type I error and a Type II error of 9 is deemed acceptable, then we may proceed with using the decision rule δ_τ implied by $\tau = 1.28$. If a relative loss of 9 is still deemed too large, then α may be set higher than 10%, e.g. $\alpha = 20\%$, and the procedure repeated until an α is found that implies a reasonable relative loss between the two types of errors.

We note that, unfortunately, it is in practice often the case that a hypothesis test is performed without consideration to the probability of making a Type II error and the potential losses incurred. Indeed, it has become convention to simply set α to some, rather arbitrary, value such as 1% or 10%, but most often 5%, and then conduct the test essentially ignoring the probability of making a Type II error. This is a simple approach that is easy to implement, which may account for its ubiquity. However, it, of course, ignores the impact that a choice of hypothesis might have in the real world.

16.4 t-test and nuisance parameters

In this section, we extend the test of two simple hypotheses to the case where there is a nuisance parameter present. We illustrate this situation using a setup where $Y \sim \left(\mu, \sigma^2\right)$ and the hypotheses are

$$
\begin{aligned}
H_0 &: \quad \mu = \mu_{H_0}, \\
H_1 &: \quad \mu = \mu_{H_1}.
\end{aligned}
$$

The difference to Section 16.3 is the presence of the parameter σ^2, the value of which is unknown but unrestricted by the hypotheses. Thus, σ^2 is a nuisance parameter in the sense of Definition 15.2.

We shall use the sample average $\overline{Y}$ as the estimator of μ and the sample variance S^2 as the estimator of σ^2. With these estimators, the distribution of the t-test statistics is given in the next theorem.

Theorem 16.3 (t-test statistics distribution with normal population and unknown variance) *Assume the population is $Y \sim N\left(\mu, \sigma^2\right)$. Let the estimator of μ be the sample average $\overline{Y}$ given by*

$$\overline{Y} = \frac{1}{n} \sum_{i=1}^{n} \widetilde{Y}_i,$$

and let the estimator of σ^2 be the sample variance given by

$$S^2 = \frac{1}{n-1} \sum_{i=1}^{n} \left(\widetilde{Y}_i - \overline{Y}\right)^2.$$

With a simple random sample, the distribution of the t-test statistics (16.3) is

$$t_{test} \sim t(n-1, \lambda),$$

where $t(n-1, \lambda)$ is the non-central t-distribution (Section 8.6.2) with $n-1$ degrees of freedom and non-centrality parameter

$$\lambda = \frac{\mu - \mu_0}{\sqrt{\sigma^2/n}}.$$

Under the null hypothesis $\mu = \mu_{H_0}$, the non-centrality parameter is $\lambda = 0$, and, therefore, the t_{test} is t-distributed with $n-1$ degrees of freedom.

The t-distribution with many degrees of freedom is quite similar to the standard normal $N(0,1)$ distribution. Likewise, the non-central t-distribution with many degrees of freedom and non-centrality parameter λ is quite similar to the normal distribution $N(\lambda, 1)$. Hence, when the sample size n is large, the non-central t-distribution with $(n-1)$ degrees of freedom and non-centrality parameter λ may be accurately approximated by the normal distribution $N(\lambda, 1)$.

Example 16.6 *Let the population be represented by $Y \sim N\left(\mu, \sigma^2\right)$ with σ^2 unknown. Suppose the hypotheses under consideration are*

$$
\begin{aligned}
H_0 &: \quad \mu = 0, \\
H_1 &: \quad \mu = 1.
\end{aligned}
$$

Assume a simple random sample $\widetilde{Y}$ of size n is available, let the estimator of μ be the sample average $\overline{Y}$, and the estimator of σ^2 be the sample variance S^2. This setting is similar to the one in Example 16.1, except that here the variance is assumed to be unknown. Under H_0, i.e. assuming that $\mu = \mu_{H_0} = 0$, then

$$t_{test} = \frac{\overline{Y} - 0}{\sqrt{\dfrac{S^2}{n}}} \sim t(n-1).$$

If the loss function is unknown, we focus initially on the Type I error. Let $\alpha \in (0,1)$. We wish to find the threshold τ such that α is the significance level of the test when using the decision rule δ_τ, given in Definition 16.3. According to Theorem 16.2, this is obtained by setting τ equal to the $(1-\alpha)$-quantile of the t-distribution with $n-1$ degrees of freedom. If $\alpha = 0.05$ and $n = 100$, this quantile is 1.66. This number should be compared with the analogous quantile in the case of the normal distribution, which was 1.64, see Example 16.4. Recall that as n increases, the t-distribution with $n-1$ degrees of freedom approaches the normal distribution, so the 0.95 quantile from the t-distribution will become closer and closer to 1.64 as the sample size n increases.

We thus have the decision rule

$$\delta_\tau \left(\widetilde{Y}\right) = \begin{cases} H_0 & \text{if } t_{test} \leq 1.66, \\ H_1 & \text{if } t_{test} > 1.66, \end{cases}$$

which is constructed such that the t-test has significance level $\alpha = 0.05$.

We may also calculate the probability of committing a Type II error. If H_1 *is true, then* $\mu = 1$ *and the non-centrality parameter of Theorem 16.3 is* $\lambda = \frac{\mu - \mu_0}{\sqrt{\sigma^2/n}}$. *Since the variance* σ^2 *is unknown then* λ *is unknown. We can, however, estimate* λ *by plugging in our estimator* S^2 *of* σ^2, *arriving at*

$$\widehat{\lambda} = \frac{\mu - \mu_0}{\sqrt{S^2/n}}.$$

If, for instance, our estimator of the variance is $S^2 = 50$, *then* $\widehat{\lambda} = \frac{1-0}{\sqrt{50/100}} = \sqrt{2} = 1.41$. *Hence, under* H_1, *the* $t_{test} \sim t(n-1, \lambda) \approx t(n-1, \widehat{\lambda}) = t(99, 1.41)$. *The probability of committing a Type II error is, in this case,*

$$\Pr\left(Type\ II\ error\ ;\ \mu_{H_1} = 1\right) = \Pr\left(\delta_\tau\left(\widetilde{Y}\right) = H_0\ ;\ \mu_{H_1} = 1\right) = \Pr\left(t_{test} \leq 1.66\right)$$
$$\approx \Pr\left(t_{approx} \leq 1.66\right),$$

where $t_{approx} \sim t(99, 1.41)$. *The CDF of the non-central t-distribution is highly complicated, so computer software is generally needed to evaluate it. Using such software, we find*

$$\Pr\left(Type\ II\ error\ ;\ \mu_{H_1} = 1\right) \approx \Pr\left(t_{approx} \leq 1.66\right) = 0.5948.$$

Hence, there is approximately a 59.5% probability of committing a Type II error in this case. Had we used the normal distribution in place of the non-central distribution to approximate the probability of a Type II error, then we would have found

$$\Pr\left(Type\ II\ error\ ;\ \mu_{H_1} = 1\right) \approx \Pr\left(Z \leq 1.66\right) = 0.5964,$$

where $Z \sim N(1.41, 1)$, *showing that the result from using the* $t(99, 1.41)$ *distribution is very close to the result from using the* $N(1.41, 1)$ *distribution.*

These probabilities of a Type II error can be compared to the one found in Example 16.4, under the assumption of a known variance. Here, the test statistics is normally distributed, which yielded a threshold value of $\tau = 1.64$ *and a resulting probability of a Type II error equal to 59%. Again, this illustrates that, when* $n = 100$, *basing a hypothesis test on the t distribution will often not be much different than basing it on the normal distribution, at least as long as the significance level* α *is not too small.*

In Example 16.6, the distribution of the test statistics does not depend on the nuisance parameter under the null, i.e. $t_{test} \sim t(n-1)$ under H_0. The distribution of the test statistics might very well depend on the nuisance parameter under the alternative, however. Again, this is the case in Example 16.6, where $t_{test} \sim t(n-1, \lambda)$ under H_1, where the non-centrality parameter λ depends on σ^2, see Theorem 16.3. When the distribution of the test statistic depends on nuisance parameters under the alternative, we need to make assumptions about the value of the nuisance parameter, e.g. by estimating it, to calculate, say, the power of the test. This was done in Example 16.6, where we estimated σ^2 using the sample variance S^2, which allowed us to approximate the power of the test, by plugging in S^2 in place of σ^2 in $\lambda = (\mu - \mu_0)/\sqrt{\sigma^2/n}$.

In the above example, the distribution of the t-test statistics only depends on the nuisance parameter under the alternative. There are many cases where the distribution of the test statistics also depends on a nuisance parameter under the null hypothesis. In these cases, it is important that we describe how we deal with them, that is, what values we use for the nuisance parameters when discussing probabilities of Type I and Type II errors. We will see an example of this when discussing the likelihood ratio test later in Section 16.8.

16.5 Hypotheses with multiple values of the parameter of interest

So far, only one value of the parameter of interest has satisfied a hypothesis whether or not nuisance parameters are present. In this section, we consider the situation where multiple values of the parameter of interest satisfy a hypothesis. This has consequences for how a test may be constructed. Here, we consider how such a construction can be done using the t-test statistics.

16.5.1 Multiple values of the parameter of interest under the alternative hypothesis

We initially consider the case where only one value of the parameter of interest θ is allowed under the null. That is, we suppose the null hypothesis is given as

$$H_0 : \; \theta = \theta_{H_0}.$$

We will consider three different forms of the alternative hypothesis, namely,

$$\begin{aligned}
H_1 &: & \theta < \theta_{H_0}, \\
H_1 &: & \theta > \theta_{H_0}, \\
H_1 &: & \theta \neq \theta_{H_0}.
\end{aligned}$$

The first two are called *one-sided* alternatives, while the latter is called a *two-sided* alternative. The decision rule used for a t-test depends on the form of the alternative hypothesis, as the following definition makes clear.

Definition 16.4 (t-test decision rule of composite hypotheses) *Assume the null hypothesis $H_0 : \theta = \theta_{H_0}$. Let t_{test} be the t-test statistics (16.3) and $\tau, \tau_1, \tau_2 \in \mathbb{R}$ numbers such that $\tau_1 < \tau_2$. Then the **t-test decision rule** $\delta_\tau()$ **of a composite hypothesis** is defined as follows.*

1. If $H_1 : \theta < \theta_{H_0}$, then

$$\delta_\tau\left(\widetilde{Y}\right) = \begin{cases} H_0 & \text{if } \; t_{test} \geq \tau, \\ H_1 & \text{if } \; t_{test} < \tau. \end{cases}$$

2. If $H_1 : \theta > \theta_{H_0}$, then

$$\delta_\tau\left(\widetilde{Y}\right) = \begin{cases} H_0 & \text{if } \; t_{test} \leq \tau, \\ H_1 & \text{if } \; t_{test} > \tau. \end{cases}$$

3. If $H_1 : \theta \neq \theta_{H_0}$, then

$$\delta_{\tau_1, \tau_2}\left(\widetilde{Y}\right) = \begin{cases} H_0 & \text{if} & t_{test} \in [\tau_1, \tau_2], \\ H_1 & \text{otherwise.} \end{cases}$$

As the definition shows, in the case of a one-sided alternative, we reject H_0 in favor of H_1 if the t-test statistics is sufficiently small when $H_1 : \theta < \theta_{H_0}$, or sufficiently large when $H_1 : \theta > \theta_{H_0}$. In the case of the two-sided alternative, we reject H_0 in favor of H_1 if the t-test statistics is either sufficiently large or sufficiently small, i.e. if it is outside the interval $[\tau_1, \tau_2]$.

If the loss function is known, we may choose the thresholds by considering the risk function, see Section 15.6. If the loss function is unknown, we follow the approach in Section 15.7 and put initial focus on the Type I error. In particular, we let $\alpha \in (0, 1)$ and design our decision rule such that it has significance level α. In the one-sided cases, we find the same solution as in Theorem 16.2 above. In the two-sided case, the solution should be altered accordingly. The following theorem contains all three cases.

Theorem 16.4 (Constructing a t-test with specific significance level; Theorem 16.2, continued) *Let $\alpha \in (0, 0.5]$ be a number and Θ a parameter space. Consider the null hypothesis*

$$H_0 : \theta = \theta_{H_0}$$

for some $\theta_{H_0} \in \Theta$. Assume that the t-test statistics t_{test} in (16.3) is a continuous random variable and the decision rules δ_τ and δ_{τ_1, τ_2} are given by Definition 16.4. Let q_p be the p-quantile of the distribution of t_{test} assuming $\theta = \theta_{H_0}$. Then the t-test has significance level α if

1. $H_1 : \theta < \theta_{H_0}$ and $\tau = q_\alpha$.

2. $H_1 : \theta > \theta_{H_0}$ and $\tau = q_{1-\alpha}$.

3. If $H_1 : \theta_{H_1} \neq \theta_{H_0}$, $\tau_1 = q_{\alpha/2}$ and $\tau_2 = q_{1-\alpha/2}$.

Further, if $\widehat{\theta}$ is a consistent estimator of θ and $\widehat{Var}(\widehat{\theta})$ is such that $\lim_{n \to} \frac{\widehat{Var}(\widehat{\theta})}{Var(\widehat{\theta})} = 1$, then the t-test is consistent.

If the distribution of t_{test} under H_0 is symmetric around zero, then $q_{\alpha/2} = -q_{1-\alpha/2}$, implying that the thresholds τ_1 and τ_2 in Theorem 16.4 can be expressed as $\tau_1 = -q_{1-\alpha/2}$ and $\tau_2 = q_{1-\alpha/2}$.

Once the threshold τ, or thresholds τ_1 and τ_2, have been chosen to ensure that the t-test has significance level α, the probability of a Type II error, $\beta(\theta)$, can be calculated similarly as in the previous section, the difference being that $\beta(\theta)$ can be evaluated for all θ relevant for H_1.

Example 16.7 *Let the population be $Y \sim N\left(\mu, \sigma^2\right)$ with σ^2 known. Suppose the hypotheses under consideration are*

$$
\begin{aligned}
H_0 &: \quad \mu = 0, \\
H_1 &: \quad \mu \neq 0.
\end{aligned}
$$

Assume a simple random sample $\widetilde{Y}$ of size n is available and let the estimator of μ be the sample average $\overline{Y}$. As we saw in Example 16.1, we have that

$$t_{test} = \frac{\overline{Y} - 0}{\sqrt{\frac{\sigma^2}{n}}} \sim N\left(\frac{\mu}{\sqrt{\frac{\sigma^2}{n}}}, 1\right).$$

Under the null hypothesis with $\mu = 0$ we have $t_{test} \sim N(0,1)$. Under the alternative hypothesis with $\mu \neq 0$, then

$$t_{test} \sim N\left(\frac{\mu}{\sqrt{\frac{\sigma^2}{n}}}\,,\,1\right) = N\left(\sqrt{n}\,\frac{\mu}{\sqrt{\sigma^2}}\,,\,1\right).$$

Under H_0, $t_{test} \sim N(0,1)$, which is a symmetric distribution around zero. According to Theorem 16.4, to ensure that $\alpha \in (0, 0.5)$ is the significance level associated to the t-test, we may set $\tau_1 = -z_{1-\alpha/2}$ and $\tau_2 = z_{1-\alpha/2}$, where $z_{1-\alpha/2} = \Phi^{-1}(1 - \alpha/2)$ is the $(1 - \alpha/2)$-quantile of the standard normal distribution.

Using the properties of the normal distribution, the probability of committing a Type II error becomes

$$
\begin{aligned}
\beta(\mu) &= \Pr\left(Type\ II\ error\ ;\ \mu\right) \\
&= \Pr\left(\delta_\tau\left(\widetilde{Y}\right) = H_0\ ;\ \mu\right) \\
&= \Pr\left(t_{test} \in [\tau_1, \tau_2]\ ;\ \mu\right) \\
&= \Pr\left(t_{test} \leq \tau_2\ ;\ \mu\right) - \Pr\left(t_{test} \leq \tau_1\ ;\ \mu\right) \\
&= \Phi\left(z_{1-\alpha/2} - \frac{\mu}{\sqrt{\frac{\sigma^2}{n}}}\right) - \Phi\left(-z_{1-\alpha/2} - \frac{\mu}{\sqrt{\frac{\sigma^2}{n}}}\right),
\end{aligned}
$$

for $\mu \neq 0$. The associated power function is $\pi(\mu) = 1 - \beta(\mu)$.

As a numerical example, let $\sigma^2 = 1$ and $\alpha = 0.05$. Then $\tau_2 = z_{1-\alpha/2} = z_{0.975} = \Phi^{-1}(0.975) = 1.96$ is the 0.975-quantile of the standard normal distribution. If a random sample of size $n = 100$ is collected, and it turns out that the realized sample mean is, say, $\overline{y} = 0.17$, then we can calculate the realized test statistic

$$t_{test} = \frac{\overline{y} - 0}{\sqrt{\frac{\sigma^2}{n}}} = \frac{0.17 - 0}{\sqrt{\frac{1}{100}}} = 1.7.$$

Since $t_{test} = 1.7 \in [-1.96, 1.96]$ we cannot reject H_0, i.e. H_0 is selected. The interpretation of this is that even though the sample mean $\overline{y} = 0.17$ is different from the value of μ under H_0, i.e. from $\mu = 0$, then this difference might simply be because of sampling uncertainty. The conclusion is that the data do not give us grounds to reject H_0 when $\alpha = 5\%$ is used as a significance level. Note, however, that if one had used, say, $\alpha = 10\%$ as a significance level, thus accepting a higher probability of committing a Type I error when H_0 is true, the acceptance region for the test would have been $[-q_{1-\alpha/2}, q_{1-\alpha/2}] = [-q_{0.95}, q_{0.95}] = [-1.64, 1.64]$, and in this case $t_{test} = 1.7 \notin [-1.64, 1.64]$. Hence, if the test is performed using an $\alpha = 10\%$ significance level, then H_0 is rejected and H_1 selected.

16.5.2   Multiple values of the parameter of interest under both hypotheses

We now consider the case where also the null hypothesis permits multiple values of the parameter of interest. Let $\theta_{H_0} \in \mathbb{R}$ be a number. We consider one of the following two pairs of hypotheses

$$
\begin{aligned}
H_0 &: \quad \theta \geq \theta_{H_0}, \\
H_1 &: \quad \theta < \theta_{H_0},
\end{aligned}
$$

or

$$H_0 \; : \; \theta \leq \theta_{H_0},$$
$$H_1 \; : \; \theta > \theta_{H_0},$$

and the t-test statistics given in (16.3). As above, we base our decision rule on the alternative hypothesis and hence use the one-sided decision rules given in the Definition 16.4. If the loss function is known, we can choose the threshold τ based on minimizing the risk. If the loss function is unknown, we again begin with a focus on the Type I error. Let $\alpha \in (0,1)$. We wish to select τ such that α is a significance level for the test. Since H_0 is now composite, the parameter space Θ_{H_0} relevant under H_0 contains more than one value, which must be accounted for. Recall the definition of α being a significance level, namely that (Definition 15.12)

$$\mathrm{Pr}\left(\textit{Type I error} \; ; \; \theta\right) = \mathrm{Pr}\left(\delta_\tau\left(\tilde{Y}\right) = H_1 \; ; \; \theta\right) \leq \alpha,$$

for all $\theta \in \Theta_{H_0}$. For instance, in the case where $H_0 : \theta \geq \theta_{H_0}$ and $H_1 : \theta < \theta_{H_0}$, we have

$$\mathrm{Pr}\left(\delta_\tau\left(\tilde{Y}\right) = H_1 \; ; \; \theta\right) = \mathrm{Pr}\left(t_{test} < \tau \; ; \; \theta\right),$$

and hence we must choose τ such that

$$\mathrm{Pr}\left(t_{test} < \tau \; ; \; \theta\right) \leq \alpha,$$

for all $\theta \geq \theta_{H_0}$. Fortunately, this problem has a straight forward solution in most cases. Indeed, it will often be the case that the probability $\mathrm{Pr}\left(t_{test} < \tau \; ; \; \theta\right)$ is a decreasing function of θ, meaning that

$$\mathrm{Pr}\left(t_{test} < \tau \; ; \; \theta\right) \leq \mathrm{Pr}\left(t_{test} < \tau \; ; \; \theta_{H_0}\right), \tag{16.7}$$

for all $\theta \geq \theta_{H_0}$. Intuitively, this is because the closer θ is to the region defined by Θ_{H_1}, the larger will be the probability of choosing H_1, i.e. of committing a Type I error. Indeed, for $\theta \geq \theta_{H_0}$ this probability is made maximally large for $\theta = \theta_{H_0}$. The upshot is that if we can choose τ such that

$$\mathrm{Pr}\left(t_{test} < \tau \; ; \; \theta_{H_0}\right) \leq \alpha,$$

then (16.7) implies that

$$\mathrm{Pr}\left(t_{test} < \tau \; ; \; \theta\right) \leq \alpha,$$

for all $\theta \geq \theta_{H_0}$, i.e. that α is a significance level for the decision rule. In fact, in practice it will often be possible to choose τ such that $\mathrm{Pr}\left(t_{test} < \tau \; ; \; \theta_{H_0}\right) = \alpha$. For instance, when t_{test} is a continuous random variable, this is accomplished by setting τ equal to the α-quantile of the distribution of t_{test}, assuming that $\theta = \theta_{H_0}$

The case $H_0 : \theta \leq \theta_{H_0}$ and $H_1 : \theta > \theta_{H_0}$ works similarly to the above. Here, we have

$$\mathrm{Pr}\left(\textit{Type I error} \; ; \; \theta\right) = \mathrm{Pr}\left(\delta_\tau\left(\tilde{Y}\right) = H_1 \; ; \; \theta\right) = \mathrm{Pr}\left(t_{test} > \tau \; ; \; \theta\right),$$

and it will often be possible to choose τ such that $\alpha = \mathrm{Pr}\left(t_{test} > \tau \; ; \; \theta_{H_0}\right) \geq \mathrm{Pr}\left(t_{test} > \tau \; ; \; \theta\right)$ for all $\theta \leq \theta_{H_0}$. Indeed, when t_{test} is a continuous random variable, this is accomplished by setting τ equal to the $(1 - \alpha)$-quantile of the distribution of the t_{test}, assuming that $\theta = \theta_{H_0}$.

The discussion above shows that also in the case where there are multiple values of θ under both hypotheses, then a consistent hypothesis test with a given significance level can still be obtained by using the thresholds suggested in Theorem 16.2. This is summarized in the following theorem.

Theorem 16.5 (Constructing a t-test with specific significance level)
Let $\alpha \in (0,1)$ be a number and Θ a parameter space. Let $\theta_{H_0} \in \Theta$. Assume that the t-test statistics t_{test} in (16.3) is a continuous random variable. Furthermore, assume that

$$\Pr\left(\frac{\widehat{\theta} - \theta_{H_0}}{\sqrt{\widehat{Var}\left(\widehat{\theta}\right)}} > c \; ; \; \theta\right), \tag{16.8}$$

is increasing in θ for any $c \in \mathbb{R}$, where θ is the true value. Let q_p be the p-quantile of the distribution of the t-test statistics under the assumption that θ_{H_0} is the true value of θ, and let $\delta_\tau()$ be as in Definition 16.3. Then the t-test has significance level α if

1. $H_0 : \theta \geq \theta_{H_0}$, $H_1 : \theta_{H_1} < \theta_{H_0}$, and $\tau = q_\alpha$.

2. $H_0 : \theta \leq \theta_{H_0}$, $H_1 : \theta_{H_1} > \theta_{H_0}$, and $\tau = q_{1-\alpha}$.

Further, if $\widehat{\theta}$ is a consistent estimator of θ and $\widehat{Var}(\widehat{\theta})$ is such that $\lim_{n\to} \frac{\widehat{Var}(\widehat{\theta})}{Var(\widehat{\theta})} = 1$, then the t-test is a consistent test.

In Theorem 16.5, the assumption that the expression (16.8) is increasing in θ means that, if the true value θ is increased, then the distribution of the t-test statistics is shifted to the right.

Example 16.8 *Let the population be $Y \sim N\left(\mu, \sigma^2\right)$ with σ^2 known. Suppose the hypotheses under consideration are*

$$\begin{aligned} H_0 &: \quad \mu \leq 0, \\ H_1 &: \quad \mu > 0. \end{aligned}$$

Assume a simple random sample $\widetilde{Y}$ of size n is available and let the estimator of μ be the sample average $\overline{Y}$. As we saw in Example 16.1, we have that

$$t_{test} = \frac{\overline{Y} - 0}{\sqrt{\frac{\sigma^2}{n}}} \sim N\left(\frac{\mu}{\sqrt{\frac{\sigma^2}{n}}} \; , \; 1\right).$$

Note that, contrary to Example 16.1, the value of μ under H_0 is not fixed in this example, but may be any value such that $\mu \leq 0$. For $\mu \leq 0$, we have

$$\Pr\left(Type\ I\ error \; ; \; \mu\right) = \Pr\left(\delta_\tau\left(\widetilde{Y}\right) = H_1 \; ; \; \mu\right) = \Pr\left(t_{test} > \tau \; ; \; \mu\right) = 1 - \Phi\left(\tau - \sqrt{n}\frac{\mu}{\sigma^2}\right).$$

Since $\Phi(x)$ is an increasing function in x, we conclude that $\Pr\left(Type\ I\ error \; ; \; \mu\right)$ is increasing in μ. Indeed, for $\mu = 0$, then $\Pr\left(Type\ I\ error \; ; \; \mu = 0\right) = 1 - \Phi(\tau) > 0$, while, if $\mu \to -\infty$, then $\Pr\left(Type\ I\ error \; ; \; \mu\right) \to 1 - \Phi(\infty) = 1 - 1 = 0$.
Let $\alpha \in (0,1)$. The discussion above shows that if we choose τ such that

$$\Pr\left(Type\ I\ error \; ; \; \mu = 0\right) = 1 - \Phi(\tau) = \alpha,$$

then α will be a significance level of the decision rule, in the sense of Definition 15.12. The solution is to set $\tau = z_{1-\alpha} = \Phi^{-1}(1 - \alpha)$, the $(1 - \alpha)$-quantile of the standard normal distribution.

Comparing Theorem 16.5 with Theorem 16.2 illustrates an important and useful concept for t-tests: When the null and alternative hypotheses are composite, e.g. $H_0 : \theta \geq \theta_{H_0}$ and $H_1 : \theta < \theta_{H_0}$, then the threshold τ used for the decision rule is the same as the one used when testing the simple null versus the composite alternative, e.g. $H_0 : \theta = \theta_{H_0}$ and $H_1 : \theta < \theta_{H_0}$. Indeed, this choice will ensure that the significance level of the test is α, also in the case of a composite null hypothesis.

16.6 The p-value of a hypothesis test

Above, we have discussed how to choose between hypotheses using a decision rule that divides the real line into two regions, the acceptance region and the critical region, and selects a hypothesis based on which region the realized test statistic in questions belongs to, see, for instance, Definition 16.4. In this section, we explore an alternative, but equivalent, decision rule which is often used in practice. The decision rule relies on a so-called *p-value*. The calculation of a p-value is slightly more involved than the calculation of an acceptance/critical region, but the decision rule is correspondingly easier to apply. We first give the definition of a p-value, which is intimately linked to a decision rule $\delta()$ through the corresponding test statistic T.

Definition 16.5 (p-value) *The **p-value** of a hypothesis test, associated with the test statistic T, is the probability of observing an outcome of the test statistic at least as extreme as T, assuming that the null hypothesis is true.*

The definition of the p-value may appear slightly cryptic, in that it not only depends on the test statistic T, but also on "an outcome of the test statistic at least as extreme as T, assuming that the null hypothesis is true". What it means is that the p-value is the probability of observing a value of the realized test statistic t, or a value that would be more unlikely than t, if the null hypothesis is true. The following example illustrates how this works in practice in the case of a one-sided alternative hypothesis.

Example 16.9 *Consider the setting from Example 16.8. The test statistic is t_{test}. Values considered more extreme than t_{test} are defined by the alternative hypothesis $H_1 : \theta > 0$, i.e. values of t_{test} that are very large. Further, according to Definition 16.5, a p-value is calculated assuming that the null hypothesis is true. Hence, for a realized value t of the t-test statistic, the p-value of this test is*

$$p\text{-}value = \Pr\left(t_{test} > t \; ; \; \mu_{H_0} = 0\right).$$

Recall that, under the null with $\mu_{H_0} = 0$, then $t_{test} \sim N(0,1)$. Using the properties of the normal distribution, we have

$$p\text{-}value = \Pr\left(t_{test} > t \; ; \; \mu_{H_0} = 0\right) = 1 - \Phi(t).$$

As a numerical example, suppose that the realized t-test statistic, obtained from the realized sample $\tilde{y}$, is $t = 1.7$. Then the p-value is

$$p\text{-}value = 1 - \Phi(1.7) = 1 - 0.9554 = 0.0446.$$

In the case of a two-sided alternative hypothesis, i.e. $H_1 : \theta \neq \theta_{H_0}$, we need to consider that outcomes of the test statistics that are extreme, from the point of view of the null hypothesis, are both very small and very large values. Suppose for instance, that $t_{test} \sim N(0,1)$ under the null and that the outcome of the test statistic was $t = -1.7$. Then values more improbable, under the null, are values of the test statistic smaller than $t = -1.7$, but also values larger than $|t| = |-1.7| = 1.7$. The following example illustrates.

Example 16.10 *Consider the setting from Example 16.7. The test statistic is t_{test}. Values more extreme than t_{test} will be defined by the alternative hypothesis $H_1 : \theta \neq 0$, i.e. values of t_{test} that are very small or very large. Further, according to Definition 16.5, a p-value* *is calculated assuming that the null hypothesis is true. Hence, for a realized value t of the t-test statistic, the p-value of this test is*

$$\text{p-value} = \Pr\left(t_{test} < -|t| \; ; \; \mu_{H_0}\right) + \Pr\left(t_{test} > |t| \; ; \; \mu_{H_0}\right).$$

Recall that, under the null, $t_{test} \sim N(0,1)$. Using the properties of the normal distribution, we have

$$
\begin{aligned}
\text{p-value} \;&=\; \Pr\left(t_{test} < -|t| \; ; \; \mu_{H_0}\right) + \Pr\left(t_{test} > |t| \; ; \; \mu_{H_0}\right) \\
&=\; \Phi(-|t|) + 1 - \Phi(|t|) \\
&=\; 2 \cdot \Phi(-|t|).
\end{aligned}
$$

As a numerical example, suppose that the realized t-test statistic is $t = 1.7$. Then the p-value is

$$\text{p-value} = 2 \cdot \Phi(-1.7) = 2 \cdot 0.0446 = 0.0891.$$

The p-value is completely determined from the decision rule and the test statistic, hence there is no new information in the p-value as such. However, it lends itself to a simple definition of a decision rule.

Definition 16.6 (Decision rule based on p-value) *Let $\alpha \in (0,1)$ be a number and $p \in (0,1)$ a p-value in the sense of Definition 16.5. Then the **decision rule based on a p-value** is defined by the decision rule*

$$
\delta\left(\widetilde{Y}\right) = \begin{cases} H_0 & if \quad p \geq \alpha, \\ H_1 & if \quad p < \alpha. \end{cases}
$$

In words, we select H_0 if the p-value is above α, while we reject H_0 (select H_1) if the p-value is below α.

Example 16.11 (Example 16.10, continued) *We saw that when the realized random sample results in a test statistics of $t = 1.7$, then the p-value of the test is 8.9%. From the p-value decision rule (Definition 16.6) we see that H_0 cannot be rejected if $\alpha = 5\%$, but it can be rejected if $\alpha = 10\%$. This is also what we found using the equivalent decision rule δ_τ in Example 16.7.*

We want to stress that the decision rule based on the p-value will, by construction, always result in the same choice of hypothesis as the decision rule on which it is based. You are asked to prove this in Problem 16.10.2. But since the p-value is a number between zero and one, and since it can readily be compared to any significance level α of the readers' choice, it is common in the academic literature to report p-values instead of test statistics and critical regions.

Lastly, we provide a result that provides insight into the p-value concept. For simplicity, we state the result for the specific case of a normally distributed test statistic. However, the result applies more generally. For instance, in the case where the distribution of the statistic is approximated by the normal distribution (Section 16.7.2), the same result holds, approximately.

Theorem 16.6 *Assume that,*

$$t_{test} \sim N\left(\sqrt{n} \cdot \frac{\theta - \theta_{H_0}}{\sqrt{\sigma^2}}, 1\right),$$

where $\sigma^2 > 0$, and n is the sample size of the sample used to calculate t_{test}. Consider one of the following three setups:

i) $H_0 : \theta = \theta_{H_0}$, $H_1 : \theta > \theta_{H_0}$, *with p-value* $= 1 - \Phi(t_{test})$.

ii) $H_0 : \theta = \theta_{H_0}$, $H_1 : \theta < \theta_{H_0}$, *with p-value* $= \Phi(t_{test})$.

iii) $H_0 : \theta = \theta_{H_0}$, $H_1 : \theta \neq \theta_{H_0}$, *with p-value* $= 2 \cdot \Phi(-|t_{test}|)$.

   Then, in all cases i)–iii),

1. *Under H_0, then p-value $\sim U(0,1)$.*

2. *Under H_1, then p-value $\to 0$ as $n \to \infty$.*

The first part of the theorem shows that if the null hypothesis being tested is true, then the outcome of the p-value is distributed as a uniform random variable on the interval $(0,1)$. This also shows that α is a significance level for the t-test, since, according to Theorem 16.6,

$$\Pr(Type\ I\ error\ ;\ \theta_{H_0}) = \Pr(p\text{-}value < \alpha\ ;\ \theta_{H_0}) = \Pr(U(0,1) < \alpha) = \alpha.$$

The second part of the theorem expresses that the test is consistent (Definition 15.7). Note that, while the null hypotheses stated in Theorem 16.6 are simple, the theorem also applies when the null hypothesis is composite.

16.6.1   p-value pitfalls

It is important to stress some of the things about which the p-value does not say anything. First, and foremost, the p-value is not the probability that H_0 is true. For the approach considered here, called a frequentist approach, the true value of the parameter θ is non-random, so concepts such as "the probability of a hypothesis being true" or "more or less likely values of a parameter θ" are not well-defined. To consider such statements, the Bayesian approach can be used, see Chapter 19. The p-value simply expresses the probability of observing the sample, or samples which would have produced more extreme values of the test statistics, under the assumption that the null hypothesis is true (Definition 16.5). That is it. Hence, this probability cannot be used to infer the probability that any of the hypotheses are true, but, as we saw, it does lend itself to an easy way of choosing between hypotheses, through the decision rule in Definition 16.6.

Secondly, even if the p-value is high under H_0, meaning the data observed is compatible with a true parameter value under H_0, the data may also be quite compatible with a true parameter value under H_1. This would happen if the power of the test is low at the true parameter under the alternative. The p-value does not contain information about power and, thus, cannot be used to say what happens if an alternative is true.

16.7   Estimation of the t-test statistics distribution

In order to construct a decision rule that satisfies the significance level and to calculate its power function, the distribution of the test statistics is needed. In the examples considered so far, it has been possible to derive the distribution of the test statistics because, for a given parameter value θ, the distribution of the population has been assumed sufficiently known to derive the distribution of the test statistics. In case we do not have sufficient knowledge to calculate the sampling distribution of the test statistics, we need to estimate the sampling distribution of the test statistics. In this section, we consider estimation of the sampling distribution of the t-test statistics, first using the bootstrap method and then using asymptotic arguments. This is analogous to the procedure for constructing confidence intervals for a parameter, when the population distribution is unknown, see Chapter 14.

16.7.1 Bootstrap estimation of the t-test statistics distribution

When performing a hypothesis test, we need the distribution of the test statistics both under the null hypothesis and under the alternative hypothesis. For illustration, consider the null hypothesis $H_0 : \theta = \theta_{H_0}$ against the alternative hypothesis $H_1 : \theta = \theta_{H_1}$ for $\theta_{H_0}, \theta_{H_1} \in \mathbb{R}$ with $\theta_{H_0} \neq \theta_{H_1}$. As we will see below, the distribution of the t-test statistics

$$t_{test} = \frac{\widehat{\theta} - \theta_{H_0}}{\sqrt{\widehat{Var}\left(\widehat{\theta}\right)}}, \tag{16.9}$$

can be estimated by the bootstrap. To determine a critical value for the t-test, or a p-value, we can use the bootstrap distribution under the null, and to determine the power of the t-test, we need to know the bootstrap distribution under the alternative. If we find the bootstrap distribution of (16.9), however, we do not know whether this distribution is under the null hypothesis or under the alternative hypothesis, because we do not know which of these hypotheses is actually true. Therefore, we need to impose more structure to make sure that the bootstrap distribution estimates the distribution of the t-test under the null hypothesis when this is needed, and that it estimates the distribution under the alternative hypothesis when this is needed. To impose this structure, we focus on the distribution of the bootstrap estimator $\widehat{\theta}^*$ of θ.

The sample can be thought of as the bootstrap population. Since $\widehat{\theta}$ is estimated on the sample, and that sample is the bootstrap population, we may assume that the true value of θ in the bootstrap population is $\widehat{\theta}$. Thus, when we draw random samples from the bootstrap population to estimate θ, we know that the true value of θ in the bootstrap population is $\widehat{\theta}$.

When we need the distribution of the bootstrap estimator $\widehat{\theta}^*$ to have θ_{H_0} as the true value, we can relocate the distribution of $\widehat{\theta}^*$ by subtracting the bootstrap population true value $\widehat{\theta}$. This implies that the bootstrap distribution of $\widehat{\theta}^* - \widehat{\theta}$ is located at 0. To get this located at θ_{H_0}, add θ_{H_0}. Thus, $\left(\widehat{\theta}^* - \widehat{\theta} + \theta_{H_0}\right)$ can be used to represent the bootstrap distribution of $\widehat{\theta}$ located at θ_{H_0}. The bootstrap test statistics under H_0, with $\theta = \theta_{H_0}$, is thus

$$t_{test}^* = \frac{\left(\widehat{\theta}^* - \widehat{\theta} + \theta_{H_0}\right) - \theta_{H_0}}{\sqrt{\widehat{Var}^*\left(\widehat{\theta}^*\right)}} = \frac{\widehat{\theta}^* - \widehat{\theta}}{\sqrt{\widehat{Var}^*\left(\widehat{\theta}^*\right)}},$$

where $\widehat{\theta}^*$ is the estimator of θ based on a sample from the bootstrap population and $\widehat{Var}^*\left(\widehat{\theta}^*\right)$ is an estimator of the variance of $\widehat{\theta}^*$.

Similarly to the bootstrap distribution of $\widehat{\theta}$ under the null hypothesis, the bootstrap distribution of $\widehat{\theta}$ under the alternative hypothesis can be obtained by relocating the bootstrap distribution of $\widehat{\theta}^*$ to the bootstrap distribution of $\widehat{\theta}^* - \widehat{\theta} + \theta_{H_1}$. Then the bootstrap test statistics under the alternative, with $\theta = \theta_{H_1}$, is

$$t_{test}^* = \frac{\left(\widehat{\theta}^* - \widehat{\theta} + \theta_{H_1}\right) - \theta_{H_0}}{\sqrt{\widehat{Var}^*\left(\widehat{\theta}^*\right)}} = \frac{\widehat{\theta}^* - \widehat{\theta} + \left(\theta_{H_1} - \theta_{H_0}\right)}{\sqrt{\widehat{Var}^*\left(\widehat{\theta}^*\right)}}.$$

A non-parametric bootstrap of the distribution of the t-test statistics at a particular value θ_H can be constructed using the following algorithm:

Algorithm 16.7.1 (Non-parametric bootstrap of t-test statistics) *Let θ_H be the true value for which the sampling distribution of the t-test statistics is wanted.*

1. *Resample with replacement from $\left(\widetilde{Y}_1, \ldots, \widetilde{Y}_n\right)$ to get a bootstrap sample $\left(\widetilde{Y}_1^*, \ldots, \widetilde{Y}_n^*\right)$.*

2. *Estimate θ and $V\left(\widehat{\theta}\right)$ based on the bootstrap sample. Denote the bootstrap-based estimates by $\widehat{\theta}^*$ and $\widehat{Var}^*\left(\widehat{\theta}^*\right)$.*

3. *Calculate the bootstrap t-test statistics t_{test}^* as*

$$t_{test}^* = \frac{\widehat{\theta}^* - \widehat{\theta} + (\theta_H - \theta_{H_0})}{\sqrt{\widehat{Var}^*\left(\widehat{\theta}^*\right)}}.$$

4. *Repeat steps 1 to 3 B times to obtain B instances of the bootstrapped t-test statistics, $(t_{test,1}^*, \ldots, t_{test,B}^*)$. The empirical distribution function of these numbers is an approximation of the distribution of t_{test} at θ_H.*

The bootstrap estimate of the p-value can be obtained by evaluating the empirical distribution of the bootstrap t-test statistics t_{test}^* for $\theta_H = \theta_{H_0}$ at the realized value t_{test}. Alternatively, the critical value τ needed to construct the t-test can be estimated by the appropriate quantile of the empirical distribution of the bootstrap t-test statistics t_{test}^* for $\theta_H = \theta_{H_0}$. The following example illustrates.

Example 16.12 *Let the population be given by Y with $\mu = E(Y)$ and $\sigma^2 = Var(Y) < \infty$. Suppose the hypotheses under consideration are*

$$\begin{aligned} H_0 &: \quad \mu = 0, \\ H_1 &: \quad \mu = 1. \end{aligned}$$

Assume a simple random sample $\widetilde{Y}$ of size n is available, let the estimator of μ be the sample average $\overline{Y}$, and the estimator of σ^2 be the sample variance S^2. Note that this example is similar to Example 16.6, except that we do not assume anything about the distribution of Y other than finite variance.

The t-test statistic is

$$t_{test} = \frac{\overline{Y} - 0}{\sqrt{\frac{S^2}{n}}}.$$

Recall that the decision rule for the t-test for these hypotheses is (Definition 16.3)

$$\delta_\tau\left(\widetilde{Y}\right) = \left\{ \begin{array}{ll} H_0 & if \quad t_{test} \leq \tau, \\ H_1 & if \quad t_{test} > \tau, \end{array} \right.$$

where τ is the threshold for the test. Let $\alpha \in (0,1)$. We seek to choose τ such that α is a significance level for the test. As we saw above, the solution is to set τ equal to the $(1-\alpha)$-quantile of the distribution of t_{test} under H_0. However, since we do not know the distribution of the population Y, we do not know the distribution of t_{test}, and therefore cannot select τ on this basis.

We may instead approximate the distribution of t_{test} using the non-parametric bootstrap technique outlined above. That is, we follow the steps in the Algorithm 16.7.1 with $\mu_H = \mu_{H_0} = 0$ to obtain the bootstrap sample $t^* = (t_1^*, \ldots, t_B^*)$, which represents the bootstrap distribution of t_{test} under H_0. Then, set τ equal to the $(1-\alpha)$-quantile of the bootstrap data $t^* = (t_1^*, \ldots, t_B^*)$. The bootstrapped p-value can be found as the fraction of bootstrapped test statistics t_i^* that exceed the realized test statistics t, i.e.

$$p\text{-}value^* = \frac{1}{B} \sum_{i=1}^{B} I(t_i^* > t),$$

where $I()$ is the indicator function and t is the realized test statistics, calculated from the realized sample $\widetilde{y}$.

The bootstrap distribution of the t-test under an alternative, say, $\theta = \theta_{H_1}$, is based on

$$t_{test,H_1}^* = \frac{\widehat{\theta}^* - \widehat{\theta} + (\theta_{H_1} - \theta_{H_0})}{\sqrt{\widehat{Var}^* \left(\widehat{\theta}^* \right)}}.$$

Rewrite to get

$$t_{test,H_1}^* = \frac{\left(\widehat{\theta}^* - \widehat{\theta} \right)}{\sqrt{\widehat{Var}^* \left(\widehat{\theta}^* \right)}} + \frac{(\theta_{H_1} - \theta_{H_0})}{\sqrt{\widehat{Var}^* \left(\widehat{\theta}^* \right)}} = \frac{\left(\widehat{\theta}^* - \widehat{\theta} \right)}{\sqrt{\widehat{Var}^* \left(\widehat{\theta}^* \right)}} + \widehat{\lambda}^* = t_{test,H_0}^* + \widehat{\lambda},$$

where

$$t_{test,H_0}^* = \frac{\left(\widehat{\theta}^* - \widehat{\theta} \right)}{\sqrt{\widehat{Var}^* \left(\widehat{\theta}^* \right)}}$$

and

$$\widehat{\lambda}^* = \frac{(\theta_{H_1} - \theta_{H_0})}{\sqrt{\widehat{Var}^* \left(\widehat{\theta}^* \right)}}$$

resembles the non-centrality parameter of Theorem 16.3. Notice, $\widehat{\lambda}^*$ is different for each bootstrap sample. As an approximation, it is possible to estimate $\widehat{\lambda}^*$ by

$$\widehat{\lambda} = \frac{(\theta_{H_1} - \theta_{H_0})}{\sqrt{\widehat{Var} \left(\widehat{\theta} \right)}}.$$

This makes it slightly faster to evaluate the bootstrap distribution for different values of θ_{H_1} since the bootstrap algorithm only needs to be run for $\theta = \theta_{H_0}$. Then bootstrap distributions at other θ_{H_1}'s are obtainable by relocating the bootstrap distribution for $\theta = \theta_{H_0}$ by $\widehat{\lambda}$.

16.7.2 Asymptotic estimation of the t-test statistics distribution

If the distribution of the t-test statistics is unknown, it might be possible to approximate it when the sample size n is large. In particular, if a Central Limit Theorem applies to $\widehat{\theta}$, it will often be the case that the t-test statistics is approximately normally distributed. The next theorem presents this result in the case of using the sample average as an estimator of the mean.

Theorem 16.7 (Asymptotic distribution of t-test statistics) *Assume the population is Y with $E(Y) = \mu$ and $Var(Y) = \sigma^2 < \infty$. Let the estimator of μ be the sample average $\overline{Y}$ given by*

$$\overline{Y} = \frac{1}{n} \sum_{i=1}^{n} \widetilde{Y}_i.$$

Let S^2 be the sample variance and $\mu_{H_0} \in \mathbb{R}$ a number. If $\widetilde{Y}$ is a simple random sample, then the asymptotic distribution of the t-test statistics (16.3) is

$$t_{test} = \frac{\overline{Y} - \mu_{H_0}}{\sqrt{\frac{S^2}{n}}} \overset{A}{\sim} N\left(\frac{\mu - \mu_{H_0}}{\sqrt{\frac{\sigma^2}{n}}}, 1\right).$$

In particular, if $\mu = \mu_{H_0}$, then $t_{test} \overset{A}{\sim} N(0, 1)$.

As discussed in earlier chapters, this approximation becomes better as n increases. Theorem 16.7 suggests that when the sample size is large, we may construct a hypothesis test in the same way as when t_{test} is normally distributed, see e.g. Examples 16.1–16.4, with the difference being that the test should be interpreted as approximate, in that sense that the theoretical properties of the test hold as the sample size n increases toward infinity. The following example illustrates.

Example 16.13 *Let the population be given by Y with $\mu = E(Y)$ and $\sigma^2 = Var(Y) < \infty$. Suppose the hypotheses under consideration are*

$$\begin{aligned} H_0 &: \quad \mu = 0, \\ H_1 &: \quad \mu = 1. \end{aligned}$$

Assume a simple random sample $\widetilde{Y}$ of size n is available, let the estimator of μ be the sample average $\overline{Y}$, and the estimator of σ^2 be the sample variance S^2. Note that this example is similar to Example 16.6, except that we do not assume anything about the distribution of Y except for finite variance.

Under H_0, i.e. assuming that $\mu = \mu_{H_0} = 0$, then

$$t_{test} = \frac{\overline{Y} - \mu_{H_0}}{\sqrt{\frac{S^2}{n}}} \overset{A}{\sim} N(0, 1),$$

i.e. t_{test} is approximately $N(0, 1)$ distributed, and the quality if this approximation increases as the sample size n increases. Similarly, under H_1, i.e. assuming that $\mu = \mu_{H_1} = 1$,

$$t_{test} \overset{A}{\sim} N\left(\frac{1}{\sqrt{\frac{\sigma^2}{n}}}, 1\right) = N\left(\sqrt{n}\frac{1}{\sqrt{\sigma^2}}, 1\right).$$

That is, we are in precisely the same setting as Examples 16.1, 16.2, 16.3, and 16.4, except for two differences. First, the variance σ^2 is unknown and must be estimated, e.g. by S^2. Secondly, while the normal distribution can be used for calculation of the threshold and the probabilities of committing Type I and Type II errors, precisely as in Examples 16.1, 16.2, 16.3, and 16.4, all conclusions have to be interpreted in an approximate sense, since the test statistic is only approximately normally distributed. As usual, we can expect the approximation to become better as the sample size n increases. In Problem 16.10.5 you are asked to explore how good this approximation is in a specific case.

Although asymptotic arguments can often be used to construct an approximate hypothesis test using the normal distribution, properties of the test might, for finite sample sizes n, not be precisely as specified by the normal distribution. For instance, even if the test is designed to have significance level α, it might not be the case that α is actually a significance level for the test, because this property is only ensured to hold for an infinite sample size. Consequently, when applied in practice with a finite sample, the test might be under- or over-sized, see the discussion in Section 15.7.

16.8 Likelihood ratio test of simple hypotheses

We close this chapter by considering another approach to constructing a hypothesis test called the *likelihood ratio test* (LR test). Where the t-test builds on the parameter values and how "close" they are to satisfying the hypotheses, the likelihood ratio test builds on the probability (or density, in the case of continuous random variables) of the sample under the different hypotheses. We consider the LR test here to show that there are other ways to construct a hypothesis test than the t-test, and not least because it has certain optimal properties. In addition, the LR test is the basis for a generalized LR test, which is a widely used for hypothesis testing in non-linear models. This will be discussed further in Chapter 18.

The likelihood ratio compares the "likelihood" of the data under the two hypotheses and uses this comparison to choose between the hypotheses. Assume the distribution of the population is known except for the parameter θ. Let the null and alternative hypotheses both be simple, and given by

$$
\begin{aligned}
H_0 &: \quad \theta = \theta_{H_0}, \\
H_1 &: \quad \theta = \theta_{H_1},
\end{aligned}
$$

with $\theta_{H_0}, \theta_{H_1} \in \mathbb{R}$. We next define the *likelihood ratio, LikR*, as the ratio of the probability, or density, of the random sample if H_0 is true over the probability, or density, of the random sample if H_1 is true.

Definition 16.7 (Likelihood ratio) *Let $\widetilde{Y}$ be a random sample and $f_{\widetilde{Y}}\left(\widetilde{y}; \theta\right)$ the sample distribution. Then the **likelihood ratio** between $\theta = \theta_{H_0}$ and $\theta = \theta_{H_1}$ is defined as*

$$
LikR\left(\widetilde{Y}\right) = \frac{f_{\widetilde{Y}}\left(\widetilde{Y}; \theta_{H_0}\right)}{f_{\widetilde{Y}}\left(\widetilde{Y}; \theta_{H_1}\right)}.
$$

The higher the likelihood ratio, the more likely it is that the random sample came from a population satisfying H_0, compared to a population satisfying H_1.

The likelihood ratio can be used to construct a decision rule. Since a higher value of the likelihood ratio implies that it is more likely that the random sample came from a population satisfying H_0, compared to a population satisfying H_1, a natural decision rule is one where "large" values of the likelihood ratio leads to choosing H_0 and "small" values of the likelihood ratio leads to choosing H_1. As in the case of the t-test, what is "large" and "small" will depend on the distribution of the likelihood ratio.

It turns out that it can be complicated to study the distribution of the likelihood ratio $LikR$. Instead, a non-linear transformation of $LikR$ given by

$$LR = -2 \cdot \ln\left(LikR\left(\widetilde{Y}\right)\right) \;=\; -2 \cdot \ln\left(\frac{f_{\widetilde{Y}}\left(\widetilde{Y};\theta_{H_0}\right)}{f_{\widetilde{Y}}\left(\widetilde{Y};\theta_{H_1}\right)}\right)$$

$$= -2 \cdot \left(\ln f_{\widetilde{Y}}\left(\widetilde{Y};\theta_{H_0}\right) - \ln f_{\widetilde{Y}}\left(\widetilde{Y};\theta_{H_1}\right)\right) \quad (16.10)$$

turns out to have a sampling distribution, which is easier to study. LR is called the *likelihood ratio test statistic*. A higher value of the likelihood ratio implies a lower value of the likelihood ratio test statistics LR, and, thus, a lower value of LR for a realized random sample implies that it is more likely that this sample came from a population for which $\theta = \theta_{H_0}$, compared to a population for which $\theta = \theta_{H_1}$.

A decision rule based on the likelihood ratio is called a *likelihood ratio test*, and it is defined next.

Definition 16.8 (Likelihood ratio test) *Assume the two simple hypotheses $H_0 : \theta = \theta_{H_0}$ and $H_1 : \theta = \theta_{H_1}$. Let $\tau \in \mathbb{R}$ be a number, $\widetilde{Y}$ a random sample, and $f_{\widetilde{Y}}\left(\widetilde{y};\theta\right)$ be the sample distribution. Then the **likelihood ratio test** is defined by the decision rule*

$$\delta_\tau\left(\widetilde{Y}\right) = \begin{cases} H_0 & if \;\; LR \leq \tau, \\ H_1 & if \;\; LR > \tau, \end{cases}$$

where the likelihood ratio test statistic LR is given in Equation (16.10).

If the loss function is known, we may choose τ by considering the risk function $Risk_{\delta_\tau}(\theta)$, see Section 15.6. If the loss function is unknown, then we follow the approach in Section 15.7 and put initial focus on the Type I error. Specifically, we let $\alpha \in (0,1)$ and seek a decision rule $\delta_\tau()$ such that α is the significance level for the decision rule (Definition 15.12). Since Θ_{H_0} consists of only one element, namely θ_{H_0}, this can be obtained by choosing τ such that

$$\alpha \geq \Pr\left(Type\ I\ error\ ;\ \theta_{H_0}\right) = \Pr\left(\delta_\tau(\widetilde{Y}) - II_1\ ;\ \theta_{H_0}\right) \;=\; \Pr\left(LR > \tau\ ;\ \theta_{H_0}\right)$$

$$= 1 - \Pr\left(LR \leq \tau\ ;\ \theta_{H_0}\right),$$

i.e. such that

$$\Pr\left(LR \leq \tau\ ;\ \theta_{H_0}\right) \geq 1 - \alpha.$$

When the distribution of the LR test statistics is continuous, as will often be the case, we can choose the threshold τ such that the inequality above holds with equality, i.e. such that

$$\Pr\left(LR \leq \tau\ ;\ \theta_{H_0}\right) = 1 - \alpha. \quad (16.11)$$

In this case, the solution is to set τ equal to the $(1 - \alpha)$-quantile of the distribution of LR under H_0, i.e. the quantile is to be calculated assuming that $\theta = \theta_{H_0}$. This quantile will, in general, depend on the distribution of the random sample $\widetilde{Y}$.

The following example illustrates how to determine the critical value τ such that the likelihood ratio test has significance level α in the case where the random sample $\widetilde{Y}$ is normally distributed.

Example 16.14 (LR test of two simple hypotheses) *Let $Y \sim N\left(\mu,\sigma^2\right)$, where σ^2 is assumed known. Suppose the null hypothesis is $H_0 : \mu = 0$ and alternative hypothesis is*

$H_1 : \mu = 1$. Let $\alpha \in (0, 1)$ be the significance level and assume a simple random sample $\widetilde{Y}$ of size n is available.

Using the PDF of the normal distribution (Definition 8.8), the likelihood ratio is

$$
\begin{aligned}
LikR\left(\widetilde{Y}\right) &= \frac{f_{\widetilde{Y}}\left(\widetilde{Y}; \mu_{H_0}\right)}{f_{\widetilde{Y}}\left(\widetilde{Y}; \mu_{H_1}\right)} \\[2mm]
&= \frac{\displaystyle\prod_{i=1}^{n} \frac{1}{\sqrt{2\pi}\sigma} \exp\left(-\frac{\widetilde{Y}_i^2}{2\cdot\sigma^2}\right)}{\displaystyle\prod_{i=1}^{n} \frac{1}{\sqrt{2\pi}\sigma} \exp\left(-\frac{(\widetilde{Y}_i-1)^2}{2\cdot\sigma^2}\right)} \\[2mm]
&= \frac{\left(\frac{1}{\sqrt{2\pi}\sigma}\right)^n \exp\left(-\frac{1}{2\cdot\sigma^2}\sum_{i=1}^{n}\widetilde{Y}_i^2\right)}{\left(\frac{1}{\sqrt{2\pi}\sigma}\right)^n \exp\left(-\frac{1}{2\cdot\sigma^2}\sum_{i=1}^{n}\left(\widetilde{Y}_i-1\right)^2\right)}.
\end{aligned}
$$

Cancel the first term in the denominator and numerator and take the logarithm on both sides to get

$$
\begin{aligned}
\ln\left(LikR\left(\widetilde{Y}\right)\right) &= -\frac{1}{2\cdot\sigma^2}\sum_{i=1}^{n}\widetilde{Y}_i^2 + \frac{1}{2\cdot\sigma^2}\sum_{i=1}^{n}\left(\widetilde{Y}_i-1\right)^2 \\[2mm]
&= \frac{1}{2\cdot\sigma^2}\sum_{i=1}^{n}\left(-\widetilde{Y}_i^2 + \left(\widetilde{Y}_i-1\right)^2\right) \\[2mm]
&= \frac{1}{2\cdot\sigma^2}\sum_{i=1}^{n}\left(-\widetilde{Y}_i^2 + \widetilde{Y}_i^2 + 1 - 2\widetilde{Y}_i\right) \\[2mm]
&= \frac{1}{2\cdot\sigma^2}\sum_{i=1}^{n}\left(1 - 2\widetilde{Y}_i\right).
\end{aligned}
$$

This implies that the likelihood ratio test statistics is

$$
LR = -2\cdot\ln\left(LikR\left(\widetilde{Y}\right)\right) = \frac{1}{\sigma^2}\sum_{i=1}^{n}\left(-1 + 2\widetilde{Y}_i\right) = \frac{-1}{\sigma^2}n + \frac{2}{\sigma^2}\sum_{i=1}^{n}\widetilde{Y}_i. \tag{16.12}
$$

Recall that, to control the probability of making a Type I error, we have to derive the distribution of the LR test statistics under the assumption that H_0 is true, i.e. under the assumption that $\mu = \mu_{H_0} = 0$. This distribution can be derived using the distribution of $\widetilde{Y}$. Indeed, assuming that H_0 is true, we get, according to the distribution of a sum of independent normally distributed $N(0, \sigma^2)$ random variables,

$$
LR = \frac{-1}{\sigma^2}n + \frac{2}{\sigma^2}\sum_{i=1}^{n}\widetilde{Y}_i \sim N\left(\frac{-1}{\sigma^2}n, \frac{4}{\sigma^2}n\right).
$$

Thus, from (16.11), we know that the critical value τ should be set equal to the $(1-\alpha)$-quantile of the $N\left(\frac{-1}{\sigma^2}n, \frac{4}{\sigma^2}n\right)$ distribution. The solution to this is (Problem 16.10.1)

$$
\tau = z_{1-\alpha}\cdot\sqrt{\frac{4}{\sigma^2}n} - \frac{1}{\sigma^2}n,
$$

where $z_{1-\alpha} = \Phi^{-1}(1-\alpha)$ is the $(1-\alpha)$-quantile of the $N(0,1)$ distribution. For example, if $\alpha = 0.05$, $n = 100$, and $\sigma^2 = 50$, then $z_{0.95} = \Phi^{-1}(0.95) = 1.64$ and

$$\tau = 1.64 \cdot \sqrt{\frac{4}{50}100} - \frac{1}{50}100 = 2.64.$$

By construction, the threshold τ will imply that the LR test has a probability of a Type I error equal to the significance level α.

Note that, contrary to the case of the t-test studied above, the distribution of the LR test statistics under the null hypothesis may depend on the parameters of the model. For instance, in Example 16.14 we saw that, under H_0, $LR \sim N\left(-n/\sigma^2, 4n/\sigma^2\right)$, such that the distribution of the test statistics under the null depends on the sample size n and the population variance σ^2. Contrast this with the t-test in Example 16.1, where $t_{test} \sim N(0,1)$ under the null. The reason for this difference is that the t-test statistics (16.3) is defined as a standardized random variable, which results in the t-test statistics being (asymptotically) pivotal, i.e. its distribution is free from parameters of the model. As we have seen, this is not the case for the LR test statistics (16.10).

Once the critical value τ has been chosen such that the likelihood ratio test has significance level α, we can proceed to examine the probability of committing a Type II error. This probability is given by

$$\beta(\theta_{H_1}) = \Pr\left(Type\ II\ error\ ;\ \theta_{H_1}\right) = \Pr\left(\delta\left(\widetilde{Y}\right) = H_0\ ;\ \theta_{H_1}\right) = \Pr\left(LR \leq \tau\ ;\ \theta_{H_1}\right).$$

The next example illustrates the calculation of the probability of a Type II error of a likelihood ratio test in the case where the random sample $\widetilde{Y}$ is normally distributed.

Example 16.15 (Example 16.14, continued) *To calculate the probability of a Type II error, the distribution of the LR test statistics is needed assuming $H_1 : \theta_{H_1} = 1$ is true. Under H_1, the distribution of $\widetilde{Y}_i$ is*

$$\widetilde{Y}_i \sim N\left(1, \sigma^2\right).$$

This implies that the LR test statistics is distributed as

$$LR = \frac{-1}{\sigma^2}n + \frac{2}{\sigma^2}\sum_{i=1}^{n}\widetilde{Y}_i \sim N\left(\frac{-1}{\sigma^2}n + \frac{2}{\sigma^2}n, \frac{4}{\sigma^2}n\right) = N\left(\frac{1}{\sigma^2}n, \frac{4}{\sigma^2}n\right).$$

Using the properties of the normal distribution, the probability of a Type II error is

$$\Pr\left(Type\ II\ error\ ;\ \theta_{H_1}\right) = \Pr\left(LR \leq \tau\ ;\ \theta_{H_1}\right) = \Phi\left(\frac{\tau - \frac{1}{\sigma^2}n}{\sqrt{\frac{4}{\sigma^2}n}}\right).$$

As a numerical example, if $\alpha = 0.05$, $n = 100$, and $\sigma^2 = 50$, then $\tau = 2.64$, and

$$\Pr\left(Type\ II\ error\ ;\ \theta_{H_1}\right) = \Phi\left(\frac{2.64 - \frac{1}{50}100}{\sqrt{\frac{4}{50}100}}\right) = \Phi\left(0.23\right) = 0.59.$$

In terms of the power function, $\pi(\theta_{H_1}) = \pi(1) = 1 - \Pr\left(Type\ II\ error\ ;\ \theta_{H_1}\right) = 1 - 0.59 = 0.41$.

Similarly to the case of the t-test, we may also define our likelihood ratio test decision rule (Definition 16.8) in terms of a p-value. Indeed, both the definition of the p-value (Definition 16.5), and the decision rule based upon the p-value (Definition 16.6), are valid for general hypothesis tests and are not restricted to the t-test. Following Definition 16.5, the p-value of the likelihood ratio test decision rule of Definition 16.8 becomes:

$$p\text{-}value = \Pr\left(LR > lr \; ; \; \theta_{H_0}\right), \tag{16.13}$$

where LR is the likelihood ratio test statistics and lr is the realized value of the likelihood ratio test statistics. Thus, just as was the case for calculation of the threshold value τ, studied above, it is necessary to know the distribution of the LR test statistics to calculate the p-value.

Example 16.16 (Example 16.14, continued) *We know that*

$$LR \sim N\left(\frac{-1}{\sigma^2}n, \frac{4}{\sigma^2}n\right)$$

under $H_0 : \mu = 0$. *Let* $lr \in \mathbb{R}$ *be a realized value of the LR test statistics. Then, by* (16.13) *and the properties of the normal distribution, the p-value is*

$$p\text{-}value = \Pr\left(LR > lr \; ; \; \mu = 0\right) = 1 - \Phi\left(\frac{lr + \frac{1}{\sigma^2}n}{\sqrt{\frac{4}{\sigma^2}n}}\right).$$

The likelihood ratio test of two simple hypotheses can be proven to be the optimal decision rule, in the sense that it is the test with significance level α that has the lowest probability of a Type II error or, equivalently, the highest power. This result follows from the so-called *Neyman-Pearson Lemma*. In the context of Example 16.15, the Neyman-Pearson Lemma implies that it is impossible to construct any decision rule with a Type I error at most 5% which simultaneously has power higher than 41%. This implies a ratio of probability of Type II error to Type I error of $0.59/0.05 = 11.8$. This ratio can only become lower by increasing the significance level since no decision rule with significance level α has a lower probability of a Type II error.

Above, we considered the case where both hypotheses H_0 and H_1 are simple. In practice, however, many hypotheses are composite. In Chapter 18, after having discussed maximum likelihood estimation, we will show how to test two composite hypotheses using a generalized version of the likelihood ratio test statistics.

16.9 Proofs

Proof of Theorem 16.1 *The loss under H_0 is $\pi_{\delta_\tau}(\mu_{H_0}) \cdot L_{01}$ and the loss under H_1 is $(1 - \pi_{\delta_\tau}(\mu_{H_1})) \cdot L_{10}$. We need to find the value of τ for which*

$$\max\left(\pi_{\delta_\tau}(\mu_{H_0}) \cdot L_{01}, (1 - \pi_{\delta_\tau}(\mu_{H_1})) \cdot L_{10}\right)$$

is smallest. Suppose $\mu_{H_0} < \mu_{H_1}$. Then $\pi_{\delta_\tau}(\mu_{H_0})$ is decreasing from 1 to 0 in τ, and, thus, $(1 - \pi_{\delta_\tau}(\mu_{H_1}))$ is increasing from 0 to 1 in τ, the smallest maximum is obtained when

$$\pi_{\delta_\tau}(\mu_{H_0}) \cdot L_{01} = (1 - \pi_{\delta_\tau}(\mu_{H_1})) \cdot L_{10} \Rightarrow$$
$$\frac{(1 - \pi_{\delta_\tau}(\mu_{H_1}))}{\pi_{\delta_\tau}(\mu_{H_0})} = \frac{L_{01}}{L_{10}}.$$

Proof of Theorem 16.2 *Suppose first that $\theta_{H_1} < \theta_{H_0}$. The proof that*

$$\Pr\left(Type\ I\ error\ ;\ \theta_{H_0}\right) = \Pr\left(\delta_\tau\left(\widetilde{Y}\right) = H_1\ ;\ \theta_{H_0}\right) = \alpha,$$

follows the arguments given immediately above Theorem 16.2. Indeed, since $\tau = q_\alpha$ is the α-quantile of t_{test}, we have

$$\Pr\left(\delta_\tau\left(\widetilde{Y}\right) = H_1\ ;\ \theta_{H_0}\right) = \Pr(t_{test} \le \tau\ ;\ \theta_{H_0}) = \alpha,$$

by the definition of the α-quantile of a continuous random variable, see Definition 5.7 and the discussion following this definition. To prove the second part of the theorem, note that because $\widehat{\theta}$ and $\widehat{Var}(\widehat{\theta})$ are, by assumption, consistent estimators, then under H_1 where $\theta = \theta_{H_1}$,

$$plim_{n\to\infty}\left(\widehat{\theta} - \widehat{\theta}_{H_0}\right) = \theta - \theta_{H_0} = \theta_{H_1} - \theta_{H_0} < 0$$

and

$$plim_{n\to\infty}\ \widehat{Var}(\widehat{\theta}) = 0.$$

Therefore, under H_1 where $\theta = \theta_{H_1}$,

$$plim_{n\to\infty}\ t_{test} = plim_{n\to\infty}\ \frac{\widehat{\theta} - \theta_{H_0}}{\widehat{Var}(\widehat{\theta})} = -\infty,$$

and, hence,

$$\lim_{n\to\infty}\Pr\left(\delta_\tau\left(\widetilde{Y}\right) = H_1\ ;\ \theta_{H_1}\right) = \lim_{n\to\infty}\Pr\left(t_{test} \le \tau\ ;\ \theta_{H_1}\right) = 1.$$

This concludes the proof in the case $\theta_{H_1} < \theta_{H_0}$. The case $\theta_{H_1} > \theta_{H_0}$ follows similar arguments. We omit the details.

Proof of Theorem 16.3 *We first recall the definition of the non-central t-distribution, see also Section 8.6.2. Let $Z \sim N(0,1)$ and $W \sim \chi^2(\nu)$ be independent random variables. Then*

$$T = \frac{Z + \lambda}{\sqrt{W/\nu}} \tag{16.14}$$

has a non-central t-distribution $t(\nu, \lambda)$. To see that the distribution of the random variable t_{test} can be represented by a non-central t-distribution, we prove that there are random variables Z and W such that t_{test} can be written as (16.14).

Note first that it is the case that a scaled version of S^2 has the χ^2 distribution. More precisely,

$$W = S^2 \frac{n-1}{\sigma^2} \sim \chi^2(n-1).$$

Also, we may define Z to be a standard normal random variable via

$$Z = \frac{\overline{Y} - \mu}{\sqrt{\sigma^2/n}} \sim N(0,1).$$

Since $\overline{Y}$ and S^2 are statistically independent random variables, we may conclude that also W and Z are statistically independent. Recall the definition of the t-test statistics

$$t_{test} = \frac{\overline{Y} - \mu_{H_0}}{\sqrt{S^2/n}}.$$

By re-arranging the expression for W above, we may write $S^2 = W \frac{\sigma^2}{n-1}$. Inserting this into the expression for t_{test} and adding $0 = \mu - \mu$ in the numerator, we get

$$t_{test} = \frac{\overline{Y} - \mu_{H_0}}{\sqrt{S^2/n}} = \frac{\overline{Y} - \mu + \mu - \mu_{H_0}}{\sqrt{\frac{W}{n}\frac{\sigma^2}{n-1}}} = \frac{\frac{\overline{Y}-\mu}{\sqrt{\sigma^2/n}} + \frac{\mu-\mu_{H_0}}{\sqrt{\sigma^2/n}}}{\sqrt{\frac{W}{n-1}}} = \frac{Z + \lambda}{\sqrt{\frac{W}{n-1}}},$$

where

$$\lambda = \frac{\mu - \mu_{H_0}}{\sqrt{\sigma^2/n}},$$

which completes the proof.

Proof of Theorem 16.4 *Using that*

$$\begin{aligned}
\Pr\left(\delta_{\tau_1,\tau_2}\left(\widetilde{Y}\right) = H_1 \; ; \; \theta\right) &= \Pr\left(t_{test} \in [\tau_1, \tau_2]^c \; ; \; \theta\right) \\
&= \Pr\left(t_{test} \leq \tau_1 \; ; \; \theta\right) + \Pr\left(t_{test} \geq \tau_2 \; ; \; \theta\right),
\end{aligned}$$

the proof follows similar arguments as the proof of Theorem 16.2. We omit the details.

Proof of Theorem 16.5 *The proof combines the arguments given immediately above Theorem 16.5 with the arguments in the proof of Theorem 16.2. We omit the details.*

Proof of Theorem 16.6 *Consider the case ii) $H_0 : \theta = \theta_{H_0}$, $H_1 : \theta < \theta_{H_0}$, with p-value $= \Phi(t_{test})$. The other cases i) and iii) follow using similar arguments.*

To prove the first part of the theorem, note that, under H_0 where $\theta = \theta_{H_0}$, then, by assumption $t_{test} \sim N(0,1)$. But since Φ is the CDF of the $N(0,1)$ distribution, the probability integral transform (Theorem 11.6) implies that p-value $= \Phi(t_{test}) \sim U(0,1)$, as we wanted to show.

To prove the second part of the theorem, note first that by the properties of the normal distribution, we can write

$$t_{test} = \sqrt{n}\frac{\theta - \theta_{H_0}}{\sqrt{\sigma^2}} + Z,$$

where $Z \sim N(0,1)$. Also, under H_1 where $\theta < \theta_{H_0}$, then

$$\sqrt{n}\frac{\theta - \theta_{H_0}}{\sqrt{\sigma^2}} \to -\infty,$$

as $n \to \infty$. We conclude that p-value $= \Phi(t_{test}) \to \Phi(-\infty) = 0$, as $n \to \infty$, as we wanted to show.

Proof of Theorem 16.7 *Note that, by adding $0 = \mu - \mu$ in the numerator, we can write*

$$t_{test} = \frac{\overline{Y} - \mu_{H_0}}{\sqrt{S^2/n}} = \frac{\overline{Y} - \mu}{\sqrt{S^2/n}} + \frac{\mu - \mu_{H_0}}{\sqrt{S^2/n}}.$$

By the Feasible Central Limit Theorem (Theorem 13.5), the first term satisfies

$$\frac{\overline{Y} - \mu}{\sqrt{S^2/n}} \overset{A}{\sim} N(0,1).$$

Since S^2 is a consistent estimator of σ^2, the continuous mapping theorem (Theorem 13.2) implies that

$$\mathrm{plim}_{n\to\infty} \frac{\mu - \mu_{H_0}}{\sqrt{S^2}} = \frac{\mu - \mu_{H_0}}{\sqrt{\sigma^2}}.$$

Combining this with Slutsky's theorem (Theorem 13.4), we have that

$$t_{test} \overset{A}{\sim} N\left(\frac{\mu - \mu_{H_0}}{\sqrt{\frac{\sigma^2}{n}}}, \ 1 \right),$$

as we wanted to show.

16.10 Exercises

Problem 16.10.1 *Let $X \sim N(\mu, \sigma^2)$ and $\alpha \in (0,1)$ be a number. Show that the α quantile of the distribution of X is given by*

$$q_\alpha = \mu + z_\alpha \sigma,$$

where z_α is the α quantile of the standard normal distribution $N(0,1)$.

Problem 16.10.2 *Let $Y \sim N(\mu, \sigma^2)$ and assume that σ^2 is known. Consider testing the hypotheses $H_0 : \mu = \mu_{H_0}$ and $H_1 : \mu > \mu_{H_0}$, where $\mu_{H_0} \in \mathbb{R}$ is some number, using a t-test.*

1. *Show that the decision rule based on a critical region for the t-test statistic (Definition 16.4) is equivalent to the decision rule based on a p-value (Definition 16.6). (Hint: To show that the decision rules are equivalent, you can show that the two decision rules always give the same answer. For instance, prove that if one decision rule outputs H_0 if and only if the other decision rule also outputs H_0.)*

2. *Show that the two decision rules are also equivalent in case the alternative hypothesis is $H_1 : \mu < \mu_{H_0}$.*

3. *Show that the two decision rules are also equivalent in case the alternative hypothesis is $H_1 : \mu \neq \mu_{H_0}$.*

Problem 16.10.3 *Consider a country with two political candidates, A and B. A polling institute wishes to assess whether the population prefers candidate A or candidate B. Since the population of voters is very large in this country, it is infeasible to poll every voter. Instead, the institute gathers a simple random sample $\widetilde{Y}$ of size $n = 1000$, where $\widetilde{Y}_i = 1$ if the i'th person in the sample prefers candidate A, and $\widetilde{Y}_i = 0$ if the i'th person in the sample prefers candidate B. Let $p \in (0,1)$ be the (unknown) fraction in the population that prefers candidate A.*

We wish to test whether $p > 50\%$ (A is preferred by the population) or $p < 50\%$ (B is preferred by the population). Candidate A is the incumbent of the political position on vote, and is seen to be quite popular. Hence, the polling institute judges that it needs to have extra high confidence if they are to conclude that $p < 50\%$. This suggest that are interested in testing the hypotheses

$$\begin{aligned} H_0 \ &: \ p \geq 50\%, \\ H_1 \ &: \ p < 50\%, \end{aligned}$$

with a particular focus on controlling the Type I error, i.e. of controlling the probability of concluding $p < 50\%$ in cases where it is actually the case that $p \geq 50\%$.

After gathering their sample, the polling institute finds that 480 of the persons asked favored candidate A, while 520 favored candidate B.

1. *Use the realized sample to estimate p and calculate a 95% confidence interval for p. (Hint: You may use the approximation to the normal distribution when calculating the confidence interval.)*

2. *Conduct a t-test of the above hypotheses at the significance level $\alpha = 5\%$. Calculate also the p-value of the test. What is the conclusion? (Hint: You may use the approximation to the normal distribution when conducting the t-test.)*

3. *Calculate the power function of the test $\pi(p)$ and plot $\pi(p)$ as a function of p for $p \in [0, 1]$. (Hint: Again you may use the approximation to the normal distribution.)*

Problem 16.10.4 *A medicine company produces a well-tested and reliable drug. It is known that this drug will have a beneficial effect on a patient in 75% of cases. The company has recently developed an alternative drug, and want to test whether the new drug is better than the old drug. Because the old drug is known to be reliable, the company only wants to distribute the new drug if there is very good evidence for the new drug being better than the old drug, in the sense that it has a beneficial effect in more than 75% of cases. The company plans to administer the new drug to a simple random sample of size $n = 3000$ and assess how many of these the new drug has a beneficial effect on.*

1. *Propose null and alternative hypotheses, H_0 and H_1, that can be used to test whether the new drug is preferred to the old drug. Propose also a significance level α for the test.*

2. *Define the t-test statistics that can be used to test the H_0 and H_1 from the first part. Calculate and plot the power function.*

3. *The medicinal company collects the random sample and finds that the drug has a positive effect on 2305 out of the 3000 patients on which the drug was administered. Use this realized random sample to conduct the hypothesis test. Calculate also the p-value of the test. Based on your results, should the medical company distribute the new drug instead of the old drug?*

Problem 16.10.5 *In this exercise, you are asked to conduct a Monte Carlo study investigating the effects of using the normal distribution for conducting a t-test, in the case where the underlying population is non-normal, see also Example 16.13 and the discussion surrounding it. Monte Carlo investigations of the properties of a hypothesis test is often performed when a new hypothesis test is suggested. It allows the researcher to assess whether the test can be trusted when applied in practice.*

Let $Y \sim Exp(\lambda)$ with $\lambda = 1/2$. Recall that $\mu = E(Y) = \lambda^{-1} = 2$. We assume that the population distribution is unknown to the researcher. The goal is to test a hypotheses of the form

$$H_0 \quad : \quad \mu = 2,$$
$$H_1 \quad : \quad \mu \neq 2.$$

Note that, in this case, the null hypothesis is true.

1. *Simulate a random sample $\widetilde{Y}$ of size $n = 10$ from Y.*

2. *Use the random sample to perform the above hypothesis test using a t-test with significance level $\alpha = 5\%$, where the distribution of the t-test statistics is approximated using asymptotic methods (Section 16.7.2).*

3. *Record whether H_0 is rejected or whether it is selected by the test.*

4. *Repeat the above steps 1–3 $M = 10,000$ times. Calculate the fraction of these M times where a Type I error has been committed, i.e. where the null hypothesis has rejected.*

5. *Repeat steps 1–4 for different samples sized, $n = 50$, $n = 100$, $n = 200$, $n = 1000$.*

6. *Plot the fraction of times a Type I error has been committed as a function of the sample size n. Compare with $\alpha = 5\%$.*

Note: Recall that the size *of a test is equal to the probability of committing a Type I error when the null is true. In this example, the α is called the* nominal size *of the test, since this is the significance level chosen and the size that the test has, theoretically, as n becomes large. Conversely, the numbers calculated in step 4 are called the* actual size *of the test. These numbers are Monte Carlo estimates of the true probabilities of committing a Type I error, taking into account that the test was designed based on the approximation to the normal distribution, which might be poor when the sample size n is small.*

Problem 16.10.6 *Consider the same setting as in Problem 16.10.5, but now investigate the power of the test using Monte Carlo simulations. Such an analysis can be used to analyze the finite sample properties of the test when the null hypothesis is false.*

1. *Go through the same steps as in Problem 16.10.5, but now assume that the population is $Exp(\lambda)$ with $\lambda = 1$. Note that now the null hypothesis $H_0 : \mu = 2$ is false.*

2. *Record again the number of times, over $M = 10,000$ Monte Carlo simulations the null was rejected. The fraction of rejections of H_0 over these $10,000$ simulations is an estimate of the power of the test when $\mu = 1$.*

3. *Re-do the analysis with $\mu = 1/4$ and $\mu = 2$. Compare the estimated power function $\pi(\mu)$ for $\mu = 1/4, 1, 2$ and comment on how the power changes as the sample size n increased.*

4. *Calculate the theoretical power of the test for $\mu = 1/4, 1, 2$, using the approximation to the normal distribution. Compare the theoretical probabilities to the estimated ones.*

Problem 16.10.7 *Re-do Problem 16.10.5 using the bootstrap approach (Section 16.7.1) to approximate the distribution of the t-test statistics in Step 2, instead of the asymptotic approach.*

Problem 16.10.8 *Re-do Problem 16.10.6 using the bootstrap approach (Section 16.7.1) to approximate the distribution of the t-test statistics, instead of the asymptotic approach.*

Problem 16.10.9 *Derive an expression for the power of the LR test statistics in Example 16.14. Hint, derive the distribution of the LR test using the expression in equation (16.12) and assuming $\widetilde{y}_i \sim N\left(1, \sigma^2\right)$. Then evaluate the probability of an outcome with this distribution less than τ_{LR}.*

Problem 16.10.10 *Let $Y \sim Ber(p)$, where $p \in (0,1)$. For instance, Y may represent the distribution of a population of yes/no voters on a referendum in a population. Let $p_0, p_1 \in (0,1)$ be distinct numbers and consider the simple hypotheses*

$$H_0 : p = p_0,$$
$$H_1 : p = p_1.$$

Assume a simple random sample $\widetilde{Y}$ of size n is available.

1. Derive the LR test statistic and show that it can be represented as

$$LR = -2 \cdot n \cdot \ln\left(\frac{1-p_0}{1-p_1}\right) - 2\left(\ln\left(\frac{p_0}{1-p_0}\right) - \ln\left(\frac{1-p_1}{p_1}\right)\right) \cdot X,$$

where $X \sim Bin(n,p)$.

2. What is the value of LR when $p_0 = p_1$? What is the interpretation of this, in terms of using LR as a test statistics?

3. Let $\alpha \in (0,1)$. Explain how you can choose a critical value τ, such that the likelihood ratio test has significance level α.

4. Suppose now that $p_0 = 0.50$, $p_1 = 0.42$, $\alpha = 5\%$, and $n = 100$. Determine τ.

5. Suppose that the realized sample $\tilde{y}$ of size $n = 100$ is such that $\sum_{i=1}^{100} \tilde{y}_i = 45$. Conduct the likelihood ratio test at a $\alpha = 5\%$ level. Do you reject H_0?

17

Maximum likelihood estimation

17.1 Introduction

Consider a random sample obtained from some population. In Chapter 10, we saw that if
we know the population distribution and the sampling mechanism, we can use this to derive
the distribution of the sample. The distribution of the sample will, in general, depend on
population parameters of the population distribution. In this chapter and the next, we will
study an estimator of these population parameters called the *maximum likelihood* (ML)
estimator. The ML estimator proposes to estimate the population parameters as those
values for which the probability, or *likelihood*, of the data is maximized. That is, the ML
estimate is defined as the values of the parameters for which the probability/density of the
sample at hand is largest. This estimation approach turns out to have several nice properties.
For example, the ML estimator is asymptotically efficient in a large class of estimators. The
ML approach does, however, require that one correctly specifies the functional form of the
distribution for the population. In this chapter, the ML estimator is defined and its basic
statistical properties, such as bias, variance, and consistency, are discussed. In the next
chapter, inference with the ML estimator will be considered.

In Section 17.3, when considering estimation of multiple parameters, we make use of
partial derivatives, a concept which extends the derivative of a function of one variable to
the derivate of a function of multiple variables. You can learn about partial derivatives in
Appendix B.3.

17.2 Maximum likelihood estimation of one parameter

To do ML estimation, it is necessary to specify the distribution function of the population
characteristics of interest. In this section, we consider the case of one characteristics, represented by the random variable Y. The distribution of Y is assumed known except for a
parameter θ. To make this explicit, we denote the distribution by $f(y; \theta)$, where the notation "$; \theta$" is used to indicate the specification of f through the parameter θ. We will take
$f(y; \theta)$ to be a probability mass function (PMF) if Y is discrete or a probability density
function (PDF) if Y is continuous. The next example considers the case of a discrete random
variable. The example will continue throughout this chapter.

Example 17.1 (Bernoulli distribution) *Suppose the population characteristics of in-*
terest is individuals choice to vote "yes" or "no" in a referendum. This implies that the
population characteristics is Bernoulli distributed. Let the random variable representing the
population be

$$Y = \begin{cases} 0 & \text{if} \quad \text{"no",} \\ 1 & \text{if} \quad \text{"yes".} \end{cases}$$

DOI: 10.1201/9781003591191-17

The PMF for a Bernoulli distribution Ber(θ) is

$$f(y;\theta) = \begin{cases} 1 - \theta & if \quad y = 0, \\ \theta & if \quad y = 1, \end{cases}$$

where $\theta \in [0,1]$ is a parameter denoting the fraction of "yes" in the population. To ease notation in what follows, we will use that the PMF for a Ber(θ) distribution may be written more compactly as

$$f(y;\theta) = (1-\theta)^{1-y} \cdot \theta^y \quad for \quad y = 0 \quad or \quad y = 1.$$

For example, if $y = 1$, then

$$f(1;\theta) = (1-\theta)^{1-1} \cdot \theta^1 = (1-\theta)^0 \cdot \theta^1 = 1 \cdot \theta = \theta,$$

and, similarly, if $y = 0$, then

$$f(0;\theta) = (1-\theta)^{1-0} \cdot \theta^0 = (1-\theta)^1 \cdot \theta^0 = (1-\theta) \cdot 1 = 1 - \theta.$$

Note that we have used the convention that $0^0 = 1$, which ensures that the expressions are valid also for $\theta = 0$ and $\theta = 1$,

The overall approach to ML estimation is the following. Consider a population characteristic Y with PDF/PMF $f(y;\theta_0)$. We will assume that the functional form of the distribution is known, that is, we know $f(y;\theta)$. Sometimes we say that $f(y;\theta)$ is a *class* of distributions, where each member is specified by a specific value of θ. What we do not know is the value of θ that correctly characterizes the population Y, namely θ_0. The value θ_0 is called the *true* value of θ. Based on a sample, the purpose is to estimate θ_0 using the knowledge that the distribution of Y is in the class of the distributions specified by $f(y;\theta)$.

Example 17.2 (Bernoulli, continued) *The class of Bernoulli distributions is*

$$f(y;\theta) = (1-\theta)^{1-y} \cdot \theta^y, \quad y = 0, 1,$$

for $\theta \in [0,1]$. Suppose the fraction of "yes" in the population is 0.6. Then $f(y;0.6)$ is the "true" or correct distribution in the class of Bernoulli distributions. Thus, $\theta_0 = 0.6$.

To further explain how ML estimation works, let $(\widetilde{Y}_1,\ldots,\widetilde{Y}_n)$ be a sample with n observations of some population characteristic, represented by the random variable Y. Assume that Y is distributed as $f(y;\theta_0)$, where the distribution function $f(y;\theta)$ is known but the true value of θ, that is θ_0, is unknown. Our goal is to use the sample to estimate θ_0. ML estimation suggests to estimate θ_0 by the value $\widehat{\theta}$ of θ that maximizes the so-called likelihood of the data $(\widetilde{Y}_1,\ldots,\widetilde{Y}_n)$. To this end, the next section introduces the *likelihood function*, which is a function of θ and the data.

17.2.1 Likelihood function of one parameter

The likelihood function is founded on the sample distribution. Let $(\widetilde{Y}_1,\ldots,\widetilde{Y}_n)$ be a sample with n observations. The sample distribution is denoted as

$$f_{\widetilde{Y}_1,\ldots,\widetilde{Y}_n}(y_1,\ldots,y_n),$$

where f is a PDF if the sample consists of continuous random variables or a PMF if the sample consists of discrete random variables.

As discussed in Chapter 10, to derive the sample distribution, it is necessary to make assumptions about how the sample is drawn. Here, we assume that the sample is a simple random sample of a given characteristic, represented by the population random variable Y with distribution $f(y; \theta_0)$. This means that all the observations are iid, that is, statistically independent and identically distributed according to $f(y; \theta_0)$. In particular, $\widetilde{Y}_i$ is statistically independent of $\widetilde{Y}_j$ and $f_{\widetilde{Y}_i}(y; \theta_0) = f_{\widetilde{Y}_j}(y; \theta_0) = f_Y(y; \theta_0)$ for all $i \neq j$. For a simple random sample, the sample distribution can be written as

$$f_{\widetilde{Y}_1,\ldots,\widetilde{Y}_n}(y_1,\ldots,y_n; \theta_0) = f_{\widetilde{Y}_1}(y_1; \theta_0) \cdots f_{\widetilde{Y}_n}(y_n; \theta_0) = \prod_{i=1}^{n} f_Y(y_i; \theta_0),$$

where the first equality sign follows from statistical independence and the second equality sign follows from identical marginal distributions. As discussed above, the value θ_0 is unknown. We will therefore also consider the sample distribution for any value θ, denoted as

$$f(y_1,\ldots,y_n; \theta) = f_{\widetilde{Y}_1,\ldots,\widetilde{Y}_n}(y_1,\ldots,y_n; \theta) = \prod_{i=1}^{n} f_Y(y_i; \theta), \tag{17.1}$$

where we have dropped the "$\widetilde{Y}_1,\ldots,\widetilde{Y}_n$" subscript on $f(y_1,\ldots,y_n; \theta)$ for notational convenience.

Example 17.3 (Bernoulli, continued) *Recall that the PMF of a Bernoulli distribution* $Ber(\theta)$ *is given by*

$$f(y; \theta) = (1 - \theta)^{1-y} \cdot \theta^y.$$

With a simple random sample $(\widetilde{Y}_1,\ldots,\widetilde{Y}_n)$, *the sample distribution is*

$$f(y_1,\ldots,y_n; \theta) = \prod_{i=1}^{n} (1 - \theta)^{1-y_i} \cdot \theta^{y_i},$$

for $\theta \in [0, 1]$. *For example, if* $n = 3$ *and* $(\widetilde{y}_1, \widetilde{y}_2, \widetilde{y}_3) = (1, 0, 0)$, *then the probability of obtaining this sample is*

$$\begin{aligned} f(1, 0, 0; \theta) &= (1 - \theta)^{1-1} \cdot \theta^1 \cdot (1 - \theta)^{1-0} \cdot \theta^0 \cdot (1 - \theta)^{1-0} \cdot \theta^0 \\ &= \theta \cdot (1 - \theta) \cdot (1 - \theta). \end{aligned}$$

The likelihood function is very similar to the sample distribution. The difference is that the sample distribution is a function of $y_1,\ldots,y_n$ for fixed values of the parameter value θ, whereas the likelihood function is a function of θ for each sample. The definition is next.

Definition 17.1 (Likelihood function, one parameter) *Let* $f(y_1,\ldots,y_n; \theta_0)$ *be the sample distribution of the sample* $\left(\widetilde{Y}_1,\ldots,\widetilde{Y}_n\right)$. *Then the **likelihood function** $L\left(\theta; \left(\widetilde{Y}_1,\ldots,\widetilde{Y}_n\right)\right)$ is a function of θ, defined by*

$$L\left(\theta; \left(\widetilde{Y}_1,\ldots,\widetilde{Y}_n\right)\right) = a \cdot f\left(\left(\widetilde{Y}_1,\ldots,\widetilde{Y}_n\right); \theta\right),$$

where $a > 0$ can be chosen to be any positive number.
For a realized sample $(\widetilde{y}_1,\ldots,\widetilde{y}_n)$, the realized likelihood function is

$$L\left(\theta; (\widetilde{y}_1,\ldots,\widetilde{y}_n)\right) = a \cdot f\left((\widetilde{y}_1,\ldots,\widetilde{y}_n); \theta\right).$$

The notation of writing "$; \left(\widetilde{Y}_1, \ldots, \widetilde{Y}_n\right)$" or "$; (\widetilde{y}_1, \ldots, \widetilde{y}_n)$" after θ in the likelihood function is to make it explicit that the likelihood function depends on the sample.

The multiplication of the sample distribution by a positive constant a allows the likelihood function to be equal to the sample distribution, in case $a = 1$, but also to any function that is positively proportional to the sample distribution function. Sometime it is convenient to work with a likelihood function that is proportional to the sample distribution function. Unless otherwise mentioned, we will set $a = 1$.

Example 17.4 (Bernoulli, continued) *Consider again a simple random sample $(\widetilde{Y}_1, \ldots, \widetilde{Y}_n)$ from a $Ber(\theta)$ distribution. We saw in Example 17.3 that the sample distribution is*

$$f(y_1, \ldots, y_n; \theta) = \prod_{i=1}^{n}(1-\theta)^{1-y_i} \cdot \theta^{y_i},$$

while the sample distribution evaluated at the realized sample $(\widetilde{y}_1, \widetilde{y}_2, \widetilde{y}_3) = (1, 0, 0)$ is

$$f(1, 0, 0; \theta) = \theta \cdot (1-\theta) \cdot (1-\theta).$$

For this realized sample, the realized likelihood function is

$$L(\theta; (1, 0, 0)) = f(1, 0, 0; \theta) = \theta \cdot (1-\theta) \cdot (1-\theta).$$

It is important to stress the difference between the likelihood function and the realized likelihood function. For a fixed value of θ, the likelihood function $L\left(\theta; \left(\widetilde{Y}_1, \ldots, \widetilde{Y}_n\right)\right)$ is a random variable and is used when we derive theoretical properties of the ML estimator. When seen as a function of θ, $L\left(\theta; \left(\widetilde{Y}_1, \ldots, \widetilde{Y}_n\right)\right)$ is called a random function. In comparison, the realized likelihood function $L(\theta; (\widetilde{y}_1, \ldots, \widetilde{y}_n))$ is calculated for a given realized sample $(\widetilde{y}_1, \ldots, \widetilde{y}_n)$. It is a deterministic function of θ. The realized likelihood function is used when data are available and the maximum likelihood estimate, to be defined in the next section, is to be calculated.

Example 17.5 (Bernoulli, continued) *Suppose $n = 2$ and $Y \sim Ber(0.6)$. Then the possible outcomes of the likelihood function are*

$$L(\theta; (\widetilde{y}_1, \widetilde{y}_2)) = \begin{cases} (1-\theta) \cdot (1-\theta) & \text{if} & (\widetilde{y}_1, \widetilde{y}_2) = (0, 0), \\ (1-\theta) \cdot \theta & \text{if} & (\widetilde{y}_1, \widetilde{y}_2) = (0, 1) \text{ or } (\widetilde{y}_1, \widetilde{y}_2) = (1, 0), \\ \theta \cdot \theta & \text{if} & (\widetilde{y}_1, \widetilde{y}_2) = (1, 1). \end{cases}$$

When written as a random function, the likelihood function is, using that $\widetilde{Y}_1, \widetilde{Y}_2 \sim Ber(0.6)$ are statistically independent,

$$L(\theta; (\widetilde{Y}_1, \widetilde{Y}_2)) = \begin{cases} (1-\theta) \cdot (1-\theta) & \text{with probability } 0.16, \\ (1-\theta) \cdot \theta & \text{with probability } 0.48, \\ \theta \cdot \theta & \text{with probability } 0.36. \end{cases}$$

This last expression shows the probability that the likelihood function will have a particular functional form in θ. For example, the functional form $L(\theta) = \theta \cdot \theta = \theta^2$ is realized with probability $0.6^2 = 0.36$, namely, when getting the sample $(\widetilde{y}_1, \widetilde{y}_2) = (1, 1)$.

It is often easier to work with the logarithm of the likelihood function, for instance when we shortly need to maximize it. We therefore devote a definition to the logarithm of the likelihood function. This function is called the *log-likelihood function*.

Definition 17.2 (Log-likelihood function, one parameter) *Let* $L\left(\theta; \left(\widetilde{Y}_1, \ldots, \widetilde{Y}_n\right)\right)$ *be a likelihood function. Then the **log-likelihood function** $l\left(\theta; \left(\widetilde{Y}_1, \ldots, \widetilde{Y}_n\right)\right)$ is a function of θ, defined by*

$$l\left(\theta; \left(\widetilde{Y}_1, \ldots, \widetilde{Y}_n\right)\right) = \ln\left(L\left(\theta; \left(\widetilde{Y}_1, \ldots, \widetilde{Y}_n\right)\right)\right).$$

The main reason why the log-likelihood function is often much easier to work with than the likelihood function itself, is the property of the logarithm that the logarithm of the product is the sum of the logarithms, i.e. $\ln(x \cdot y) = \ln x + \ln y$ when $x, y > 0$. To see why this simplifies things, consider the case where we have a simple random sample. Taking logarithms of the sample distribution (17.1) yields the log-likelihood function

$$l\left(\theta; \left(\widetilde{Y}_1, \ldots, \widetilde{Y}_n\right)\right) = \ln\left(L\left(\theta; \left(\widetilde{Y}_1, \ldots, \widetilde{Y}_n\right)\right)\right) = \ln\left(\prod_{i=1}^{n} f_Y(y_i; \theta)\right) = \sum_{i=1}^{n} \ln(f_Y(y_i; \theta)).$$

$$(17.2)$$

In calculations, as well as in numerical evaluations, it is often much easier to work with the sum rather than the product.

Example 17.6 (Bernoulli, continued) *The log-likelihood function for a simple random sample from a Bernoulli population $Ber(\theta)$ is*

$$
\begin{aligned}
l\left(\theta; \left(\widetilde{Y}_1, \ldots, \widetilde{Y}_n\right)\right) &= \ln\left(\prod_{i=1}^{n}(1-\theta)^{1-\widetilde{Y}_i} \cdot \theta^{\widetilde{Y}_i}\right) \\
&= \sum_{i=1}^{n}\left((1-\widetilde{Y}_i)\ln(1-\theta) + \widetilde{Y}_i\ln(\theta)\right) \\
&= \ln(1-\theta)\sum_{i=1}^{n}(1-\widetilde{Y}_i) + \ln(\theta)\sum_{i=1}^{n}\widetilde{Y}_i.
\end{aligned}
$$

It can be further simplified by setting

$$n_1 = \sum_{i=1}^{n}\widetilde{Y}_i,$$

and

$$n_0 = \sum_{i=1}^{n}(1-\widetilde{Y}_i) = n - \sum_{i=1}^{n}\widetilde{Y}_i = n - n_1,$$

where n_0 and n_1 are the number of observations with $\widetilde{Y}_i$ equal to 0 and 1, respectively. Thus, $n = n_0 + n_1$ and

$$l\left(\theta; \left(\widetilde{Y}_1, \ldots, \widetilde{Y}_n\right)\right) = \ln(1-\theta) \cdot n_0 + \ln(\theta) \cdot n_1.$$

17.2.2 Maximum likelihood estimator of one parameter

The maximum likelihood estimator is defined as the random variable $\widehat{\theta}$ which maximizes the likelihood function $L\left(\theta; \left(\widetilde{Y}_1, \ldots, \widetilde{Y}_n\right)\right)$, when viewed as a function of θ defined in a parameter space Θ. The definition is next.

Definition 17.3 (Maximum likelihood estimator, one parameter) *Let* $L\left(\theta;\left(\widetilde{Y}_1,\ldots,\widetilde{Y}_n\right)\right)$ *be the likelihood function and* $\Theta \subseteq \mathbb{R}$ *a set of real numbers. Then the* **maximum likelihood (ML) estimator** $\widehat{\theta}$ *of one parameter* θ_0 *is*

$$\widehat{\theta} = \underset{\theta \in \Theta}{\arg\max}\, L\left(\theta;\left(\widetilde{Y}_1,\ldots,\widetilde{Y}_n\right)\right),$$

where we recall that " $\arg\max_{\theta \in \Theta} L\left(\theta;\left(\widetilde{Y}_1,\ldots,\widetilde{Y}_n\right)\right)$ *" returns the value of* θ *in* Θ *that maximizes* $L\left(\theta;\left(\widetilde{Y}_1,\ldots,\widetilde{Y}_n\right)\right)$.

For a realized sample $(\widetilde{y}_1,\ldots,\widetilde{y}_n)$, *the* **maximum likelihood (ML) estimate** $\widehat{t}$ *of* θ_0 *is*

$$\widehat{t} = \underset{\theta \in \Theta}{\arg\max}\, L\left(\theta;(\widetilde{y}_1,\ldots,\widetilde{y}_n)\right).$$

It is important to stress that the ML estimator $\widehat{\theta}$ is a random variable and, thus, has a distribution. It is this distribution we need in order to pass judgement on the quality of the ML estimator. The ML estimate $\widehat{t}$ is the realized value of the ML estimator $\widehat{\theta}$ for a realized sample. Note that, in practice, we will often abuse notation slightly and write $\widehat{\theta}$ for both the ML estimator and the realized ML estimate $\widehat{t}$.

We can replace the likelihood function in Definition 17.3 with the log-likelihood function and obtain the same solution $\widehat{\theta}$. The reason is that taking a strictly increasing function like $\ln()$ to a function does not change the location of an extremum. At the maximizer $\widehat{\theta}$, the value of the likelihood function $L(\widehat{\theta};(\widetilde{Y}_1,\ldots,\widetilde{Y}_n))$ will be different from the value of the log-likelihood function $\ln\left(L\left(\widehat{\theta};(\widetilde{Y}_1,\ldots,\widetilde{Y}_n)\right)\right)$, but since our interest is in the location of the maximum, namely $\widehat{\theta}$, and not the value of the likelihood function itself, it does not matter in principle whether we use the likelihood function or the log-likelihood function for obtaining $\widehat{\theta}$. Both in practice and for theoretical calculations, however, it turns out that it is often more convenient to work with the log-likelihood function.

17.2.3 Maximum likelihood estimator with differentiable likelihood function

In many practical cases, the ML estimator $\widehat{\theta}$ is obtained by numerically maximizing the log-likelihood function $l\left(\theta;\left(\widetilde{Y}_1,\ldots,\widetilde{Y}_n\right)\right)$, e.g. using computer software. In some cases, however, it is possible to obtain an analytical closed-form solution for $\widehat{\theta}$. In the case where the log-distribution function $\ln f(y;\theta)$ is differentiable in θ, this may be achieved by solving the first-order condition, i.e. by finding the value of θ such that the derivative of the log-likelihood function is equal to zero. The derivative of the log-likelihood function plays a prominent role in ML estimation theory. It is also called the *score function*. The definition is given next.

Definition 17.4 (Score function, one parameter) *Let* $(\widetilde{Y}_1,\ldots,\widetilde{Y}_n)$ *be a random sample of* Y *and assume that the distribution function* $f(y;\theta)$ *of* Y *is differentiable in* θ *for all* y. *The* **score function** $s\left(\theta;\left(\widetilde{Y}_1,\ldots,\widetilde{Y}_n\right)\right)$ *is defined by*

$$s\left(\theta;\left(\widetilde{Y}_1,\ldots,\widetilde{Y}_n\right)\right) = \frac{dl\left(\theta;\left(\widetilde{Y}_1,\ldots,\widetilde{Y}_n\right)\right)}{d\theta},$$

where $l\left(\theta;\left(\widetilde{Y}_1,\ldots,\widetilde{Y}_n\right)\right) = \ln\left(L\left(\theta;\left(\widetilde{Y}_1,\ldots,\widetilde{Y}_n\right)\right)\right)$ *is the log-likelihood function.*

Using the score function, the first-order condition for an optimum $\widehat{\theta}$ of the likelihood function is the value $\widehat{\theta}$ that solves the equation

$$s\left(\widehat{\theta};\left(\widetilde{Y}_1,\ldots,\widetilde{Y}_n\right)\right) = 0.$$

The next example illustrates a case where solving the first-order condition can result in a closed-form solution of the ML estimator

Example 17.7 (Bernoulli, continued) *The log-likelihood function for the Bernoulli distribution with a simple random sample is*

$$l\left(\theta;\left(\widetilde{Y}_1,\ldots,\widetilde{Y}_n\right)\right) = \ln(1-\theta)\cdot n_0 + \ln(\theta)\cdot n_1.$$

This function is differentiable in θ for $\theta \in (0,1)$. For any θ in this interval, the score function is

$$\begin{aligned}
s\left(\theta;\left(\widetilde{Y}_1,\ldots,\widetilde{Y}_n\right)\right) &= \frac{d\left(\ln(1-\theta)\cdot n_0 + \ln(\theta)\cdot n_1\right)}{d\theta} \\
&= -\frac{1}{1-\theta}n_0 + \frac{1}{\theta}n_1.
\end{aligned}$$

The first-order condition for an optimum is

$$s\left(\widehat{\theta};\left(\widetilde{Y}_1,\ldots,\widetilde{Y}_n\right)\right) = 0.$$

Insert for the score function

$$-\frac{1}{1-\widehat{\theta}}n_0 + \frac{1}{\widehat{\theta}}n_1 = 0.$$

Rearrange to get

$$-\widehat{\theta}\cdot n_0 + \left(1-\widehat{\theta}\right)\cdot n_1 = 0,$$

and then solve for $\widehat{\theta}$

$$\widehat{\theta} = \frac{n_1}{n_0+n_1} = \frac{n_1}{n}.$$

Thus, the ML estimator $\widehat{\theta}$ of θ_0 is equal to the fraction of "1"s in the sample. This ML estimator can also be written as a sample average of the $\widetilde{Y}_i$'s,

$$\widehat{\theta} = \frac{n_1}{n} = \frac{\sum_{i=1}^n \widetilde{Y}_i}{n} = \frac{1}{n}\sum_{i=1}^n \widetilde{Y}_i = \overline{Y}.$$

Notice how $\widehat{\theta}$ is a random variable because it depends on the random sample.

On a technical note, when finding the optimum of a function using the first-order condition, we must be careful to check that the solution is a maximum, and not a minimum, and that there is not a maximum on the boundary of the parameter space. For more on optimizing functions, see Appendix B.

17.2.4   Maximum likelihood estimator with non-differentiable likelihood function

There are cases of interest where the distribution function $f(y;\theta)$ is not differentiable in θ. Then the maximization of the likelihood function must be solved by other means than the first-order conditions, most often by numerically maximizing the log-likelihood function. In some cases, however, it is possible to obtain an analytical closed-form solution for the ML estimator, even when the distribution function is non-differentiable in θ. The next example illustrates.

Example 17.8 (Uniform distribution) *Consider a population characteristics Y which is evenly distributed over the population. The lowest value of the characteristics is 0 and the highest value is θ_0. Suppose θ_0 is unknown and the object of interest.*

The distribution matching such a population is the uniform distribution $Y \sim U(0, \theta_0)$, where the density function for a uniform distribution $U(0, \theta)$ is

$$f(y;\theta) = \begin{cases} \frac{1}{\theta} & \text{if } 0 \leq y \leq \theta, \\ 0 & \text{otherwise.} \end{cases}$$

This can also be expressed as

$$f(y;\theta) = \frac{1}{\theta} \cdot I\left(0 \leq y \leq \theta\right),$$

where $I()$ is the indicator function.

Assume a realized simple random sample $(\widetilde{y}_1, \ldots, \widetilde{y}_n)$. Then the sample distribution and the likelihood function are

$$L(\theta) = f(\widetilde{y}_1, \ldots, \widetilde{y}_n; \theta) = \prod_{i=1}^{n} \frac{1}{\theta} \cdot I\left(0 \leq \widetilde{y}_i \leq \theta\right) = \frac{1}{\theta^n} \cdot \prod_{i=1}^{n} I\left(0 \leq \widetilde{y}_i \leq \theta\right).$$

Note that if θ is such that it is larger than all of the realized observations, then all the indicator functions evaluate to one, i.e.

$$I\left(0 \leq \widetilde{y}_i \leq \theta\right) = 1 \text{ if } \theta \geq \widetilde{y}_i$$

implying

$$L(\theta) = \frac{1}{\theta^n} \text{ if } \theta \geq \widetilde{y}_i \text{ for all } i.$$

Conversely, if θ is such that there is at least one $\widetilde{y}_i$ which is larger than θ, then

$$I\left(0 \leq \widetilde{y}_i \leq \theta\right) = 0 \text{ if } \theta < \widetilde{y}_i$$

implying that

$$L(\theta) = 0 \text{ if } \theta < \widetilde{y}_i \text{ for some } i.$$

Let $\widetilde{y}_{(n)} = \max_{i \in \{1, \ldots, n\}} \widetilde{y}_i$ be the largest value in the realized sample $\widetilde{y}_1, \ldots, \widetilde{y}_n$. The value $\widetilde{y}_{(n)}$ is called the nth-order statistics, or the largest order statistics, or extreme order statistics. The above discussion shows that

$$L(\theta) = \begin{cases} 0 & \text{if } \theta < \widetilde{y}_{(n)}, \\ \frac{1}{\theta^n} & \text{if } \theta \geq \widetilde{y}_{(n)}. \end{cases}$$

Thus, the likelihood function $L(\theta)$ is not differentiable around the point $\theta = \widetilde{y}_{(n)}$.

Even though the likelihood function is not differentiable, we may solve the maximization problem by considering the structure of the likelihood function. Clearly, the likelihood function is not maximized for any $\theta < \widetilde{y}_{(n)}$, since for these values we have $L(\theta) = 0$. Hence, the maximum value must be attained for some $\theta \geq \widetilde{y}_{(n)}$, where

$$L(\theta) = \frac{1}{\theta^n}.$$

This is a decreasing function in θ, meaning that it is maximized for the smallest possible value of θ, i.e. for $\theta = \widetilde{y}_{(n)}$. That is, the ML estimate is

$$\widehat{\theta} = \widetilde{y}_{(n)} = \max_{i \in \{1,\dots,n\}} \widetilde{y}_i.$$

In conclusion, the ML estimator of the highest possible value that Y can take is equal to the highest value observed in the sample.

17.3 Maximum likelihood estimation of multiple parameters

The sampling distribution may rely on more than one parameter, so, in this section, we generalize the ML estimator of one parameter to multiple parameters. Most of the concepts from Section 17.2 can be generalized without change, except that the parameter θ should be viewed as a set of K parameters $\theta_1, \dots, \theta_K$, ordered from θ_1 to θ_K. This can be written as

$$\theta = (\theta_1, \dots, \theta_K).$$

Such an ordered set is also called a *vector*. To be precise, θ is called a K-dimensional vector because it has K elements. Therefore, the resulting class of distributions $f(y; \theta_1, \dots, \theta_K)$ depending on the K parameters can be written in short as

$$f(y; \theta) = f(y; \theta_1, \dots, \theta_K).$$

We have chosen to refer to $\theta = (\theta_1, \dots, \theta_K)$ as a vector because this is the common notation in the context of multiple parameters. We do not, however, use any mathematical theory concerning vectors. That is, a K-dimensional vector can, for the present purposes, simply be viewed as a container of K different numbers, parameters, or random variables. In Volume II of this book, we shall make use of the mathematical theory for vectors to analyze complex models. In the context of this volume, these considerations are not important because the focus is on understanding the fundamental concepts of maximum likelihood estimation and this can be done without considering the mathematical analysis of vectors.

17.3.1 Likelihood function of multiple parameters

The likelihood function with multiple parameters is defined next.

Definition 17.5 (Likelihood function, multiple parameters)

*Let $f(y_1, \dots, y_n; \theta_1, \dots, \theta_K)$ be the sample distribution of the random sample $\left(\widetilde{Y}_1, \dots, \widetilde{Y}_n\right)$. Then the **likelihood function** $L\left(\theta_1, \dots, \theta_K; \left(\widetilde{Y}_1, \dots, \widetilde{Y}_n\right)\right)$ is a function of $\theta_1, \dots, \theta_K$, defined by*

$$L\left(\theta_1, \dots, \theta_K; \left(\widetilde{Y}_1, \dots, \widetilde{Y}_n\right)\right) = a \cdot f\left(\left(\widetilde{Y}_1, \dots, \widetilde{Y}_n\right); \theta_1, \dots, \theta_K\right),$$

where $a > 0$ can be chosen to be any positive number.

For a realized sample $(\tilde{y}_1, \ldots, \tilde{y}_n)$, the realized likelihood function is

$$L(\theta_1, \ldots, \theta_K; (\tilde{y}_1, \ldots, \tilde{y}_n)) = a \cdot f((\tilde{y}_1, \ldots, \tilde{y}_n); \theta_1, \ldots, \theta_K).$$

The next example illustrates the likelihood function for a normal distribution with two parameters.

Example 17.9 (Normal distribution) *Assume the population distribution is*

$$Y \sim N(\mu_0, \sigma_0^2),$$

where $\mu_0 \in \mathbb{R}$ and $\sigma_0^2 > 0$ are unknown parameters. The aim is to estimate μ_0 and σ_0^2 using the ML estimator.

The parameter vector is

$$\theta = (\mu, \sigma^2),$$

and the true value of the parameter vector is

$$\theta_0 = (\mu_0, \sigma_0^2).$$

The class of density functions for the population is the class of normal distributions, given by

$$f(y; \mu, \sigma^2) = \frac{1}{\sqrt{2\pi\sigma^2}} \exp\left(-\frac{1}{2}\frac{(y-\mu)^2}{\sigma^2}\right).$$

Assume a simple random sample $\left(\tilde{Y}_1, \ldots, \tilde{Y}_n\right)$ is available. This implies the likelihood function

$$L\left(\mu, \sigma^2; \left(\tilde{Y}_1, \ldots, \tilde{Y}_n\right)\right) = \prod_{i=1}^{n} f(\tilde{Y}_i; \mu, \sigma^2) = \left(\frac{1}{\sqrt{2\pi\sigma^2}}\right)^n \prod_{i=1}^{n} \exp\left(-\frac{1}{2}\frac{(\tilde{Y}_i - \mu)^2}{\sigma^2}\right).$$

As mentioned for the case of one parameter, it is often easier to work with the logarithm of the likelihood function. This leads to the definition of the log-likelihood function.

Definition 17.6 (Log-likelihood function, multiple parameters)
*Let $L\left(\theta_1, \ldots, \theta_K; \left(\tilde{Y}_1, \ldots, \tilde{Y}_n\right)\right)$ be a likelihood function. Then the **log-likelihood function** $l\left(\theta_1, \ldots, \theta_K; \left(\tilde{Y}_1, \ldots, \tilde{Y}_n\right)\right)$ is defined by*

$$l\left(\theta_1, \ldots, \theta_K; \left(\tilde{Y}_1, \ldots, \tilde{Y}_n\right)\right) = \ln\left(L\left(\theta_1, \ldots, \theta_K; \left(\tilde{Y}_1, \ldots, \tilde{Y}_n\right)\right)\right).$$

Example 17.10 (Normal distribution, continued) *The log-likelihood function for the normal distribution with a simple random sample is*

$$
\begin{aligned}
l\left(\mu, \sigma^2; \left(\tilde{Y}_1, \ldots, \tilde{Y}_n\right)\right) &= \ln\left(L\left(\mu, \sigma^2; \left(\tilde{Y}_1, \ldots, \tilde{Y}_n\right)\right)\right) \\
&= \sum_{i=1}^{n} \ln\left(f(\tilde{Y}_i; \mu, \sigma^2)\right) \\
&= \sum_{i=1}^{n}\left[-\frac{1}{2}\ln(2\pi) - \frac{1}{2}\ln(\sigma^2) - \frac{1}{2}\frac{(\tilde{Y}_i - \mu)^2}{\sigma^2}\right] \\
&= -\frac{n}{2}\cdot\ln(2\pi) - \frac{n}{2}\ln(\sigma^2) - \frac{1}{2}\frac{1}{\sigma^2}\sum_{i=1}^{n}(\tilde{Y}_i - \mu)^2,
\end{aligned}
$$

where it is used that $\frac{1}{\sqrt{2\pi\sigma^2}} = (2\pi\sigma^2)^{-0.5}$ and $\sum_{i=1}^{n} 1 = n$.

17.3.2 Maximum likelihood estimator of multiple parameters

The maximum likelihood estimator $\widehat{\theta} = (\widehat{\theta}_1, \ldots, \widehat{\theta}_K)$ is the value of $\theta = (\theta_1, \ldots, \theta_K)$ that maximizes the likelihood function. The definition is next.

Definition 17.7 (Maximum likelihood estimator, multiple parameters)
*Let $L\left(\theta_1, \ldots, \theta_K; \left(\widetilde{Y}_1, \ldots, \widetilde{Y}_n\right)\right)$ be the likelihood function and $\Theta \subseteq \mathbb{R}^K$ a subset of $\mathbb{R}^K$, the K-dimensional real numbers. Then the **maximum likelihood (ML) estimator** $\widehat{\theta}$ of θ_0 is*

$$\widehat{\theta} = \left(\widehat{\theta}_1, \ldots, \widehat{\theta}_K\right) = \underset{(\theta_1,\ldots,\theta_K)\in\Theta}{\arg\max} \; L\left(\theta_1, \ldots, \theta_K; \left(\widetilde{Y}_1, \ldots, \widetilde{Y}_n\right)\right).$$

*The **maximum likelihood (ML) estimate** $\widehat{t}$ of θ_0 is*

$$\widehat{t} = \left(\widehat{t}_1, \ldots, \widehat{t}_K\right) = \underset{(\theta_1,\ldots,\theta_K)\in\Theta}{\arg\max} \; L\left(\theta_1, \ldots, \theta_K; \left(\widetilde{y}_1, \ldots, \widetilde{y}_n\right)\right).$$

The ML estimator $\widehat{\theta} = \left(\widehat{\theta}_1, \ldots, \widehat{\theta}_K\right)$ is a K-dimensional vector of random variables. Therefore, $\widehat{\theta} = \left(\widehat{\theta}_1, \ldots, \widehat{\theta}_K\right)$ has a K-dimensional multivariate distribution. The ML estimate is a K-dimensional vector of numbers. As in the case of a single parameter, we will often abuse notation slightly and write $\widehat{\theta}$ for both the ML estimator and the ML estimate $\widehat{t}$.

17.3.3 ML estimator of multiple parameters with differentiable likelihood function

For a likelihood model with one parameter, the score function is a single function, whereas for a likelihood function with K parameters, the score function consists of K functions. A function that consists of more than one function is called a *vector function*. For example, a function $g(x)$ consisting of the three functions x, x^2, and x^3 is a 3-dimensional vector function and it can be written as

$$g(x) = \left(x, x^2, x^3\right).$$

If $x = 2$, then $g(2) = \left(2, 2^2, 2^3\right) = (2, 4, 8)$. Each function in the vector score function is a partial derivative of the log-likelihood function. The next definition extends Definition 17.4 of the score function to a score function of a likelihood function of K parameters.

Definition 17.8 (Score function, multiple parameters) *Let $(\widetilde{Y}_1, \ldots, \widetilde{Y}_n)$ be a random sample of Y and assume that the log distribution function $\ln\left(f(y; \theta)\right)$ of Y is differentiable in θ for all y. The **score function** $s\left(\theta_1, \ldots, \theta_K; \left(\widetilde{Y}_1, \ldots, \widetilde{Y}_n\right)\right)$ (in short $s\left(\theta_1, \ldots, \theta_K\right)$) is a K-dimensional vector function defined as*

$$s\left(\theta_1, \ldots, \theta_K\right) = \left(s_1\left(\theta_1, \ldots, \theta_K\right), \ldots, s_K\left(\theta_1, \ldots, \theta\right)\right),$$

where s_k, for $k = 1, \ldots, K$, is the k'th partial derivative with respect to the log-likelihood function, i.e.

$$s_k\left(\theta_1, \ldots, \theta_K\right) = \frac{\partial \, l\left(\theta_1, \ldots, \theta_K; \left(\widetilde{Y}_1, \ldots, \widetilde{Y}_n\right)\right)}{\partial \theta_k}.$$

The next example illustrates the derivation of the score function for the normal distribution, which have $K = 2$ parameters.

Example 17.11 (Normal distribution, continued) *Each function $s_k()$ in the vector score function $s()$ for a $N(\mu, \sigma^2)$ population is*

$$s_1\left(\mu, \sigma^2\right) = \frac{\partial l\left(\mu, \sigma^2\right)}{\partial \mu} = \frac{1}{\sigma^2}\sum_{i=1}^{n}(\tilde{Y}_i - \mu)$$

$$s_2\left(\mu, \sigma^2\right) = \frac{\partial l\left(\mu, \sigma^2\right)}{\partial \sigma^2} = -\frac{n}{2}\frac{1}{\sigma^2} + \frac{1}{2}\frac{1}{(\sigma^2)^2}\sum_{i=1}^{n}(\tilde{Y}_i - \mu)^2.$$

Hence,

$$s\left(\mu, \sigma^2\right) = \left(s_1\left(\mu, \sigma^2\right), s_2\left(\mu, \sigma^2\right)\right) = \left(\frac{1}{\sigma^2}\sum_{i=1}^{n}(\tilde{Y}_i - \mu) \; , \; -\frac{n}{2}\frac{1}{\sigma^2} + \frac{1}{2}\frac{1}{(\sigma^2)^2}\sum_{i=1}^{n}(\tilde{Y}_i - \mu)^2\right).$$

The first-order conditions for an optimum $\widehat{\theta}$ of the likelihood function is the value $\widehat{\theta} = \left(\widehat{\theta}_1, ..., \widehat{\theta}_K\right)$ that solves the K equations

$$s_k\left(\widehat{\theta}_1, \ldots, \widehat{\theta}_K\right) = \frac{\partial\, l\left(\widehat{\theta}_1, \ldots, \widehat{\theta}_K\right)}{\partial \theta_k} = 0,$$

for $k = 1, \ldots, K$. This may also be written in vector notation as

$$s\left(\widehat{\theta}_1, \ldots, \widehat{\theta}_K\right) = \underbrace{(0, \ldots, 0)}_{K\ 0\text{'s}}.$$

The next example shows how solving the first-order conditions can result in a closed-form solution to the ML estimator.

Example 17.12 (Normal distribution, continued) *The ML estimator $\left(\widehat{\mu}, \widehat{\sigma^2}\right)$ solves the first-order conditions*

$$s\left(\widehat{\mu}, \widehat{\sigma^2}\right) = (0, 0).$$

Insert for the score vector function to get

$$\left(\frac{1}{\widehat{\sigma^2}}\sum_{i=1}^{n}(\tilde{Y}_i - \widehat{\mu}) \; , \; -\frac{n}{2}\frac{1}{\widehat{\sigma^2}} + \frac{1}{2}\frac{1}{\left(\widehat{\sigma^2}\right)^2}\sum_{i=1}^{n}(\tilde{Y}_i - \widehat{\mu})^2\right) = (0, 0).$$

This represents the two equations

$$\frac{1}{\widehat{\sigma^2}}\sum_{i=1}^{n}(\tilde{Y}_i - \widehat{\mu}) = 0$$

and

$$-\frac{n}{2}\frac{1}{\widehat{\sigma^2}} + \frac{1}{2}\frac{1}{\left(\widehat{\sigma^2}\right)^2}\sum_{i=1}^{n}(\tilde{Y}_i - \widehat{\mu})^2 = 0.$$

First note that it must be the case that $\widehat{\sigma^2} > 0$ for these equations to hold. Use this to solve the first equation for $\widehat{\mu}$

$$\widehat{\mu} = \frac{1}{n}\sum_{i=1}^{n}\tilde{Y}_i.$$

Then insert into the second equation and solve for $\widehat{\sigma^2}$

$$\widehat{\sigma^2} = \frac{1}{n}\sum_{i=1}^{n}(\tilde{Y}_i - \widehat{\mu})^2.$$

The ML estimator of the mean is the sample average $\overline{Y}$. The estimator of the variance is slightly different from the sample variance S^2, which we recall is given by

$$S^2 = \frac{1}{n-1}\sum_{i=1}^{n}(\tilde{Y}_i - \widehat{\mu})^2.$$

The difference is the division by $(n-1)$ rather than with n. Their relationship is

$$\widehat{\sigma^2} = \frac{n-1}{n}S^2.$$

Hence, the ML estimator of the variance is always smaller than the sample variance. The difference between the two estimators will go to zero if the sample size n increases.

17.3.4 ML estimation with several population characteristics

A joint distribution of several population characteristics can be estimated by ML. This can be done without any changes of the framework outlined in Section 17.3, except that $f(y;\theta_1,\ldots,\theta_K)$ is replaced by $f(y_1,\ldots,y_q;\theta_1,\ldots,\theta_K)$, where q is the number of population characteristics.

The next example illustrates a case with $q = 2$ population characteristics, where their joint distribution of the characteristics is specified with $K = 5$ unknown parameters.

Example 17.13 (Multiple normal population characteristics) *Consider two population characteristics Y_1 and Y_2. For example, Y_1 could be height and Y_2 could be weight. Then $q = 2$. Assume Y_1 and Y_2 are jointly normally distributed according to*

$$Y = (Y_1, Y_2) \sim N(\mu_{01}, \mu_{02}, \sigma_{01}^2, \sigma_{02}^2, \sigma_{012}),$$

where $\mu_{01}, \mu_{02} \in \mathbb{R}$ and $\sigma_{01}^2, \sigma_{02}^2 > 0$ are the means and variances of Y_1 and Y_2, respectively, while σ_{012} is the covariance between Y_1 and Y_2. Thus, $\theta_0 = (\mu_{01}, \mu_{02}, \sigma_{01}^2, \sigma_{02}^2, \sigma_{012})$ is the vector of true parameter values, which we would like to estimate by ML estimation.

The PDF of the bivariate normal distribution is (Definition 9.1)

$$f(y;\theta) = \frac{1}{\sqrt{(2\pi)^2 \cdot (\sigma_1^2\sigma_2^2 - \sigma_{12}^2)}}$$

$$\cdot \exp\left(-\frac{1}{2}\frac{(y_1 - \mu_1)^2\,\sigma_2^2 + (y_2 - \mu_2)^2\,\sigma_1^2 - 2\,(y_1 - \mu_1)\,(y_2 - \mu_2)\,\sigma_{12}}{(\sigma_1^2\sigma_2^2 - \sigma_{12}^2)}\right).$$

Assume a simple random sample $\left(\tilde{Y}_1,\ldots,\tilde{Y}_n\right)$, where $\tilde{Y}_i = (\tilde{Y}_{1i}, \tilde{Y}_{2i})$ is a 2-dimensional random vector for each i. After some derivations, the log-likelihood function can be written as

$$l(\theta) = \sum_{i=1}^{n}\ln\left(f(\tilde{Y}_i;\theta)\right)$$

$$= -n\ln(2\pi) - \frac{n}{2}\left(\sigma_1^2\sigma_2^2 - \sigma_{12}^2\right)$$

$$-\frac{1}{2}\sum_{i=1}^{n}\left(-\frac{1}{2}\frac{\left(\tilde{Y}_1 - \mu_1\right)^2\sigma_2^2 + \left(\tilde{Y}_2 - \mu_2\right)^2\sigma_1^2 - 2\left(\tilde{Y}_1 - \mu_1\right)\left(\tilde{Y}_2 - \mu_2\right)\sigma_{12}}{(\sigma_1^2\sigma_2^2 - \sigma_{12}^2)}\right).$$

The first order conditions for a maximum of the log-likelihood function can be found by setting the score function equal to zero and solving the resulting five equations in five unknowns. By doing so, the solution can be written as

$$\widehat{\mu}_1 = \frac{1}{n} \sum_{i=1}^{n} \widetilde{Y}_{1i},$$

$$\widehat{\mu}_2 = \frac{1}{n} \sum_{i=1}^{n} \widetilde{Y}_{2i},$$

$$\widehat{\sigma_1^2} = \frac{1}{n} \sum_{i=1}^{n} \left(\widetilde{Y}_{1i} - \widehat{\mu}_1 \right)^2,$$

$$\widehat{\sigma_2^2} = \frac{1}{n} \sum_{i=1}^{n} \left(\widetilde{Y}_{2i} - \widehat{\mu}_2 \right)^2,$$

and

$$\widehat{\sigma_{12}} = \frac{1}{n} \sum_{i=1}^{n} \left(\widetilde{Y}_{1i} - \widehat{\mu}_1 \right) \cdot \left(\widetilde{Y}_{2i} - \widehat{\mu}_2 \right).$$

It can be seen that the ML estimator of μ_{01} and σ_{01}^2 do not depend on the realized values of the characteristic Y_2, and, similarly, the ML estimators μ_{02} and σ_{02}^2 do not depend on the realized values of the characteristic Y_1. Only the ML estimator of the covariance σ_{012} involves the observations for both characteristics.

In Problem 17.8.3 you are asked to derive the ML estimator of θ in a related, but simplified, setup.

The following example illustrates a case also with $q = 2$ population characteristics Y_1 and Y_2, but where the specification of the population is done by specifying the marginal distribution of Y_1 and the conditional distribution of Y_2 given Y_1.

Example 17.14 (Two population characteristics) *Let Y_1 be a binary variable that equals 1 if a randomly selected individual likes fish, and 0 if not, and let Y_2 be a binary variable that equals 1 if that individual likes chips, and 0 if not. Suppose individuals who like fish have a different preference for chips than those who do not like fish. Hence, Y_1 and Y_2 are not statistically independent.*

We model the statistical dependence between Y_1 and Y_2 by specifying the conditional distribution of Y_2 given Y_1. For those who like fish, $Y_1 = 1$, assume the probability they like chips is τ_1. For those who do not like fish, $Y_1 = 0$, assume the probability they like chips is τ_2. This can be modeled by a Bernoulli distribution written as

$$Y_2 | Y_1 = y_1 \sim Ber(\tau_1 y_1 + \tau_2 (1 - y_1)),$$

or, equivalently,

$$f_{Y_2|Y_1}(y_2|y_1; \tau_1, \tau_2) = (\tau_1 y_1 + \tau_2(1 - y_1))^{y_2} \cdot (1 - (\tau_1 y_1 + \tau_2(1 - y_1)))^{1-y_2}.$$

Assume the probability someone likes fish is γ. This can be modeled by the Bernoulli distribution

$$Y_1 \sim Ber(\gamma),$$

or, equivalently,

$$f_{Y_1}(y_1; \gamma) = \gamma^{y_1} \cdot (1 - \gamma)^{1-y_1}.$$

The joint distribution of Y_1 and Y_2 is

$$f_{Y_1,Y_2}(y_1,y_2;\theta) = f_{Y_1,Y_2}(y_1,y_2;\gamma,\tau_1,\tau_2) = f_{Y_2|Y_1}(y_2|y_1;\tau_1,\tau_2) \cdot f_{Y_1}(y_1;\gamma),$$

where $\theta = (\gamma,\tau_1,\tau_2)$.

Assume a simple random sample of the pairs $\left(\tilde{Y}_{1i}, \tilde{Y}_{2i}\right)$, $i = 1,\ldots,n$. Then the log-likelihood function is

$$
\begin{aligned}
l(\gamma,\tau_1,\tau_2) &= \sum_{i=1}^{n} \ln\left(f_{Y_1,Y_2}(\tilde{Y}_{1i},\tilde{Y}_{2i};\gamma,\tau_1,\tau_2)\right) \\
&= \sum_{i=1}^{n} \ln\left(f_{Y_2|Y_1}(\tilde{Y}_{2i}|\tilde{Y}_{1i};\tau_1,\tau_2)\right) + \sum_{i=1}^{n} \ln\left(f_{Y_1}(\tilde{Y}_{1i};\gamma)\right).
\end{aligned}
$$

Maximizing this log-likelihood function can be done by maximizing the first term with respect to τ_1 and τ_2, and the second term with respect to γ. The reason is that the first term does not depend on γ and, for a similar reason, the second term does not depend on τ_1 and τ_2.

The maximization of the first term can be found using the first-order conditions for τ_1, τ_2. To this end, we first derive the log-likelihood function. Insert for f to get

$$
\begin{aligned}
\sum_{i=1}^{n} \ln\left(f_{Y_2|Y_1}(\tilde{Y}_{2i}|\tilde{Y}_{1i};\tau_1,\tau_2)\right) &= \sum_{i=1}^{n} \tilde{Y}_{2i} \ln\left(\tau_1\tilde{Y}_{1i} + \tau_2(1-\tilde{Y}_{1i})\right) \\
&\quad + \sum_{i=1}^{n} \left(1-\tilde{Y}_{2i}\right)\ln\left(1 - \left(\tau_1\tilde{Y}_{1i} + \tau_2(1-\tilde{Y}_{1i})\right)\right).
\end{aligned}
$$

The first-order condition with respect to τ_1 is thus

$$\sum_{i=1}^{n} \tilde{Y}_{2i}\tilde{Y}_{1i}\frac{1}{\tau_1\tilde{Y}_{1i} + \tau_2(1-\tilde{Y}_{1i})} - \sum_{i=1}^{n} \left(1-\tilde{Y}_{2i}\right)\tilde{Y}_{1i}\frac{1}{1 - \tau_1\tilde{Y}_{1i} - \tau_2(1-\tilde{Y}_{1i})} = 0. \qquad (17.3)$$

The first-order condition with respect to τ_2 is

$$\sum_{i=1}^{n} \tilde{Y}_{2i}\left(1-\tilde{Y}_{1i}\right)\frac{1}{\tau_1\tilde{Y}_{1i} + \tau_2(1-\tilde{Y}_{1i})} - \sum_{i=1}^{n} \left(1-\tilde{Y}_{2i}\right)\left(1-\tilde{Y}_{1i}\right)\frac{1}{1 - \tau_1\tilde{Y}_{1i} - \tau_2(1-\tilde{Y}_{1i})} = 0. \qquad (17.4)$$

Equation (17.3) can be written as

$$\sum_{i:\tilde{Y}_{1i}=1} \tilde{Y}_{2i}\frac{1}{\tau_1} - \sum_{i:\tilde{Y}_{1i}=1} \left(1-\tilde{Y}_{2i}\right)\frac{1}{1-\tau_1} = 0,$$

since all terms in summation with $\tilde{Y}_{1i} = 0$ equals 0. The solution to this equation, and thus the ML estimator, is

$$\hat{\tau}_1 = \frac{1}{\sum_{i:\tilde{Y}_{1i}=1} 1} \sum_{i:\tilde{Y}_{1i}=1} \tilde{Y}_{2i}.$$

In words, $\hat{\tau}_1$ equals the average of $\tilde{Y}_{2i}$ for all those observations with $\tilde{Y}_{1i} = 1$.

Similarly, Equation (17.4) can be written as

$$\sum_{i:\tilde{Y}_{1i}=0} \tilde{Y}_{2i}\frac{1}{\tau_2} - \sum_{i:\tilde{Y}_{1i}=0} \left(1-\tilde{Y}_{2i}\right)\frac{1}{1-\tau_2} = 0,$$

since all terms in summation with $\widetilde{Y}_{1i} = 1$ equals 0. The solution to this equation, and thus the ML estimator, is

$$\widehat{\tau}_2 = \frac{1}{\sum\limits_{i:\widetilde{Y}_{1i}=0} 1} \sum_{i:\widetilde{Y}_{1i}=0} \widetilde{Y}_{2i}.$$

In words, $\widehat{\tau}_2$ equals the average of $\widetilde{Y}_{2i}$ for all those observations with $\widetilde{Y}_{1i} = 0$.

The maximization of the second term can be found using the first-order condition for γ. To this end, insert for f to get

$$\sum_{i=1}^{n} \ln\left(f_{Y_1}(\widetilde{Y}_{1i}; \gamma)\right) = \sum_{i=1}^{n} \widetilde{Y}_{1i} \ln(\gamma) + \sum_{i=1}^{n} \left(1 - \widetilde{Y}_{1i}\right) \ln(1 - \gamma).$$

Since this is the Bernoulli distribution, the ML estimator is

$$\widehat{\gamma} = \frac{1}{n} \sum_{i=1}^{n} \widetilde{Y}_{1i}.$$

Hence, the ML estimator of θ in the joint distribution $f_{Y_1,Y_2}(y_1, y_2; \theta)$ of Y_1 and Y_2, is $\widehat{\theta} = (\widehat{\gamma}, \widehat{\tau}_1, \widehat{\tau}_2)$.

17.4　ML estimation with time series

The ML estimation framework is applicable generally i.e. also in cases where the random sample is not a simple random sample. In this section, we discuss ML estimation with time series, where the observations typically are not independent.

Suppose the population $Y = (Y_1, \ldots, Y_T)$ is a time series and $\widetilde{Y} = (\widetilde{Y}_1, \ldots, \widetilde{Y}_T)$ is a time series sample from Y. The dependence between the elements in a time series imply that the sample distribution is not a product of the marginal distributions, meaning that Equations (17.1) and (17.2) do not hold. Using the so-called *prediction-decomposition*, which is an application of the chain rule of probabilities (see Theorem 4.1, formula (10.2), and Problem 17.8.5), the sample distribution of $(\widetilde{Y}_1, \ldots, \widetilde{Y}_T)$ can instead be written as

$$f_{\widetilde{Y}_1,\ldots,\widetilde{Y}_T}(y_1, \ldots, y_T; \theta) = f_{Y_1}(y_1; \theta) \cdot \prod_{t=2}^{T} f_{Y_t|Y_1,\ldots,Y_{t-1}}(y_t|y_1, \ldots, y_{t-1}; \theta). \tag{17.5}$$

Thus, the sample distribution can be evaluated if the initial distribution $f_{Y_1}(y_1; \theta)$ is specified along with the conditional distributions $f_{Y_t|Y_1,\ldots,Y_{t-1}}(y_t|y_1, \ldots, y_{t-1}; \theta)$ for $t = 2, \ldots, T$. Specifying these conditional distributions in the time series case, where Y_t depends on $Y_{t-1}, \ldots, Y_1$, is in many cases easier than specifying the full distribution of $(Y_1, \ldots, Y_T)$.

There is an important class of time series where the conditional distribution of Y_t given Y_{t-1} is independent of $Y_{t-2}, \ldots, Y_1$. Intuitively speaking, for this type of time series, once Y_{t-1} is known, the previous values $Y_{t-2}, \ldots, Y_1$ hold no extra information about Y_t. A time series with this behavior is said to possess the *Markov property*. The definition is given next.

Definition 17.9 (Markov Property) *Let $Y = (Y_1, \ldots, Y_T)$ be a time series. We say that Y has the **Markov property** if*

$$f_{Y_t|Y_1,\ldots,Y_{t-1}}(y_t|y_1, \ldots, y_{t-1}) = f_{Y_t|Y_{t-1}}(y_t|y_{t-1})$$

for all t and all values of $y_t, \ldots, y_1$.

For a time series with the Markov property, the sample distribution (17.5) simplifies to

$$f_{\widetilde{Y}_1,\ldots,\widetilde{Y}_T}(y_1,\ldots,y_T;\theta) = f_{Y_1}(y_1;\theta) \cdot \prod_{t=2}^{T} f_{Y_t|Y_{t-1}}(y_t|y_{t-1};\theta),$$

with log-likelihood function

$$
\begin{aligned}
l\left(\theta;\left(\widetilde{Y}_1,\ldots,\widetilde{Y}_T\right)\right) &= \ln\left(L\left(\theta;\left(\widetilde{Y}_1,\ldots,\widetilde{Y}_T\right)\right)\right) \\
&= \ln\left(f_{Y_1}(y_1) \cdot \prod_{t=2}^{T} f_{Y_t|Y_{t-1}}(y_t|y_{t-1};\theta)\right) \\
&= \ln(f_{Y_1}(y_1)) + \sum_{t=2}^{T} \ln(f_{Y_t|Y_{t-1}}(y_t|y_{t-1};\theta)).
\end{aligned}
$$

These calculations show that to do ML estimation when the sample is a time series with the Markov property, it is only necessary to specify the initial distribution $f_{Y_1}(y_1;\theta)$ along with the conditional distributions $f_{Y_t|Y_{t-1}}(y_t|y_{t-1};\theta)$ for $t = 2,\ldots,T$. Hence, we avoid having to specify the entire conditional distributions $f_{Y_t|Y_1,\ldots,Y_{t-1}}(y_t|y_1,\ldots,y_{t-1};\theta)$ for all t.

Example 17.15 (ML estimation for autoregressive time series) *Consider a time series population Y, modeled by the so-called* autoregressive process of order one,

$$Y_t = \phi_0 \cdot Y_{t-1} + \varepsilon_t, \quad t \geq 2,$$

where $\varepsilon_t \sim N(0,1)$ is an iid sequence and $\phi_0 \in (-1,1)$ is the unknown parameter to be estimated. In this case, Y possesses the Markov property: Conditional on Y_{t-1}, Y_t is statistically independent of $Y_{t-2},\ldots,Y_1$, due to the sequence ε_t being iid. Since ε_t is normally distributed, then the conditional distribution of Y_t given Y_{t-1} is normally distributed. Further,

$$
\begin{aligned}
E(Y_t|Y_{t-1}) &= E(\phi_0 \cdot Y_{t-1} + \varepsilon_t|Y_{t-1}) = \phi_0 \cdot Y_{t-1} + 0 = \phi_0 \cdot Y_{t-1}, \\
Var(Y_t|Y_{t-1}) &= Var(\phi_0 \cdot Y_{t-1} + \varepsilon_t|Y_{t-1}) \\
&= Var(\phi_0 \cdot Y_{t-1}|Y_{t-1}) + Var(\varepsilon_t|Y_{t-1}) \\
&= 0 + 1 \\
&= 1,
\end{aligned}
$$

where we used that, conditional on Y_{t-1}, then $\phi_0 \cdot Y_{t-1}$ is a constant, and also that ε_t is statistically independent of Y_{t-1}. Summarizing, it holds that,

$$Y_t|Y_{t-1} = y_{t-1} \sim N(\phi_0 \cdot y_{t-1}, 1),$$

so that the conditional PMF of Y_t is

$$f_{Y_t|Y_{t-1}}(y_t|y_{t-1}) = \frac{1}{\sqrt{2\pi}} \cdot \exp\left(-\frac{1}{2}(y_t - \phi_0 \cdot y_{t-1})^2\right),$$

for $t = 2,\ldots,T$. These considerations specify the conditional distributions. We also need to specify the distribution of the initial value Y_1. Assume that

$$Y_1 \sim N\left(0, \frac{1}{1-\phi_0^2}\right).$$

The PDF of Y_1 is thus

$$f_{Y_1}(y_1) = \frac{1}{\sqrt{2\pi(1-\phi_0^2)^{-1}}} \cdot \exp\left(-\frac{1}{2}\frac{y_1^2}{(1-\phi_0^2)^{-1}}\right) = \frac{\sqrt{1-\phi_0^2}}{\sqrt{2\pi}} \cdot \exp\left(-\frac{1}{2}y_1^2(1-\phi_0^2)\right).$$

Using this, the log-likelihood function becomes

$$\begin{aligned}
l\left(\phi; \left(\widetilde{Y}_1, \ldots, \widetilde{Y}_T\right)\right) &= \ln(f_{Y_1}(y_1;\phi)) + \sum_{t=2}^{T} \ln(f_{Y_t|Y_{t-1}}(y_t|y_{t-1};\phi)) \\
&= -\frac{1}{2}\ln(2\pi) + \frac{1}{2}\ln(1-\phi^2) - \frac{1}{2}\widetilde{Y}_1^2(1-\phi^2) \\
&\quad + \sum_{t=2}^{T}\left(-\frac{1}{2}\ln(2\pi) - \frac{1}{2}(\widetilde{Y}_t - \phi\cdot\widetilde{Y}_{t-1})^2\right) \\
&= -\frac{T}{2}\ln(2\pi) + \frac{1}{2}\ln(1-\phi^2) - \frac{1}{2}\widetilde{Y}_1^2(1-\phi^2) - \frac{1}{2}\sum_{t=2}^{T}(\widetilde{Y}_t - \phi\cdot\widetilde{Y}_{t-1})^2.
\end{aligned}$$

The first term in the log-likelihood function, $-\frac{T}{2}\ln(2\pi)$, does not depend on the parameter ϕ, and can thus be ignored when deriving the ML estimator $\widehat{\phi}$. The second and third terms make up the contribution from the initial value Y_1, and the last term is the contribution from the remaining values $\widetilde{Y}_2, \ldots, \widetilde{Y}_T$. The first order condition results in a non-linear function of ϕ which is difficult to solve analytically. Instead, one may use computer software to numerically maximize the log-likelihood function to find the ML estimate $\widehat{\phi}$.

The non-linearity of the first order condition arises from the contribution of the term from the initial value Y_1. Had it been assumed that this was a constant, e.g. $Y_1 = 0$, or that the distribution of Y_1 did not depend on ϕ, e.g. $Y_1 \sim N(0,1)$, then it would have been possible to solve the first order condition to arrive at a closed form expression for the ML estimator $\widehat{\phi}$. You are asked to do this in Problem 17.8.6 at the end of this chapter.

17.5   Properties of maximum likelihood estimators

The following sections explore some statistical properties of ML estimators. In particular, Section 17.5.1 studies the bias and variance of the ML estimator, while Section 17.5.2 considers consistency of the ML estimator. Inference using the ML estimator will be studied in Chapter 18.

17.5.1   Bias and variance of the ML estimator

Except for a few cases, ML estimators are biased for finite sample sizes n. The few exceptions are typically characterized by the ML estimator having a closed form solution. In addition, it is typically not possible to explicitly derive the variance of the ML estimator unless it has a closed-form solution. Instead, we can estimate the bias and variance of the ML estimators by using estimates of their sampling distributions, either by bootstrap or asymptotic approaches.

One of the few cases where we can derive the bias and variance of the ML estimator analytically is the case of a normal population with a simple random sample. This case is investigated in the next example.

Example 17.16 (Normal distribution, continued) *Let $Y \sim N\left(\mu_0, \sigma_0^2\right)$ and assume a simple random sample $\left(\widetilde{Y}_1, \ldots, \widetilde{Y}_n\right)$ is available. According to Example 17.12, the ML estimator of the mean μ_0 is*

$$\widehat{\mu} = \frac{1}{n} \sum_{i=1}^{n} \widetilde{Y}_i.$$

The expected value of $\widehat{\mu}$ is

$$E\left(\widehat{\mu}\right) = E\left(\frac{1}{n} \sum_{i=1}^{n} \widetilde{Y}_i\right) = \frac{1}{n} \sum_{i=1}^{n} E\left(\widetilde{Y}_i\right) = \frac{1}{n} \sum_{i=1}^{n} \mu_0 = \mu_0$$

and, thus,

$$\operatorname{Bias}\left(\widehat{\mu}\right) = E\left(\widehat{\mu}\right) - \mu_0 = \mu_0 - \mu_0 = 0.$$

According to Example 17.12, the ML estimator of the variance σ_0^2 is

$$\widehat{\sigma^2} = \frac{1}{n} \sum_{i=1}^{n} \left(\widetilde{Y}_i - \widehat{\mu}\right)^2,$$

The expected value of $\widehat{\sigma^2}$ is

$$E\left(\widehat{\sigma^2}\right) = E\left(\frac{1}{n} \sum_{i=1}^{n} \left(\widetilde{Y}_i - \widehat{\mu}\right)^2\right) = \frac{1}{n} \sum_{i=1}^{n} E\left(\left(\widetilde{Y}_i - \widehat{\mu}\right)^2\right). \qquad (17.6)$$

Consider the term in the summation, and rewrite

$$
\begin{aligned}
E\left(\left(\widetilde{Y}_i - \widehat{\mu}\right)^2\right) &= E\left(\left(\left(\widetilde{Y}_i - \mu_0\right) - \left(\widehat{\mu} - \mu_0\right)\right)^2\right) \\
&= E\left(\left(\widetilde{Y}_i - \mu_0\right)^2\right) + E\left(\left(\widehat{\mu} - \mu_0\right)^2\right) - 2E\left(\left(\widetilde{Y}_i - \mu_0\right) \cdot \left(\widehat{\mu} - \mu_0\right)\right).
\end{aligned}
$$

The first term is

$$E\left(\left(\widetilde{Y}_i - \mu_0\right)^2\right) = \sigma_0^2.$$

The second term is

$$
\begin{aligned}
E\left(\left(\widehat{\mu} - \mu_0\right)^2\right) &= E\left(\left(\frac{1}{n} \sum_{i=1}^{n} \widetilde{Y}_i - \mu_0\right)^2\right) = \frac{1}{n^2} E\left(\left(\sum_{i=1}^{n} \left(\widetilde{Y}_i - \mu_0\right)\right)^2\right) \\
&= \frac{1}{n^2} E\left(\sum_{i=1}^{n} \left(\widetilde{Y}_i - \mu_0\right)^2 + \sum_{i=1}^{n} \sum_{j=1, j \neq i}^{n} \left(\widetilde{Y}_i - \mu_0\right)\left(\widetilde{Y}_j - \mu_0\right)\right) \\
&= \frac{1}{n^2}\left(n \cdot \sigma_0^2 + 0\right) \\
&= \frac{1}{n} \cdot \sigma_0^2.
\end{aligned}
$$

The third term is

$$
E\left(\left(\tilde{Y}_i - \mu_0\right) \cdot (\hat{\mu} - \mu_0)\right) = E\left(\left(\tilde{Y}_i - \mu_0\right) \cdot \left(\frac{1}{n}\sum_{j=1}^{n}\left(\tilde{Y}_j - \mu_0\right)\right)\right)
$$

$$
= E\left(\frac{1}{n}\sum_{j=1}^{n}\left(\tilde{Y}_i - \mu_0\right)\left(\tilde{Y}_j - \mu_0\right)\right)
$$

$$
= \frac{1}{n}\cdot\sigma_0^2.
$$

Inserting this into (17.6) yields

$$
E\left(\widehat{\sigma^2}\right) = \frac{1}{n}\sum_{i=1}^{n}E\left(\left(\tilde{Y}_i - \hat{\mu}\right)^2\right)
$$

$$
= \frac{1}{n}\sum_{i=1}^{n}\left(\sigma_0^2 + \frac{1}{n}\cdot\sigma_0^2 - 2\cdot\frac{1}{n}\cdot\sigma_0^2\right)
$$

$$
= \frac{1}{n}\sum_{i=1}^{n}\left(\sigma_0^2 - \frac{1}{n}\cdot\sigma_0^2\right)
$$

$$
= \sigma_0^2 - \frac{1}{n}\cdot\sigma_0^2.
$$

This implies that the bias of $\widehat{\sigma^2}$ is

$$
Bias(\widehat{\sigma^2}) = \left(\sigma_0^2 - \frac{1}{n}\cdot\sigma_0^2\right) - \sigma_0^2 = -\frac{1}{n}\sigma_0^2.
$$

Hence, the ML estimator of the variance of a normal distribution is downward biased. For a small sample size, the downward bias is relatively large.

The variance of the ML estimator $\hat{\mu}$ has been derived earlier since the ML estimator of the mean is the same estimator as the sample average. Hence,

$$
V\left(\hat{\mu}\right) = \sigma_0^2 \cdot \frac{1}{n}.
$$

The precision of the ML estimator $\hat{\mu}$ of μ_0 measured by MSE is (Theorem 11.1)

$$
MSE\left(\hat{\mu}\right) = Bias(\hat{\mu})^2 + Var(\hat{\mu}) = 0^2 + \sigma_0^2 \cdot \frac{1}{n} = \sigma_0^2 \cdot \frac{1}{n}.
$$

The variance of the ML estimator $\widehat{\sigma^2}$ can be shown to be

$$
V\left(\widehat{\sigma^2}\right) = \frac{2\sigma_0^4}{n-1}\left(\frac{n-1}{n}\right)^2 = 2\sigma_0^4 \cdot \frac{(n-1)}{n^2}.
$$

The precision of the ML estimator $\widehat{\sigma^2}$ of σ_0^2 measured by MSE is (Theorem 11.1)

$$
MSE\left(\widehat{\sigma^2}\right) = Bias\left(\widehat{\sigma^2}\right)^2 + Var\left(\widehat{\sigma^2}\right) = \sigma_0^4 \cdot \frac{1}{n^2} + 2\sigma_0^4 \cdot \frac{(n-1)}{n^2}.
$$

17.5.2 Consistency of the ML estimator

Under quite general assumptions, the ML estimator is a consistent estimator of the parameters of a population distribution. Consistency of ML estimators must generally be established by justifying a link between choosing the parameter values that makes the sample most likely, and the true population values of the parameters. This link can be established by the so-called *information inequality*. The information inequality is stated in the next theorem.

Theorem 17.1 (Information inequality) *Let $f(y; \theta)$ be a class of distributions indexed by a K-dimensional vector $\theta \in \Theta$ and assume $f(y; \theta_0)$ is the distribution of Y, where $\theta_0 \in \Theta$. Then, for all $\theta \in \Theta$,*

$$E\left(\ln\left(f\left(Y; \theta_0\right)\right)\right) - E\left(\ln\left(f\left(Y; \theta\right)\right)\right) \geq 0. \tag{17.7}$$

Equivalently, $\theta = \theta_0$ is a solution to the following maximization problem

$$\max_{\theta \in \Theta} E\left(\ln\left(f\left(Y; \theta\right)\right)\right). \tag{17.8}$$

The expected value in (17.8) is unknown. By the analog principle, we can often estimate an expected value by a sample average. The sample average of the expected value in (17.8) is

$$\widehat{E}\left(\ln\left(f\left(Y; \theta\right)\right)\right) = \frac{1}{n} \sum_{i=1}^{n} \ln\left(f\left(\widetilde{Y}_i; \theta\right)\right). \tag{17.9}$$

The right-hand side is the log-likelihood function with a simple random sample. Therefore, the ML estimator is solving the empirical version of (17.8). Typically, to prove that the right-hand side of (17.9) converges to the expected value in (17.8), a Law of Large Numbers can be invoked.

The next theorem states a set of conditions under which the ML estimator is consistent.

Theorem 17.2 (Consistency of the ML estimator) *Let $\widetilde{Y} = (\widetilde{Y}_1, \ldots, \widetilde{Y}_n)$ be a simple random sample of Y and suppose that Y has PDF/PMF $f(y; \theta_0)$. Assume the following.*

1. Θ is a compact set such that $\theta_0 \in \Theta$.

2. $f(y; \theta)$ is continuous in θ for all values of y.

3. $\Pr\left(f(Y; \theta) \neq f(Y; \theta_0)\right) > 0$ for all $\theta \neq \theta_0$.

4. $E(\sup_{\theta \in \Theta} |\ln f(Y; \theta)|) < \infty$.

Then the ML estimator $\widehat{\theta}$, given in Definition 17.7, is a consistent estimator of θ_0.

We briefly discuss the four conditions in Theorem 17.2. The first condition states that the sample space Θ which we maximize the likelihood function over should be compact (i.e. closed and bounded). In the case where θ is a single parameter, then we will often have that Θ is a closed interval, i.e. $\Theta = [a, b]$ for some $a < b$. If θ is a vector of parameters, we can e.g. consider Θ being a hyper-rectangle so that $\theta_i \in \Theta_i = [a_i, b_i]$ with $a_i < b_i$ for all i. Notice that even though it is a requirement that Θ is closed and bounded, the condition does not restrict the size of Θ as long as the size is finite. Hence, in practice, if it is difficult to assess which values θ_0 can plausibly take, we may simply choose Θ to be very large. For instance, if θ is a single parameter, we may set $\Theta = [-M, M]$, where M is some large number. In this way, the condition is formally satisfied, but in practical terms, we search over an unbounded set Θ. We note that consistency of the ML estimator can be proved without requiring that

Θ is compact, i.e. the first condition of Theorem 17.2 can be weakened by slightly altering the other conditions.

The second condition in Theorem 17.2 is a *regularity condition* stating that the distribution function $f(y;\theta)$ is continuous in θ. This is a very weak condition which will be satisfied in most practical cases. However, there are exceptions, see e.g. Example 17.8.

The third condition in Theorem 17.2 is an *identification condition* for ML estimation. Together with the information inequality (Theorem 17.1) it implies that the expected value of the log-likelihood function is uniquely maximized at $\theta = \theta_0$. If it had been the case that the likelihood function is also maximized by some other value of θ, say $\theta_1 \neq \theta_0$, then θ is not identified by ML estimation: Two different values of θ imply the same maximum of the expected likelihood function. The third condition in Theorem 17.2 is made to avoid this issue, by stating that we cannot have $f(Y;\theta) = f(Y;\theta_0)$ with probability one unless $\theta = \theta_0$.

The fourth condition in Theorem 17.2 is a technical condition ensuring that the expected likelihood function is finite no matter which $\theta \in \Theta$ is considered. Similar to the third condition, this condition is also rather weak and will be satisfied by most models encountered in practice.

Example 17.17 (Normal distribution, continued) *Continuing Example 17.16, where $Y \sim N(\mu_0, \sigma_0^2)$, we show using Theorem 17.2 that the ML estimator $\widehat{\theta} = (\widehat{\mu}, \widehat{\sigma^2})$, as given in Example 17.16, is a consistent estimator of $\theta_0 = (\mu_0, \sigma_0^2)$. We note that in this case, consistency can also be established by other methods, e.g. by invoking the Law of Large Numbers.*

First, we let the parameter space Θ be a compact set, say $\Theta = (\mu, \sigma^2 : \mu \in [-1000, 1000],$ $\sigma^2 \in [0.0001, 1000])$. This is a closed and bounded, and hence compact, set, so it satisfies the first condition of Theorem 17.2. Note that we have now explicitly assumed that the true value $\theta_0 = (\mu_0, \sigma_0^2)$ is in the interior of Θ, i.e. that $-1000 \leq \mu_0 \leq 1000$ and $0.0001 \leq \sigma_0^2 \leq 1000$. If this assumption is questionable for the problem at hand, the bounds in Θ should be widened. The PDF relevant for the normal distribution is

$$f(y;\theta) = \frac{1}{\sqrt{2\pi\sigma^2}} \cdot \exp\left(-\frac{1}{2\sigma^2}(y-\mu)^2\right).$$

For $\theta = (\mu, \sigma^2) \in \Theta$, $f(y;\theta)$ is continuous. The only problematic point would be if $\sigma^2 = 0$, but this value is not included in the parameter space Θ. This verifies the second condition of Theorem 17.2.

To see that the third condition is verified, note that

$$\ln f(Y;\theta) = -\frac{1}{2}\ln(2\pi) - \frac{1}{2}\ln(\sigma^2) - \frac{1}{2\sigma^2}(Y-\mu)^2.$$

Using this, along with the fact that Y is normally distributed, we can conclude that for $\theta \neq \theta_0$

$$\Pr\left(\ln(f(Y;\theta)) \neq \ln(f(Y;\theta_0))\right) \;=\; \Pr\left(-\frac{1}{2}\ln\left(\frac{\sigma^2}{\sigma_0^2}\right) - \frac{1}{2}\left(\frac{(Y-\mu)^2}{\sigma^2} - \frac{(Y-\mu_0)^2}{\sigma_0^2}\right) \neq 0\right)$$

$$> \;\; 0.$$

Since the logarithm is a strictly increasing function, this implies that $\Pr\left(f(Y;\theta) \neq f(Y;\theta_0)\right) > 0$ for $\theta \neq \theta_0$, showing that the third condition is satisfied.

To verify the fourth condition, consider

$$|\ln f(y;\theta)| \;=\; \left| -\frac{1}{2}\ln(2\pi) - \frac{1}{2}\ln(\sigma^2) - \frac{1}{2\sigma^2}(y-\mu)^2 \right|$$

$$\leq \;\; \frac{1}{2}\ln(2\pi) + \frac{1}{2}|\ln(\sigma^2)| + \frac{1}{2\sigma^2}(y-\mu)^2,$$

where we used the triangle inequality, $|x + y| \leq |x| + |y|$. The supremum of a sum is always smaller than the sum of the suprema, thus

$$\sup_{\theta \in \Theta} |\ln f(y; \theta)| \leq \frac{1}{2} \ln(2\pi) + \sup_{\theta \in \Theta} \frac{1}{2} |\ln(\sigma^2)| + \sup_{\theta \in \Theta} \frac{1}{2\sigma^2} (y - \mu)^2 .$$

Since

$$\sup_{\theta \in \Theta} \frac{1}{2} |\ln(\sigma^2)| \leq |\ln(0.0001)| + \ln(1000)$$

and

$$\sup_{\theta \in \Theta} \frac{1}{2\sigma^2} (y - \mu)^2 \quad \leq \quad \sup_{\theta \in \Theta} \frac{1}{2\sigma^2} \left(y^2 + \mu^2 + 2|\mu| \cdot |y| \right)$$

$$\leq \quad \frac{1}{0.0001} \left(y^2 + 1000^2 + 2 \cdot 1000 \cdot |y| \right) ,$$

we can conclude that

$$E(\sup_{\theta \in \Theta} |\ln f(Y; \theta)|) \quad \leq \quad E \left(\frac{1}{2} \ln(2\pi) + |\ln(0.0001)| + \ln(1000) \right.$$

$$\left. + \frac{1}{0.0001} \left(Y^2 + 1000^2 + 2 \cdot 1000 \cdot |Y| \right) \right)$$

$$= \quad \frac{1}{2} \ln(2\pi) + |\ln(0.0001)| + \ln(1000)$$

$$+ \frac{1}{0.0001} \left(E(Y^2) + 1000^2 + 2 \cdot 1000 \cdot E(|Y|) \right)$$

$$< \quad \infty,$$

because $E(Y^2)$ and $E(|Y|)$ are finite when Y is normally distributed. In conclusion, the fourth condition of Theorem 17.2 also holds. We can conclude that the ML estimator $\widehat{\theta} = (\widehat{\mu}, \widehat{\sigma^2})$ is a consistent estimator of $\theta_0 - (\mu_0, \sigma_0^2)$.

Although the technical conditions underlying the theoretical properties of the ML estimator, such as consistency (Theorem 17.2) and asymptotic normality (studied in the next chapter, Theorem 18.1), appear complicated, they are, in fact, quite mild. That is, they are satisfied for most parametric distributions encountered in practical applications. This makes the ML estimation framework very powerful, in particular in cases where the population distribution is known except for a finite number of parameters θ.

17.6 Maximum-likelihood estimation viewed in an information context

We finish the introduction of the ML estimator by an alternative view on how it can be characterized. The ML estimator can be characterized as finding the distribution closest to the population distribution based on the sample. The closeness between two distributions $f(y; \theta)$ and $f(y; \theta_0)$ is going to be defined using the so-called *Kullback-Leibler divergence*. It is defined next.

Definition 17.10 (Kullback-Leibler divergence) *Let the distribution of Y be $f(y;\theta_0)$. The **Kullback-Leibler divergence** $KL(f(Y;\theta),f(Y;\theta_0))$ between the distributions $f(y;\theta)$ and $f(y;\theta_0)$ is defined by*

$$KL(f(Y;\theta),f(Y;\theta_0)) = E(\ln(f(Y;\theta_0))) - E(\ln(f(Y;\theta))).$$

It is seen that the left-hand side of the information inequality (17.7) is the Kullback-Leibler divergence. Therefore, Theorem 17.1 implies that the Kullback-Leibler divergence is non-negative for all θ. We say that the lower the Kullback-Leibler divergence, the closer $f(y;\theta)$ is to $f(y;\theta_0)$ in the Kullback-Leibler sense. It can be shown that the ML estimator is equivalent to the value of θ that minimizes the estimated Kullback-Leibler divergence between the class of population distributions and the true population distribution.

The Kullback-Leibler divergence of the distributions $f(Y;\theta)$ and $f(Y;\theta_0)$ can be interpreted as a risk function $R(f(Y;\theta))$ for using $f(Y;\theta)$ to approximate $f(Y;\theta_0)$. Since $E(\ln(f(Y;\theta_0)))$ is constant, the term $-E(\ln(f(Y;\theta)))$ measures the risk though it is not calibrated to be 0 when $\theta = \theta_0$. Thus,

$$-E(\ln(f(Y;\theta)))$$

can serve as a risk function. The estimate of this risk function with a simple random sample is

$$\widehat{R}(f(Y;\theta)) = -\frac{1}{n}\sum_{i=1}^{n}\ln\left(f\left(\widetilde{Y}_i;\theta\right)\right).$$

In this light, the ML estimator minimizes this estimated risk function.

Sometimes the risk function is defined as 2 times $\widehat{R}(f(Y;\theta))$. The minimum value of this risk function, obtained by evaluating it at the ML estimator, is called the *deviance*,

$$deviance = -2\frac{1}{n}\sum_{i=1}^{n}\ln\left(f\left(\widetilde{Y}_i;\widehat{\theta}\right)\right).$$

17.7   Proofs

Proof of Theorem 17.1 *Note first that due to the so-called Jensen's inequality, it holds that $E(\phi(X)) \leq \phi(E(X))$ when X is a random variable and ϕ is a concave function. Using the properties of the logarithm, we may write*

$$E(\ln(f(Y;\theta))) - E(\ln(f(Y;\theta_0))) = E\left(\ln\left(\frac{f(Y;\theta)}{f(Y;\theta_0)}\right)\right).$$

Since $\ln()$ is a concave function, we may use Jensen's inequality, as discussed above, to conclude that

$$E\left(\ln\left(\frac{f(Y;\theta)}{f(Y;\theta_0)}\right)\right) \leq \ln\left(E\left(\frac{f(Y;\theta)}{f(Y;\theta_0)}\right)\right).$$

Suppose Y is a continuous random variable. The case where Y is discrete works analogously with sums instead of integrals. Since Y has the continuous distribution given by $f(y;\theta_0)$, then, for any function $g()$, it holds that (Theorem 5.1)

$$E(g(Y)) = \int g(y)f(y;\theta_0)dy.$$

Applying this to the function $g(y) = f(y; \theta)/f(y; \theta_0)$ yields

$$E\left(\frac{f(Y; \theta)}{f(Y; \theta_0)}\right) = \int \frac{f(y; \theta)}{f(y; \theta_0)} f(y; \theta_0)\, dy = \int f(y; \theta)\, dy = 1,$$

where the last equality follows because $f(y; \theta)$ is a PDF and hence integrates to one. Putting it all together, we get

$$E\left(\ln\left(f\left(Y; \theta\right)\right)\right) - E\left(\ln\left(f\left(Y; \theta_0\right)\right)\right) \le \ln\left(E\left(\frac{f(Y; \theta)}{f(Y; \theta_0)}\right)\right) = \ln(1) = 0,$$

which concludes the proof.

Proof of Theorem 17.2 *We briefly sketch the main intuition behind this proof, as a full treatment is beyond the scope of this book. We refer to e.g. Amemiya (1985) for the full details.*

As discussed above, the third condition in the theorem, together with the information inequality, ensures that the value of θ that maximizes $E(\ln(Y; \theta))$ is unique and equal to θ_0. Because of the assumptions of a simple random sample the log-likelihood function divided by the sample size will be close to $E\left(\ln\left(f(Y; \theta)\right)\right)$ when the sample size is large. The second condition, i.e. the continuity property, then ensures that, when the sample size is large, the maximizer of the log-likelihood function $\widehat{\theta}$ is close to the maximizer of $E\left(\ln\left(f(Y; \theta)\right)\right)$, that is, θ_0. The remaining conditions ensure that this closeness between $\widehat{\theta}$ and θ_0 improves as the sample size increases in such a way that $\widehat{\theta}$ converges in probability to θ_0.

17.8 Exercises

Problem 17.8.1 *Consider a simple random sample $\widetilde{Y}_1, \ldots, \widetilde{Y}_n$ of the population random variable Y. Assume that Y has the exponential distribution with parameter λ_0, $Y \sim Exp(\lambda_0)$ with $\lambda_0 > 0$, i.e. the PDF of Y is*

$$f(y; \lambda_0) = \lambda_0 e^{-\lambda_0 y}, \quad y > 0.$$

1. *Write up the log-likelihood function and the score function.*

2. *Solve the first order condition for the score function to arrive at the ML estimator $\widehat{\lambda}$ of λ_0.*

3. *Show that the derivative of the score function is always negative and conclude that $\widehat{\lambda}$ indeed maximizes the likelihood function.*

Problem 17.8.2 *Consider a simple random sample $\widetilde{Y}_1, \ldots, \widetilde{Y}_n$ of the population random variable Y. Assume that Y has the Poisson distribution with parameter λ_0, $Y \sim Poi(\lambda_0)$ with $\lambda_0 > 0$, i.e. the PDF of Y is*

$$f(y; \lambda_0) = \frac{\lambda_0^y}{y!} e^{-\lambda_0}, \quad y = 0, 1, \ldots.$$

1. *Write up the log-likelihood function and the score function.*

2. *Solve the first order condition for the score function to arrive at the ML estimator $\widehat{\lambda}$ of λ_0.*

3. *Show that the derivative of the score function is always negative and conclude that $\widehat{\lambda}$ indeed maximizes the likelihood function.*

Problem 17.8.3 *Let the two continuous random variables Y_1 and Y_2 be jointly normally distributed according to the bivariate normal distribution*

$$(Y_1, Y_2) \sim N(0, 0, 1, 1, \rho_0),$$

where $\rho_0 \in (-1, 1)$ is the parameter of interest. Hence, Y_1 and Y_2 are marginally distributed as standard normal random variables and $Corr(Y_1, Y_2) = \rho_0$.

Assume a simple random sample of the pairs $\left(\widetilde{Y}_{1i}, \widetilde{Y}_{2i}\right)$, $i = 1, \ldots, n$.

1. *Write up the log-likelihood function and the score function as functions of ρ.*

2. *Solve the first order condition for the score function to arrive at the ML estimator $\widehat{\rho}$ of ρ_0.*

Problem 17.8.4 *Consider a simple random sample $\widetilde{Y}_1, \ldots, \widetilde{Y}_n$ of the population random variable Y. Assume that Y has the t-distribution with parameter ν_0, $Y \sim t(\nu_0)$ with $\nu_0 > 0$. The PDF of the $t(\nu)$ distribution is*

$$f(y; \nu) = \frac{\Gamma\left(\frac{\nu+1}{2}\right)}{\sqrt{\pi\nu}\,\Gamma\left(\frac{\nu}{2}\right)} \left(1 + \frac{x^2}{\nu}\right)^{-\frac{\nu+1}{2}}, \quad y \in \mathbb{R}, \tag{17.10}$$

where for $x > 0$, $\Gamma(x) = \int_0^\infty s^{x-1} e^{-s} ds$ denotes the so-called Gamma function. *Since the PDF (17.10) is quite complicated, it is not possible to solve the first order condition and arrive at an ML estimate $\widehat{\nu}$ is closed form. Instead, to estimate ν from a sample using the ML estimator, the log-likelihood function must be maximized numerically using computer software.*

1. *Write up the log-likelihood function.*

2. *Simulate a simple random sample of size $\widetilde{Y}_1, \ldots, \widetilde{Y}_n$ of size $n = 20$ from the $t(\nu_0)$ with $\nu_0 = 5$. Optimize the log-likelihood function numerically to arrive at an ML estimate $\widehat{\nu}$.*

3. *Conduct a Monte Carlo study by repeating the previous question $M = 10,000$ times. Use the M Monte Carlo estimates of ν_0 to calculate Monte Carlo estimates of the bias and variance of the ML estimator $\widehat{\nu}$.*

Problem 17.8.5 *Let $Y_1, \ldots, Y_t$ be random variables. Prove that Equation (17.5) holds. (Hint: Use the definition of conditional PMF/PDF repeatedly.)*

Problem 17.8.6 *Consider the population time series Y generated by*

$$Y_t = \phi_0 \cdot Y_{t-1} + \epsilon_t, \quad t \geq 2,$$

where $\epsilon_t \sim N(0, \sigma_0^2)$ is an iid sequence, $\phi_0 \in (-1, 1)$, and $\sigma_0 > 0$. Suppose that the initial value Y_1 is standard normal, i.e. $Y_1 \sim N(0, 1)$. Note that this setting is similar to that in Example 17.15, with two differences: (1) the distribution of the initial value does not depend on the parameters and (2) the error sequence ϵ_t is here distributed as $N(0, \sigma^2)$, whereas it had the $N(0, 1)$ distribution in Example 17.15. Thus, in the setting of this problem, there are two unknown parameters, $\theta_0 = (\phi_0, \sigma_0^2)$.

Assume a time series sample $\widetilde{Y} = (\widetilde{Y}_1, \ldots, \widetilde{Y}_T)$ is available.

1. *Write up the log-likelihood function as a function of $\theta = (\phi, \sigma^2)$.*

2. *Write up the score function as a function of $\theta = (\phi, \sigma^2)$.*

3. *Solve the first order condition for the score function to arrive at the ML estimator $\widehat{\theta}$ of θ_0.*

Problem 17.8.7 *Consider a model similar to Problem 17.8.6*

$$Y_t = \phi_0 \cdot Y_{t-1} + \epsilon_t, \quad t \geq 2,$$

but now assume that $\epsilon_t \sim N(0,1)$ and $Y_1 \sim N\left(0, \frac{1}{1-\phi_0^2}\right)$ as in Example 17.15. As discussed in Example 17.15 this change of the distribution of Y_1 implies that the ML estimator $\widehat{\phi}$ is not available in closed form; in particular, it is not equal to the one derived in Problem 17.8.6. In this exercise, you are asked to conduct a Monte Carlo study to compare the ML estimator of this model where $Y_1 \sim N\left(0, \frac{1}{1-\phi_0^2}\right)$ and with the estimator derived in Problem 17.8.6. Note that when $Y_1 \sim N\left(0, \frac{1}{1-\phi_0^2}\right)$, then the estimator derived in Problem 17.8.6 can be viewed as the ML estimator that ignores the first observation Y_1. (Alternatively, it can be viewed as a conditional ML estimator, where we condition on the first observation Y_1, see Volume II of this book.)

1. *Set $\phi_0 = 0.75$. Simulate a time series $\widetilde{Y}_1, \ldots, \widetilde{Y}_T$ of size $T = 10$ from the model suggested in this exercise, i.e. where $Y_1 \sim N\left(0, \frac{1}{1-\phi_0^2}\right)$. Maximize the log-likelihood function numerically to arrive at an ML estimate $\widehat{\phi}$. (Note: The log-likelihood function is derived in Example 17.15.) Use also the estimator derived in Problem 17.8.6 on the sample to arrive at the alternative estimate $\widetilde{\phi}$.*

2. *Conduct a Monte Carlo study by repeating the previous question $M = 10,000$ times. Use the M Monte Carlo estimates of ϕ_0 to calculate Monte Carlo estimates of the bias and variance of the ML estimators $\widehat{\phi}$ and $\widetilde{\phi}$. What can you conclude?*

3. *Repeat the Monte Carlo study, but this time for a sample of size $T = 1000$. What can you conclude?*

18

Inference with maximum likelihood estimation

18.1 Introduction

To examine the quality of an estimator, the sampling distribution of the estimator is needed. In general, we cannot derive a closed form solution for the sampling distribution of the ML estimator. Instead, we may estimate the sampling distribution by either a bootstrap distribution or an asymptotic distribution. Section 18.2 briefly considers the bootstrap approach while Section 18.3 considers the asymptotic approach. Section 18.4 is devoted to studying certain efficiency properties of the ML estimator. Sections 18.5 and 18.6 show how the estimated sampling distribution of the ML estimator can be used to construct confidence intervals and perform hypothesis tests, respectively.

The theoretical results given in this chapter rely on certain regularity conditions, as stated in Theorem 18.1. These conditions are sufficient, but not necessary, for many of the results given in this chapter. Thus, for any given theorem, there may exist weaker conditions under which the theorem holds. To be parsimonious with regard to assumptions, we have not sought to state each theorem with minimal conditions. As mentioned in the previous chapter, although the technical conditions stated below appear complicated, they are, in fact, quite mild. That is, they are satisfied for most parametric distributions encountered in practical applications.

In Chapter 17, we introduced the ML estimator for a single parameter $\theta_0 \in \mathbb{R}$ and then extended the setup to include the ML estimator with K parameters. As we saw, this is doable without needing a mathematical treatment of vectors. In this chapter, we only present the results on conducting inference for the ML estimator in the case of a single parameter $\theta_0 \in \mathbb{R}$. The reason is that, although the intuitions and ideas of the ML estimator are similar in the one-dimensional and K-dimensional case, the mathematical theory of the latter relies heavily on vector and matrix theory. Therefore, we focus on the case of a single parameter here, and postpone the discussion of inference in the K-dimensional case to Volume II of this book.

18.2 Bootstrap distribution of the ML estimator

The ML estimator is based on a known class of distributions $f(y; \theta)$, among which one is the unknown population distribution, namely $f(y; \theta_0)$. This framework matches the framework of the parametric bootstrap, where the population distribution is assumed known except for the parameters θ, see Section 13.2.2. Thus, the parametric bootstrap is ideally suited for estimating the sampling distribution of the ML estimator.

DOI: 10.1201/9781003591191-18

To be precise, to estimate the sampling distribution of the ML estimator $\widehat{\theta}$, we can use as bootstrap population the distribution $f\left(y; \widehat{\theta}\right)$. When the ML estimator $\widehat{\theta}$ is consistent for θ_0 (Theorem 17.2), $f(y; \widehat{\theta})$ will, in general, be a consistent estimator for $f(y; \theta_0)$. Then the bootstrap can be implemented according to the procedure in Section 13.2.3. In some cases, it may be difficult to draw observations from the distribution given by $f()$. In Chapter C in the appendix, we study various methods for drawing observations from any distribution.

18.3 Asymptotic distribution of the ML estimator of one parameter

Under certain regularity conditions, a central limit-type result will apply to the ML estimator, meaning that its sampling distribution approaches a normal distribution as the sample size increases toward infinity. As usual, we will take this as justification for using the normal distribution as an approximation of the exact distribution of the ML estimator when the sample size is large. In this section, we present the asymptotic distribution for the ML estimator of one parameter. It turns out that the correct normal distribution to use depends on the first and second derivative of the log-likelihood function.

18.3.1 Asymptotic distribution with one parameter

In what follows, we will for a differentiable function $f()$ use the notation $\frac{df(x)}{dx}\big|_{x=x_0}$ to mean the derivative of $f(x)$ evaluated at $x = x_0$. The asymptotic distribution of the ML estimator in the case of one parameter is given in the next theorem.

Theorem 18.1 (Asymptotic distribution of the ML estimator, one parameter)
Let $\widetilde{Y} = (\widetilde{Y}_1, \ldots, \widetilde{Y}_n)$ be a simple random sample of Y and suppose that Y has either a PDF or PMF $f(y; \theta_0)$, where $\theta_0 \in \mathbb{R}$ is a single parameter. Assume that the conditions for consistency of the ML estimator (Theorem 17.2) hold, along with the following.

1. *θ_0 is in the interior of Θ.*

2. *$f(y; \theta)$ is twice continuously differentiable in θ for all values of y.*

3. *The function $f(y; \theta)$ is such that we may interchange integration and differentiation, i.e. $\frac{d}{d\theta} \int f(y; \theta) dy = \int \frac{d}{d\theta} f(y; \theta) dy$ and $\frac{d^2}{d\theta^2} \int f(y; \theta) dy = \int \frac{d^2}{d\theta^2} f(y; \theta) dy$.*

4. *$E\left(\sup_{\theta \in N} \left|\frac{\partial^2}{\partial \theta^2} f(Y; \theta)\right|\right) < \infty$ for a neighborhood N of θ_0.*

5. *$E\left(\frac{d^2}{d\theta^2} \ln f(Y; \theta)\big|_{\theta=\theta_0}\right) \neq 0$.*

Then the asymptotic distribution of the ML estimator $\widehat{\theta}$ of θ_0 is

$$\widehat{\theta} \overset{A}{\sim} N\left(\theta_0, \frac{J}{H^2} \cdot \frac{1}{n}\right), \tag{18.1}$$

as $n \to \infty$, where

$$J = E\left(\left(\frac{d \ln \left(f(Y; \theta)\right)}{d\theta}\bigg|_{\theta=\theta_0}\right)^2\right), \tag{18.2}$$

and

$$H = E\left(\left.\frac{d^2 \ln\left(f(Y;\theta)\right)}{d\theta^2}\right|_{\theta=\theta_0}\right). \tag{18.3}$$

We briefly discuss the five conditions in Theorem 18.1. The first condition states that the true value of the parameter θ_0 should be in the interior of the compact parameter space Θ. If θ_0 is not in the interior of Θ, i.e. if θ_0 is on the boundary of the parameter space, the asymptotic distribution of the ML estimator is typically different than that stated in Theorem 18.1. The second and third conditions are rather weak regularity conditions on the distribution function $f(y;\theta)$ that are satisfied by most distribution functions of interest. It is possible to give higher-level conditions on $f(y;\theta)$ ensuring that the third condition holds. The fourth condition is a technical condition, ensuring that a Law of Large Numbers result can be applied to the score function when evaluated at the ML estimator $\hat{\theta}$. Lastly, the fifth condition ensures that $H \neq 0$ so that the variance $\frac{J}{H^2} \cdot \frac{1}{n}$ in (18.1) is well defined.

Recall that the score function $s(\theta; \widetilde{Y}_1, \ldots, \widetilde{Y}_n)$ is the derivative of the log-likelihood function. Thus, for a simple random sample, we may equivalently write (18.2) and (18.3) as (Problem 18.8.1)

$$J = E\left(s\left(\theta_0; (\widetilde{Y}_1, \ldots, \widetilde{Y}_n)\right)^2\right),$$

and

$$H = E\left(\left.\frac{ds\left(\theta; (\widetilde{Y}_1, \ldots, \widetilde{Y}_n)\right)}{d\theta}\right|_{\theta=\theta_0}\right),$$

respectively. The term J is called the *Jacobian* and H is called the *Hessian*. When θ_0 is a single parameter, J and H are numbers. In Volume II of this book, when studying the ML estimator of the K-dimensional parameter θ_0, the Jacobian and Hessian will be matrices.

The next example illustrates the calculations needed to specify the asymptotic distribution for the case of a Bernoulli distribution.

Example 18.1 (Bernoulli, continued) *The PMF of the $Ber(\theta)$ distribution is*

$$f(y;\theta) = (1-\theta)^{1-y} \cdot \theta^y, \quad y = 0, 1.$$

Taking the logarithm of f gives

$$\ln\left(f(y;\theta)\right) = (1-y) \cdot \ln\left(1-\theta\right) + y \cdot \ln\left(\theta\right).$$

The first derivative of $\ln\left(f(y;\theta)\right)$ with respect to θ is

$$\frac{d\ln\left(f(y;\theta)\right)}{d\theta} = -\frac{1-y}{1-\theta} + \frac{y}{\theta},$$

and the second derivative is

$$\frac{d^2 \ln\left(f(y;\theta)\right)}{d\theta^2} = -\frac{1-y}{(1-\theta)^2} - \frac{y}{\theta^2}.$$

The first derivative squared is

$$\begin{aligned}
\left(\frac{d\ln\left(f(y;\theta)\right)}{d\theta}\right)^2 &= \frac{(1-y)^2}{(1-\theta)^2} + \frac{y^2}{\theta^2} - 2 \cdot \frac{1-y}{1-\theta} \cdot \frac{y}{\theta} \\
&= \frac{1-y}{(1-\theta)^2} + \frac{y}{\theta^2} - 2 \cdot \frac{1-y}{1-\theta} \cdot \frac{y}{\theta} \\
&= \frac{1-y}{(1-\theta)^2} + \frac{y}{\theta^2},
\end{aligned}$$

where the second equation follows because $y^2 = y$ and $(1 - y)^2 = (1 - y)$ for $y = 0, 1$, and the third equality follows because $(1 - y) \cdot y = 0$ for $y = 0, 1$.

Now, taking the expectation of the second derivative with respect to the distribution of Y yields

$$E\left(\frac{d^2 \ln\left(f(Y;\theta)\right)}{d\theta^2}\right) = E\left(-\frac{1-Y}{(1-\theta)^2} - \frac{Y}{\theta^2}\right) = -\frac{1-\theta_0}{(1-\theta)^2} - \frac{\theta_0}{\theta^2},$$

where we used that since $Y \sim Ber(\theta_0)$, then

$$E(Y) = \theta_0.$$

Similarly, take the expectation of the first derivative squared

$$E\left(\left(\frac{d \ln\left(f(Y;\theta)\right)}{d\theta}\right)^2\right) = E\left(\frac{1-Y}{(1-\theta)^2} + \frac{Y}{\theta^2}\right) = \frac{1-\theta_0}{(1-\theta)^2} + \frac{\theta_0}{\theta^2}.$$

Now evaluate the two expectations in θ_0 :

$$
\begin{aligned}
J &= E\left(\left(\frac{d \ln\left(f(Y;\theta)\right)}{d\theta}\bigg|_{\theta=\theta_0}\right)^2\right) \\
&= \frac{1-\theta_0}{(1-\theta_0)^2} + \frac{\theta_0}{\theta_0^2} \\
&= \frac{1}{1-\theta_0} + \frac{1}{\theta_0} \\
&= \frac{\theta_0}{(1-\theta_0)\theta_0} + \frac{1-\theta_0}{(1-\theta_0)\theta_0} \\
&= \frac{1}{(1-\theta_0)\,\theta_0},
\end{aligned}
$$

and

$$
\begin{aligned}
H &= E\left(\frac{d^2 \ln\left(f(Y;\theta)\right)}{d\theta^2}\bigg|_{\theta=\theta_0}\right) \\
&= -\frac{1-\theta_0}{(1-\theta_0)^2} - \frac{\theta_0}{\theta_0^2} \\
&= -\frac{1}{1-\theta_0} - \frac{1}{\theta_0} \\
&= -\frac{\theta_0}{(1-\theta_0)\theta_0} - \frac{1-\theta_0}{(1-\theta_0)\theta_0} \\
&= -\frac{1}{(1-\theta_0)\,\theta_0}.
\end{aligned}
$$

In conclusion, by Theorem 18.1, the asymptotic distribution of the ML estimator $\widehat{\theta}$ is

$$\widehat{\theta} \overset{A}{\sim} N\left(\theta_0, \frac{J}{H^2}\frac{1}{n}\right) = N\left(\theta_0, \frac{(1-\theta_0)\theta_0}{n}\right).$$

Comparing J and H in Example 18.1 shows that $J = -H$. This equality between the expected value of the square of the score and minus the expected value of the derivative of the score is a general result, known as the *information equality*. It is stated next.

Theorem 18.2 (Information equality) *Suppose the conditions of Theorem 18.1 hold. Then*

$$
E\left(\left(\left.\frac{d\ln\left(f(Y;\theta)\right)}{d\theta}\right|_{\theta=\theta_0}\right)^2\right) = -E\left(\left.\frac{d^2\ln\left(f(Y;\theta)\right)}{d\theta^2}\right|_{\theta=\theta_0}\right)
$$

or, using the J and H notation,

$$
J = -H.
$$

A consequence of the information equality is that the asymptotic distribution of the ML estimator can be written as

$$
\widehat{\theta} \overset{A}{\sim} N\left(\theta_0, \frac{1}{-H}\cdot\frac{1}{n}\right) = N\left(\theta_0, \frac{1}{J}\cdot\frac{1}{n}\right),
$$

cf. Theorem 18.1.

By definition of the maximum likelihood estimator, the score function will necessarily equal zero when evaluated at the ML estimator, i.e.

$$
s\left(\widehat{\theta}; \left(\widetilde{Y}_1,\ldots,\widetilde{Y}_n\right)\right) = 0.
$$

The next result states that the expected value of the score is zero at the true parameter θ_0.

Theorem 18.3 *Suppose the conditions of Theorem 18.1 hold. Then*

$$
E\left(\left.\frac{d\ln\left(f(Y;\theta)\right)}{d\theta}\right|_{\theta=\theta_0}\right) = 0,
$$

and, as a consequence,

$$
E\left(s\left(\theta_0; \left(\widetilde{Y}_1,\ldots,\widetilde{Y}_n\right)\right)\right) = 0.
$$

A consequence of Theorem 18.3 is that, since the expected value of the score function $s()$ evaluated at θ_0 equals 0, then J is also the variance of the score function when evaluated at θ_0.

18.3.2   Estimation of the variance of the asymptotic distribution

In order to use the asymptotic distribution of the ML estimator in practice, an estimator of the variance matrix is needed. Recall that the asymptotic distribution of the ML estimator is (Theorem 18.1)

$$
\widehat{\theta} \overset{A}{\sim} N\left(\theta_0, \frac{J}{H^2}\cdot\frac{1}{n}\right),
$$

which by the information equality (Theorem 18.2) is equivalent to

$$
\widehat{\theta} \overset{A}{\sim} N\left(\theta_0, \frac{1}{J}\cdot\frac{1}{n}\right) = N\left(\theta_0, \frac{1}{-H}\cdot\frac{1}{n}\right). \tag{18.4}
$$

These relations show that estimators of the asymptotic variance of the ML estimator should either estimate J, H, or J/H^2. The next definition proposes different estimators of J and H.

Definition 18.1 (Estimators of the Jacobian and Hessian, one parameter)
Suppose $\ln\left(f(y;\theta)\right)$ is twice continuously differentiable in θ for all y and let $\widehat{\theta}$ be the ML estimator of the parameter $\theta_0 \in \mathbb{R}$. The following quantities are estimators of either the Jacobian J or the Hessian H.

1. **Expected information estimators:**

$$\widehat{J}_{EX} \;=\; E_Y\left(\left(\left.\frac{d\ln\left(f(Y;\theta)\right)}{d\theta}\right|_{\theta=\widehat{\theta}}\right)^2\right),$$

$$\widehat{H}_{EX} \;=\; E_Y\left(\left(\left.\frac{d^2\ln\left(f(Y;\theta)\right)}{d\theta^2}\right|_{\theta=\widehat{\theta}}\right)\right),$$

where $E_Y()$ denotes that the expectation is to be taken with respect to the random variable or vector Y, and not with respect to $\widehat{\theta}$.

2. **Outer product of the scores estimator:**

$$\widehat{J}_{OP} = \frac{1}{n}\sum_{i=1}^{n}\left(\left.\frac{d\ln\left(f(\widetilde{Y}_i;\theta)\right)}{d\theta}\right|_{\theta=\widehat{\theta}}\right)^2.$$

3. **Observed Hessian estimator:**

$$\widehat{H}_{OH} = \frac{1}{n}\sum_{i=1}^{n}\left.\frac{d^2\ln\left(f(\widetilde{Y}_i;\theta)\right)}{d\theta^2}\right|_{\theta=\widehat{\theta}}.$$

The expected information estimators, $\widehat{J}_{EX}$ and $\widehat{H}_{EX}$, are calculated analytically using the distribution function $f(y;\theta)$. If such calculations are possible, these estimators are very useful. The latter two estimators, $\widehat{J}_{OP}$ and $\widehat{H}_{OH}$, use the sample to directly estimate the Jacobian and Hessian, respectively. Also in this case, it is necessary to evaluate derivatives of the log distribution function. However, if the ML estimator is found by numerically maximizing the log-likelihood using computer software, it is often possible to numerically calculate the Hessian as well. More precisely, when numerical optimization is performed using an approach that numerically evaluates the derivatives of the log-likelihood function, the software may output the Observed Hessian estimator $\widehat{H}_{OH}$, meaning that H can be estimated directly from the sample without the need for calculating derivatives. This can be very useful in complex problems where analytic calculation of the derivatives of the log distribution function is difficult or impossible.

The following theorem states that under certain regularity conditions, the estimators given in Definition 18.1 are consistent estimators of either J or H.

Theorem 18.4 (Consistency of Jacobian and Hessian estimators, one parameter) *Suppose the conditions of Theorem 18.1 hold. Then*

$$\plim_{n\to\infty} \widehat{J}_{EX} = \plim_{n\to\infty} \widehat{J}_{OP} = J,$$

and

$$\plim_{n\to\infty} \widehat{H}_{EX} = \plim_{n\to\infty} \widehat{H}_{OH} = H.$$

Summing up, Theorem 18.1 shows that the asymptotic distribution of the ML estimator $\widehat{\theta}$ is

$$\widehat{\theta} \stackrel{A}{\sim} N\left(\theta_0, V\cdot\frac{1}{n}\right),$$

where, due to the information equality (Theorem 18.2), the *asymptotic variance V of $\widehat{\theta}$* can be written as

$$V = \frac{J}{H^2} = \frac{1}{J} = -\frac{1}{H}.$$

Theorem 18.4, together with the continuous mapping theorem (Theorem 13.2), imply that the following

$$\widehat{V}_{EX} = \frac{1}{\widehat{J}_{EX}} = -\frac{1}{\widehat{H}_{EX}},$$

$$\widehat{V}_{OP} = \frac{1}{\widehat{J}_{OP}},$$

$$\widehat{V}_{OH} = -\frac{1}{\widehat{H}_{OH}},$$

$$\widehat{V}_{SW} = \frac{\widehat{J}_{OP}}{\widehat{H}_{OH}^2},$$

are all consistent estimators of the asymptotic variance matrix V. Note that the "SW" in $\widehat{V}_{SW}$ stands for "sandwich", a name that derives after the form $J/H^2 = H^{-1}JH^{-1}$, where J is "sandwiched" between a pair of H^{-1}s.

Example 18.2 (Bernoulli, continued) *The asymptotic distribution of the ML estimator $\widehat{\theta}$ was shown above to be*

$$\widehat{\theta} \overset{A}{\sim} N\left(\theta_0, (1-\theta_0)\theta_0 \frac{1}{n}\right).$$

The asymptotic variance is

$$V = (1-\theta_0)\theta_0.$$

An estimator of V is found by plugging in the estimate $\widehat{\theta}$ in place of θ_0,

$$\widehat{V} = (1-\widehat{\theta})\widehat{\theta}.$$

This is the expected information variance estimator, $\widehat{V}_{EX}$.

    For the Bernoulli population, the variance estimators based on either $\widehat{H}_{OH}$ or $\widehat{J}_{OP}$ give the same result as $\widehat{V}_{EX}$. For instance, for $\widehat{J}_{OP}$, this can be seen as follows,

$$
\begin{aligned}
\widehat{J}_{OP} &= \frac{1}{n}\sum_{i=1}^{n}\left(\frac{1-\widetilde{Y}_i}{\left(1-\widehat{\theta}\right)^2} + \frac{\widetilde{Y}_i}{\widehat{\theta}^2}\right) \\
&= \frac{\frac{1}{n}\sum_{i=1}^{n} 1-\widetilde{Y}_i}{\left(1-\widehat{\theta}\right)^2} + \frac{\frac{1}{n}\sum_{i=1}^{n}\widetilde{Y}_i}{\widehat{\theta}^2} \\
&= \frac{1-\widehat{\theta}}{\left(1-\widehat{\theta}\right)^2} + \frac{\widehat{\theta}}{\widehat{\theta}^2} \\
&= \frac{1}{\left(1-\widehat{\theta}\right)\widehat{\theta}},
\end{aligned}
$$

and, thus,

$$\widehat{V}_{OP} = \frac{1}{\widehat{J}_{OP}} = \left(1-\widehat{\theta}\right)\widehat{\theta} = \widehat{V}_{EX}.$$

Although it is not the case in Example 18.2, the four estimators $\widehat{V}_{EX}$, $\widehat{V}_{OH}$, $\widehat{V}_{OP}$, $\widehat{V}_{SW}$ are, in general, different. Typically, $\widehat{V}_{EX}$ and $\widehat{V}_{OH}$ are the more precise estimators of the variance of the ML estimator. The "sandwich" estimator $\widehat{V}_{SW}$ is mostly relevant in cases where misspecification of the distribution of the population characteristic is suspected. This case is briefly discussed in the next section.

18.3.3 ML estimation in case of misspecification

Throughout these chapters on ML estimation, it is assumed that the model $f(y;\theta)$ is correctly specified, in the sense that the population characteristic has the distribution $f(y;\theta_0)$ with $\theta_0 \in \Theta$. That is, we are assuming that the functional form of the distribution of Y is known and that the only unknown is the parameter θ_0 which we are trying to estimate. In practice, however, we might not know the true functional form of the distribution of Y. If we are using the class of functions $f(y;\theta)$ to do ML estimation but the true distribution of Y is not among these, say Y is distributed according to the PMF, or PDF, $g(y;\theta_0)$ where $g() \neq f()$, then we say that the model of Y is *misspecified*. ML estimation in the case of misspecification is known as *quasi maximum likelihood estimation* and can be understood in terms of minimizing the Kullback-Leibler divergence (Section 17.6). In this case, the results studied in these chapters do not hold. It turns out, however, that the ML estimator does possess desirable properties in some misspecified cases. Indeed, it can be shown that under certain conditions, consistency and even a asymptotic distribution result like (18.1) may nevertheless hold even when the model is misspecified.

Although we will not go into further detail regarding quasi maximum likelihood estimation in this book, we note that the sandwich estimator can be useful in cases of misspecification. Importantly, in this case, the information equality (Theorem 18.2) does not hold, which means that $J \neq -H$ and hence J/H^2 does not reduce to $-1/H$. In case of misspecification, because of the violation of the information equality, the variance of the ML estimator should be estimated using the sandwich estimator $\widehat{V}_{SW}$, since all the other estimators studied above rely on the simplification provided by the information equality.

18.4 Efficiency of the ML estimator

In this section, we consider results on how precise any estimator, not just the ML estimator, can possibly be. Results of this type are called *efficiency bounds*. Efficiency bounds can be derived for certain classes of estimators, such as unbiased estimators. They can also be derived based on the asymptotic distributions of estimators. In that case, they are called *asymptotic efficiency bounds*.

We compare the performance of the ML estimator to the efficiency bounds. As we will see, it turns out that the ML estimator is asymptotically efficient. In other words, in large samples the ML estimator makes efficient use of the information in terms of estimating the unknown parameters.

18.4.1 The Cramér-Rao lower bound

It is possible to define measures that quantify how much information a distribution contains about its parameters. This is useful knowledge to assess how precisely a parameter may be estimated. One such measure is the *Fisher information*, defined next

Definition 18.2 (Fisher information) *Let $f(y; \theta_0)$ be the distribution of Y. Assume that $\ln f(y; \theta)$ is continuously differentiable in θ for all y. Then the **Fisher information** $I(\theta_0)$ is defined as*

$$I(\theta_0) = E\left(\left(\left.\frac{\partial \ln\left(f(Y; \theta)\right)}{\partial \theta}\right|_{\theta=\theta_0}\right)^2\right).$$

Due to the information matrix equality (Theorem 18.2), when $\ln\left(f(y; \theta)\right)$ is twice continuously differentiable in θ, the Fisher information is also equal to the negative of the expected value of the second derivative of $\ln\left(f(Y; \theta_0)\right)$, i.e.

$$I(\theta_0) = -E\left(\left.\frac{\partial^2 \ln\left(f(Y; \theta)\right)}{\partial \theta^2}\right|_{\theta=\theta_0}\right).$$

Further, from (18.4), which is a consequence of Theorem 18.1 and the information matrix equality, we see that the asymptotic variance of the ML estimator is equal to the inverse Fisher matrix, i.e.

$$\widehat{\theta} \overset{A}{\sim} N\left(\theta_0, I(\theta_0)^{-1} \cdot \frac{1}{n}\right).$$

The second derivative of a function describes the curvature of the function and, hence, the Fisher information is linked to the curvature in θ of the logarithm to the distribution function $f(y; \theta)$ around $\theta = \theta_0$. A function without curvature has Fisher information equal to 0.

Intuitively, if $\ln f(y; \theta)$ has a lot of curvature around $\theta = \theta_0$, then it is easier to pinpoint the maximum, θ_0, than in a case where $\ln f(y; \theta)$ is very flat around $\theta = \theta_0$, i.e. where the curvature is low. For this reason, we can typically estimate a parameter from a distribution more precisely for a given sample size if the Fisher information is large (high curvature) compared to a case, with the same sample size, but with a lower Fisher information (low curvature).

Example 18.3 (Bernoulli, continued) *The Fisher information for a Bernoulli distribution $Ber(\theta_0)$ is*

$$I(\theta_0) = \frac{1}{1 - \theta_0} + \frac{1}{\theta_0} = \frac{1}{(1 - \theta_0)\,\theta_0}.$$

It can be seen that the Fisher information is lowest in case $\theta_0 = 0.5$, where $I(0.5) = 4$ and largest for values of θ_0 close to 0 or 1. Therefore, we can estimate θ_0 more precisely for a given sample size if e.g. θ_0 is close to 0 or 1 rather than close to 0.5. The intuition behind this is that if θ_0 is, say, very small, e.g. $\theta_0 = 0.00001$, then with a very high probability almost all elements in a sample will be zero, thus indicating a very low value θ_0. It is in this sense that such a sample contains a lot of information about the unknown parameter θ_0.

Recall that the precision of an estimator is defined using a risk function. In what follows, we assume the squared loss function and, thus, the risk function is the mean squared error (MSE; Definition 11.6) function. Efficiency of an estimator is then assessed by comparing the MSE of the estimator to the MSEs of other estimators in a given class. If the class of estimators under consideration are unbiased estimators, then the MSE may be assessed using the variance of the estimator.

For a given sample size, it is possible to find a lower bound for the variance of an estimator among estimators which are unbiased. The result is called the *Cramér-Rao lower bound* and is stated in the next theorem.

Theorem 18.5 (Cramér-Rao lower bound) *Suppose the conditions of Theorem 18.1 hold. Let $\widehat{\theta}$ be an unbiased estimator of the single parameter θ_0, that is, $E(\widehat{\theta}) = \theta_0$. Then*

$$Var\left(\widehat{\theta}\right) \geq I(\theta_0)^{-1} \cdot \frac{1}{n}, \tag{18.5}$$

where $I(\theta_0)$ is the Fisher information.

Note that while Theorem 18.5 relies on the regularity conditions of Theorem 18.1, it is not a statement about the ML estimator. The theorem holds for any unbiased estimator where the underlying distribution function of the population satisfies the conditions.

As Theorem 18.5 shows, the Cramér-Rao lower bound of the inverse information matrix divided by the sample size provides a lower bound on the variance of any unbiased estimator. We say that an unbiased estimator with variance equal to the Cramér-Rao lower bound is *efficient* among unbiased estimators.

It is worth emphasizing several properties of the Cramér-Rao lower bound (18.5). Firstly, it does not depend on the estimator $\widehat{\theta}$. It only requires that the class of estimators consists exclusively of unbiased estimators satisfying the quite mild conditions of Theorem 18.1. Secondly, the Cramér-Rao lower bound does not guarantee that there exists an unbiased estimator with a variance equal to the Cramér-Rao lower bound. In case no estimator in the class achieves the Cramér-Rao lower bound, there may still be an estimator which has a variance not larger than any other estimator in the class. An estimator which has the lowest variance among all unbiased estimators, is called a *Uniform Minimum Variance Unbiased* (UMVU) *estimator*. If an unbiased estimator achieves the Cramér-Rao lower bound, it is necessarily an UMVU estimator. Although there are some special cases where the ML estimator is an UMVU estimator, the ML estimator is, in general, biased in finite samples and will therefore not be an UMVU estimator. When the sample size increases toward infinity, however, it turns out that the ML estimator, in the limit is unbiased and achieves the Cramér-Rao lower bound. This is studied in the next section.

18.4.2   Asymptotic efficiency

As we saw in Chapter 17, the ML estimator is, in general, biased in finite samples. Therefore, Theorem 18.5 does not apply to the ML estimator in general. It is possible, however, to use the Cramér-Rao lower bound on biased estimators as long as these estimators are consistent and asymptotically normal. This result is given in the following theorem.

Theorem 18.6 (Asymptotic lower bound) *Suppose the conditions of Theorem 18.1 hold. Assume $\widehat{\theta}$ is a consistent estimator of θ_0 and that it has the asymptotic distribution*

$$\widehat{\theta} \overset{A}{\sim} N\left(\theta_0, V \cdot \frac{1}{n}\right),$$

where $V > 0$ is some number. Then there exists a lower bound on the asymptotic variance V of $\widehat{\theta}$ given by

$$V \geq (I(\theta_0))^{-1}, \tag{18.6}$$

where $I(\theta_0)$ is the Fisher information.

If it is the case that the number V in Theorem 18.6 fulfils

$$V = (I(\theta_0))^{-1}, \quad for\ every\ \theta_0 \in \Theta,$$

then the asymptotic variance V of the estimator $\widehat{\theta}$ satisfies (18.6) with equality, i.e. the variance of the estimator $\widehat{\theta}$ achieves the Cramér-Rao lower bound as the sample size increases toward infinity. In this case, we say that the estimator $\widehat{\theta}$ is *asymptotically efficient.*

On a technical note, that $\widehat{\theta}$ is consistent with an asymptotic normal distribution implies that $\widehat{\theta}$ is asymptotically unbiased, where we define asymptotically unbiased as the expected value of the asymptotic distribution of $\widehat{\theta}$ being equal to θ_0. Hence, with an asymptotically unbiased estimator, the inequality in (18.6) is similar to the Cramér-Rao lower bound for unbiased estimators. The difference is that the result in Theorem 18.6 is given using the asymptotic distribution and, thus, provides a definition of asymptotic efficiency.

Similarly to the situation studied above, we remark that Theorem 18.6 is not a statement about the ML estimator. The theorem holds for any estimator that is consistent and asymptotically normal, as long as the underlying distribution function of the population satisfies the conditions of Theorem 18.1. However, Equation (18.4), which is a consequence of Theorem 18.1, together with Definition 18.2, shows that the asymptotic variance of the ML estimator equals the Cramér-Rao lower bound. Hence, the ML estimator is asymptotically efficient. This result is one of the main attractions of the ML estimator because it implies that the ML estimator efficiently uses the information in a large sample.

18.5 Confidence intervals

Confidence intervals based on ML estimators can be constructed using the approach laid out in Chapter 14. In particular, the parametric bootstrap approach suggested for the ML estimator in Section 18.2 lends itself directly to constructing bootstrap confidence intervals as in Section 14.5. Similarly, the asymptotic normality of the ML estimator (Theorem 18.1) lends itself directly to constructing approximate confidence intervals as in Section 14.6.

The following sections briefly summarize how these confidence intervals may be constructed in the case of the ML estimator. We refer to Chapter 14 for further details and theoretical justifications behind the confidence intervals.

18.5.1 Bootstrap confidence intervals

In ML estimation, we assume that the distribution of Y is known apart from the parameters to be estimated. This implies that the parametric bootstrap can be applied since the bootstrap population can be estimated by $f\left(y;\widehat{\theta}\right)$. We shall use the studentized approach to construct the confidence interval. Thus, for an $\alpha \in (0, 0.5]$, the $(1-\alpha)$-bootstrap confidence interval for $\widehat{\theta}$ is

$$\widehat{CI}^*_{1-\alpha,student} = \left[\widehat{\theta} + t^*_{\alpha/2} \cdot \sqrt{\widehat{V}/n} \; , \; \widehat{\theta} + t^*_{1-\alpha/2} \cdot \sqrt{\widehat{V}/n}\right],$$

where $\widehat{V}$ is an estimator of the asymptotic variance V of $\widehat{\theta}$, see Section 18.3.2, and $t^*_{\alpha/2}$, $t^*_{1-\alpha/2}$ are the $\alpha/2$ and $(1-\alpha/2)$ quantiles, respectively, in the bootstrap distribution of the bootstrap statistics

$$t^*_b = \frac{\widehat{\theta}^* - \widehat{\theta}}{\sqrt{\widehat{V}^*/n}}.$$

The bootstrap statistics is calculated by drawing observations $\widetilde{Y}_i^*$, $i = 1, \ldots, n$, from the distribution $f\left(y; \widehat{\theta}\right)$, and then calculating $\widehat{\theta}^*$ and $\widehat{V}^*$ using the ML estimator to form the statistics t_b^*.

18.5.2 Asymptotic confidence intervals

Since the ML estimator $\widehat{\theta}$ is asymptotically normally distributed (Theorem 18.1), we can construct an asymptotic confidence interval for $\widehat{\theta}$ using Theorem 14.3. Thus, for an $\alpha \in (0, 0.5]$, an approximate $(1 - \alpha)$-confidence interval for $\widehat{\theta}$ is

$$\widehat{CI}_{1-\alpha,asymp} = \left[\widehat{\theta} + q_{\alpha/2} \cdot \sqrt{\widehat{V}/n} \, , \, \widehat{\theta} + q_{1-\alpha/2} \cdot \sqrt{\widehat{V}/n}\right],$$

where q_p is the p-quantile of the standard normal $N(0, 1)$ distribution and $\widehat{V}$ is an estimator of the asymptotic variance V of $\widehat{\theta}$, see Section 18.3.2. Recall that for the standard normal distribution, we have $q_{\alpha/2} = -q_{1-\alpha/2}$, so that we can also write the confidence interval as

$$\widehat{CI}_{1-\alpha,asymp} = \left[\widehat{\theta} - q_{1-\alpha/2} \cdot \sqrt{\widehat{V}/n} \, , \, \widehat{\theta} + q_{1-\alpha/2} \cdot \sqrt{\widehat{V}/n}\right].$$

18.6 Hypothesis testing

In this section, we consider how to test hypotheses about the parameter θ_0 in the distribution of Y when using the ML estimator $\widehat{\theta}$ as estimator of θ_0. Consider the following null hypothesis and two-sided alternative,

$$\begin{aligned} H_0 &: \quad \theta_0 = \theta_{H_0}, \\ H_1 &: \quad \theta_0 \neq \theta_{H_0}, \end{aligned} \tag{18.7}$$

where θ_{H_0} is a number in the interior of the parameter set Θ. In the following sections, we present three different ways of conducting the test of these hypotheses, based on ML estimation. Then, in Section 18.6.4, we comment on the differences and similarities between the three approaches.

18.6.1 The generalized likelihood ratio test

In this subsection, we generalize the likelihood ratio test of two simple hypotheses (Definition 16.8) to the more practically relevant test of (18.7). The new test is called the *generalized likelihood ratio* (GLR) *test*. It is constructed using the GLR test statistics, defined next.

Definition 18.3 (Generalized likelihood ratio test statistics) *Let $\theta_{H_0} \in \mathbb{R}$ be a number in the interior of the parameter space Θ and $\widetilde{Y}$ a random sample of Y with distribution $f(y; \theta_0)$. The **generalized likelihood ratio (GLR) test statistics** is defined as*

$$GLR = -2 \cdot \left(l\left(\theta_{H_0}; \widetilde{Y}\right) - l\left(\widehat{\theta}; \widetilde{Y}\right)\right), \tag{18.8}$$

where $l\left(\theta; \widetilde{Y}\right) = \ln\left(L(\theta; \widetilde{Y})\right)$ is the log-likelihood function and $\widehat{\theta}$ is the ML estimator of θ_0.

Note that since θ_{H_0} is in Θ and because $\widehat{\theta}$, by definition, maximizes the log-likelihood function, then it is necessarily the case that

$$l\left(\widehat{\theta};\widetilde{Y}\right) \geq l\left(\theta_{H_0};\widetilde{Y}\right).$$

Therefore, the generalized likelihood ratio GLR is always non-negative, i.e. $GLR \geq 0$.

We can think of the GLR as measuring the "distance" between the likelihood of the data under $H_0 : \theta = \theta_{H_0}$ and the maximized likelihood under the alternative $H_1 : \theta \neq \theta_{H_0}$. If this distance is large, it suggests that H_1 is true, because this indicates that the restriction imposed by H_0 implies a lower likelihood. To determine what is "large", we need the sampling distribution of the GLR test statistics, and, typically, we need to estimate this sampling distribution, e.g. using bootstrap or asymptotic approaches. The next theorem presents the asymptotic distribution of the GLR test statistics.

Theorem 18.7 (Asymptotic distribution of the GLR test statistics) *Consider the GLR test statistic (18.8) and the hypotheses given in (18.7). Suppose the conditions of Theorem 18.1 hold. Then the asymptotic distribution of the GLR test statistics is*

$$GLR \overset{A}{\sim} \chi^2(1, \lambda),$$

as $n \to \infty$, where $\chi^2(1, \lambda)$ is the non-central chi-square distribution (Section 8.6.1) with one degree of freedom and non-centrality parameter λ, given by

$$\lambda^2 = \frac{(\theta_0 - \theta_{H_0})^2}{Var\left(\widehat{\theta}\right)}.$$

Under the null hypothesis, $\lambda = 0$, i.e. the GLR test statistics is χ^2 distributed with 1 degree of freedom, whereas under the alternative hypothesis, $\lambda^2 \to \infty$, as $n \to \infty$.

The sampling distribution of the GLR test statistics can also be estimated by a bootstrap approach. As discussed above, since the population distribution with ML estimation is assumed known, except for the unknown parameter θ_0, we can use a parametric bootstrap. In particular, we can apply a Monte Carlo experiment to derive the distribution of the GLR test statistics for a given value θ_H of θ. The distribution under H_0 is found by setting $\theta_H = \theta_{H_0}$ and under H_1 by setting $\theta_H \neq \theta_{H_0}$. The algorithm is as follows.

Algorithm 18.6.1 (Bootstrap distribution of the GLR test statistics)
Let $B \geq 1$ denote the number of bootstrap samples.

1. *For each $b = 1, \ldots, B$, draw a bootstrap sample $\left(\widetilde{Y}_i^*\right)$, $i = 1, \ldots, n$ from $f(y; \theta_H)$. Then calculate GLR_b^* using (18.8) and the bootstrap sample.*

2. *The bootstrap distribution of the GLR test statistics at $\theta = \theta_H$ is the empirical distribution of GLR_b^*, $b = 1, \ldots, B$.*

The bootstrap distribution of the GLR test statistics under H_0 is found for $\theta_H = \theta_{H_0}$. To calculate the power of the GLR test using the bootstrap distribution, one approach is to first calculate the critical value of the test using a bootstrap distribution satisfying $H_0 : \theta_0 = \theta_{H_0}$, i.e. using $\theta_H = \theta_{H_0}$. Then generate the bootstrap distribution at an alternative $\theta_H \neq \theta_{H_0}$ of interest and calculate the rejection probability using the critical value found with the bootstrap distribution under H_0.

18.6.2 The Wald test

Another test of the hypotheses (18.7) is the so-called *Wald test*. Compared to the GLR test, which we saw relied on the distance between the likelihood under H_0 and H_1, the Wald test compares the ML estimator $\widehat{\theta}$ directly with the hypothesized value of θ under the null, i.e. with θ_{H_0}. The definition of the Wald test statistics is given next.

Definition 18.4 (Wald test statistics with ML estimator) *Let $\theta_{H_0} \in \mathbb{R}$ satisfy H_0. The **Wald test statistics** is defined as*

$$Wald = \frac{(\widehat{\theta} - \theta_{H_0})^2}{\widehat{Var}(\widehat{\theta})},$$

(18.9)

where $\widehat{\theta}$ is the ML estimator of θ_0 and $\widehat{Var}(\widehat{\theta})$ is an estimator of $Var(\widehat{\theta})$.

The idea behind the Wald test is to measure how close the null hypothesis is to being satisfied by the ML estimator. To decide how close is "close", we need the distribution of the Wald test statistics.

Theorem 18.8 (Asymptotic distribution of the Wald test statistics) *Consider the Wald test statistic (18.9) and the hypotheses given in (18.7). Suppose the conditions of Theorem 18.1 hold and let $\widehat{Var}(\widehat{\theta}) = \widehat{V}/n$, where $\widehat{V}$ is one of the estimators of the asymptotic variance of $\widehat{\theta}$ given in Section 18.3.2. Then the asymptotic distribution of the Wald test statistics is*

$$Wald \overset{A}{\sim} \chi^2(1, \lambda),$$

as $n \to \infty$, where $\chi^2(1, \lambda)$ is the non-central chi-square distribution (Section 8.6.1) with one degree of freedom and non-centrality parameter λ, given by

$$\lambda^2 = \frac{(\theta_0 - \theta_{H_0})^2}{Var\left(\widehat{\theta}\right)}.$$

Under the null hypothesis, $\lambda = 0$, i.e. the Wald test statistics is χ^2 distributed with 1 degree of freedom, whereas under the alternative hypothesis, $\lambda^2 \to \infty$, as $n \to \infty$.

Notice that, the Wald test statistics is asymptotic pivotal under H_0, i.e. asymptotically it does not rely on the value of the parameter of the model. This is a result of the studentization in (18.9).

Theorem 18.8 shows that the asymptotic distribution of the Wald test statistics is the same as the asymptotic distribution of the GLR test statistics in Theorem 18.7. This implies that both tests have the same power function based on the asymptotic distributions. Their exact sampling distributions in finite samples, however, need not be the same. Typically, for a given level of Type I error, the GLR test tends to be more powerful than the Wald test.

The sampling distribution of the Wald test statistics can also be estimated by a bootstrap approach. The algorithm is the same as the Algorithm 18.6.1, except the GLR test statistics is replaced by the Wald test statistics.

18.6.3 The Lagrange multiplier test

A third test of the hypotheses (18.7) is the *Lagrange multiplier* (LM) *test*. Compared to the tests above, the LM test only relies on the value of θ under the null. Therefore, with only one parameter, it does not require calculation of the ML estimator. The definition of the LM test statistics is given next.

Definition 18.5 (Lagrange multiplier test statistics with ML estimator)
*Let $\theta_{H_0} \in \mathbb{R}$ be a number in the interior of the parameter space Θ and $\widetilde{Y}$ a random sample of Y with distribution $f(y; \theta_0)$. The **Lagrange multiplier (LM) test statistics** is defined as*

$$LM = \frac{\left(\left. \frac{dl(\theta; \widetilde{Y})}{d\theta} \right|_{\theta = \theta_{H_0}} \right)^2}{\widehat{Var} \left(\left. \frac{dl(\theta; \widetilde{Y})}{d\theta} \right|_{\theta = \theta_{H_0}} \right)}, \tag{18.10}$$

where $l\left(\theta; \widetilde{Y}\right) = \ln\left(L(\theta; \widetilde{Y})\right)$ is the log-likelihood function and $\widehat{Var}\left(\left. \frac{dl(\theta; \widetilde{Y})}{d\theta} \right|_{\theta = \theta_{H_0}} \right)$ is a consistent estimator of $Var\left(\left. \frac{dl(\theta; \widetilde{Y})}{d\theta} \right|_{\theta = \theta_{H_0}} \right).$

The idea behind the LM test is to measure the slope of the log distribution function at the value of θ specified under H_0, i.e. at θ_{H_0}. Since the slope (derivative) of the log-likelihood function is also denoted the score (Definition 17.4), the LM test is sometimes also called the *score test*. We know from the first order condition attached to the ML estimator that, if H_0 is true, the score should be close to 0 for $\theta = \theta_{H_0}$, which is the value of the slope at the optimum.

The next theorem presents the asymptotic distribution of the LM test statistics.

Theorem 18.9 (Asymptotic distribution of the LM test statistics) *Consider the LM test statistic (18.10) and the hypotheses given in (18.7). Suppose the conditions of Theorem 18.1 hold. Then the asymptotic distribution of the LM test statistics is*

$$LM \overset{A}{\sim} \chi^2(1, \lambda),$$

as $n \to \infty$, where $\chi^2(1, \lambda)$ is the non-central chi-square distribution (Section 8.6.1) with one degree of freedom and non-centrality parameter λ, given by

$$\lambda^2 = \frac{(\theta_0 - \theta_{H_0})^2}{Var\left(\widehat{\theta}\right)}.$$

Under the null hypothesis, $\lambda = 0$, i.e. the LM test statistics is χ^2 distributed with 1 degree of freedom, whereas under the alternative hypothesis, $\lambda^2 \to \infty$, as $n \to \infty$.

Notice that in (18.10), the LM test statistics is asymptotic pivotal under H_0, i.e. asymptotically it does not rely on the value of the parameter of the model. This is due to the studentization of the score in (18.10).

Theorem 18.9 shows that the asymptotic distribution of the LM test statistics is the same as the asymptotic distribution of the GLR test statistics in Theorem 18.7 and the Wald test statistics in Theorem 18.8. Their exact sampling distributions in finite samples, however, need not be the same. Typically, for a given level of Type I error, the GLR test tends to be more powerful that the LM test. If it is too complicated to compute the ML estimator, then the LM test can be used.

The sampling distribution of the LM test statistics can also be estimated by a bootstrap distribution. The algorithm is the same as the Algorithm 18.6.1, except the GLR test statistics is replaced by the LM test statistics.

18.6.4 The trinity of tests

The discussion in Sections 18.6.1–18.6.3 has suggested three different ways to test the hypotheses (18.7) using maximum likelihood estimation, namely the generalized likelihood ratio test (GLR; Definition 18.3), the Wald test (Wald; Definition 18.4), and the Lagrange multiplier test (LM; Definition 18.5). Together, they are sometimes referred to as the *trinity of tests*. As we saw, the GLR test compares the likelihood of the data assuming $\theta = \theta_{H_0}$ (i.e. under H_0) with the likelihood assuming $\theta = \widehat{\theta}$ (i.e. under H_1), whereas the Wald test directly compares the ML estimator $\widehat{\theta}$ to the value of θ under H_0, i.e. to θ_{H_0}. The LM test, in contrast, does not rely on the ML estimator $\widehat{\theta}$ itself, but instead evaluates the score function at the value of θ under H_0, i.e. at $\theta = \theta_{H_0}$. Hence, the choice of which test to use may be informed by how difficult or costly it is to obtain the ML estimator and whether the score function can be calculated analytically. Although the finite sample distribution of the three test statistics may be different, which may in turn result in different bootstrap distributions, they are *asymptotically equivalent* in the sense that they share the same asymptotic distribution as given in Theorems 18.7, 18.8, and 18.9. In particular, they are all asymptotically χ^2-distributed with one degree of freedom under H_0.

18.7 Proofs

Proof of Theorem 18.1 *We provide a sketch of the proof, with the main purpose being to illustrate how the asymptotic distribution of the ML estimator depends on the derivatives of the log-likelihood function.*

The point of departure is the mathematical result known as the mean value theorem *(MVT). The MVT states how a differentiable function can be represented by its derivative. To be precise let $g(x)$ be a differentiable function and $x_0 \in \mathbb{R}$. Then the MVT states that the function $g(x)$ can be written as*

$$g(x) = g(x_0) + \frac{dg(x^*)}{dx}(x - x_0),$$

where x^* is a number between x and x_0. Here we abuse notation slightly and write $\frac{dg(x^*)}{dx}$ for $\frac{dg(x)}{dx}\Big|_{x=x^*}$.

Choose as the $g()$ function in the MVT the score function and choose x_0 to be θ_0. By the MVT, we may represent the score function evaluated at the point $\theta = \widehat{\theta}$ as

$$s(\widehat{\theta}) = s(\theta_0) + \frac{ds(\theta^*)}{d\theta}(\widehat{\theta} - \theta_0),$$

with θ^* being a number between $\widehat{\theta}$ and θ_0, and

$$s(\theta) = \sum_{i=1}^{n} \frac{d}{d\theta} \ln\left(f(\widetilde{Y}_i; \theta)\right)$$

the score function where we for notational convenience have suppressed the dependence of the score function on $\widetilde{Y}$. The first-order condition for the ML estimator implies that $s(\widehat{\theta}) = 0$. Then solve for $(\widehat{\theta} - \theta_0)$ in the above to get

$$(\widehat{\theta} - \theta_0) = -\left(\frac{ds(\theta^*)}{d\theta}\right)^{-1} s(\theta_0),$$

which implies

$$\widehat{\theta} = \theta_0 - \left(\frac{ds(\theta^*)}{d\theta}\right)^{-1} s(\theta_0).$$

By multiplying the second term by $1 = n \cdot \frac{1}{n}$, we may write

$$\widehat{\theta} = \theta_0 - \left(\frac{1}{n}\frac{ds(\theta^*)}{d\theta}\right)^{-1} \frac{1}{n}s(\theta_0). \tag{18.11}$$

Notice that $\frac{1}{n}s(\theta)$ and $\frac{1}{n}\frac{ds(\theta)}{d\theta}$ are sample averages of $\frac{d}{d\theta}\ln\left(f(\widetilde{Y}_i;\theta)\right)$ and $\frac{d^2}{d\theta^2}\ln\left(f(\widetilde{Y}_i;\theta)\right)$, respectively. The fourth condition of the theorem, along with the fact that $\widetilde{Y}$ is a simple random sample, means that we may apply Law of Large Numbers-type result to conclude that

$$\underset{n\to\infty}{plim}\ \frac{1}{n}\frac{ds(\theta^*)}{d\theta} = E\left(\frac{d^2\ln\left(f(Y;\theta)\right)}{d\theta^2}\bigg|_{\theta=\theta_0}\right) = H,$$

where we also used the fact that because θ^* lies between θ_0 and $\widehat{\theta}$ and because $\widehat{\theta}$ is consistent for θ_0, i.e. $\underset{n\to\infty}{plim}\ \widehat{\theta} = \theta_0$, then $\underset{n\to\infty}{plim}\ \theta^* = \theta_0$. Note also that the fifth condition implies that $H \neq 0$, meaning that H^{-1} is well-defined. Similarly, we may apply a Central Limit Theorem to conclude that

$$\frac{1}{n}s(\theta_0) \overset{A}{\sim} N\left(0, J \cdot \frac{1}{n}\right).$$

This latter result relies on the fact that

$$E\left(\frac{1}{n}s(\theta_0)\right) = E\left(\frac{d}{d\theta}\ln\left(f(Y;\theta_0)\right)\right) = 0$$

and, thus, because $E\left(s(\theta_0)\right) = 0$ (Theorem 18.3),

$$Var\left(\frac{1}{n}s(\theta_0)\right) = E\left(\left(\frac{d}{d\theta}\ln\left(f(Y;\theta)\right)\bigg|_{\theta=\theta_0}\right)^2\right) = J.$$

Let $Q \sim N(0, J)$. Combining the above with (18.11), we conclude from Slutsky's theorem (Theorem 13.4), the continuous mapping theorem (Theorem 13.2), and the properties of the normal distribution, that

$$\widehat{\theta} = \theta_0 - \left(\frac{1}{n}\frac{ds(\theta^*)}{d\theta}\right)^{-1} \frac{1}{n}s(\theta_0) \overset{A}{=} \theta_0 - H^{-1}Q \sim N\left(\theta_0, H^{-2}J \cdot \frac{1}{n}\right),$$

as we wanted to show.

Proof of Theorem 18.2   *Suppose that Y is a continuous random variable such that $f(y;\theta)$ is a PDF. The case where Y is a discrete random variable follows similar arguments as those below, except that sums are used instead of integrals.*

    We have that

$$\frac{d\ln(f(y;\theta))}{d\theta} = \frac{1}{f(y;\theta)}\frac{df(y;\theta)}{d\theta}, \tag{18.12}$$

so that

$$E\left(\left(\frac{d\ln(f(Y;\theta))}{d\theta}\Big|_{\theta=\theta_0}\right)^2\right) = E\left(\left(\frac{1}{f(y;\theta_0)}\right)^2\left(\frac{df(y;\theta)}{d\theta}\Big|_{\theta=\theta_0}\right)^2\right)$$

$$= \int\left(\frac{1}{f(y;\theta_0)}\right)^2\left(\frac{df(y;\theta)}{d\theta}\Big|_{\theta=\theta_0}\right)^2 f(y;\theta_0)dy$$

$$= \int\frac{1}{f(y;\theta_0)}\left(\frac{df(y;\theta)}{d\theta}\Big|_{\theta=\theta_0}\right)^2 dy.$$

Also,

$$\frac{d^2\ln(f(y;\theta))}{d\theta^2} = -\frac{1}{f(y;\theta)^2}\left(\frac{df(y;\theta)}{d\theta}\right)^2 + \frac{1}{f(y;\theta)}\frac{d^2f(y;\theta)}{d\theta^2},$$

so that

$$E\left(\frac{d^2\ln(f(Y;\theta))}{d\theta^2}\Big|_{\theta=\theta_0}\right) = E\left\{-\frac{1}{f(Y;\theta_0)^2}\left(\frac{df(Y;\theta)}{d\theta}\Big|_{\theta=\theta_0}\right)^2\right.$$

$$\left.+\frac{1}{f(Y;\theta_0)}\frac{d^2f(Y;\theta)}{d\theta^2}\Big|_{\theta=\theta_0}\right\}$$

$$= \int\left\{-\frac{1}{f(y;\theta_0)^2}\left(\frac{df(y;\theta)}{d\theta}\Big|_{\theta=\theta_0}\right)^2\right.$$

$$\left.+\frac{1}{f(y;\theta_0)}\frac{d^2f(y;\theta)}{d\theta^2}\Big|_{\theta=\theta_0}\right\}f(y;\theta_0)dy$$

$$= -\int\frac{1}{f(y;\theta_0)}\left(\frac{df(y;\theta)}{d\theta}\Big|_{\theta=\theta_0}\right)^2 dy + \int\frac{d^2f(y;\theta)}{d\theta^2}\Big|_{\theta=\theta_0}dy$$

$$= -\int\frac{1}{f(y;\theta_0)}\left(\frac{df(y;\theta)}{d\theta}\Big|_{\theta=\theta_0}\right)^2 dy + \frac{d^2}{d\theta^2}\int f(y;\theta)dy|_{\theta=\theta_0}$$

$$= -\int\frac{1}{f(y;\theta_0)}\left(\frac{df(y;\theta)}{d\theta}\Big|_{\theta=\theta_0}\right)^2 dy + \frac{d^2}{d\theta^2}1\Big|_{\theta=\theta_0}$$

$$= -\int\frac{1}{f(y;\theta_0)}\left(\frac{df(y;\theta)}{d\theta}\Big|_{\theta=\theta_0}\right)^2 dy,$$

where we in the third-to-last equality interchanged differentiation and integration, which is possible under the conditions of the theorem, and in the second-to-last equality used that $f(y;\theta)$ is a PDF and hence integrates to one. The last equality follows because the derivative of a constant is zero. From the derivations above, we see that

$$E\left(\left(\frac{d\ln(f(Y;\theta))}{d\theta}\Big|_{\theta=\theta_0}\right)^2\right) = -E\left(\frac{d^2\ln(f(Y;\theta))}{d\theta^2}\Big|_{\theta=\theta_0}\right),$$

as we wanted to show.

Proof of Theorem 18.3 *Suppose that Y is a continuous random variable such that $f(y;\theta)$ is a PDF. The case where Y is a discrete random variable follows similar arguments as those below, except that sums are used instead of integrals.*

Using (18.12), we get

$$E\left(\frac{d\ln(f(Y;\theta))}{d\theta}\right) = E\left(\frac{1}{f(Y;\theta)}\frac{df(Y;\theta)}{d\theta}\right) = \int \frac{1}{f(y;\theta)}\frac{df(y;\theta)}{d\theta}f(y;\theta_0)dy.$$

Evaluating this in $\theta = \theta_0$ yields

$$E\left(\frac{d\ln(f(Y;\theta))}{d\theta}\bigg|_{\theta=\theta_0}\right) = \int \frac{1}{f(y;\theta_0)}\frac{df(y;\theta)}{d\theta}\bigg|_{\theta=\theta_0} f(y;\theta_0)dy = \int \frac{df(y;\theta)}{d\theta}\bigg|_{\theta=\theta_0} dy.$$

Under the conditions of the theorem, we may interchange integration and differentiation, yielding

$$\begin{aligned}
E\left(\frac{d\ln(f(Y;\theta))}{d\theta}\bigg|_{\theta=\theta_0}\right) &= \int \frac{df(y;\theta)}{d\theta}\bigg|_{\theta=\theta_0} dy \\
&= \frac{d}{d\theta}\int f(y;\theta)dy\bigg|_{\theta=\theta_0} \\
&= \frac{d}{d\theta}\int 1\,dy\bigg|_{\theta=\theta_0} \\
&= 0,
\end{aligned}$$

where we used that $f(y;\theta)$ is a PDF and hence integrates to one. This proves the first part of the theorem. The second part follows using the first part along with the assumption of a simple random sample,

$$\begin{aligned}
E\left(s\left(\theta_0;\widetilde{Y}\right)\right) &= E\left(\sum_{i=1}^{n}\frac{d\ln(f(\widetilde{Y}_i;\theta))}{d\theta}\bigg|_{\theta=\theta_0}\right) \\
&= \sum_{i=1}^{n}E\left(\frac{d\ln(f(\widetilde{Y}_i;\theta))}{d\theta}\bigg|_{\theta=\theta_0}\right) \\
&= \sum_{i=1}^{n}0 \\
&= 0,
\end{aligned}$$

which completes the proof.

Proof of Theorem 18.4 *The proof of this theorem relies on the consistency of the ML estimation, i.e. plim$_{n\to\infty}\widehat{\theta} = \theta_0$. For the expected information estimators, we use that under* *the conditions of the theorem, we may interchange limits and expectations, implying that*

$$
\begin{aligned}
\underset{n\to\infty}{plim}\ \widehat{J}_{EX} \;&=\; \underset{n\to\infty}{plim}\, E_Y\left(\left(\left.\frac{d\ln\left(f(Y;\theta)\right)}{d\theta}\right|_{\theta=\widehat{\theta}}\right)^2\right)\\[2mm]
&=\; E_Y\left(\underset{n\to\infty}{plim}\left(\left.\frac{d\ln\left(f(Y;\theta)\right)}{d\theta}\right|_{\theta=\widehat{\theta}}\right)^2\right)\\[2mm]
&=\; E_Y\left(\left(\left.\frac{d\ln\left(f(Y;\theta)\right)}{d\theta}\right|_{\theta=\theta_0}\right)^2\right)\\[2mm]
&=\; J.
\end{aligned}
$$

Similar arguments imply that

$$
\underset{n\to\infty}{plim}\ \widehat{H}_{EX} = E\left(\left.\frac{d^2\ln\left(f(Y;\theta)\right)}{d\theta^2}\right|_{\theta=\theta_0}\right) = H.
$$

For the outer product of the scores estimator, we use again that $\widehat{\theta}$ is consistent for θ_0, along with the fact that the conditions of the theorem imply that a Law of Large Numbers may be applied to conclude

$$
\underset{n\to\infty}{plim}\ \widehat{J}_{OP} = \underset{n\to\infty}{plim}\ \frac{1}{n}\sum_{i=1}^{n}\left(\left.\frac{d\ln\left(f(\widetilde{Y}_i;\theta)\right)}{d\theta}\right|_{\theta=\widehat{\theta}}\right)^2 = E\left(\left(\left.\frac{d\ln\left(f(Y;\theta)\right)}{d\theta}\right|_{\theta=\theta_0}\right)^2\right) = J.
$$

A similar argument shows applies to the observed Hessian estimator,

$$
\underset{n\to\infty}{plim}\ \widehat{H}_{OH} = E\left(\left.\frac{d^2\ln\left(f(Y;\theta)\right)}{d\theta^2}\right|_{\theta=\theta_0}\right) = H.
$$

This concludes the proof.

Proof of Theorem 18.5　*The proof of the Cramér-Rao lower bound relies on a neat trick. Let X and Z be random variables such that $E(Z) = 0$ and let $\rho = Corr(X,Z)$. Then, by the definition of the correlation,*

$$
\rho = \frac{Cov(X,Z)}{\sqrt{Var(X)\cdot Var(Z)}} = \frac{E(X\cdot Z)}{\sqrt{Var(X)\cdot Var(Z)}},
$$

where we used that $Cov(X,Z) = E(X\cdot Z)$ when either X or Z has mean zero. Re-arranging, we have

$$
Var(X) = \frac{E(X\cdot Z)^2}{\rho^2\cdot Var(Z)}.
$$

Since $|\rho| \le 1$, we have

$$
Var(X) = \frac{E(X\cdot Z)^2}{\rho^2\cdot Var(Z)} \ge \frac{E(X\cdot Z)^2}{Var(Z)}.
$$

Now, we apply this result with $X = \widehat{\theta}$ and $Z = s\left(\theta_0; \widetilde{Y}\right)$, which is valid because that $E\left(s\left(\theta_0; \widetilde{Y}\right)\right) = 0$ (Theorem 18.3), to conclude

$$Var(\widehat{\theta}) \geq \frac{E\left(\widehat{\theta} \cdot s\left(\theta_0; \widetilde{Y}\right)\right)^2}{Var\left(s\left(\theta_0; \widetilde{Y}\right)\right)}.$$

Using the definition of the score function, that $\widetilde{Y}$ is a simple random sample, and $E\left(\left.\frac{d\ln(f(Y;\theta))}{d\theta}\right|_{\theta=\theta_0}\right) = 0$ (Theorem 18.3), we have

$$
\begin{aligned}
Var\left(s\left(\theta_0; \widetilde{Y}\right)\right) &= Var\left(\left.\sum_{i=1}^{n} \frac{d\ln\left(f(\widetilde{Y}_i; \theta)\right)}{d\theta}\right|_{\theta=\theta_0}\right) \\
&= \sum_{i=1}^{n} Var\left(\left.\frac{d\ln\left(f(\widetilde{Y}_i; \theta)\right)}{d\theta}\right|_{\theta=\theta_0}\right) \\
&= \sum_{i=1}^{n} E\left(\left(\left.\frac{d\ln\left(f(\widetilde{Y}_i; \theta)\right)}{d\theta}\right|_{\theta=\theta_0}\right)^2\right) \\
&= n \cdot I(\theta_0),
\end{aligned}
$$

where $I(\theta_0)$ is the Fisher information (Definition 18.2). Further, letting $f_{\widetilde{Y}}()$ denote the sample distribution, we have

$$
\begin{aligned}
E\left(\widehat{\theta} \cdot s\left(\theta_0; \widetilde{Y}\right)\right) &= E\left(\left.\widehat{\theta} \cdot \frac{d\ln\left(f_{\widetilde{Y}}(\widetilde{Y}; \theta)\right)}{d\theta}\right|_{\theta=\theta_0}\right) \\
&= E\left(\left.\widehat{\theta} \cdot \frac{1}{f_{\widetilde{Y}}(\widetilde{Y}; \theta_0)} \cdot \frac{d\left(f_{\widetilde{Y}}(\widetilde{Y}; \theta)\right)}{d\theta}\right|_{\theta=\theta_0}\right) \\
&= \int \widehat{\theta} \cdot \frac{1}{f_{\widetilde{Y}}(\widetilde{y}; \theta_0)} \cdot \left.\frac{df_{\widetilde{Y}}(\widetilde{y}; \theta)}{d\theta}\right|_{\theta=\theta_0} \cdot f_{\widetilde{Y}}(\widetilde{y}; \theta_0) d\widetilde{y} \\
&= \int \widehat{\theta} \cdot \left.\frac{df_{\widetilde{Y}}(\widetilde{y}; \theta)}{d\theta}\right|_{\theta=\theta_0} d\widetilde{y} \\
&= \left.\frac{d}{d\theta} \int \widehat{\theta} \cdot f_{\widetilde{Y}}(\widetilde{y}; \theta) d\widetilde{y}\right|_{\theta=\theta_0} \\
&= \left.\frac{d}{d\theta} E_\theta\left(\widehat{\theta}\right)\right|_{\theta=\theta_0},
\end{aligned}
$$

where we use the notation $E_\theta()$ to indicate that the expectation is taken with respect to the distribution $f_{\widetilde{Y}}(y; \theta)$. Now, since $\widehat{\theta}$ is unbiased by assumption, we have $E_\theta\left(\widehat{\theta}\right) = \theta$ and, thus,

$$E\left(\widehat{\theta} \cdot s\left(\theta_0; \widetilde{Y}\right)\right) = \left.\frac{d}{d\theta}\theta\right|_{\theta=\theta_0} = 1.$$

Putting it all together, we conclude that

$$Var(\widehat{\theta}) \geq \frac{E\left(\widehat{\theta} \cdot s\left(\theta_0; \widetilde{Y}\right)\right)^2}{Var\left(s\left(\theta_0; \widetilde{Y}\right)\right)} = \frac{1^2}{n \cdot I(\theta_0)} = I(\theta_0)^{-1} \cdot \frac{1}{n},$$

as we wanted to prove.

Proof of Theorem 18.6 *The proof follows the same steps as the proof of Theorem 18.5, except that all statement are made asymptotically. We skip the details for brevity.*

Proof of Theorems 18.7, 18.8, 18.9 *We sketch the proof of Theorem 18.8; the proofs of Theorems 18.7 and 18.9 are beyond the scope of this book.*
By adding and subtracting θ_0, we may write the Wald statistics as (18.9)

$$Wald = \frac{(\widehat{\theta} - \theta_{H_0})^2}{\widehat{Var}(\widehat{\theta})} = \frac{(\widehat{\theta} - \theta_0 + \theta_0 - \theta_{H_0})^2}{\widehat{Var}(\widehat{\theta})}$$

By Theorem 18.1, we have that

$$\widehat{\theta} - \theta_0 + \theta_0 - \theta_{H_0} \overset{A}{\sim} N\left(\theta_0 - \theta_{H_0}, Var(\widehat{\theta})\right).$$

By the properties of the normal distribution, this shows that, since $\widehat{Var}\left(\widehat{\theta}\right)$ is a consistent estimator for $Var\left(\widehat{\theta}\right)$,

$$\frac{\widehat{\theta} - \theta_{H_0}}{\sqrt{\widehat{Var}\left(\widehat{\theta}\right)}} = \frac{\widehat{\theta} - \theta_0 + \theta_0 - \theta_{H_0}}{\sqrt{\widehat{Var}\left(\widehat{\theta}\right)}} \overset{A}{\sim} N\left(\frac{\theta_0 - \theta_{H_0}}{Var\left(\widehat{\theta}\right)}, 1\right),$$

and, hence,

$$Wald \overset{A}{\sim} \chi^2(1, \lambda),$$

where $\chi^2(1, \lambda)$ is the non-central χ^2 distribution with one degree of freedom and with non-centrality parameter λ given by

$$\lambda^2 = \frac{\theta_0 - \theta_{H_0}}{Var\left(\widehat{\theta}\right)},$$

as we wanted to show. Further, under H_0 then $\theta_0 - \theta_{H_0} = 0$, implying that $\lambda = 0$. To see how λ^2 behaves under H_1 for large sample sizes, recall that since $\widehat{\theta}$ is a consistent estimator, then

$$Var\left(\widehat{\theta}\right) \to 0,$$

as $n \to \infty$. Thus, under H_1, where $\theta_0 - \theta_{H_0} \neq 0$, we have that

$$\lambda^2 = \frac{\theta_0 - \theta_{H_0}}{Var\left(\widehat{\theta}\right)} \to \infty,$$

as $n \to \infty$. This concludes the proof.

18.8   Exercises

Problem 18.8.1 *Let $\widetilde{Y}_1, \ldots, \widetilde{Y}_n$ be a simple random sample from the random variable Y with distribution function $f(y; \theta_0)$, with $\theta_0 \in \mathbb{R}$. Assume that $\ln(f(y; \theta))$ is twice continuously differentiable in θ for all y. Let*

$$s(\theta; \widetilde{Y}_1, \ldots, \widetilde{Y}_n) = \sum_{i=1}^{n} \ln\left(f(\widetilde{Y}_i; \theta)\right)$$

be the score function.

1. *Prove that*

$$E\left(s(\theta; \widetilde{Y}_1, \ldots, \widetilde{Y}_n)^2\right) = J,$$

   where J is given in Equation (18.2).

2. *Prove that*

$$E\left(\left.\frac{ds(\theta; \widetilde{Y}_1, \ldots, \widetilde{Y}_n)}{d\theta}\right|_{\theta = \theta_0}\right) = H,$$

   where H is given in Equation (18.3).

Problem 18.8.2 *Consider a simple random sample $\widetilde{Y}_1, \ldots, \widetilde{Y}_n$ of the population random variable Y. Assume that Y has the exponential distribution with parameter λ_0, $Y \sim Exp(\lambda_0)$ with $\lambda_0 > 0$, i.e. the PDF of Y is*

$$f(y; \lambda_0) = \lambda_0 e^{-\lambda_0 y}, \quad y > 0.$$

1. *Derive the Jacobian J (18.2) and Hessian H (18.3).*

2. *Based on the simple random sample $\widetilde{Y}_1, \ldots, \widetilde{Y}_n$, suggest estimators of the asymptotic variance $V = J/H^2$ in (18.1). In particular, state the forms of $\widehat{V}_{EX}, \widehat{V}_{OP}, \widehat{V}_{OH}, \widehat{V}_{SW}$.*

3. *Set $\lambda_0 = 2$ and $n = 100$. Using $\lambda_0 = 2$, calculate the true asymptotic variance V in (18.1). Conduct a Monte Carlo study to assess the performance of the estimators $\widehat{V}_{EX}, \widehat{V}_{OP}, \widehat{V}_{OH}, \widehat{V}_{SW}$ by calculating the Monte Carlo root mean squared error of each of the estimators.*

4. *Repeat the Monte Carlo study with $n = 5000$.*

Problem 18.8.3 *Consider the same setup as in Problem 18.8.2. In this exercise, you are asked to implement a parametric bootstrap approach that can be used to estimate the variance of the ML estimator.*

1. *Set $\lambda_0 = 2$ and $n = 20$. Using $\lambda_0 = 2$, calculate the true asymptotic variance V in (18.1). Conduct a Monte Carlo study to assess the performance of the parametric bootstrap by calculating the Monte Carlo root mean squared error of the bootstrap estimate of V. As is Problem 18.8.2, implement also the estimators $\widehat{V}_{EX}, \widehat{V}_{OP}, \widehat{V}_{OH}, \widehat{V}_{SW}$. Compare their performance with that of the bootstrap approach.*

2. *Repeat the Monte Carlo study with $n = 5000$. Comment on the relative performance of the estimators when $n = 20$ versus $n = 5000$.*

Problem 18.8.4 *Consider the same setup as in Problems 18.8.2 and 18.8.3. In this exercise, you are asked to construct 95% confidence intervals for λ_0 and calculate their empirical coverage rates.*

1. *Set $n = 20$ and consider a 95% confidence interval. Conduct a Monte Carlo study and for each simulation calculate a 95% confidence interval using a bootstrap approach (Section 18.5.1) and a 95% confidence interval using an asymptotic approach (Section 18.5.2). In particular, use the Monte Carlo study to calculate the empirical coverage rate of the two confidence intervals and compare with the nominal coverage rate of $\alpha = 95\%$.*

2. *Repeat the Monte Carlo study with $n = 5000$. Comment on the relative performance of the approaches when $n = 20$ versus $n = 5000$.*

Problem 18.8.5 *Consider the same setup as in Problems 18.8.2–18.8.4. Set $\lambda_0 = 2$ and consider the hypotheses*

$$
\begin{aligned}
H_0 : \lambda_0 &= 2, \\
H_1 : \lambda_0 &\neq 2.
\end{aligned}
$$

In this exercise, you are asked to implement the trinity of tests and investigate their empirical (finite sample) size and compare it to a parametric bootstrap approach, by testing H_0 versus H_1.

1. *Set $n = 20$ and consider a $\alpha = 5\%$ significance level. Conduct a Monte Carlo study to assess the performance of four different tests of H_0: The GLR test, the Wald test, the LM test, and a test based on a parametric bootstrap approach. In particular, use the Monte Carlo study to calculate the empirical rejection rate of the four tests and compare with the nominal significance level $\alpha = 5\%$.*

2. *Repeat the Monte Carlo study with the sample sizes $n = 50, 100, 200, 400, 800, 1600, 3200$. For each test, plot the empirical rejection rates as a function of sample size n. Comment on the relative performance of the approaches when the sample size n is small versus when the sample size n is large.*

Problem 18.8.6 *Consider the same setup as in Problem 18.8.5. Set $\lambda_0 = 2$. In this exercise, you are asked to implement the trinity of tests and investigate their empirical (finite sample) power and compare it to a parametric bootstrap approach, by testing H_0 versus H_1, where*

$$
\begin{aligned}
H_0 : \lambda_0 &= 1.75, \\
H_1 : \lambda_0 &\neq 1.75.
\end{aligned}
$$

1. *Set $n = 20$ and consider a $\alpha = 5\%$ significance level. Conduct a Monte Carlo study to assess the performance of four different tests of H_0: The GLR test, the Wald test, the LM test, and a test based on a parametric bootstrap approach. In particular, use the Monte Carlo study to calculate the empirical rejection rate of the four tests.*

2. *Repeat the Monte Carlo study with the sample sizes $n = 50, 100, 200, 400, 800, 1600, 3200$. For each test, plot the empirical rejection rates as a function of sample size n. Comment on the relative performance of the approaches when the sample size n is small versus when the sample size n is large.*

Problem 18.8.7 *Consider the time series population modeled by the process given by*

$$Y_t = \phi_0 \cdot Y_{t-1} + \epsilon_t, \quad t \geq 2,$$

where $\epsilon_t \sim N(0,1)$ and $Y_1 \sim N\left(0, \frac{1}{1-\phi_0^2}\right)$ as in Example 17.15 in Chapter 17.

1. *Set $\phi_0 = 0.75$. Simulate a time series sample $\widetilde{Y}_1, \ldots, \widetilde{Y}_T$ of size $T = 10$ from this population. Maximize the log-likelihood function numerically to arrive at an ML estimate $\widehat{\phi}$. Use a parametric bootstrap approach to estimate the variance of $\widehat{\phi}$.*

2. *Repeat the previous question but with a sample size $T = 1000$.*

3. *Conduct a Monte Carlo simulation study that compares the ML estimator ϕ_0 when $T = 10$ and $T = 1000$ in terms of bias and MSE.*

19

Bayesian methods

19.1 Introduction

In the preceding chapters, we have been working under the so-called *frequentist paradigm*. For the purposes of this book, this has mainly meant that population parameters θ have been treated as unknown, but fixed, quantities. We have seen how to make use of data to estimate and perform inference on θ in this paradigm. The *Bayesian paradigm*, in contrast, treats population parameters θ as random variables with probability distributions. The probability distributions of θ reflect our beliefs about the likely values of the parameters. The goal in this paradigm is to synthesize prior knowledge and available data to arrive at a probability distribution representing our complete knowledge of the problem, which can subsequently be used in statistical analyses and for making decisions.

The prior knowledge we have about a problem is represented by a probability distribution over the possible values of the parameters. Since it is knowledge we have obtained before observing a random sample, this distribution over the parameters is called the "prior distribution". The random sample may alter our beliefs of the likely values of the parameters. The alteration due to the random sample is made through a link between the random sample and the parameters. This link is the likelihood function, as introduced in Chapter 17, since the likelihood function is the function mapping parameter values to the distribution of the random sample.

The prior distribution and the likelihood function can be combined by Bayes' theorem (Theorem 4.4) to provide the distribution of the beliefs over the parameters, incorporating both our prior beliefs and the information contained in the sample. This distribution is called the "posterior distribution". It is different from the prior distribution if the random sample alters the beliefs represented by the prior distribution. In general, the larger the sample size, the more influence the information in the sample has on the posterior distribution relative to the prior distribution.

19.2 The prior and posterior distributions

Let θ_0 be the parameter of interest. Our beliefs of the likely values of θ_0 is represented by the random variable θ. The *prior distribution* of θ is represented by the function $p(\theta)$. If θ is a discrete random variable, then $p(\theta)$ is a PMF, whereas if θ is a continuous random variable, $p(\theta)$ is a PDF. Note that we, with slight abuse of notation, use the Greek letter "θ" both to represent the random variable and the real value used as input to the distributions, e.g. in $p(\theta)$.

We seek to update our prior belief on the true value θ_0, represented by the prior distribution $p(\theta)$, by incorporating the information contained in a random sample. Let

$\widetilde{Y} = (\widetilde{Y}_1, \ldots, \widetilde{Y}_n)$ be the random sample and $\widetilde{y} = (\widetilde{y}_1, \ldots, \widetilde{y}_n)$ a realization of this random sample. If θ is the true value of θ_0, the sample distribution is

$$f_{\widetilde{Y}|\theta}(\widetilde{y} \mid \theta) = f_{\widetilde{Y}_1, \ldots, \widetilde{Y}_n|\theta}(\widetilde{y}_1, \ldots, \widetilde{y}_n \mid \theta).$$

In what follows, we will simply refer to $f_{\widetilde{Y}|\theta}(\widetilde{y} \mid \theta)$ as the sample distribution. Considered as a function of θ, it is the likelihood of the data. We update our beliefs of θ_0 by incorporating the information from the random sample into the probability distribution of θ. Formally, this is done by conditioning on the information in the sample. That is, we need to calculate

$$f_{\theta|\widetilde{Y}}(\theta \mid \widetilde{y}) = f_{\theta|\widetilde{Y}_1, \ldots, \widetilde{Y}_n}(\theta \mid \widetilde{y}_1, \ldots, \widetilde{y}_n).$$

This probability can be calculated using Bayes' theorem (Theorem 4.4)

$$f_{\theta|\widetilde{Y}}(\theta \mid \widetilde{y}) = \frac{f_{\widetilde{Y}|\theta}(\widetilde{y} \mid \theta) \cdot p(\theta)}{f_{\widetilde{Y}}(\widetilde{y})}, \tag{19.1}$$

where

$$f_{\widetilde{Y}}(\widetilde{y}) = \int f_{\widetilde{Y}|\theta}(\widetilde{y} \mid \theta) \cdot p(\theta) \, d\theta,$$

in case θ is a continuous random variable, or

$$f_{\widetilde{Y}}(\widetilde{y}) = \sum_{\theta} f_{\widetilde{Y}|\theta}(\widetilde{y} \mid \theta) \cdot p(\theta),$$

in case θ is a discrete random variable. Notice, $f_{\widetilde{Y}}(\widetilde{y})$ is not the distribution of $\widetilde{Y}$ as such, but instead our belief of the distribution of $\widetilde{Y}$, based on the prior distribution $p(\theta)$ of θ.

The distribution $f_{\theta|\widetilde{Y}}(\theta \mid \widetilde{y})$ of θ given the random sample $\widetilde{Y}$, is called the *posterior distribution*, and it is defined next.

Definition 19.1 (Posterior distribution) *Let $p(\theta)$ be a prior distribution and $f_{\widetilde{Y}|\theta}(\widetilde{y} \mid \theta)$ the sample distribution. Then $f_{\theta|\widetilde{Y}}(\theta \mid \widetilde{y})$ given by (19.1) is called the* **posterior distribution** *of θ.*

The posterior distribution is the basic output of the Bayesian method. Whereas the frequentist method typically builds on optimization of an objective function, the Bayesian approach builds on updating the prior distribution using the sample to arrive at the posterior distribution, incorporating both the prior information and information from the sample.

Next is an example to illustrate how to calculate a posterior distribution.

Example 19.1 *Consider a random variable Y assumed to be Bernoulli distributed, $Y \sim Ber(\theta_0)$. The number $\theta_0 \in (0, 1)$, i.e. the probability of "success", is the parameter of interest. Let θ be a continuous random variable, defined over the possible values of θ_0, i.e. $\theta \in (0, 1)$. Let $\widetilde{Y} = (\widetilde{Y}_1, \ldots, \widetilde{Y}_n)$ be a simple random sample from Y. From Example 17.3, the sample distribution is*

$$f_{\widetilde{Y}|\theta}(\widetilde{y} \mid \theta) = \prod_{i=1}^{n} \theta^{\widetilde{y}_i} (1 - \theta)^{\widetilde{y}_i} = \theta^{\sum_{i=1}^{n} \widetilde{y}_i} (1 - \theta)^{\sum_{i=1}^{n}(1 - \widetilde{y}_i)}.$$

Suppose we, a priori, consider all values in $(0, 1)$ equally likely. In this case, we may use the $U(0, 1)$ distribution as prior distribution, i.e.

$$p(\theta) = 1, \quad \theta \in (0, 1).$$

Then the posterior distribution (19.1) is

$$f_{\theta|\widetilde{Y}}(\theta \mid \widetilde{y}) = \frac{f_{\widetilde{Y}|\theta}(\widetilde{y} \mid \theta) \cdot p(\theta)}{f_{\widetilde{Y}}(\widetilde{y})} = \frac{\theta^{\sum_{i=1}^{n} \widetilde{y}_i} (1-\theta)^{\sum_{i=1}^{n}(1-\widetilde{y}_i)} \cdot 1}{\int_0^1 \theta^{\sum_{i=1}^{n} \widetilde{y}_i} (1-\theta)^{\sum_{i=1}^{n}(1-\widetilde{y}_i)} \cdot 1 \, d\theta}, \quad \theta \in (0,1).$$

Consider the case with only $n = 1$ observation $\widetilde{y}$. If that observation is $\widetilde{y} = 1$, then the posterior distribution is

$$f_{\theta|\widetilde{Y}}(\theta \mid \widetilde{y} = 1) = \frac{\theta}{\int_0^1 \theta \, d\theta} = \frac{\theta}{\left[\frac{1}{2}\theta^2\right]_0^1} = 2\theta, \quad \theta \in (0,1).$$

Conversely, if that observation is $\widetilde{y} = 0$, then the posterior distribution is

$$f_{\theta|\widetilde{Y}}(\theta \mid \widetilde{y} = 0) = \frac{1-\theta}{\int_0^1 (1-\theta) \, d\theta} = \frac{1-\theta}{\left[\theta - \frac{1}{2}\theta^2\right]_0^1} = 2(1-\theta), \quad \theta \in (0,1).$$

Since θ is the probability of obtaining $Y = 1$, an observation of $\widetilde{y} = 1$ gives a posterior distribution with higher probability for values of θ larger than $1/2$ compared to the prior distribution. In other words, the observation of $\widetilde{y}$ "pulls" the prior distribution toward values of θ, which makes it more likely to obtain that observation. This is illustrated in Figure 19.1, where the prior $p(\theta)$ and posterior $f_{\theta|\widetilde{Y}}(\theta \mid \widetilde{y} = 1)$ are shown.

Suppose there are two observations $\widetilde{y} = (\widetilde{y}_1, \widetilde{y}_2)$. Depending on the sample, the posterior distribution is

$$\begin{aligned}
f_{\theta|\widetilde{Y}}(\theta \mid \widetilde{y} &= (0,0)) = \frac{(1-\theta)^2}{\int_0^1 (1-\theta)^2 \, d\theta} = \frac{1-\theta}{\left[\theta + \frac{1}{3}\theta^3 - \theta^2\right]_0^1} = 3(1-\theta)^2, \\
f_{\theta|\widetilde{Y}}(\theta \mid \widetilde{y} &= (0,1)) = \frac{(1-\theta)\theta}{\int_0^1 (1-\theta)\theta \, d\theta} = \frac{(1-\theta)\theta}{\left[\frac{1}{2}\theta^2 - \frac{1}{3}\theta^3\right]_0^1} = 6(1-\theta)\theta, \\
f_{\theta|\widetilde{Y}}(\theta \mid \widetilde{y} &= (1,0)) = f_{\theta|\widetilde{Y}}(\theta \mid \widetilde{y} = (0,1)) = 6(1-\theta)\theta, \\
f_{\theta|\widetilde{Y}}(\theta \mid \widetilde{y} &= (1,1)) = \frac{\theta^2}{\int_0^1 \theta^2 \, d\theta} = \frac{\theta^2}{\left[\frac{1}{3}\theta^3\right]_0^1} - 3\theta^2,
\end{aligned}$$

for $\theta \in (0,1)$. Figure 19.1 shows the posterior $f_{\theta|\widetilde{Y}}(\theta \mid \widetilde{y} = (1,1))$, which is the posterior arising from a sample of size $n = 2$ with $\widetilde{y} = (1,1)$. Compared to the case of $n = 1$ and $\widetilde{y} = 1$, i.e. $f_{\theta|\widetilde{Y}}(\theta \mid \widetilde{y} = 1)$, the posterior has even more mass concentrated at the higher values of θ. Conversely, if the second observation is $\widetilde{y}_2 = 0$ instead of $\widetilde{y}_2 = 1$, then the posterior $f_{\theta|\widetilde{Y}}(\theta \mid \widetilde{y} = (1,0))$ becomes symmetric around $\theta = 0.5$, like the prior distribution. Note, however, that in this case, more mass of the posterior distribution is centered around $\theta = 1/2$, compared to $\theta = \pm 1$, because the presence of both a "success" $(Y = 1)$ and a "failure" $(Y = 0)$ in the sample make the extreme values of θ_0 less likely.

19.3 Examples of prior distributions

To make the Bayesian approach operational, we need to specify the prior distribution and the sample distribution or likelihood function. The choice of sample distribution can be guided by similar principles as those discussed in Chapter 17 when we studied maximum

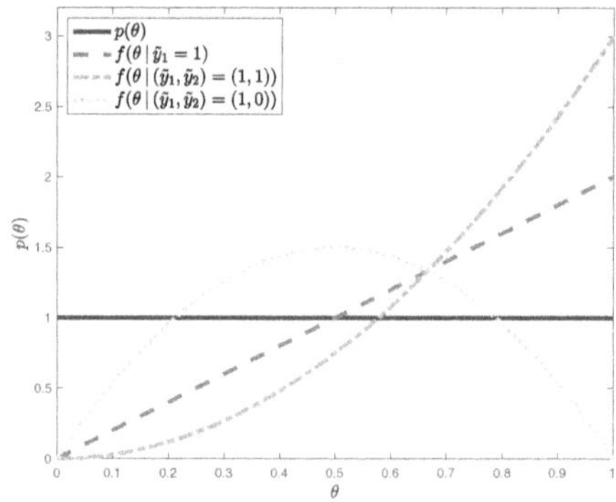

FIGURE 19.1

Prior distribution, $p(\theta)$, and posterior distributions, $f_{\theta|\widetilde{Y}}(\theta\,|\widetilde{y})$, of Example 19.1.

likelihood estimation. In particular, for a given distribution of the population and a given sampling mechanism, we may derive the sample distribution, as discussed in Chapter 10.

The choice of prior distribution should reflect our beliefs of the likely values of the parameter before observing the sample, including how certain we are of these beliefs. An important special case is when have no prior information about the parameter of interest. There are several approaches to represent this situation, going under the name of *non-informative priors* (or *diffuse* or *flat* priors). In the following, we consider various priors and how they may be interpreted.

19.3.1 The degenerate prior

An extreme situation is when we are completely sure about the value of a parameter. For example, if we are certain that the true value of θ_0 is equal to the value a, this implies the prior distribution

$$p(\theta) = \begin{cases} 1 & \text{if} \quad \theta = a, \\ 0 & \text{otherwise.} \end{cases}$$

This is a degenerate distribution. It implies that the posterior is

$$f_{\theta|\widetilde{Y}}(\theta\mid\widetilde{y}) = \frac{f_{\widetilde{Y}|\theta}(\widetilde{y}\mid\theta)\cdot p(\theta)}{\sum_{\theta} f_{\widetilde{Y}|\theta}(\widetilde{y}\mid\theta)\cdot p(\theta)} = \begin{cases} 1 & \text{if} \quad \theta = a, \\ 0 & \text{otherwise,} \end{cases}$$

since $p(\theta) = 0$ for any other value of θ than a. Hence, no matter the sample, the posterior distribution equals the prior distribution. In other words, the sample has no influence on our beliefs.

19.3.2 The uniform prior

The uniform distribution can be used as a non-informative prior when it is known that the parameter values belongs to a bounded interval $\theta \in [a, b]$, where $b > a$. Then choosing $\theta \sim U(a, b)$ as prior distribution makes the density of any value of θ in the interval the same. This prior distribution is given by

$$p(\theta) = \frac{1}{b - a}, \quad \theta \in [a, b].$$

For an illustration of the $U(a,b)$ density with $a = 0$ and $b = 1$, see the top left panel in Figure 19.2. This prior was used in Example 19.1 for the parameter θ_0, in order to capture the situation where we had no a priori reasons to favor any values in $(0,1)$ over others.

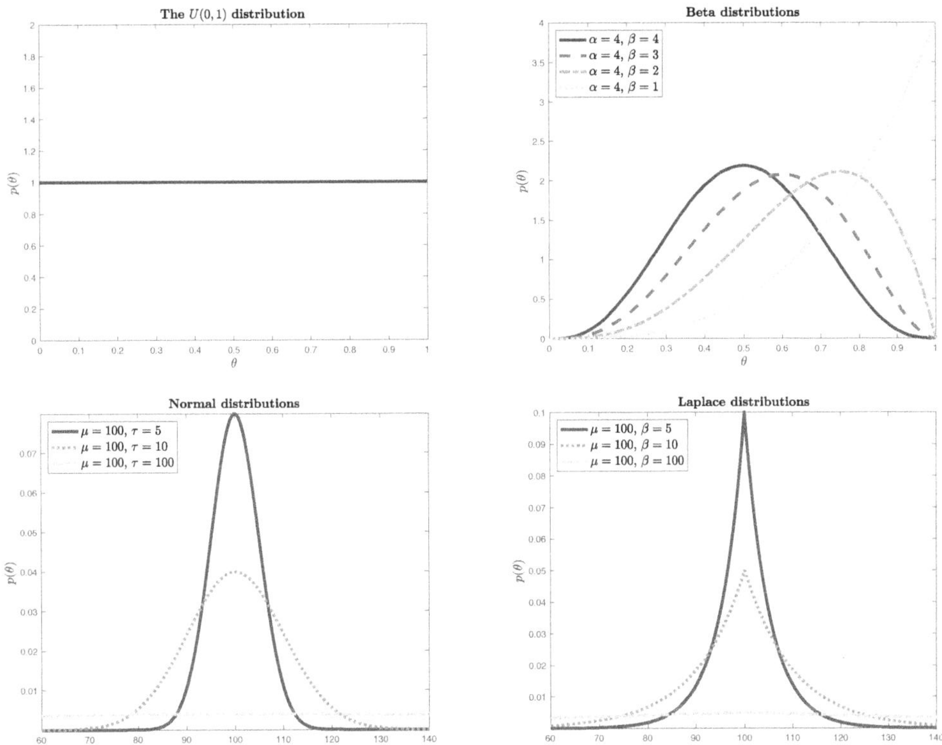

FIGURE 19.2
Examples of prior distributions. Top left: PDF $p(\theta)$ of the uniform distribution $U(a,b)$ with $a = 0$, $b = 1$. Top right: PDF $p(\theta)$ of the beta distribution $Beta(\alpha,\beta;a,b)$ with $a = 0$, $b = 1$, and α,β as given in the plot. Bottom left: PDF $p(\theta)$ of the normal distribution $N(\mu,\tau^2)$ with $\mu = 100$ and τ as given in the plot. Bottom right: PDF $p(\theta)$ of the Laplace distribution with $\mu = 100$ and β as given in the plot.

19.3.3 The beta prior

Suppose it is known that the parameter value θ_0 is in some interval $[a,b]$, $a < b$, and that we believe that some values in $[a,b]$ are more likely than others. In this case, a beta distribution (Definition 8.7) may be used as prior.

The beta distribution, $\theta \sim Beta(\alpha,\beta;a,b)$ for $\alpha,\beta > 0$, has four parameters. Recall that when $\alpha = \beta = 1$, the beta distribution $Beta(1,1;a,b)$ reduces to the uniform distribution $U(a,b)$, but for other values of α,β, we may specify informative prior beliefs on the likely values of θ_0. To illustrate this, let $a = 0$ and $b = 1$, meaning that θ is restricted to the unit interval $[0,1]$. If we, for instance, expect that values of θ around 0.5 are more likely than values close to 0 and 1, we may specify this using $\alpha = \beta \neq 1$. If we instead believe that, say, higher values of θ are more likely than lower values, this may be specified using $\alpha > \beta$. The top right panel of Figure 19.2 plots the PDF of the beta distribution for various values of α,β in the case where $a = 0$ and $b = 1$. We see how choosing $\alpha = \beta = 4$ leads to a symmetric distribution centered around 0.5, while varying α and β such that $\alpha > \beta$ results in distributions with relatively more mass located at the larger values of θ.

19.3.4   The normal prior

When the values of θ are not bounded, we may use a PDF with support over all the real numbers. One popular choice is the normal distribution.

Suppose that $\theta \sim N(\mu, \tau^2)$ is used as prior distribution, i.e.

$$p(\theta) = \frac{1}{\sqrt{2\pi\tau^2}} \exp\left(-\frac{1}{2\tau^2}(\theta - \mu)^2\right), \quad \theta \in \mathbb{R}.$$

This is a prior with a bell shape centered around the value μ. We may think of μ as the prior "best guess" of θ_0. Choosing τ^2 small implies that the distribution is concentrated around μ, representing high prior belief that θ_0 is close to the value μ. Conversely, larger values of τ^2 implies that the distribution is more spread out around μ, representing a correspondingly lower prior belief in the best guess μ. Indeed, choosing τ^2 very large implies that the bell shape becomes flat, essentially representing a non-informative prior.

To illustrate this, let $\mu = 100$ be our best guess of θ_0 before observing any data. The bottom left panel of Figure 19.2 shows the PDFs of the normal prior distribution $N(\mu, \tau^2)$ for $\tau = 5, 10, 100$. We see that, for $\tau = 5$ and $\tau = 10$, the prior distribution is centered around $\mu = 100$, with the lower value of τ indicating a correspondingly higher belief in the prior best-guess $\mu = 100$. Indeed, the figure shows that the $N(100, 5^2)$ prior puts essentially all the prior beliefs in values of θ in the interval $[85,\ 115]$, while the $N(100, 10^2)$ prior results in putting a substantial belief in values outside this interval as well. Conversely, choosing τ very large, here $\tau = 100$, leads to a prior distribution that is essentially flat, i.e. non-informative.

19.3.5   The Laplace prior

The normal distribution has very "light" tails in the sense that the PDF behaves as e^{-x^2/τ^2} as $|x| \to \infty$. This means that a normal $N(\mu, \tau^2)$ prior implies a very low prior belief in values of θ far from the initial best guess μ. This may not be appropriate in situations where the researcher cannot, a priori, rule out "extreme" values of θ_0.

A prior distribution with heavier tails than the normal distribution, and thus more easily accommodating extreme outcomes away from the best guess, can be specified using the so-called *Laplace distribution*. The PDF of the Laplace distribution is given by

$$p(\theta) = \frac{1}{2\beta} \exp\left(-\frac{|\theta - \mu|}{\beta}\right), \quad \theta \in \mathbb{R},$$

where $\mu \in \mathbb{R}$ and $\beta > 0$. The mean of the Laplace distribution is μ and the scale of the distribution is determined by β. Choosing β small will mean that the distribution is concentrated around μ, representing high prior belief that θ_0 is close to the value μ. Conversely, larger values of β implies that the distribution is more spread out around μ, representing a correspondingly lower prior belief in the best guess μ. Indeed, choosing β very large implies that the bell shape becomes flat, essentially representing a non-informative prior.

To illustrate this, let $\mu = 100$ be our best guess of θ_0 before observing any data. The bottom right panel of Figure 19.2 shows the PDFs of the Laplace prior distribution for $\beta = 5, 10, 100$. We see that, for $\beta = 5$ and $\beta = 10$, the prior distribution is centered around $\mu = 100$, with the lower value of β indicating a correspondingly higher belief in the prior best-guess $\mu = 100$. Compared to the normal distribution (bottom left panel), the decay of the tails toward 0 of the Laplace distribution is much slower. Choosing β very large, here $\beta = 100$, leads to a prior distribution that is essentially flat, i.e. non-informative.

19.3.6 Improper priors

It turns out that a prior does not necessarily have to be a proper distribution. When this is the case, the prior is called an *improper prior*. For example, consider a prior for θ being a positive constant $p(\theta) = c$ for all $\theta \in (-\infty, \infty)$. Then

$$\int_{-\infty}^{\infty} p(\theta)\, d\theta = \int_{-\infty}^{\infty} c\, d\theta = \infty,$$

and, thus, $p(\theta)$ is not a density function. But it may be an improper prior according to the following definition.

Definition 19.2 (Improper prior) *Let $f_{\widetilde{Y}|\theta}(\widetilde{y} \mid \theta)$ be the sample distribution and $p(\theta)$ a function such that $\int p(\theta)d\theta = \infty$ if θ is continuous or $\sum p(\theta) = \infty$ if θ is discrete. Then $p(\theta)$ is called an **improper prior** if and only if the posterior $f_{\theta|\widetilde{Y}}(\theta \mid \widetilde{y})$, defined by*

$$f_{\theta|\widetilde{Y}}(\theta \mid \widetilde{y}) = \frac{f_{\widetilde{Y}|\theta}(\widetilde{y} \mid \theta) \cdot p(\theta)}{f_{\widetilde{Y}}(\widetilde{y})},$$

is a valid PMF or PDF, where $f_{\widetilde{Y}}(\widetilde{y})$ is either the sum or the integral of the numerator taken over all θ where $p(\theta) > 0$.

To check if $p(\theta)$ is an improper prior, it is sufficient to check if

$$f_{\widetilde{Y}}(\widetilde{y}) = \int_{-\infty}^{\infty} f_{\widetilde{Y}|\theta}(\widetilde{y} \mid \theta) \cdot p(\theta)\, d\theta$$

equals a positive number.

Example 19.2 *Consider a population given by the exponential distribution with density*

$$f(y|\lambda) = \lambda \exp(-\lambda y), \quad y \geq 0,$$

where λ is a positive number. With a simple random sample $\widetilde{Y} = (\widetilde{Y}_1, \ldots, \widetilde{Y}_n)$, the sample distribution is

$$f(\widetilde{y}|\lambda) = \lambda^n \exp(-\lambda(\widetilde{y}_1 + \ldots + \widetilde{y}_n)).$$

Let $p(\lambda) = c$, where $c > 0$ is a constant, be the candidate for an improper prior. To verify that $p(\lambda) = c$ is an improper prior, check if

$$f_{\widetilde{Y}}(\widetilde{y}) = \int_0^{\infty} f(\widetilde{y}|\lambda)\, p(\lambda)\, d\lambda$$

exists. It can be shown that

$$\int_0^{\infty} f(\widetilde{y}|\lambda)\, p(\lambda)\, d\lambda = \int_0^{\infty} \lambda^n \exp(-\lambda(\widetilde{y}_1 + \ldots + \widetilde{y}_n))\, c\, d\lambda = c\frac{n!}{(\widetilde{y}_1 + \ldots + \widetilde{y}_n)^{n+1}}.$$

This is a finite number and, thus, $p(\lambda) = c$ is an improper prior for an exponential distributed population and a simple random sample. Note that the actual value of $c > 0$ does not matter, because it cancels out when calculating the posterior distribution (19.1) using Bayes' theorem.

The improper priors may allow us to put equal weight on all values of θ when the support of θ is unbounded as shown in Example 19.2. Hence, this improper prior serves as the non-informative prior analogous to the uniform prior when the support of θ is unbounded.

19.3.7 Hyperparameters of a prior distribution

As the examples above show, the prior distribution will often depend on certain parameters, e.g. α and β in the beta distribution, μ and τ^2 in the normal distribution, and μ and β in the Laplace distribution. These parameters must be chosen by the statistician in a way to best reflect her prior knowledge. Since they are parameters introduced by the statistician, and not a feature of the original problem, they are often called *hyperparameters*. In practice, to choose the values of the hyperparameters, we may choose values such that the prior distribution represents what we have in mind as prior beliefs.

Although we will not go into detail with it here, methods for choosing the hyperparameters in a data-driven way also exists. The resulting Bayesian approach goes under the name of *empirical Bayes*.

19.4 Conjugate priors

In this section, we consider two particular cases where the prior distribution and the posterior distributions belongs to the same family of distributions. For a given likelihood function, a prior distribution which results in a posterior distribution of the same form, is called a *conjugate prior distribution*. One case of a conjugate prior is for a normal population and the other case is for a Bernoulli population.

19.4.1 Normal prior and normal population

In case the population is normally distributed, the prior distribution is normal, and that the sample is a simple random sample, then the posterior distribution is a normal distribution. This implies that if we do Bayesian updating several times, e.g. if we get new samples every month, then the posterior distributions for each update stays within the same normal family of distributions. The next theorem contains this result.

Theorem 19.1 (Conjugate normal prior) *Assume the prior distribution and the population distributions are*

$$\text{Prior} \quad : \quad \theta \sim N(\mu, \tau^2),$$
$$\text{Population} \quad : \quad Y \sim N(\theta_0, \sigma_0^2),$$

where $\theta_0 \in \mathbb{R}$, $\sigma_0^2 > 0$ and $\mu \in \mathbb{R}, \tau^2 > 0$ are hyperparameters. Assume σ_0^2 is known. Furthermore, assume a simple random sample $\widetilde{Y}_1, \ldots, \widetilde{Y}_n$. Then the posterior distribution is

$$\text{Posterior} : \theta | \widetilde{Y} \sim N(\mu_{po}, \tau_{po}^2),$$

where

$$\tau_{po}^2 \;=\; \frac{1}{n}\sigma_0^2 \frac{1}{\left(1 + \frac{1}{n}\frac{\sigma_0^2}{\tau^2}\right)},$$

$$\mu_{po} \;=\; \frac{1}{\left(1 + \frac{1}{n}\frac{\sigma_0^2}{\tau^2}\right)}\left(\overline{y} + \frac{1}{n}\mu\frac{\sigma_0^2}{\tau^2}\right),$$

and where $\overline{y}$ is the sample average

$$\overline{y} = \frac{1}{n}\sum_{i=1}^{n}\widetilde{y}_i.$$

From the expressions in Theorem 19.1, we observe that the larger the sample size n, the closer μ_{po} is to the sample average $\overline{y}$, i.e.

$$\mu_{po} = \frac{1}{\left(1 + \frac{1}{n}\frac{\sigma_0^2}{\tau^2}\right)} \left(\overline{y} + \frac{1}{n}\mu\frac{\sigma_0^2}{\tau^2}\right) \approx \frac{1}{(1+0)}(\overline{y} + 0) = \overline{y}$$

for n large. Thus, the influence of the prior distribution disappears for larger and larger n. That is, the information in the sample determines the posterior distribution for very large n. Recall that, by the Law of Large Numbers (Theorem 11.4), for a simple random sample, the sample average $\overline{y}$ is a consistent estimator of the population mean μ_0. Thus, for large n, the posterior distribution has a mean close to the population mean.

As discussed in Section 19.3.4, the more trust we have in the prior best-guess μ, the lower we may choose the value of the variance τ^2 in the prior. In this case, it can be seen that the initial best-guess μ dominates μ_{po} compared to the sample average $\overline{y}$. That is, for small τ^2, $\frac{1}{n}\frac{\sigma_0^2}{\tau^2}$ is large, and, thus,

$$\mu_{po} = \frac{1}{\left(1 + \frac{1}{n}\frac{\sigma_0^2}{\tau^2}\right)} \left(\overline{y} + \frac{1}{n}\mu\frac{\sigma_0^2}{\tau^2}\right) \approx \frac{\left(\overline{y} + \frac{1}{n}\mu\frac{\sigma_0^2}{\tau^2}\right)}{\frac{1}{n}\frac{\sigma_0^2}{\tau^2}} = \frac{\overline{y}}{\frac{1}{n}\frac{\sigma_0^2}{\tau^2}} + \mu \approx \mu.$$

The following example illustrates how Theorem 19.1 may be used in practice.

Example 19.3 (Conjugate normal prior) *Consider the setting of Theorem 19.1 and suppose that $\sigma_0^2 = 100$ and our prior on θ_0 is given by*

$$\theta \sim N(\mu, \tau^2),$$

with $\mu = 100$ and τ equal to either 5, 10, or 100. Suppose also that we observe the outcome of a simple random sample of size $n = 10$ such that $\overline{y} = 122.75$. The left panel of Figure 19.3 shows the three prior distributions and the three resulting posterior distributions

$$Posterior : \theta|\widetilde{Y} \sim N(\mu_{po}, \tau_{po}^2),$$

where μ_{po} and τ_{po}^2 are given in Theorem 19.1. The values of μ_{po} are $\mu_{po} = 116.25$, 120.68, and 122.73 for the three posteriors, respectively. The figure shows how the prior distribution is "dragged" toward the sample mean $\overline{y}$ to arrive at a posterior distribution. The more non-informative the prior, i.e. the higher τ is, the closer the corresponding posterior distribution will be to the sample mean.

We note that in the more realistic case where both μ_0 and σ_0^2 are unknown, still with a normal population and a simple random sample, there also exists a conjugate prior. It is the so-called *inverse Gamma distribution* on σ^2 and a normal distribution, conditional on σ^2, on μ.

19.4.2 Beta prior and Bernoulli population

Another case in which the prior and the posterior distributions are within the same family of distribution, is with a Bernoulli population, a beta prior, and a simple random sample. The next theorem contains the result.

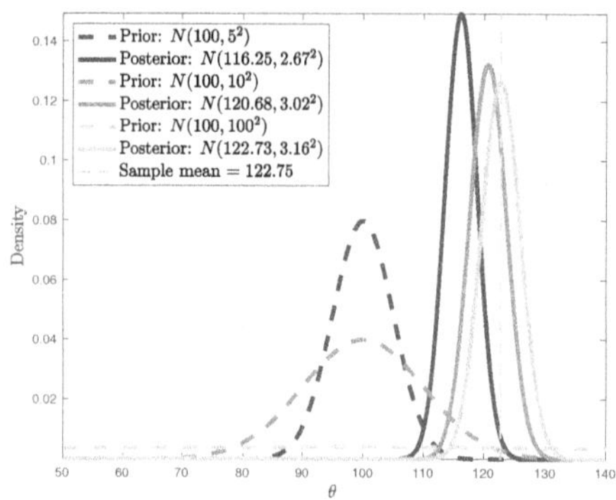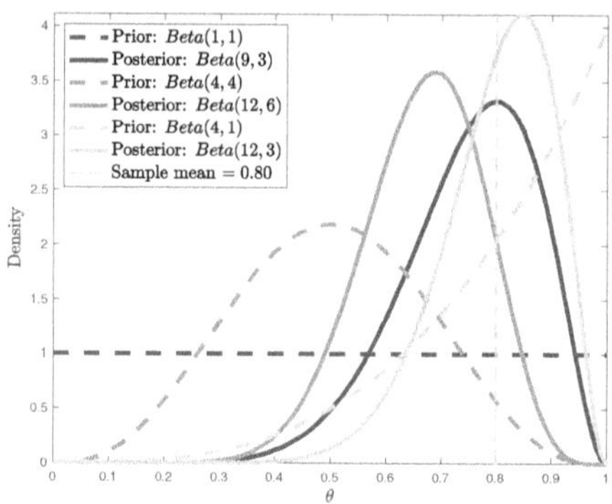

FIGURE 19.3

Left: Normal priors (dashed) and posteriors (solid), as given in Example 19.3. The vertical dashed line denotes the sample mean $\overline{y} = 122.75$. Right: Beta priors (dashed) and posteriors (solid), as given in Example 19.4. The vertical dashed line denotes the sample mean $\overline{y} = 0.80$.

Theorem 19.2 (Conjugate beta prior) *Assume the prior distribution and the population distributions are*

$$\begin{aligned} Prior \quad &: \quad \theta \sim Beta(\alpha, \beta), \\ Population \quad &: \quad Y \sim Ber(\theta_0), \end{aligned}$$

where $\theta_0 \in (0,1)$ and $\alpha, \beta > 0$ are hyperparameters. Assume a simple random sample $\widetilde{Y}_1, \ldots, \widetilde{Y}_n$. Then the posterior distribution is

$$Posterior : \theta|\widetilde{Y} \sim Beta(\alpha_{po}, \beta_{po}),$$

where

$$\alpha_{po} = \alpha + \sum_{i=1}^{n} \widetilde{Y}_i,$$

and

$$\beta_{po} = \beta + n - \sum_{i=1}^{n} \widetilde{Y}_i.$$

The following example illustrates how Theorem 19.2 may be used in practice.

Example 19.4 (Conjugate beta prior) *Consider the following sets of values for the hyperparameters $(\alpha, \beta) = (1,1)$, $(4,4)$, and $(4,1)$. Recall from Section 19.3.3 that $(\alpha, \beta) = (1,1)$ implies that the beta distribution is a uniform distribution on $(0,1)$, i.e. a non-informative prior.*

Suppose that $n = 10$ and that the sample is such that $\sum_{i=1}^{10} \widetilde{Y}_i = 8$, i.e. there are 8 "successes" in the 10 Bernoulli trials $\widetilde{Y}_1, \ldots, \widetilde{Y}_{10}$.

The right panel of Figure 19.3 shows the three prior distributions and the three resulting posterior distributions

$$Posterior : \theta|\widetilde{Y} \sim Beta(\alpha_{po}, \beta_{po}),$$

where α_{po} and β_{po} are given in Theorem 19.2.

Example 19.5 *Example 19.1, in which the posterior distribution is derived directly for a Bernoulli population and a $U(0,1)$ prior, is an example of the conjugate beta prior since the distribution $Beta(1,1;0,1)$ is the same as the $U(0,1)$ distribution. For example, with $n = 1$, the posterior distribution is either $Beta(2,1;0,1)$ or $Beta(1,2;0,1)$ for $\widetilde{y} = 1$ or $\widetilde{y} = 0$, respectively.*

Choosing a conjugate normal prior or a conjugate beta prior may ease computation considerably, since the posterior distribution is generally easier to handle and evaluate. In particular, this may result in closed-form expressions for the mean and other moments, and for quantiles of the posterior distribution. In general, however, it is important to choose priors, populations, and sampling methods that are good fit for the application at hand. That is, the choices of these should be based on perceived correctness, not computational expediency. The situation where the posterior distribution is difficult or impossible to derive analytically, e.g. because the prior is not conjugate, is discussion in Section 19.6 below.

19.5 Reparametrization and the Jeffreys prior

In some situations, it may be natural to consider different parametrizations of a problem. For instance, in analyzing the personal annual incomes of the working population of a country, this analysis could either be done in terms of the incomes themselves or, alternatively, in terms of the log-incomes. As we will see, different parametrizations will not alter the outcome of a Bayesian analysis. That is, the posterior distribution arrived at using different parametrizations are equivalent, if the prior distributions in the different parametrizations are chosen accordingly (Theorem 19.3). However, if the priors are not chosen in this way, then the resulting posterior distributions may differ.

There is a particular choice of prior, called the "Jeffreys prior", which resolves this problem. As we will see below, if the prior is specified to be the Jeffreys prior, then the posterior distributions arising from any parametrization will be equivalent.

19.5.1 Reparametrization

The same population may be parameterized many different ways. For example, $Y \sim N\left(\mu, \sigma^2\right)$ is the same population as $Y \sim N\left(\ln(\tau), \sigma^2\right)$ for $\mu = \ln(\tau)$. If we choose the prior distribution without considering how the sample distribution is parameterized, then it typically results in posterior distributions, which are not equivalent. In order for the posterior distributions to be equivalent, the prior distributions must be related in a particular manner, which is given in the following theorem.

Theorem 19.3 (Equivalence of parametrizations) *Let $g(\theta)$ be a differentiable and one-to-one function and let $\tau = g(\theta)$. Assume the sample distribution $f_{\widetilde{y}|\tau}\left(\widetilde{y}|\tau\right)$ is a reparametrization of $f_{\widetilde{y}|\theta}\left(\widetilde{y}|\theta\right)$, in the sense that*

$$f_{\widetilde{y}|\tau}\left(\widetilde{y}|\tau\right) = f_{\widetilde{y}|\theta}\left(\widetilde{y}|\theta\right) \quad \text{when } \tau = g(\theta).$$

Let $p_\theta(\theta)$ be the prior distribution of θ. Then the posterior distributions $f_{\theta|\widetilde{Y}}\left(\theta|\widetilde{y}\right)$ and $f_{\tau|\widetilde{y}}\left(\tau|\widetilde{y}\right)$ with $\tau = g(\theta)$ are equivalent if and only if the prior distribution $p_\tau(\tau)$ of τ is given by

$$p_\tau(\tau) = a \cdot p_\theta\left(g^{-1}(\tau)\right) \cdot \left| \frac{dg^{-1}(\tau)}{d\tau} \right|,$$

where $a > 0$ is a constant.

The value of the positive constant $a > 0$ in Theorem 19.3 is immaterial for the subsequent Bayesian analysis, because it cancels out when applying Bayes' theorem to derive the posterior as in (19.1). When possible, it is customary to set a such that $p_\tau(\tau)$ becomes a proper density, i.e. such that $\int p_\tau(\tau)d\tau = 1$.

Using the result in Theorem 19.3, the next example demonstrates that the interpretation of a prior may be quite different depending on how the sample distribution is parameterized.

Example 19.6 *Consider two researchers, Alice and Bob, who have the same belief on the distribution of the population. Suppose Alice specifies the population distribution as $Y \sim N(\mu, 1)$, i.e.*

$$f_{\widetilde{Y}|\mu}(\widetilde{y}|\mu) = \phi(\widetilde{y} - \mu), \quad \mu \in \mathbb{R},$$

where $\phi()$ denotes the PDF of the standard normal distribution. Bob specifies the population distribution as $Y \sim N(\ln(\tau), 1)$, i.e.

$$f_{\widetilde{Y}|\mu}(\widetilde{y}|\mu) = \phi(\widetilde{y} - \ln(\tau)), \quad \tau > 0.$$

The two specifications are equivalent when

$$\mu = \ln(\tau) \quad \text{or} \quad \tau = \exp(\mu).$$

Suppose that both researchers want to impose non-informative priors on their respective parameters μ and τ. For simplicity, assume we know that $\mu \in [0, 10]$ and, thus, $\tau \in [1, \exp(10)]$. Then Alice specifies the uniform prior

$$p_A(\mu) = \frac{1}{10}, \quad \mu \in [0, 10],$$

and Bob specifies the uniform prior

$$p_B(\tau) = \frac{1}{\exp(10) - 1}, \quad \tau \in [1, \exp(10)].$$

Alice obtains the posterior distribution

$$f_{\mu|\widetilde{Y}}(\mu|\widetilde{y}) = f_{\widetilde{Y}|\mu}(\widetilde{y}|\mu) \cdot p_A(\mu) = \phi(\widetilde{y} - \mu) \cdot \frac{1}{10},$$

while Bob obtains the posterior distribution

$$f_{\tau|\widetilde{Y}}(\tau|\widetilde{y}) = f_{\widetilde{Y}|\tau}(\widetilde{y}|\tau) \cdot p_B(\tau) = \phi(\widetilde{y} - \ln(\tau)) \cdot \frac{1}{\exp(10) - 1}.$$

To compare the two posterior distributions, rewrite $f_{\mu|\widetilde{Y}}(\mu|\widetilde{y})$ as a function of τ. Denote this distribution $h(\tau|\widetilde{y})$. By change of variables (Theorem 2.14),

$$h(\tau|\widetilde{y}) = f_{\mu|\widetilde{Y}}(\ln(\tau)|\widetilde{y}) \left| \frac{d\ln(\tau)}{d\tau} \right| = \phi(\widetilde{y} - \ln(\tau)) \cdot \frac{1}{10} \cdot \frac{1}{\tau}.$$

It is seen that $h(\tau|\widetilde{y})$ is not equal to $f_{\tau|\widetilde{Y}}(\tau|\widetilde{y})$. That is, even though both researchers specify the same population and sample distribution, imposing a uniform prior on both μ and τ does not lead to the same posterior distribution.

According to Theorem 19.3, for the two researchers to get to the same posterior distribution, when Alice imposes a uniform prior, then Bob should have imposed the prior

$$p_B(\tau) = p_\mu(\exp(\tau)) \cdot \left| \frac{d\ln(\tau)}{d\tau} \right| = \frac{1}{10} \cdot \frac{1}{\tau}.$$

This is not a uniform prior. Thus, an uninformative prior in one specification of the population may not be an uninformative prior in a reparametrization of the population.

19.5.2   The Jeffreys prior

It is possible to define a prior which leads to the same posterior distributions no matter the parametrization of the sample distribution. This prior is called the *Jeffreys prior*. The Jeffreys prior is constructed using the Fisher information $I(\theta)$ of the sample distribution (Definition 18.2). In case θ is a one-dimensional parameter, the Jeffreys prior is proportional to the square root of the absolute value of the Fisher information, i.e.

$$p(\theta) = a \cdot \sqrt{|I(\theta)|},$$

where $a > 0$ is a constant. To calculate the posterior distribution, the constant a is not important because it cancels out in Bayes' theorem when deriving the posterior distribution (19.1). However, when $\int p(\theta)d\theta$ exists, we may set a such that $p(\theta)$ becomes a proper density, i.e. such that $\int p(\theta)d\theta = 1$. This is accomplished by setting

$$a = \left(\int \sqrt{|I(\theta)|}d\theta \right)^{-1}.$$

If $\int p(\theta)d\theta$ does not exist, we may be able to treat the Jeffreys prior as an improper prior (Definition 19.2).

Example 19.7 (The Jeffreys prior, Bernoulli distributed population)
Consider the Fisher information of a simple random sample of size n from a Bernoulli distribution $Y \sim Ber(\theta_0)$ for $\theta_0 \in (0,1)$. The Fisher information is given by

$$I(\theta) = n\frac{1}{\theta(1-\theta)},$$

and, thus, the Jeffreys prior is

$$p(\theta) = a \cdot \sqrt{n\frac{1}{\theta(1-\theta)}},$$

where $a > 0$ may be set to any positive constant. It can be seen that the Fisher information, and thus the prior density, is largest for θ close to 0 and θ close to 1, and lowest for $\theta = 0.5$. It is possible to show that

$$\int_0^1 p(\theta)d\theta = \int_0^1 a \cdot \sqrt{n\frac{1}{\theta(1-\theta)}}d\theta = a\sqrt{n}\pi.$$

Thus, letting $a = 1/\left(\sqrt{n}\pi\right)$ makes $p(\theta)$ a proper density.

The next example illustrates the case where the Jeffreys prior turns out to be an improper prior.

Example 19.8 (The Jeffreys prior, normally distributed population) *Let the population be $Y \sim N(\mu_0, \sigma_0^2)$, where σ_0^2 is known and μ_0 is the unknown parameter of interest. Let θ be the random variable representing our belief of μ_0. Assuming a simple random sample $\tilde{Y}$ of size n, the Fisher information is given by (Definition 18.2)*

$$I(\theta) = E\left(\left(\frac{d\ln\left(f_{\tilde{Y}|\theta}(Y;\theta) \right)}{d\theta} \right)^2 \right) = \frac{n}{\sigma_0^2}, \tag{19.2}$$

see Problem 19.10.3. Thus, the Jeffreys prior on θ is

$$p(\theta) = a \cdot \frac{n}{\sigma_0^2},$$

where $a > 0$ may be set to any positive constant. If it is known that $\mu_0 \in [b, c]$ for some numbers $b < c$, then we may set $a = \frac{\sigma_0^2}{n(c-b)}$ to arrive at a proper density for $p(\theta)$. If μ_0 may be any number in $(-\infty, \infty)$, however, then we may set, say, $a = 1$, and treat $p(\theta)$ as an improper prior.

Note that it is necessary to check if the Jeffreys prior is either a distribution or an improper prior. In case it is none of the above, the Jeffreys prior should not be used.

The Jeffreys prior is not motivated by prior beliefs about the likely parameter values. Instead, it is solely determined by the sample distribution. The sample distribution can, however, also be said to be a result of our beliefs, since we specify a class of distributions for the population. Thus, with the Jeffreys prior, solely our beliefs about the models of the population matter for the posterior distribution. In addition, the Jeffreys prior assigns the highest density to the parameter values θ for which the random sample is most informative, where "most informative" is measured by the Fisher information.

19.6   Simulation-based estimation of the posterior distribution

As we saw above in (19.1), the posterior distribution may be derived using Bayes' theorem. In some cases, most notably when the prior and likelihood are chosen such that the prior is a conjugate prior, it may be possible to derive the posterior distribution analytically. We saw illustrations of this approach in Examples 19.3 and 19.4. In general, however, the posterior distribution may be highly intractable, which prohibits analytical derivation. The following example illustrates this situation.

Example 19.9 (Deriving the posterior distribution, Laplace prior and normal sample) *Consider the case of a population Y, represented by the normal distribution $N(\theta_0, \sigma_0^2)$. Assume, for simplicity, that σ_0^2 is known. The parameter θ_0 is the unknown parameter of interest. Suppose that the prior knowledge of θ_0 is captured by the Laplace distribution (Section 19.3.5)*

$$p(\theta) = \frac{1}{2\beta} e^{-|\theta - \mu|/\beta}, \quad \theta \in \mathbb{R},$$

where $\mu \in \mathbb{R}$, $\beta > 0$ are hyperparameters.

Assume a simple random sample $\widetilde{Y} = (\widetilde{Y}_1, \ldots, \widetilde{Y}_n)$ of Y is available. The sample distribution is

$$f_{\widetilde{Y}|\theta}(\widetilde{y} \mid \theta) = \left(\frac{1}{\sqrt{2\pi\sigma_0^2}}\right)^n \exp\left(-\frac{1}{2\sigma_0^2} \sum_{i=1}^{n} (\widetilde{y}_i - \theta)^2\right).$$

According to (19.1), the posterior distribution is

$$f_{\theta|\widetilde{Y}}(\theta \mid \widetilde{y}) = \frac{f_{\widetilde{Y}|\theta}(\widetilde{y} \mid \theta) \cdot p(\theta)}{\int f_{\widetilde{Y}|\theta}(\widetilde{y} \mid \theta) \cdot p(\theta)\, d\theta}.$$

The numerator is

$$\int f_{\widetilde{Y}|\theta}(\widetilde{y} \mid \theta) \cdot p(\theta) \, d\theta = \left(\frac{1}{\sqrt{2\pi\sigma_0^2}}\right)^n \frac{1}{2\beta} \exp\left(-\frac{1}{2\sigma_0^2}\sum_{i=1}^n (\widetilde{y}_i - \theta)^2\right) \exp(-|\theta - \mu|/\beta),$$

while the denominator is

$$\int f_{\widetilde{Y}|\theta}(\widetilde{y} \mid \theta) \cdot p(\theta) \, d\theta = \left(\frac{1}{\sqrt{2\pi\sigma_0^2}}\right)^n \frac{1}{2\beta} \int \exp\left(-\frac{1}{2\sigma_0^2}\sum_{i=1}^n (\widetilde{y}_i - \theta)^2\right) \exp(-|\theta - \mu|/\beta) d\theta,$$

yielding the posterior

$$f_{\theta|\widetilde{Y}}(\theta \mid \widetilde{y}) = \frac{\exp\left(-\frac{1}{2\sigma_0^2}\sum_{i=1}^n (\widetilde{y}_i - \theta)^2\right)\exp(-|\theta - \mu|/\beta)}{\int \exp\left(-\frac{1}{2\sigma_0^2}\sum_{i=1}^n (\widetilde{y}_i - \theta)^2\right)\exp(-|\theta - \mu|/\beta)d\theta}.$$

There is no closed-form solution for the integral in the denominator. Hence, we cannot evaluate the posterior distribution analytically.

As Example 19.9 shows, a given choice of prior distribution and of the distribution of the population may result in a posterior distribution that is difficult to derive and evaluate. In particular, closed-form expressions for the PDF, CDF, for moments, and for quantiles of the posterior distribution may not be available. In other words, even though the posterior distribution is precisely described by (19.1), using the posterior distribution in practice may be challenging because it may be hard to evaluate with reasonable precision. This is in particular often a problem when there are multiple unknown parameters, i.e. when $\theta = (\theta_1, \ldots, \theta_K)$ for some $K \geq 2$. For this reason, considerable work has been done to develop tools for evaluating the posterior distribution numerically, i.e. using computer software. Monte Carlo simulation has proved to be a particularly powerful tool for Bayesian analysis. Although some of these methods are computationally cumbersome, modern computers are generally fast enough to implement the methods. Access to these methods has revolutionized Bayesian methods, making it possible to evaluate even very complicated posteriors.

Simulation from a distribution is discussed in general in Appendix C. These simulation techniques are, in particular, useful for Bayesian analysis by simulating draws from the posterior distribution. The draws can subsequently be used to approximate relevant aspects of the posterior distribution using Monte Carlo techniques. An illustration of this is given in Example C.6 in Appendix C, where the so-called "Metropolis-Hastings algorithm" is used to simulate from the posterior distribution derived in Example 19.9.

We can use draws from the posterior distribution to visualize the posterior distribution of a parameter. Suppose we have M observations, $X_1^{\#}, \ldots, X_M^{\#}$, drawn from the posterior distribution $f_{\theta|\widetilde{Y}}(\theta \mid \widetilde{y})$. An estimator of the posterior PDF may be constructed using a so-called *kernel density estimator*,

$$\widehat{f}_h(x) = \frac{1}{Mh}\sum_{i=1}^M K\left(\frac{x - X_i^{\#}}{h}\right), \quad x \in \mathbb{R},$$

where $h > 0$ is a *bandwidth parameter*, chosen by the statistician, and $K()$ is a *kernel function* satisfying certain assumptions. A convenient and often-used kernel function is the standard Gaussian PDF $K(x) = \phi(x)$. The kernel density estimator provides a smoothed visualization of the posterior density function.

19.7   Estimators based on the posterior distribution

The main outcome of Bayesian methods is the posterior distribution, but we may also have an interest in having an estimate of a population parameter or other aspects of the population distribution. In this section, we consider various estimators of such aspects, derived from the posterior distribution.

19.7.1   Point estimators derived from the posterior distribution

Similarly to the aspects of any distribution, the mean and median reveals interesting properties of the posterior distribution. In addition, they may be taken as estimates of the population parameter of interest. The choice of estimator may be justified by a particular loss function.

Let the posterior distribution be $f_{\theta|\widetilde{Y}}(\theta \mid \widetilde{y})$ and let the interest be a point estimate $\widehat{\theta}$ of the population parameter θ_0. One possibility is to set $\widehat{\theta}$ equal to the mean of the posterior distribution of θ. It is denoted the *posterior mean* of θ and given by

$$\widehat{\theta}_{Post-Mean} = \int \theta \cdot f_{\theta|\widetilde{Y}}(\theta|\widetilde{Y})d\theta.$$

Another possibility for a point estimate $\widehat{\theta}$ of θ_0 is to use the *posterior median* of θ. It is defined by a value $\widehat{\theta}_{Post-Median}$, satisfying

$$\int_{-\infty}^{\widehat{\theta}_{Post-Median}} f_{\theta|\widetilde{Y}}(\theta|\widetilde{Y})d\theta = 0.5.$$

In Bayesian analysis, sometimes the value of θ that maximizes the posterior density is used as an estimate of θ_0. It is called a *maximum a posteriori* (MAP) *estimate* and it equals the mode of the posterior density. It is given by

$$\widehat{\theta}_{MAP} = \arg\max_{\theta} f_{\theta|\widetilde{Y}}(\theta \mid \widetilde{y}).$$

Notice, to calculate the MAP estimate, it is not necessary to calculate the denominator in the expression of the posterior distribution (19.1). This implies that we can avoid calculation, for instance in case θ is continuous, of an integral of dimension equal to the number of parameters.

Example 19.10 (Example 19.3, continued) *Consider the case of a conjugate normal prior. Then*

$$\widehat{\theta}_{Post-Mean} = \widehat{\theta}_{Post-Median} = \widehat{\theta}_{MAP} = \mu_{po}.$$

From Example 19.3, the values of μ_{po} are $\mu_{po} = 116.25$, 120.68, and 122.73 for the three posteriors, respectively. These values can be contrasted with the sample mean $\overline{y} = 122.75$, which is the maximum likelihood estimate $\widehat{\theta}_{ML}$ of θ_0.

Example 19.11 (Example 19.4, continued) *In the case of a conjugate beta prior, the mean of the beta distribution (Theorem 8.7) can be used to deduce that the posterior mean is*

$$\widehat{\theta}_{Post-Mean} = \frac{\alpha_{po}}{\alpha_{po} + \beta_{po}}.$$

In Example 19.4, the values of $\widehat{\theta}_{Post-Mean}$ are 0.75, 0.67, and 0.80 for the three posterior distributions. The median of the beta distribution has a very complicated expression, but for $\alpha_{po}, \beta_{po} > 1$, which is the case here, the posterior median can be approximated by

$$\widehat{\theta}_{Post-Median} \approx \frac{\alpha_{po} - \frac{1}{3}}{\alpha_{po} + \beta_{po} - \frac{2}{3}},$$

which comes out at $0.76, 0.67$, and 0.81 for the three posterior distributions. Lastly, the mode of the beta distribution can be used to deduce that the posterior mode, for $\alpha_{po}, \beta_{po} > 1$, is

$$\widehat{\theta}_{MAP} = \frac{\alpha_{po} - 1}{\alpha_{po} + \beta_{po} - 2},$$

which comes out at $0.80, 0.69$, and 0.85 for the three posterior distributions. These values can all be contrasted with the sample mean $\overline{y} = 0.80$, which is the maximum likelihood estimate $\widehat{\theta}_{ML}$ of θ_0.

19.7.2   Interval estimators derived from the posterior distribution

Similarly to a confidence interval studied in the frequentist paradigm in Chapter 14, we can define a Bayesian analogue, called the $(1-\alpha)$-*posterior density interval*, sometimes also called the $(1-\alpha)$-*credible interval*. For $\alpha \in (0,1)$, a $(1-\alpha)$-posterior density interval is any set $PDI_{1-\alpha}$ that satisfies

$$\mathrm{Pr}_\theta(\theta \in PDF_{1-\alpha}) = \int_{\theta \in PDI_{1-\alpha}} f_{\theta|\widetilde{Y}}(\theta \mid \widetilde{y}) \, d\theta = 1 - \alpha, \tag{19.3}$$

where $\mathrm{Pr}_\theta()$ denotes the probability taken with respect to the posterior distribution $f_{\theta|\widetilde{Y}}(\theta|\widetilde{Y})$. For a given $\alpha \in (0,1)$, there are many different posterior density intervals that satisfy the condition (19.3). To choose among them, we can impose an additional criteria on the interval. For instance, we may choose the interval of θ between the $\alpha/2$-quantile and the $(1 - \alpha/2)$-quantile of the posterior distribution. Another criteria is the shortest interval. This criteria can be obtained by requiring that the posterior density must be at least as high for any θ in the interval than for any θ outside the interval. Such a $(1 - \alpha)$-posterior density interval is called the *highest $(1 - \alpha)$-posterior density interval*, denoted $HPDI_{1-\alpha}$.

Though a $(1 - \alpha)$-posterior density interval appears to be very similar to a $(1 - \alpha)$-confidence interval (Definition 14.1), they are not the same. A $(1 - \alpha)$-confidence interval is a frequentist concept, constructed such that it covers the true value θ_0 with probability $(1-\alpha)$ in the thought experiment that the statistical experiment can be repeated infinitely many times, that is, over repeated sampling of $\widetilde{y}$. In contrast, a $(1 - \alpha)$-posterior density interval is constructed directly from the posterior distribution, which is a result of the prior distribution on θ and the information in the sample $\widetilde{y}$.

19.7.3   Bayesian decision theory and loss functions

The existence of a posterior distribution over the parameters θ provides the advantage that we can evaluate the expected loss of any decision we may want to make. To illustrate this, this section considers the decision of which estimator to choose. Another decision may be to choose between various hypotheses. This is the topic of Section 19.8. We first define the optimal decision rule and then we return to the case, where an estimate of a population parameter is needed.

Let a decision rule be represented by the function $\delta(\widetilde{y})$. For example, $\delta(\widetilde{y}) = \widehat{\theta}$, that is, based on the sample $\widetilde{y}$, we choose $\widehat{\theta}$ to be the estimate of the population parameter θ_0. Let $L()$ be a loss function such that $L(\delta(\widetilde{y}), \theta)$ is the loss of making decision $\delta(\widetilde{y})$ when θ is the true population parameter. Since we have a distribution over the possible values of θ, namely the posterior distribution, we can calculate the expected loss of the decision $\delta(\widetilde{y})$. This leads to the *expected posterior loss* of $\delta(\widetilde{y})$, and it is defined next.

Definition 19.3 (Expected posterior loss) *Let $L(\delta(\widetilde{y}), \theta)$ be the loss function of making decision $\delta(\widetilde{y})$ when θ is the true population parameter, and let $f_{\theta|\widetilde{Y}}(\theta|\widetilde{y})$ be the posterior distribution of θ. Then the **expected posterior loss**, EPL, of $\delta(\widetilde{y})$ is defined as*

$$EPL(\delta(\widetilde{y})) = E_{\theta|\widetilde{Y}}(L(\delta(\widetilde{y}), \theta)|\widetilde{y}),$$

where $E_{\theta|\widetilde{Y}}(\ |\widetilde{y})$ denotes the expectation taken with respect to the posterior distribution $f_{\theta|\widetilde{Y}}(\theta|\widetilde{y})$.

In Definition 19.3, in case θ is a continuous random variable, then the expected posterior loss is calculated as

$$EPL(\delta(\widetilde{y})) = \int L(\delta(\widetilde{y}), \theta) \cdot f_{\theta|\widetilde{Y}}(\theta \mid \widetilde{y}) \, d\theta.$$

In case θ is a discrete random variable, then the expected posterior loss is calculated as

$$EPL(\delta(\widetilde{y})) = \sum_{\theta} L(\delta(\widetilde{y}), \theta) \cdot f_{\theta|\widetilde{Y}}(\theta \mid \widetilde{y}),$$

where the summation is taken over all possible values of θ.

The expected posterior loss may be used to determine an optimal decision rule by solving for the decision rule than minimizes the expected posterior loss. Such an optimal decision rule is called the *Bayes decision rule* and it is defined next.

Definition 19.4 (Bayes decision rule) *Let $EPL(\delta(\widetilde{y}))$ be the expected posterior loss of a decision rule $\delta(\widetilde{y})$. Then the **Bayes decision rule** is defined as a decision rule $\delta^*(\widetilde{y})$ that solves*

$$\delta^*(\widetilde{y}) = \arg\min_{\delta} EPL(\delta(\widetilde{y})).$$

The Bayes decision rule is the optimal decision rule, in the sense of minimizing the expected posterior loss. Returning to considering how to pick an estimator $\widehat{\theta}$ of the population parameter θ_0, we may derive the optimal estimator for a particular loss function using the Bayes decision rule. Consider, for instance, the case of a squared loss function $L(\widehat{\theta}, \theta) = (\widehat{\theta} - \theta)^2$. Then it can be shown that the Bayes decision rule is to set the estimate $\widehat{\theta}$ equal to the mean of the posterior distribution,

$$\widehat{\theta}^* = \widehat{\theta}_{Post-Mean}.$$

In case the loss function is the absolute loss $L(\widehat{\theta}, \theta) = |\widehat{\theta} - \theta|$, then it can be shown that the Bayes decision rule is to set $\widehat{\theta}$ equal to the median of the posterior distribution,

$$\widehat{\theta}^* = \widehat{\theta}_{Post-Median}.$$

19.8 Bayesian analysis of hypotheses

In this section, we consider how to analyze hypotheses with Bayesian methods. The largest difference to the frequentist analysis of hypotheses, studied in Chapters 15 and 16, is that we in the Bayesian framework have a distribution of the likely values of θ_0, represented by the random variable θ, whereas we in the frequentist framework consider θ_0 a fixed number or vector. The upshot of this is that we, with the Bayesian approach, may attach probabilities to various hypotheses being true, e.g. the probability that $\theta \in \Theta_0$ for some set Θ_0, whereas such a statement is vacuous in the frequentist paradigm. This is a great strength of the Bayesian approach to statistics, since it aligns well with how we evaluate hypotheses in our everyday life, i.e. by attempting to attach probabilities to various hypotheses being true. Further, in case we want to use the analysis of hypotheses, based on a loss function, to make a decision, we can use the Bayes decision rule (Definition 19.4).

Let Θ be a parameter space, $\theta_0 \in \Theta$ the parameter of interest, and let θ be the random variable describing our beliefs about the likely values of θ_0. Consider $K + 1$ hypotheses, $H_0, \ldots, H_K$, given by

$$H_j : \theta \in \Theta_{H_j}, \quad j = 0, \ldots, K,$$

where $\Theta_{H_0}, \ldots, \Theta_{H_K}$ are parameter sets such that $\Theta_{H_0} \cup \ldots \cup \Theta_{H_K} = \Theta$. For $K = 1$, this setting is similar to that of Chapters 16 and 16, where statistical hypothesis testing is discussed from a frequentist point of view.

Given a prior $p(\theta)$ on θ, we may derive our prior beliefs about the hypotheses being true. Suppose θ is a continuous random variable. Then,

$$\Pr(H_j \text{ true}) = \Pr\left(\theta \in \Theta_{H_j}\right) = \int p(\theta) \cdot I\left(\theta \in \Theta_{H_j}\right) d\theta, \quad j = 0, \ldots, K,$$

where $I()$ is the indicator function. Once a random sample $\tilde{y}$ is available, we may calculate the probability of a hypothesis being true based on the posterior distribution $f_{\theta|\tilde{Y}}(\theta \mid \tilde{y})$,

$$\Pr(H_j \text{ true}|\tilde{y}) = \int f_{\theta|\tilde{Y}}(\theta \mid \tilde{y}) \cdot I\left(\theta \in \Theta_{H_j}\right) d\theta, \quad j = 0, \ldots, K.$$

When analysis of hypotheses is used for decision making, we may model the various cost and gains of a decision using a loss function. Let $\delta(\tilde{y}) \in \{H_0, H_1, \ldots, H_K\}$ be a decision rule and $L()$ a loss function such that $L\left(\delta(\tilde{y}), \theta\right)$ is the loss of the decision $\delta(\tilde{y})$ when θ is the true value. The expected posterior loss of $\delta(\tilde{y})$ is

$$EPL\left(\delta(\tilde{y})\right) = \int L\left(\delta(\tilde{y}), \theta\right) f_{\theta|\tilde{Y}}(\theta| \tilde{y}) d\theta = \sum_{j=0}^{K} \int L\left(\delta(\tilde{y}), \theta\right) f_{\theta|\tilde{Y}}(\theta| \tilde{y}) I\left(\theta \in \Theta_j\right) d\theta.$$

Assume, for simplicity, there are two hypotheses H_0 and H_1 and the loss only depends on whether H_0 or H_1 is true. In that case, the loss function can be written as

$$L\left(\delta(\tilde{y}), \theta\right) = L_{01} I\left(\theta \in \Theta_0\right) + L_{10} I\left(\theta \in \Theta_1\right),$$

where $L_{01}, L_{10} \geq 0$ are the losses from taking decision H_1 when H_0 is true and vice versa. This loss function is also illustrated in the following table.

		Truth	
		H_0	H_1
Decision	$\delta(\tilde{y}) = H_0$	0	L_{10}
	$\delta(\tilde{y}) = H_1$	L_{01}	0

A similar loss function is used in Example 15.15 of Chapter 15. In this case, the expected posterior loss of $\delta\left(\widetilde{y}\right)$ is

$$EPL\left(\delta\left(\widetilde{y}\right)\right) \;=\; \int L_{01}f_{\theta\mid\widetilde{Y}}(\theta\mid\widetilde{y})I\left(\theta\in\Theta_0\right) + \int L_{10}f_{\theta\mid\widetilde{Y}}(\theta\mid\widetilde{y})I\left(\theta\in\Theta_1\right) \tag{19.4}$$

$$=\; L_{01}\cdot\Pr(H_0\text{ true}\mid\widetilde{y})\cdot I\left(\delta\left(\widetilde{y}\right)=H_1\right) + L_{10}\cdot\Pr(H_1\text{ true}\mid\widetilde{y})\cdot I\left(\delta\left(\widetilde{y}\right)=H_0\right).$$

The Bayes decision rule is the decision rule that minimizes the expected posterior loss. The solution $\delta^*\left(\widetilde{y}\right)$ to minimizing the expected posterior loss given in (19.4) is

$$\delta^*\left(\widetilde{y}\right) = \begin{cases} H_0 & \text{if } \dfrac{\Pr(H_1\text{ true}\mid\widetilde{y})}{\Pr(H_0\text{ true}\mid\widetilde{y})} < \dfrac{L_{01}}{L_{10}}, \\[2ex] H_1 & \text{if } \dfrac{\Pr(H_1\text{ true}\mid\widetilde{y})}{\Pr(H_0\text{ true}\mid\widetilde{y})} > \dfrac{L_{01}}{L_{10}}, \\[2ex] \text{indifferent} & \text{if } \dfrac{\Pr(H_1\text{ true}\mid\widetilde{y})}{\Pr(H_0\text{ true}\mid\widetilde{y})} = \dfrac{L_{01}}{L_{10}}. \end{cases}$$

The ratio of the probability of each hypothesis being true is called the *Bayes factor*. For two hypotheses H_i and H_j, the Bayes factor is

$$BF_{ij} = \frac{\Pr\left(H_i\text{ true}\mid\widetilde{y}\right)}{\Pr\left(H_j\text{ true}\mid\widetilde{y}\right)}.$$

In case we are not able or willing to specify a loss function, Bayes factors are sometimes used as the basis for choosing a hypothesis. Interpreted in a loss function context, choosing the hypothesis favored by the Bayes factor amounts to assuming that the losses are equal. For example, with two hypotheses, the Bayes decision rule is determined by the favored Bayes factor if $L_{10} = L_{01}$.

We remark that, in the frequentist approach, we needed to define decision rules and then define additional criteria to choose between the decision rules. For example, we may use the minimax criterion to choose between decision rules. Since the Bayesian approach provides a distribution over θ, namely the posterior distribution, we can directly solve for the optimal decision rule, which is the Bayes decision rule. In the frequentist approach, in contrast, it is not necessary to specify a prior distribution.

19.9   Proofs

Proof of Theorem 19.1   *The density of the prior distribution is*

$$p(\theta) = \frac{1}{\sqrt{2\pi\tau^2}}\exp\left(-\frac{(\theta-\mu)^2}{2\tau^2}\right).$$

Let the simple random sample be $\widetilde{Y} = (\widetilde{Y}_1,\ldots,\widetilde{Y}_n)$*. The sample distribution is*

$$f_{\widetilde{Y}\mid\theta}(\widetilde{y}\mid\theta) = \left(\frac{1}{\sqrt{2\pi\sigma_0^2}}\right)^n \exp\left(-\frac{1}{2\sigma_0^2}\sum_{i=1}^{n}(\widetilde{y}_i-\theta)^2\right).$$

According to (19.1), the posterior distribution is found by multiplying $f_{\widetilde{Y}\mid\theta}(\widetilde{y}\mid\theta)$ *by*

$p(\theta)$ *and dividing by* $f_{\widetilde{Y}}(\widetilde{y})$. *After some calculations, it can be shown that*

$$
\begin{aligned}
f_{\theta|\widetilde{Y}}(\theta \mid \widetilde{y}) &= \frac{1}{f_{\widetilde{Y}}(\widetilde{y})} \cdot f_{\widetilde{Y}|\theta}(\widetilde{y} \mid \theta) \cdot p(\theta) \\
&= \frac{1}{f_{\widetilde{Y}}(\widetilde{y})} \cdot \left(\frac{1}{\sqrt{2\pi\sigma_0^2}}\right)^n \exp\left(-\frac{1}{2\sigma_0^2}\sum_{i=1}^{n}(\widetilde{y}_i - \theta)^2\right) \\
&\quad \cdot \frac{1}{\sqrt{2\pi\tau^2}} \exp\left(-\frac{1}{2\tau^2}(\theta - \mu)^2\right) \\
&= c_1 \cdot \exp\left(-\frac{1}{2\sigma_0^2}\sum_{i=1}^{n}(\widetilde{y}_i - \theta)^2 - \frac{1}{2\tau^2}(\theta - \mu)^2\right),
\end{aligned}
$$

where

$$
c_1 = \frac{1}{f_{\widetilde{Y}}(\widetilde{y})} \cdot \left(\frac{1}{\sqrt{2\pi\sigma_0^2}}\right)^n \cdot \frac{1}{\sqrt{2\pi\tau^2}}.
$$

Now, raise the squares

$$
f_{\theta|\widetilde{Y}}(\theta \mid \widetilde{y}) = c_1 \exp\left(-\frac{1}{2\sigma^2}\left(\sum_{i=1}^{n}\widetilde{y}_i^2 + n\theta^2 - 2\theta\sum_{i=1}^{n}\widetilde{y}_i\right) - \frac{1}{2\tau^2}\left(\theta^2 + \mu^2 - 2\mu\theta\right)\right).
$$

Let c_2 *be*

$$
c_2 = \frac{1}{2\sigma^2}\sum_{i=1}^{n}\widetilde{y}_i^2 + \frac{1}{2\tau^2}\mu^2.
$$

Then

$$
f_{\theta|\widetilde{Y}}(\theta \mid \widetilde{y}) = c_3 \exp\left(-\frac{1}{2\sigma^2}\left(n\theta^2 - 2\theta n\overline{y}\right) - \frac{1}{2\tau^2}\left(\theta^2 - 2\mu\theta\right)\right),
$$

where

$$
c_3 = c_1 \cdot \exp(-c_2),
$$

and

$$
\sum_{i=1}^{n}\widetilde{y}_i = n\overline{y}.
$$

Let

$$
d_1 = \frac{n}{\sigma^2} \quad \text{and} \quad d_2 = \frac{1}{\tau^2}.
$$

Then write $f_{\theta|\widetilde{Y}}(\theta \mid \widetilde{y})$ *as*

$$
\begin{aligned}
f_{\theta|\widetilde{Y}}(\theta \mid \widetilde{y}) &= c_3 \exp\left(-\frac{1}{2}\left((d_1 + d_2)\theta^2 - (d_1\overline{y} - d_2\mu)\,2\theta\right)\right) \\
&= c_3 \exp\left(-\frac{1}{2}\left((d_1 + d_2)\theta^2 - \left((d_1 + d_2)\frac{d_1\overline{y} + d_2\mu}{d_1 + d_2}\right)2\theta\right)\right) \\
&= c_3 \exp\left(-\frac{1}{2}(d_1 + d_2)\left(\theta^2 - \frac{d_1\overline{y} + d_2\mu}{d_1 + d_2}2\theta\right)\right) \\
&= c_3 \exp\left(-\frac{1}{2}(d_1 + d_2)\left(\theta - \frac{d_1\overline{y} + d_2\mu}{d_1 + d_2}\right)^2 + c_4\right),
\end{aligned}
$$

where

$$
c_4 = \frac{1}{2}(d_1 + d_2)\left(\frac{d_1\overline{y} + d_2\mu}{d_1 + d_2}\right)^2.
$$

Then

$$f_{\theta|\widetilde{Y}}(\theta \mid \widetilde{y}) = c_5 \exp\left(-\frac{1}{2}(d_1 + d_2)\left(\theta - \frac{d_1\overline{y} + d_2\mu}{d_1 + d_2}\right)^2\right),$$

where

$$c_5 = c_3 \exp(-c_4).$$

Let

$$\tau_{po}^2 = \frac{1}{d_1 + d_2} = \frac{1}{\frac{n}{\sigma^2} + \frac{1}{\tau^2}}$$

and

$$\mu_{po} = \frac{d_1\overline{y} + d_2\mu}{d_1 + d_2} = \tau_{po}^2\left(\frac{n}{\sigma^2}\overline{y} + \frac{1}{\tau^2}\mu\right).$$

Then

$$f_{\theta|\widetilde{Y}}(\theta \mid \widetilde{y}) = c_5 \exp\left(-\frac{1}{2\tau_{po}^2}(\theta - \mu_{po})^2\right).$$

This is the density function of the normal distribution $N(\mu_{po}, \tau_{po}^2)$. The reason is the term inside the exponential function is the same term which would be inside the exponential term is case $\theta \sim N(\mu_{po}, \tau_{po}^2)$. Since, by construction, $f_{\theta|\widetilde{Y}}(\theta \mid \widetilde{y})$ is a density function, and c_5 is a constant not depending on θ, then $f_{\theta|\widetilde{Y}}(\theta \mid \widetilde{y})$ must be this normal density function. That is,

$$Posterior : \theta|\widetilde{y} \sim N(\mu_{po}, \tau_{po}^2),$$

as we wanted to show.

Proof of Theorem 19.2 *The proof of Theorem 19.2 is left as an exercise to the reader (Problem 19.10.4).*

Proof of Theorem 19.3 *Suppose that θ is a continuous variable. The case where θ is a discrete random variable follows similar arguments with sums instead of integrals. Using the parametrization θ and the associated prior $p_\theta(\theta)$, the posterior is given by applying Bayes' rule as in (19.1), i.e.*

$$f_{\theta|\widetilde{Y}}(\theta \mid \widetilde{y}) = \frac{f_{\widetilde{Y}|\theta}(\widetilde{y} \mid \theta)p_\theta(\theta)}{f_{\widetilde{Y}}(\widetilde{y})} = \frac{f_{\widetilde{Y}|\theta}(\widetilde{y} \mid \theta)p_\theta(\theta)}{\int f_{\widetilde{Y}|\theta}(\widetilde{y}|\theta)p_\theta(\theta)d\theta}.$$

We may reparametrize this in terms of $\tau = g(\theta)$ using a change of variables (Theorem 2.14), yielding the reparametrized posterior

$$h_{\theta|\widetilde{Y}}(\tau \mid \widetilde{y}) = f_{\theta|\widetilde{Y}}(g^{-1}(\tau)|\widetilde{y})\left|\frac{dg^{-1}(\tau)}{d\tau}\right|.$$

We insert the posterior distribution $f_{\theta|\widetilde{Y}}(\theta \mid \widetilde{y})$ to get

$$h_{\theta|\widetilde{Y}}(\tau \mid \widetilde{y}) = \frac{f_{\widetilde{Y}|\theta}(\widetilde{y} \mid g^{-1}(\tau))p_\theta(g^{-1}(\tau))\left|\frac{dg^{-1}(\tau)}{d\tau}\right|}{\int f_{\widetilde{Y}|\theta}(\widetilde{y}|\theta)p_\theta(\theta)d\theta}.$$

Note that, by the change-of-variables $\theta = g^{-1}(\tau)$, the integral in the denominator may we written as

$$\int f_{\widetilde{Y}|\theta}(\widetilde{y}|\theta)p_\theta(\theta)d\theta = \int f_{\widetilde{Y}|\theta}(\widetilde{y}|g^{-1}(\tau))p_\theta(g^{-1}(\tau))\left|\frac{dg^{-1}(\tau)}{d\tau}\right|d\tau.$$

Thus, we may write the reparametrized posterior as

$$h_{\theta|\widetilde{Y}}(\tau|\,\widetilde{y}) = \frac{f_{\widetilde{Y}|\theta}(\widetilde{y}\,|g^{-1}(\tau))p_\theta(g^{-1}(\tau))\left|\frac{dg^{-1}(\tau)}{d\tau}\right|}{\int f_{\widetilde{Y}|\theta}(\widetilde{y}|g^{-1}(\tau))p_\theta(g^{-1}(\tau))\left|\frac{dg^{-1}(\tau)}{d\tau}\right|d\tau}. \tag{19.5}$$

Consider now the parametrization τ and the associated prior $p_\tau(\tau)$. The posterior in this case is also given by applying Bayes' rule as in (19.1), i.e.

$$f_{\tau|\widetilde{Y}}(\tau|\,\widetilde{y}) = \frac{f_{\widetilde{Y}|\tau}(\widetilde{y}\,|\tau)p_\tau(\tau)}{\int f_{\widetilde{Y}|\tau}(\widetilde{y}|\tau)p_\tau(\tau)d\tau}.$$

Using the assumption that, when $\tau = g(\theta)$, $f_{\widetilde{Y}|\tau}(\widetilde{y}|\tau) = f_{\widetilde{Y}|\theta}(\widetilde{y}|\theta) = f_{\widetilde{Y}|\theta}(\widetilde{y}|g^{-1}(\tau))$, we conclude that the posterior in the τ-parametrization, $f_{\tau|\widetilde{Y}}(\tau|\,\widetilde{y})$, is equal to the reparametrized posterior in the θ-parametrization (19.5) if and only if

$$p_\tau(\tau) = a \cdot p_\theta(g^{-1}(\tau))\left|\frac{dg^{-1}(\tau)}{d\tau}\right|,$$

where $a > 0$ is a constant. This concludes the proof.

19.10 Exercises

Problem 19.10.1 *Consider a simple random sample $\widetilde{Y} = (\widetilde{Y}_1, \ldots, \widetilde{Y}_n)$ from a Bernoulli distribution $Ber(\theta)$. Suppose the prior distribution is*

$$p(\theta) = \begin{cases} 0.5 & \text{if} \quad \theta = 1/4, \\ 0.5 & \text{if} \quad \theta = 3/4. \end{cases}$$

1. *Write up the sample distribution.*

2. *Explain why a posterior distribution only can have non-negative values for θ in the support of the prior, that is, for θ for which $p(\theta) > 0$.*

3. *Derive the posterior distribution.*

4. *Consider the case of $n = 1$ observation. Calculate the posterior distributions when the observation equals 1 and when the observation equals 0. Interpret your results in a comparison to the prior distribution.*

Problem 19.10.2 *Let $\mu \in \mathbb{R}$ and $\beta > 0$. Show that the Laplace prior (Section 19.3.5)*

$$p(\theta) = \frac{1}{2\beta}\exp\left(-\frac{|\theta - \mu|}{\beta}\right), \quad \theta \in \mathbb{R},$$

is a proper PDF.

1. *Show that $p(\theta)$ is a proper PDF.*

2. *Show that the mean of the Laplace distribution is μ.*

Problem 19.10.3 *Consider the setup in Example 19.8 and show that the Fisher information is given as in (19.2).*

Problem 19.10.4 *Prove Theorem 19.2.*

Problem 19.10.5 *Let the population Y be a normal population $N(\mu, \sigma^2)$ with known variance σ^2. Assume a simple random sample $\widetilde{Y} = (\widetilde{Y}_1, \ldots, \widetilde{Y}_n)$ is available. Consider the improper prior given as a constant $p(\mu) = 1$ for all $\mu \in (-\infty, \infty)$.*

1. *Derive the posterior distribution.*

2. *Show that the posterior mean equals the sample average.*

Appendices

A

Appendix A: Probability theory

A.1 Sets and set operators

In this section, we provide some insight into sets and set operators. The reason sets are important for probability theory is that the probability model is built on sets, set operations, and set functions.

A.1.1 Representation of a set

A *set* is a collection of *elements* called *members* of the set. The elements of a set can in principle be anything, such as physical objects, mathematical functions or relations, abstract human constructs, etc. The elements may even be sets themselves. For the purpose of probability theory, the elements we will consider will most often be numbers, representing outcomes of a statistical experiment, or sets. We will denote a generic element of a set by the Greek letter ω (omega). We will typically use capital letters to denote a set and curly brackets to enclose the elements.

Definition A.1 (Set, set membership) *A **set** Ω is a collection of elements. For any element ω, it must be possible to determine whether ω is in the set Ω. If ω is an element in Ω, we say that ω belongs to Ω, which can be written mathematically as*

$$\omega \in \Omega,$$

*where "$\in$" is the relation "element in" or "belongs to". If $\omega \in \Omega$, we also say that ω is a **member** of Ω. The elements of a set are distinct, meaning that an element cannot be a member of the same set twice. In an element ω is not a member of a set Ω, then we write $\omega \notin \Omega$.*

If a set Ω consists of the elements $\omega_1, \ldots, \omega_M$, we write

$$\Omega = \{\omega_1, \ldots, \omega_M\}.$$

Here, we allow $M = \infty$, i.e. we allow for a set to consist of infinitely many members. A set without any elements is called the empty set *and it is denoted $\emptyset$, which can also be written as $\emptyset = \{\ \}$.*

Example A.1 (Simple set of numbers) *The set of the numbers 0 and 1 is*

$$A = \{0, 1\}.$$

The element (number) 0 is an element of A, i.e. $0 \in A$. The number 5 is not in A, which we write as $5 \notin A$.

DOI: 10.1201/9781003591191-A

Example A.2 (Important sets of numbers) *There are several sets of numbers which are widely used and, thus, have been given their own names and symbols. Some of those are given in the following table.*

Numbers	*Elements*	*Symbol*
Natural	$\{1, 2, 3, 4, 5, \ldots\ldots\}$	$\mathbb{N}$
Integers	$\{\ldots\ldots -3, -2, -1, 0, 1, 2, 3, \ldots\ldots\}$	$\mathbb{Z}$
Rational	$\{\ldots\ldots, \frac{a}{b}, \ldots\ldots\}, a, b \in \mathbb{Z}$	$\mathbb{Q}$
Real	$xxx.xxxxx\ldots$, *where x represents digits* $(0, \ldots, 9)$	$\mathbb{R}$

A rational number $q \in \mathbb{Q}$ is any number which can be written as the ratio of two integers. A real number is, intuitively speaking, something that can be written in decimal number form, where we allow for infinitely many decimals. This means that the real numbers include all rational numbers as well as numbers that are not rational, i.e. numbers that cannot be written as a ratio of two integers, such as π and $\sqrt{2}$.

Example A.3 (Sets of sets) *Consider the set B consisting of three elements, namely the set $A = \{0, 1\}$ from Example A.1 and the sets $\mathbb{N}$ and $\mathbb{R}$ from Example A.2,*

$$B = \{\{0, 1\}, \mathbb{N}, \mathbb{R}\}.$$

The set B is a set of sets.

Note that 0 is not an element of B, i.e. $0 \notin B$, since the element 0 is not in B. Similarly, $1 \notin B$. The set consisting of 0 and 1, however, is in B, that is, $\{0, 1\} \in B$.

When a set is an interval of numbers, we will use square brackets or parentheses depending on whether the endpoints are included in the interval.

Definition A.2 (Intervals) *Let a and b be two real numbers, $a, b \in \mathbb{R}$, with $a < b$. The **interval from** a **to** b is the set of all real numbers between a and b. Four different intervals from a to b are possible, depending on whether a and b are themselves in the interval. The following table lists the four possibilities.*

Notation	*Description*	*Type*
$[a, b]$	*All real numbers between a and b, a and b included*	*Closed*
(a, b)	*All real numbers between a and b, a and b not included*	*Open*
$[a, b)$	*All real numbers between a and b, a included and b not included*	*Half open*
$(a, b]$	*All real numbers between a and b, a not included and b included*	*Half open*

Example A.4 *Consider the set consisting of all the real numbers between 0 and 1, including 0 and 1. This is also called the* unit *interval. According to Definition A.2, this set can be written as $[0, 1]$.*

Both 0 and 1 are members of the unit interval, i.e. $0 \in [0, 1]$ and $1 \in [0, 1]$, as are the numbers 0.42 and $\pi/100$, i.e. $0.42, \pi/100 \in [0, 1]$. The number 2 is not in the interval, i.e. $2 \notin [0, 1]$.

A.1.2 Subsets

A set may be larger than another set, meaning that all the elements of the smaller set are in the larger set. This lead to the concept of a *subset*.

Definition A.3 (Subset, proper subset) *A set A is said to be a **subset** of the set B if and only if B contains all elements that are in A, i.e. if and only if*

$$\omega \in A \Rightarrow \omega \in B.$$

This is denoted

$$A \subseteq B.$$

If $A \subseteq B$ and $B \subseteq A$, meaning that the two sets A and B contain the same elements, we say that A and B are equal and write

$$A = B.$$

*If $A \subseteq B$ but there exist $\omega \in B$ such that $\omega \notin A$, then we call A a **proper subset** of B. If we want to highlight that A is a proper subset of B, we may write*

$$A \subset B.$$

Example A.5 *Let*

$$
\begin{aligned}
A &= \{2,4\}, \\
B &= \{1,2,3,4\}.
\end{aligned}
$$

Then

$$A \subseteq B$$

because all elements in A are also in B.

In Example A.5, the set A is actually a proper subset of B, so we could write $A \subset B$. Unless it is important to know that a set is a proper subset, we will most often use the notation "$\subseteq$" to denote subsets, even if they may be proper subsets.

Example A.6 (Example A.2 continued) *The sets of numbers from Example A.2 have the following relation*

$$\mathbb{N} \subseteq \mathbb{Z} \subseteq \mathbb{Q} \subseteq \mathbb{R}.$$

That is, the set of natural numbers is a subset of the set of integers, which is a subset of the set of rational numbers, which is a subset of the set of real numbers.

Example A.7 *The interval $[0, 20]$ is a subset of $\mathbb{R}$ but it is not a subset of the rationals $\mathbb{Q}$ because the interval $[0, 20]$ contains numbers that are not rational, such as π and $\sqrt{2}$.*

Example A.8 *By definition, the empty set $\emptyset$ is a subset of any set B. That is, for any set B, we have that $\emptyset \subseteq B$. This is an implication of the definition of a subset. By Definition A.3, $\emptyset$ is a subset of the set B if every element in $\emptyset$ is also in B. Since $\emptyset$ contains no elements, this definition is trivially satisfied.*

Although they may look similar, there is an important difference between an element ω and a set consisting of that element, $\{\omega\}$. For example, we can use arithmetic operators on the number 8 like adding it to another number whereas we can use set operators (Section A.1.3) on the set $\{8\}$. This is reflected in how we use the notation. For example, let Ω be a set of elements. Then, for an element ω, we can evaluate if ω is a member of Ω, i.e. whether $\omega \in \Omega$, but we cannot meaningfully ask if ω is a subset of Ω, i.e. the operation $\omega \subseteq \Omega$ is not well defined. Conversely, we can evaluate if $\{\omega\} \subseteq \Omega$ but not if $\{\omega\} \in \Omega$.

Example A.9 (Example A.1 continued) *Consider the set A from Example A.1 consisting of the two numbers 0 and 1, i.e.*

$$A = \{0, 1\}.$$

Consider also the set B consisting the two sets $\{0\}$ and $\{1\}$,

$$B = \{\{0\}, \{1\}\}.$$

The set A is a set of numbers, while B is a set of sets. That is, the elements of A are different from the elements of B. Therefore, we cannot meaningfully compare A and B. For instance, it is not the case that A equals B.

Consider the number $\omega = 0$. Then $0 \in A$ and $\{0\} \subseteq A$. Also $\{0\} \in B$ and $\{\{0\}\} \subseteq B$.

A.1.3   Set operators

In a similar manner that operations on numbers, such as addition and multiplication, are useful, operations on sets will prove useful for probability modeling. We consider set operations on one or more sets that returns sets of the same kind, i.e. with the same type of members.

The *union* operator "$\cup$" joins all elements of two sets; the *intersection* operator "$\cap$" results in a set which contains the members that two sets have in common; the *difference* operator "$\setminus$" returns the elements that are in one set but not another; the *complement* operator "c" returns all the elements that are not in a particular set. These operations are formally defined next.

Definition A.4 (Operators on sets) *Let A and B be two sets. The **union** of A and B, written $A \cup B$, is the set consisting of elements in A or in B, or in both. That is*

$$\omega \in A \cup B \iff \omega \in A \text{ or } \omega \in B.$$

*The **intersection** of A and B, written $A \cap B$, is the set consisting of elements that are in both A and B. That is*

$$\omega \in A \cap B \iff \omega \in A \text{ and } \omega \in B.$$

*The **difference** between A and B, written $A \setminus B$, is the set consisting of elements that are in A but not in B. That is*

$$\omega \in A \setminus B \iff \omega \in A \text{ and } \omega \notin B.$$

*Let Ω be a set that is larger than A, i.e. $A \subseteq \Omega$. The **complement** of A, written A^c, with respect to Ω is the set consisting of all elements in Ω that are not in A. That is*

$$\omega \in A^c \iff \omega \notin A.$$

The four set operations defined here are listed in the following table.

Name	Symbol	Definition
Union	$\cup$	$A \cup B = \{$all ω's in either A or B, or in both$\}$
Intersection	$\cap$	$A \cap B = \{$all ω's in both A and $B\}$
Difference	$\setminus$	$A \setminus B = \{$all ω's in A which are not in $B\}$
Complement	c	$A^c = \{$all ω's in Ω not in $A\}$

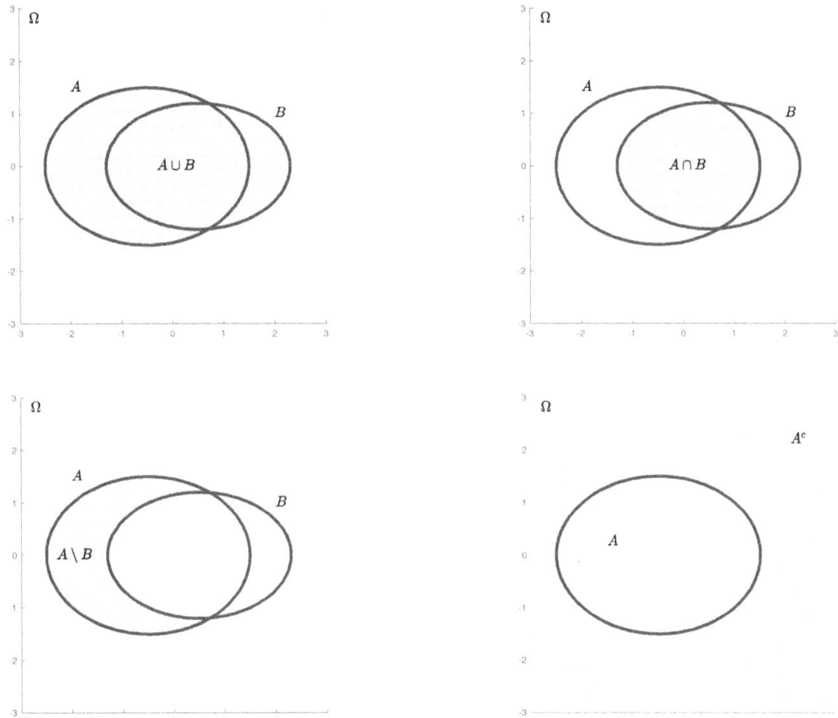

FIGURE A.1
Venn diagrams illustrating the various set operations. Shaded areas illustrate the elements included in the respective sets. Top left: Set union $A \cup B$; all the elements in A and/or B. Top right: Set intersection $A \cap B$; all the elements in A and B. Bottom left: Set difference $A \setminus B$; all the elements in A that are not in B. Bottom right: Set complement A^c; all the elements not in A.

Note that the complement operator is defined in reference to a larger set, i.e. Ω in Definition A.4. Given the definition of the difference of sets, we have that $A^c = \Omega \setminus A$.

The four set operators are illustrated in Figure A.1. These types of illustrations are called *Venn diagrams*.

Example A.10 *Let $A = \{1, 2, 3, 4\}$ and $B = \{2, 4, 6\}$. Then*

$$
\begin{aligned}
A \cup B &= \{1, 2, 3, 4, 6\}, \\
A \cap B &= \{2, 4\}, \\
A \backslash B &= \{1, 3\}, \\
B \backslash A &= \{6\}.
\end{aligned}
$$

Example A.11 (The irrational numbers) *The set of irrational numbers $\mathbb{I}$ consists of all those numbers which are real but not rational. Thus, this set can be written as the set difference $\mathbb{I} = \mathbb{R} \setminus \mathbb{Q}$. Examples of irrational numbers include π and $\sqrt{2}$.*

Some of the operations may result in a set with no elements. For example, if A and B have no elements in common, then the intersection of A and B is the empty set: $A \cap B = \emptyset$. Then A and B are called *disjoint sets*.

Definition A.5 (Disjoint sets) *The sets A and B are **disjoint** if and only if $A \cap B = \emptyset$.*

The set operators $\cup$ and $\cap$ have some of the same properties as addition and multiplication have in the case of the real numbers. Some of these are listed in the next theorem.

Theorem A.1 (Properties of union and intersection) *Let A, B, C be sets. The set operators $\cup$ and $\cap$ have the* commutative property*:*

$$\begin{aligned} A \cup B &= B \cup A, \\ A \cap B &= B \cap A. \end{aligned}$$

They have the associative property*:*

$$\begin{aligned} (A \cup B) \cup C &= A \cup (B \cup C), \\ (A \cap B) \cap C &= A \cap (B \cap C). \end{aligned}$$

They have the distributive property*:*

$$\begin{aligned} A \cup (B \cap C) &= (A \cup B) \cap (A \cup C), \\ A \cap (B \cup C) &= (A \cap B) \cup (A \cap C). \end{aligned}$$

In Theorem A.1, we have used parentheses in the usual way, i.e. operations inside parentheses should be performed first.

Combining the operators $\cup$ and $\cap$ with the complement operator, yields the useful results known as *De Morgan's laws*.

Theorem A.2 (De Morgan's laws) *Let A and B be sets. Then*

$$\begin{aligned} (A \cup B)^c &= A^c \cap B^c, \\ (A \cap B)^c &= A^c \cup B^c. \end{aligned}$$

A.1.4 Finite and infinite unions and intersections

We will need to do repeated set operations. For example, we may need to take the union of A with B and C.

Definition A.6 (Finite unions and intersections) *Let $A_1, A_2, \ldots, A_k$ be sets. The **finite union** of $A_1, \ldots, A_k$, written $\cup_{i=1}^{k} A_i = A_1 \cup \ldots \cup A_k$, is the set consisting of all the elements that are in at least one of the sets $A_1, \ldots, A_k$. That is,*

$$\omega \in \bigcup_{i=1}^{k} A_i \iff \omega \in A_i \text{ for at least one } i = 1, \ldots k.$$

*The **finite intersection** of $A_1, \ldots, A_k$, written $\cap_{i=1}^{k} A_i = A_1 \cap \ldots \cap A_k$, is the set consisting of all the elements that are in all the sets $A_1, \ldots, A_k$. That is,*

$$\omega \in \bigcap_{i=1}^{k} A_i \iff \omega \in A_i \text{ for all } i = 1, \ldots k.$$

Example A.12 *Suppose $A_i = \{0, 1, 2, \ldots, i\}$, $i = 1, \ldots, k$, and $k = 5$. That is, $A_1 = \{0, 1\}$, $A_2 = \{0, 1, 2\}$, $A_3 = \{0, 1, 2, 3\}$, $A_4 = \{0, 1, 2, 3, 4\}$, and $A_5 = \{0, 1, 2, 3, 4, 5\}$. Then*

$$\bigcup_{i=1}^{5} A_i = \{0, 1, 2, 3, 4, 5\},$$

$$\bigcap_{i=1}^{5} A_i = \{0, 1\}.$$

Sometimes we need to take unions and intersections of infinitely many sets. This is accomplished by extending the above definition of finite unions and intersections.

Definition A.7 (Infinite unions and intersections) *Let $A_1, A_2, \ldots$ be a sequence of infinitely many sets. The* **infinite union** *of $A_1, A_2, \ldots$, written $\cup_{i=1}^{\infty} A_i = A_1 \cup A_2 \cup \ldots$, is the set consisting of all the elements that are in at least one of the sets $A_1, A_2, \ldots$. That is,*

$$\omega \in \bigcup_{i=1}^{\infty} A_i \iff \omega \in A_i \text{ for at least one } i = 1, 2, \ldots.$$

The **infinite intersection** *of $A_1, A_2, \ldots$, written $\cap_{i=1}^{\infty} A_i = A_1 \cap A_2 \cap \ldots$, is the set consisting of all the elements that are in all the sets $A_1, A_2, \ldots$. That is,*

$$\omega \in \bigcap_{i=1}^{\infty} A_i \iff \omega \in A_i \text{ for all } i = 1, 2, \ldots.$$

To have a collection of infinitely many sets A_i, it is necessary to provide a formula for each A_i to know how A_i develops as i increases.

Example A.13 *Let $A_i = \{i, (i+2), (i+4)\}$. Then*

$$
\begin{aligned}
A_1 &= \{1, 3, 5\}, \\
A_2 &= \{2, 4, 6\}, \\
A_3 &= \{3, 5, 7\}, \\
&\ \ \vdots \\
A_{911} &= \{911, 913, 915\}, \\
&\ \ \vdots
\end{aligned}
$$

and, thus, A_i is an infinite collection of sets for $i = 1, 2, 3, \ldots$. We have

$$
\begin{aligned}
\bigcup_{i=1}^{\infty} A_i &= \{1, 2, 3, 4, 5, \ldots.\}, \\
\bigcap_{i=1}^{\infty} A_i &= \emptyset.
\end{aligned}
$$

Note, the infinite union here equals the set of natural numbers $\mathbb{N}$.

Example A.14 *Let $A_i = \{0, (-1)^i\}$. Then*

$$
\begin{aligned}
A_1 &= \{-1, 0\}, \\
A_2 &= \{0, 1\}, \\
A_3 &= \{-1, 0\}, \\
A_4 &= \{0, 1\}, \\
&\ \ \vdots
\end{aligned}
$$

because (-1) raised to an odd number is -1 and (-1) raised to an even number is 1. For $i = 1, 2, 3, \ldots$, $A_1, A_2, A_3, \ldots$ is an infinite collection of sets. We have

$$
\begin{aligned}
\bigcup_{i=1}^{\infty} A_i &= \{-1, 0, 1\}, \\
\bigcap_{i=1}^{\infty} A_i &= \{0\}.
\end{aligned}
$$

Example A.15 *Consider the half-open intervals $A_i = [0, 1 + 1/i)$ and the closed intervals $B_i = [0, 1 - 1/i]$ for $i = 1, 2, 3, \ldots$. We have*

$$\bigcup_{i=1}^{\infty} A_i = [0, 2), \quad \bigcup_{i=1}^{\infty} B_i = [0, 1),$$

$$\bigcap_{i=1}^{\infty} A_i = [0, 1], \quad \bigcap_{i=1}^{\infty} B_i = [0, 0] = \{0\}.$$

The infinite union of the A_i's does not contain 2 because there is no i for which A_i includes 2. The infinite union of B_i's does not include 1 because the endpoint of B_i is always less than 1: although $1 - 1/i$ gets arbitrarily close to 1, it does not attain the value 1. The infinite intersection of the A_i's does include 1 because $1 + 1/i$ is larger than 1 no matter how large an i is considered.

Strictly speaking, Definition A.7 only defines so-called *countable unions* and *countable intersections*, i.e. unions and intersections of countable sets $A_1, A_2, \ldots$. Similar operations are possible for an uncountable number of sets.

Example A.16 (Uncountable unions and intersections) *Let $\mathcal{A} = [0, 1]$ be the unit interval and consider the collection of sets $A_\alpha = (-\alpha, \alpha)$, where $\alpha \in \mathcal{A}$. The real numbers in the unit interval are uncountable, meaning that they cannot be listed as $\omega_1, \omega_2, \ldots$. Therefore, we cannot write, say, the infinite union of all the A's as $\cup_{i=1}^{\infty} A_i$. Instead, when the infinite union is with respect to uncountably many sets, we use the notation $\cup_{A \in \mathcal{A}} A$ where $\mathcal{A}$ is called an* index set. *We have*

$$\bigcup_{A \in \mathcal{A}} A = (-1, 1),$$

$$\bigcap_{A \in \mathcal{A}} A = \{0\}.$$

A.2 Measurability, random variables, and induced probability measures

Throughout the book, random variables are extensively used, as are probabilities concerning statements about random variables. The aim of this section is to formalize these concepts.

Recall from Chapter 1 that the ingredients of a probability model are the sample space Ω, an event space $\mathcal{F}$, and a probability measure P. The sample Ω (Definition 1.1) is a set consisting of all the possible outcomes that are relevant for the statistical experiment. We assume that the event space $\mathcal{F}$ is a σ-algebra (Definition 1.3) and P a probability measure (Definition 1.4), which is a function acting on sets in $\mathcal{F}$ and outputting a probability (i.e. a number between zero and one). Together, the triple $(\Omega, \mathcal{F}, P)$ forms a *probability space*. In what follows we take the probability space $(\Omega, \mathcal{F}, P)$ as given.

We are interested in how to use the probability space to assign probabilities to outcomes of statistical experiments, described by random variables. To fix ideas, let the set A be a subset of the real numbers, i.e. $A \subseteq \mathbb{R}$. Given that a random variable takes values in the real numbers, we are interested in assigning probabilities to events such as "the outcome of X is in the set A", which we write as "$X \in A$". In other words, we want to assign numerical probabilities to such statements, which we write as

$$\Pr(X \in A), \tag{A.1}$$

where $A \subseteq \mathbb{R}$. In what follows, we will formalize what is meant by such a statement. First, we may ask: Is it possible to assign such probabilities in (A.1) in a consistent manner for all sets $A \subseteq \mathbb{R}$? Here, "consistent" means that we want the probabilities to adhere to the probability calculus, as laid out in Definition 1.4 of Chapter 1. It turns out that the answer to this question is, in general, no. We therefore need to restrict the sets A used in (A.1) in a certain way. This restriction is related to the restriction of sets that the probability measure P is allowed to act on, i.e. we will require that the sets A used in $\Pr(X \in A)$ lie in a certain σ-algebra.

Assigning probabilities to sets in a consistent way requires the concept of *measurability*. This will, in turn, allow us to formally define what a *random variable* is. Lastly, these concepts can be leveraged to define probabilities concerning events of random variables, using the so-called *induced probability measure*, which will formalize what we mean by (A.1). Before delving into these concepts, we need a more thorough discussion of σ-algebras.

A.2.1 Sigma-algebras

A σ-algebra is typically constructed by specifying events of our interest. Based on these events of interest, we make sure they are included in the σ-algebra and then adding sets such that the requirements of the σ-algebra are met. This is illustrated by the next example.

Example A.17 *Consider a case where the population is $\Omega = \{0, 1, 2, 3\}$. If the interest is in the outcomes "0", "1", and "more than 1", then the events of interests can be written as*

$$
\begin{aligned}
A &= \{0\}, \\
B &= \{1\}, \\
C &= \{2, 3\}.
\end{aligned}
$$

A σ-algebra containing these events is (Problem 1.7.2)

$$\mathcal{F} = \{\{0\}, \{1\}, \{2, 3\}, \{0, 2, 3\}, \{1, 2, 3\}, \{0, 1\}, \Omega, \emptyset\}.$$

That is, $\mathcal{F}$ contains the sets A, B, and C, together with the complements of these sets and unions of all of these sets. Note, the sets $\{2\}$, $\{3\}$, $\{0, 2\}$, $\{1, 2\}$, $\{0, 3\}$, $\{1, 3\}$, $\{0, 1, 2\}$, and $\{0, 1, 3\}$ are not members of the σ-algebra because our interest is defined as being the sets A, B, and C.

When there are only a finite number of events of interest, then a σ-algebra can be constructed as in Example A.17. If the number of elements in the sample space Ω is finite, then we can choose as σ-algebra $\mathcal{F}$ the set all possible subsets of Ω. The set of all subsets of Ω is called the *power set* of Ω, sometimes denoted as $\mathcal{P}(\Omega)$ (Problem 1.7.3 asks you to prove that the power set is a σ-algebra). If Ω contains, say, K elements, then it is possible to make 2^K different subsets, that is, the power set of Ω contains 2^K elements. To construct the power set, start by including each element of Ω and then continue until all unions and complements of sets are included. The next example illustrates.

Example A.18 *Suppose the outcome of a statistical experiment can be either "low satisfaction", L, "medium satisfaction", M, or "high satisfaction", H. Then the sample space is*

$$\Omega = \{L, M, H\}.$$

Suppose we are interested in the three events $\{L\}$, $\{M\}$, and $\{H\}$. Then the σ-algebra must contain, in addition to $\{L\}$, $\{M\}$, and $\{H\}$, their complements $\{M, H\}$, $\{L, H\}$, and $\{L, M\}$, and all unions of every combination of those sets. In total this gives

$$\mathcal{F} = \{\{L\}, \{M\}, \{H\}, \{L, M\}, \{L, H\}, \{M, H\}, \{L, M, H\}, \emptyset\}.$$

This is also the power set of Ω, i.e. $\mathcal{F} = \mathcal{P}(\Omega)$, and it has $2^3 = 8$ elements.

When the sample space Ω is infinite, it may become impossible to assign probabilities to all subsets of Ω in a consistent way. Technically speaking, this may happen when Ω is *uncountably infinite*, e.g. if $\Omega = [0, 1]$ consists of all real numbers between 0 and 1. In these cases, the power set of Ω is "too large", in the sense that there will be no practically meaningful way to assign probabilities to all sets in $\mathcal{P}(\Omega)$. Therefore, in these cases, it becomes necessary to define the σ-algebra such that it does not contain all subsets of Ω, i.e. we do not define the σ-algebra $\mathcal{F}$ to be the power set of Ω. A convenient way of specifying a σ-algebra which is smaller than the power set of Ω, while still containing the subsets we are interested in, is to *generate* the σ-algebra from these subsets. These selected subsets are called a *generating set* for the σ-algebra.

Definition A.8 (Generating set) *Let $\mathcal{C}$ be a collection of subsets of Ω. The σ-algebra generated by $\mathcal{C}$, denoted by $\sigma(\mathcal{C})$, is a σ-algebra satisfying the following two conditions:*

1. $\mathcal{C} \subseteq \sigma(\mathcal{C})$.

2. If $\mathcal{B}$ is another σ-algebra such that $\mathcal{C} \subseteq \mathcal{B}$, then $\sigma(\mathcal{C}) \subseteq \mathcal{B}$.

Condition 1 in Definition A.8 ensures that the sets in $\mathcal{C}$ are also in the σ-algebra generated by $\mathcal{C}$. This is important, since $\mathcal{C}$ will usually be chosen to consist of the sets to which we are most interested in assigning probabilities. Condition 2 in Definition A.8 ensures that the σ-algebra generated by $\mathcal{C}$ is the smallest σ-algebra containing $\mathcal{C}$. In other words, the σ-algebra generated by $\mathcal{C}$, $\sigma(\mathcal{C})$, is a σ-algebra which is just big enough to contain the sets in $\mathcal{C}$. Ultimately, this will ensure that σ-algebras constructed from a suitable generating set is not "too large", in the sense discussed above.

Example A.19 *In Example A.17, the sets A, B, and C generated the σ-algebra.*

The most common situation where Ω is uncountably infinite is if it is (a subset of) the real numbers, i.e. if $\Omega \subseteq \mathbb{R}$. This will for instance be the case if the statistical experiment can have any real number in some interval as outcomes. In this case, the most useful σ-algebra $\mathcal{F}$ to choose for the probability model is almost always the so-called *Borel σ-algebra*, the definition of which is given next.

Definition A.9 (Borel σ-algebra) *Let $a < b$ and define the sample space $\Omega = [a, b] \subseteq \mathbb{R}$ as the real numbers between a and b. The **Borel σ-algebra**, denoted by $\mathcal{B}([a, b])$ is the σ-algebra generated by the open sets, i.e. generated by the sets in $\mathcal{C} = \{(c, d) : a \leq c < d \leq b\}$.*

In Definition A.9 we allow a to be minus infinity and b to be plus infinity. In particular, the case $a = -\infty$ and $b = \infty$ means that $\Omega = \mathbb{R}$. In this case, the Borel σ-algebra is denoted by $\mathcal{B}(\mathbb{R})$. A set in the Borel σ-algebra is also called a *Borel set*.

Example A.20 *Let Ω be all the real numbers in the closed interval from 0 to 5, i.e. $\Omega = [0, 5]$. Suppose we would like to be able to assign probabilities to any open interval (a, b), $0 \leq a < b \leq 5$. Hence, we will use the open intervals as generating sets, resulting in $\mathcal{F} = \mathcal{B}([0, 5])$ being the Borel σ-algebra, see Definition A.9. Using the definition of a σ-algebra (Definition 1.3), we can infer that $\mathcal{F}$ must also contain sets like*

$$(a, b)^c = [0, a] \cup [b, 5].$$

Likewise, $\mathcal{F}$ also contains all half-open and half closed intervals. To see this, let $a < c < b < d$ and note that

$$(c, d) \cap ([0, a] \cup [b, 5]) = [b, d).$$

Similarly, if $c < a < d < b$, then

$$(c, d) \cap ([0, a] \cup [b, 5]) = (c, a].$$

The closed intervals are also included in $\mathcal{F}$. Let $a < c < b < d$. Then

$$[a, b) \cup (c, d] = [a, d].$$

It could also be constructed as

$$\bigcap_{i=1}^{\infty} \left(a - \frac{1}{i}, b + \frac{1}{i} \right) = [a, b].$$

$\mathcal{F}$ also contains each element (number) in Ω because

$$\bigcap_{i=1}^{\infty} \left(a - \frac{1}{i}, a + \frac{1}{i} \right) = [a, a] = \{a\}.$$

The Borel σ-algebra, as defined above, is generated by the open intervals in $\mathbb{R}$. The following theorem shows that the definition could equally well have been in the form of closed intervals, half-open intervals, or intervals of the form $(-\infty, x]$ where $x \in \mathbb{R}$.

Theorem A.3 (Alternative generating sets for the Borel σ-algebra) *Let $\mathcal{B}(\mathbb{R})$ be the Borel σ-algebra. Then,*

$$
\begin{aligned}
\mathcal{B}(\mathbb{R}) &= \sigma((a, b] : -\infty \leq a < b < \infty) \\
&= \sigma([a, b) : -\infty < a < b \leq \infty) \\
&= \sigma([a, b] : -\infty < a \leq b < \infty) \\
&= \sigma((-\infty, x] : x \in \mathbb{R}).
\end{aligned}
$$

A.2.2 Measurability

Consider two sets Ω and Σ and the two σ-algebras $\mathcal{A}$ and $\mathcal{B}$, associated to Ω and Σ, respectively. The collections $(\Omega, \mathcal{A})$ and $(\Sigma, \mathcal{B})$ are each called *measurable spaces*.

Consider a function f with domain Ω and codomain Σ. For defining measurability, we need the concept of an *inverse image*, or *preimage*, of f.

Definition A.10 (Inverse image of function) *Let Ω and Σ be sets and consider a function $f : \Omega \to \Sigma$ with domain Ω and codomain Σ. Let $B \subseteq \Sigma$ be a subset of Σ. The **inverse image** of B under f, written as $f^{-1}(B)$, is the set of elements ω in Ω such that $f(\omega) \in B$. That is,*

$$f^{-1}(B) = \{\omega \in \Omega : f(\omega) \in B\}.$$

The measurability of a function f is defined with respect to two measurable spaces, say $(\Omega, \mathcal{A})$ and $(\Sigma, \mathcal{B})$. Specifically, if f is a function with domain Ω and codomain Σ, we say that f is measurable if for all sets B in $\mathcal{B}$, the inverse image of B under f is in $\mathcal{A}$. The formal definition is next.

Definition A.11 (Measurable function) *Let $(\Omega, \mathcal{A})$ and $(\Sigma, \mathcal{B})$ be measurable spaces and consider the function $f : \Omega \to \Sigma$ with domain Ω and codomain Σ. We say that f is $\mathcal{A}/\mathcal{B}$-measurable if and only if*

$$f^{-1}(B) = \{\omega \in \Omega : f(\omega) \in B\} \in \mathcal{A} \quad \text{for all } B \in \mathcal{B}.$$

As may be seen in Definition A.11, the concept of measurability is intimately connected to the measurable spaces under study, $(\Omega, \mathcal{A})$ and $(\Sigma, \mathcal{B})$, and in particular to the two σ-algebras, $\mathcal{A}$ and $\mathcal{B}$. Therefore, the formal designation is that a function may be "$\mathcal{A}/\mathcal{B}$-measurable". If there is no possibility for confusion, i.e. if it is understood which σ-algebras are relevant, then it is customary to simply say that a function is "measurable", instead of "$\mathcal{A}/\mathcal{B}$-measurable" i.e. not to designate the σ-algebras specifically.

The concept of measurability may be used to formally define a random variable.

A.2.3 Formal definition of a random variable

Consider a sample space Ω. We are interested in real-valued random variables, i.e. random variables that have outcomes in $\mathbb{R}$. Formally, we consider functions X with domain Ω and codomain $\mathbb{R}$. Let $\mathcal{F}$ be a σ-algebra on Ω. The discussion of measurability above shows that we also need to consider a σ-algebra on the codomain $\mathbb{R}$, and it turns out that the Borel σ-algebra $\mathcal{B}(\mathbb{R})$ (Definition A.9) is a very convenient choice. Hence, $(\Omega, \mathcal{F})$ and $(\mathbb{R}, \mathcal{B}(\mathbb{R}))$ are the measurable spaces we consider. We now define the function X to be a random variable if it is $\mathcal{F}/\mathcal{B}(\mathbb{R})$-measurable. The formal definition is given next.

Definition A.12 (Random variable) *Let Ω be a sample space and $\mathcal{F}$ a σ-algebra on Ω. Consider a function $X : \Omega \to \mathbb{R}$ with domain Ω and codomain $\mathbb{R}$. We say that X is a* **random variable** *on $(\Omega, \mathcal{F})$ if and only if X is $\mathcal{F}/\mathcal{B}(\mathbb{R})$-measurable, i.e. if and only if*

$$f^{-1}(B) = \{\omega \in \Omega : X(\omega) \in B\} \in \mathcal{F} \quad \text{for all } B \in \mathcal{B}(\mathbb{R}).$$

We began this section by asking how to assign probabilities to events such as "$X \in A$", where X is a random variable and A is a subset of the real numbers. Since our probability measure P is a function that takes sets from the σ-algebra $\mathcal{F}$ and outputs probabilities, Definition A.12 shows us the way forward. That is, the definition ensures that if A is a Borel set, that is, if $A \in \mathcal{B}(\mathbb{R})$, then the inverse image $X^{-1}(A)$ will be in $\mathcal{F}$. Since it is precisely such sets to which the probability measure P assigns probabilities, this shows that we can indeed make sense of

$$\Pr(X \in A),$$

but, in general, only for $A \in \mathcal{B}(\mathbb{R})$ and not for all conceivable subsets of $\mathbb{R}$. In particular, as discussed in Section A.2.1, if Ω is uncountable, then it may not be possible to assign probabilities to "$X \in A$" if $A \notin \mathcal{B}(\mathbb{R})$. Note, however, that the definition of the Borel σ-algebra (Definition A.9) shows that all open intervals, i.e. sets of the form $A = (a, b)$, are Borel sets. Theorem A.3 shows that closed and half-open intervals, i.e. sets of the form $A = [a, b]$, $A = (a, b]$, $A = [a, b)$, and $A = (-\infty, b]$, are also Borel sets. The definition of a σ-algebra ensures that countable unions and intersections of these sets are also Borel sets. This includes single points such as $\{a\}$ for $a \in \mathbb{R}$. These considerations show that the set of Borel sets, i.e. the Borel σ-algebra $\mathcal{B}(\mathbb{R})$, is a very rich class of sets. Indeed, it contains all the sets that we normally want to consider in practice. Hence, the restriction to only being able to assign probabilities to "$X \in A$" when $A \in \mathcal{B}(\mathbb{R})$ is not a practically relevant restriction.

The next section takes the final step and shows how to go from the probability measure P to assigning probabilities to "$X \in A$".

A.2.4 Induced probability measure

The discussion above motivates the following definition of probabilities of events concerning random variables, i.e. $\Pr(X \in A)$. The key is to use the probability measure, along with the inverse image of a random variable, to *induce* a new probability measure.

Definition A.13 (Induced probability measure) *Let $(\Omega, \mathcal{F}, P)$ be a probability space and X a random variable on $(\Omega, \mathcal{F})$. Then the* **probability measure induced by** X, *denoted P_X, is defined as*

$$P_X(A) = P\left(X^{-1}(A)\right), \quad for\ A \in \mathcal{B}(\mathbb{R}).$$

The following theorem shows that the probability measure induced by a random variable is indeed a proper probability measure.

Theorem A.4 *Let $(\Omega, \mathcal{F}, P)$ be a probability space, X a random variable on $(\Omega, \mathcal{F})$, and P_X the probability measure induced by X. Then $(\mathbb{R}, \mathcal{B}(\mathbb{R}), P_X)$ is a probability space. In particular, P_X is a probability measure in the sense of Definition 1.4.*

Sometimes the induced probability measure is also called the *push-forward measure* because it "pushes" the probability measure $P()$, which acts on subsets of Ω, into another probability measure $P_X()$, which acts on subsets of $\mathbb{R}$.

With Definition A.13 in hand, we are ready define what we mean by the probability that "$X \in A$" in (A.1). That is, for a probability space $(\Omega, \mathcal{F}, P)$ and a random variable X defined on this space, we define

$$\Pr(X \in A) = P_X(A), \quad A \in \mathcal{B}(\mathbb{R}), \tag{A.2}$$

where P_X is the probability measure induced by X.

As discussed above, the induced probability measure allows us to assign probabilities to events of the form "$X \in A$", as long as $A \in \mathcal{B}(\mathbb{R})$. A particularly useful Borel set to apply this to is a set of the form $(-\infty, x]$ where $x \in \mathbb{R}$ (recall that Theorem A.3 ensures that $(-\infty, x] \in \mathcal{B}(\mathbb{R})$), which leads to the definition of a *cumulative distribution function* (CDF) of the random variable X.

Definition A.14 (Cumulative distribution function, CDF) *Let $(\Omega, \mathcal{F}, P)$ be a probability space and X a random variable. Then the* **cumulative distribution function** *of X, denoted $F_X()$, is defined as*

$$F_X(x) = P_X((-\infty, x]), \quad x \in \mathbb{R},$$

where P_X is the probability measure induced by X.

Using the definition of $\Pr()$ in (A.2), we can also write the CDF of X as $F_X(x) = \Pr(X \le x)$ for $x \in \mathbb{R}$.

It is possible to show that if $F_X(x)$ is known for all $x \in \mathbb{R}$, then $P_X()$ can be constructed. That is, the CDF $F_X()$ and the induced probability measure $P_X()$ contain the same information about the distribution of the random variable X.

A.3   The probability limit operator

Consider a sequence of random variables $X_1, X_2, \ldots$. The following defines *convergence in probability* and the *probability limit* of such a sequence.

Definition A.15 (Convergence in probability, probability limit) *Let $X_1, X_2, \ldots$ be a sequence of random variable. We say that $X_1, X_2, \ldots$ **converges in probability** to the random variable X if for all $\epsilon > 0$,*

$$\lim_{n \to \infty} \Pr\left(|X_n - X| > \epsilon\right) = 0.$$

*If this is the case, then the random variable X is called the **probability limit** of the sequence $X_1, X_2, \ldots$, and we write*

$$\plim_{n \to \infty} X_n = X.$$

Definition A.15 provides a useful concept of the limit of a sequence of random variables, which is different to the limit of a sequence of real numbers (see Chapter B in the appendix for a discussion of limits of sequences of real numbers). In statistics, sequences of random variables are typically encountered as sequences of estimators, $\widehat{\theta}_n$, which are indexed by the sample size n. Probability limits of sequences of estimators is closely related to the concept of *consistency*, introduced in Section 11.5 of the book. Indeed, the sequence of estimators $\widehat{\theta}_1, \widehat{\theta}_2, \ldots$ is consistent for the parameter θ, in the sense of Definition 11.9, if $\plim_{n \to \infty} \widehat{\theta}_n = \theta$, i.e. if the probability limit of $\widehat{\theta}_n$ is θ.

The notation "$\plim_{n \to \infty}$" is used to stand for the *probability limit operator*: It takes the probability limit of whatever is to the right of "$\plim_{n \to \infty}$". The following theorem provides some rules of calculation with the probability limit operator when adding or multiplying sequences of random variables.

Theorem A.5 (Rules of calculation for plim) *Assume $\plim_{n \to \infty} Z_n = a_Z$ and $\plim_{n \to \infty} W_n = a_W$, where a_Z, a_W are two constants. Let also b, c be constants. Then*

1. $\plim_{n \to \infty} (bZ_n + c) = b \cdot \plim_{n \to \infty} (Z_n) + c = b \cdot a_Z + c.$

2. $\plim_{n \to \infty} (Z_n + W_n) = \left(\plim_{n \to \infty} Z_n\right) + \left(\plim_{n \to \infty} W_n\right) = a_Z + a_W.$

3. $\plim_{n \to \infty} (Z_n \cdot W_n) = \plim_{n \to \infty} Z_n \cdot \plim_{n \to \infty} W_n = a_Z \cdot a_W.$

4. $\plim_{n \to \infty} Z_n^{-1} = \left(\plim_{n \to \infty} Z_n\right)^{-1} = a_Z^{-1}$ *if* $a_Z \neq 0.$

A.4 Proofs

For brevity, we have chosen not to include proofs of the results given in this appendix. The interested reader may find the proofs in advanced textbooks that develops probability theory from a measure theoretic viewpoint.

B

Appendix B: Real analysis

This appendix provides background and additional details concerning the part of *real analysis*, i.e. the analysis of real-valued functions, used throughout the book.

B.1 Supremum and infimum of set

In this section, we formalize the concepts of *supremum* and *infimum* of a set. In Section B.2.2, we study the analogous concepts for a function. We begin with the concept of a set of numbers *bounded from above (or from below)*.

Definition B.1 (Set bounded from above, bounded from below, upper bound, lower bound) *Let $A \subseteq \mathbb{R}$ be a set of real numbers. The set A is **bounded from above** if and only if there exists a number $M \in \mathbb{R}$ such that $x \leq M$ for all $x \in A$. Such a number M is called an **upper bound** of A.*

*Conversely, the set A is **bounded from below** if and only if there exists a number $L \in \mathbb{R}$ such that $x \geq L$ for all $x \in A$. Such a number L is called a **lower bound** of A.*

Example B.1 *Consider the interval $A = (0, 1]$. This set is bounded from both above and below. To see this, apply Definition B.1 with, e.g. the upper and lower bounds $M = 10$ and $L = -5$. There are many lower and upper bounds for A, for instance, $M = 2$ and $L = 0$ are also upper and lower bounds, respectively.*

Example B.2 *Consider the interval $A = (-\infty, 100)$. This set is bounded from above but not from below. Thus, it has no lower bounds.*

With the above in place, we now define the supremum of a set A as the smallest upper bound of A.

Definition B.2 (Supremum of set) *Let $A \subseteq \mathbb{R}$ be a set of real numbers. The number $S \in \mathbb{R}$ is the **supremum** of A if and only if S is an upper bound of A and if for all upper bounds S' of A, it holds that $S \leq S'$. In this case, we write $S = \sup A$.*

Example B.3 (Example B.1 continued) *Consider again the interval from Example B.1, i.e. $A = (0, 1]$. The supremum of A is the smallest lower bound of A, i.e. $\sup A = 1$.*

Example B.4 (Example B.2 continued) *Consider again the interval from Example B.2, i.e. $A = (-\infty, 100)$. The supremum of A is the smallest lower bound of A, i.e. $\sup A = 100$.*

Note that in Example B.4, the supremum of $A = (-\infty, 100)$, namely $S = 100$, is not in A itself. This is the main reason why it is often more convenient to use the concept of supremum instead of the concept of maximum. Indeed, the maximum of $A = (-\infty, 100)$ does not exist, because no matter which value $x \in A$ you choose, there is always another value $y \in A$ such that $y > x$. In contrast, the supremum does exist, although it does not lie in A in this case.

The infimum of a set A is defined analogously to the supremum, namely as the largest lower bound of A. The formal definition is given next.

Definition B.3 (Infimum of set) *Let $A \subseteq \mathbb{R}$ be a set of real numbers. The number $I \in \mathbb{R}$ is the **infimum** of A if and only if I is a lower bound of A and if for all lower bounds I' of A, it holds that $I \geq I'$. In this case, we write $I = \inf A$.*

Example B.5 (Example B.3, continued) *Consider again the interval from Example B.1, i.e. $A = (0, 1]$. The infimum of A is the largest lower bound of A, i.e. $\inf A = 0$. Note that the infimum does not belong to the set A in this case.*

Example B.6 (Example B.4, continued) *Consider again the interval from Example B.2, i.e. $A = (-\infty, 100)$. In this case, because A does not have any lower bounds, we say that the infimum does not exist. We may also write $\inf A = -\infty$.*

B.2 Functions of one variable

This section describes functions of one variable. We will use the concept of a *limit* of a function to discuss *continuity* and *differentiability* of a function. Before that, we need some preliminaries.

B.2.1 Domain, codomain, and range

A function $f()$ of one variable assigns to each input x one output $f(x)$. The set of permissible inputs is called the *domain*, D, of the function. A set in which all outcomes of the function, is contained is called the *codomain*, C, of the function. We will only work with real functions. This means that both the domain and the codomain are subsets of the real numbers $\mathbb{R}$. A function with domain $D \subseteq \mathbb{R}$ and co-domain $C \subseteq \mathbb{R}$ is denoted as

$$f : D \to C.$$

There may be values in the codomain which are not outputs of the function for any number in the domain. The set of numbers which are an output of the function for some number in the domain is called the *range*, R, of the function.

Example B.7 *Let the function $f : D \to C$ be given by $f(x) = x^2$ with domain be $D = [-2, 10]$. Then $C = [-200, 200]$ is a codomain of f. There are many other codomains, e.g. $C = [-10, 150]$. The defining property of the codomain C is that $f(x) \in C$ for all $x \in D$, i.e. for any x in the domain D, the output of the function $f(x)$ is contained in the codomain.*

The range of the function is $R = [0, 100]$. The range of the function is necessarily a subset of any codomain of the function.

As Example B.7 illustrates, a function may have many different codomains. Often, the codomain will not be important for the subsequent analysis, and in this case, it is usual to simply specify the codomain as all the real numbers. That is, in this situation, a function $f()$ with domain D may be written as

$$f : D \to \mathbb{R},$$

or, simply,

$$f(x), \quad x \in D.$$

Example B.8 (Example B.7, continued) *We may write the function from Example B.7 as*

$$f(x) = x^2, \quad x \in [-2, 10].$$

B.2.2 Supremum and infimum of a function

The concepts of supremum and infimum of a set, studied in Section B.1, may be extended straightforwardly to functions. We first define the concept of a function being bounded from above and from below.

Definition B.4 (Function bounded from above, bounded from below, upper bound, lower bound) *Let $f : D \to \mathbb{R}$ be a function, where $D \subseteq \mathbb{R}$ is a subset of the real numbers. The function f is **bounded from above** if and only if there exists a number $M \in \mathbb{R}$ such that $f(x) \leq M$ for all $x \in D$. Such a number M is called an **upper bound** of f.*

*Conversely, the function f is **bounded from below** if and only if there exists a number $L \in \mathbb{R}$ such that $f(x) \geq L$ for all $x \in D$. Such a number L is called a **lower bound** of f.*

The supremum and infimum of a function is defined analogous to that of sets, i.e. as the smallest upper bound and the largest lower bound, respectively.

Definition B.5 (Supremum of function) *Let $f : D \to \mathbb{R}$ be a function, where $D \subseteq \mathbb{R}$ is a set of real numbers. The number $S \in \mathbb{R}$ is the **supremum** of f if and only if S is an upper bound of f and if for all upper bounds S' of f, it holds that $S \leq S'$. In this case, we write $S = \sup_{x \in D} f(x)$.*

Definition B.6 (Infimum of function) *Let $f : D \to \mathbb{R}$ be a function, where $D \subseteq \mathbb{R}$ is a set of real numbers. The number $I \in \mathbb{R}$ is the **infimum** of f if and only if I is a lower bound of f and if for all lower bounds I' of f, it holds that $I \geq I'$. In this case, we write $I = \inf_{x \in D} f(x)$.*

Example B.9 (Example B.8, continued) *Consider the function $f(x) = x^2$ defined on the interval $D = [-2, 10]$. By the above definitions, $\inf_{x \in [-2, 10]} f(x) = 0$ and $\sup_{x \in [-2, 10]} f(x) = 10^2 = 100$. That is, the minimum of $f(x)$ over D is zero (attained in $x = 0$), and the maximum over D is 100 (attained in $x = 10$) .*

Example B.10 *Consider the function $f(x) = x$ defined on the interval $D = [0, 1)$. By the above definitions, $\inf_{x \in [0,1)} f(x) = 0$ and $\sup_{x \in [0,1)} f(x) = 1$. Note that the minimum of $f(x)$ over D is zero (attained in $x = 0$), but that the maximum of $f(x)$ does not exist over D.*

B.2.3   Inverse image

For a function $f()$, instead of relating an input x to the output $f(x)$, we may ask what input values x result in a given output y, i.e. which x solves $y = f(x)$. Let $y^* \in R$ be a number in the range of a function $f()$ with domain D. There may be one or more values x in the domain of the function for which $f(x) = y^*$. Let A_{y^*} be all those values. Formally,

$$A_{y^*} = \{x \in D : f(x) = y^*\}.$$

This set is called the *inverse image* of y^*.

Example B.11 *Consider the function* $f(x) = x^2$ *with domain* $D = [-10, 10]$. *Then the inverse image of* 4 *is*

$$A_4 = \{x \in D : x^2 = 4\} = \{-2, 2\},$$

because $(-2)^2 = 4$ *and* $2^2 = 4$. *Similarly, the inverse image of* 5 *is*

$$A_5 = \{x \in D : x^2 = 5\} = \{-\sqrt{5}, \sqrt{5}\}.$$

The inverse image of 0 *is*

$$A_0 = \{x \in D : x^2 = 0\} = \{0\}.$$

The concept of an inverse image can be extended to the inverse image of a set B, as the set of all values $x \in D$ such that $f(x)$ is in B. Formally,

$$A_B = \{x \in D : f(x) \in B\}.$$

We will often write the inverse image of B as $f^{-1}(B)$.

Example B.12 (Example B.11, continued) *The inverse image of the interval* $B = [1, 9]$ *is*

$$f^{-1}([1, 9]) = A_{[1,9]} = \{x \in D : x^2 \in [1, 9]\} = [-3, -1] \cup [1, 3].$$

We can now formally define the inverse image of a function.

Definition B.7 (Inverse image of function) *Let* $f : D \to \mathbb{R}$ *be a function with domain* $D \subseteq \mathbb{R}$. *For a set* $B \subseteq \mathbb{R}$, *the **inverse image of** B is*

$$f^{-1}(B) = \{x \in D : f(x) \in B\}.$$

We note that the inverse image of a set is necessarily a subset of the domain of the function, i.e. $f^{-1}(B) \subseteq D$ for any set $B \subseteq \mathbb{R}$. We also note that it could be the case that there are no x in the domain of $f()$ such that $f(x) \in B$. In that case, the inverse image of B is the empty set $\emptyset$.

Example B.13 (Example B.12, continued) *Since* $f(x) = x^2$ *is non-negative for all* x, *the inverse image of the interval* $B = (-5, -1)$ *is*

$$f^{-1}((-5, -1)) = \{x \in D : x^2 \in (-5, -1)\} = \emptyset.$$

B.2.4 Inverse function

In case the inverse image of any number in the range of f is a singleton, i.e. consists of only one number, then we say that f has an *inverse function*. It is denoted $f^{-1}()$ and formally defined here.

Definition B.8 (Inverse function) *Let $f : D \to R$ be a function with domain $D \subseteq \mathbb{R}$ and range $R \subseteq \mathbb{R}$. Then the function $f^{-1} : R \to D$ with domain R and range D is called the* **inverse function** *of f if and only if the inverse image of f contains exactly one element for every number y in the range R of f.*

The inverse image $f^{-1}(y)$ can always be calculated for any input $y \in \mathbb{R}$. Formally, as we saw in Section B.2.3, the inverse image $f^{-1}(y)$ is a set. Recall that a function must only associate one output value with an input value. What the definition of an inverse function does is that it adds a condition for $f^{-1}()$ to be a function. In particular, it outputs real values, which can be contrasted to the inverse image which outputs sets.

Example B.14 *Consider the function $f(x) = x^2$ with domain $D = [-10, 10]$. Then there does not exist an inverse function for f because the inverse image of, say, $y = 4$ has two members*

$$f^{-1}(\{4\}) = \{-2, 2\}.$$

If we, instead, consider the function $g(x) = x^2$ with domain $D = [0, 10]$, then the inverse image for every number in the range has only one member. For example,

$$g^{-1}(\{4\}) = \{2\}.$$

Thus, in this case, the inverse function does exist. The functional form of this inverse function is given by

$$g^{-1}(y) = \sqrt{y},$$

for $y \in R = [0, 100]$.

By definition, an inverse function reverses the result of a function, taking the result back to the input value. That is, the inverse function has the property that $f^{-1}(f(x)) = x$. Likewise, if y is in the range of $f()$, then $f(f^{-1}(y)) = y$.

Example B.15 *Let $D = [0, 10]$, $R = [0, 100]$, and consider the function*

$$f : D \to R$$

given by $f(x) = x^2$. As we saw in Example B.14, $f()$ has an inverse function

$$f^{-1} : R \to D,$$

given by $f^{-1}(y) = \sqrt{y}$. For all $x \in D$, we have

$$f^{-1}(f(x)) = f^{-1}(x^2) = \sqrt{x^2} = x,$$

while, for all $y \in R$, we have

$$f(f^{-1}(y)) = f(\sqrt{y}) = (\sqrt{y})^2 = y.$$

The conditions for an inverse function to exist can also be described using the mathematical concepts of "one-to-one", "onto", and "bijection". Formally, we say a function $f()$ is *one-to-one* if there are no input values in the domain that results in the same output values:

$$f \text{ one-to-one} \iff \text{ For any } x_1, x_2 \in D \text{ with } x_1 \neq x_2, \text{ then } f(x_1) \neq f(x_2).$$

We say that a function is *onto* if the codomain equals the range of the function. If a function is both one-to-one and onto, then the function is a *bijection*. This is exactly the condition for the inverse of a function to exist: If a function is a bijection, then it has an inverse function.

B.2.5 Sequences and limits of sequences

To discuss differentiation and integration of functions, it is useful to first introduce *sequences* of numbers. For example, sometimes a sequence is presented as

$$2, 4, 6, 8, 10, \ldots.$$

Then we may instinctively know how to continue the sequence, e.g. by adding 2 to a number to get the next number in the sequence. We can write this formally as

$$x_i = 2 \cdot i,$$

where $i = 1, 2, \ldots$ denotes the location of the number in the sequence and x_i is the number corresponding to the i'th location. The sequence could also be specified iteratively as

$$x_i = x_{i-1} + 2,$$

for $i = 1, 2, \ldots$ and $x_0 = 0$. In either specification, there is a recipe defining the number at the location i in the sequence. This is formalized next in the definition of a sequence.

Definition B.9 (Sequence of numbers) *The numbers $x_1, x_2, \ldots, x_i, \ldots$ is a **sequence of numbers** if there is a function $g()$ such that for every index i, then $x_i = g(i)$. A finite sequence of m numbers is denoted $\{x_i\}_{i=1}^m$ and an infinite sequence of numbers is denoted $\{x_i\}_{i=1}^\infty$.*

Example B.16 *Here are some examples of sequences of numbers*

$$\begin{aligned}
x_i &= i &\Rightarrow\quad &\{x_i\}_{i=1}^\infty = \{1, 2, 3, 4, \ldots\}, \\
x_i &= (-1)^i &\Rightarrow\quad &\{x_i\}_{i=1}^\infty = \{-1, 1, -1, 1, -1, \ldots\ldots\}, \\
x_i &= 3 - \frac{1}{i} &\Rightarrow\quad &\{x_i\}_{i=1}^\infty = \{2, 2.5, 2.67, 2.75, \ldots\ldots\}.
\end{aligned}$$

For some infinite sequences, the numbers in the sequence approach a certain value as the index i increases. If that happens, we say the sequence has a *limit*. It is formally defined next.

Definition B.10 (Convergence of sequence, limit of sequence) *The sequence $\{x_i\}_{i=1}^\infty$ of numbers is said to **converge** to a point $L \in \mathbb{R}$ if and only if for every number $\varepsilon > 0$, there exists an $N^* \in \mathbb{N}$ such that*

$$i \geq N^* \;\Rightarrow\; |x_i - L| < \varepsilon.$$

*If this condition is satisfied, we say that the **limit** of $\{x_i\}_{i=1}^\infty$ exists and equals L. In this case, we denote the limit as*

$$\lim_{i \to \infty} x_i = L.$$

Example B.17 *Let a sequence be given by*

$$x_i = 3 - \frac{1}{i}, \quad i = 1, 2, \ldots.$$

This sequence converges to the limit $L = 3$. To prove this claim, we use the definition of a limit of a sequence. Let $\varepsilon > 0$. Then we need to check if there is a $N^* \in \mathbb{N}$ such that for all $i \geq N^*$, the numerical difference between L and x_i, i.e. $|x_i - L|$, is smaller than ε. First, we insert the definition of x_i to find

$$|x_i - 3| = \left|3 - \frac{1}{i} - 3\right| = \left|-\frac{1}{i}\right| = \frac{1}{i}.$$

This shows that if i is such that $1/i < \varepsilon$, then $|x_i - 3| < \varepsilon$. Now, let N^* be an integer such that $N^* > \frac{1}{\varepsilon}$. With this value of N^*, the definition of a limit with $L = 3$ is satisfied, i.e.

$$|x_i - 3| < \varepsilon$$

for all $i \geq N^*$. Since ε was arbitrary, this argument works for all $\varepsilon > 0$. Therefore, according to Definition B.10, we have shown that the limit of x_i exists and that the limit is

$$\lim_{i \to \infty} x_i = 3.$$

B.2.6 Limit of a function

The concept of a limit of a function is meant to describe what happens with the function values for input values in very small neighborhoods around some input value $c \in \mathbb{R}$. Therefore, to consider a limit of a function, we only want to consider input values for which a neighborhood of other potential input points exists. Such input values are called *limit points*, or *accumulation points*, and they are defined next.

Definition B.11 (Limit point) *Let $A \subseteq \mathbb{R}$ be a set of real numbers. A point $c \in A$ is a **limit point** of A if and only if for any $\varepsilon > 0$ there exist at least one $x \subset A$, $x \neq c$, such that $|x - c| < \varepsilon$.*

If a point in A is not a limit point of A, then it is called an *isolated point*.

Example B.18 *Consider the set $A = [0, 1]$. Then, all points $a \in A$ are limit points of A.*

Example B.19 *Consider the set $A = [0, 1] \cup \{2\}$. Then, all points $a \in [0, 1]$ are limit points of A, while the point $a = \{2\}$ is an isolated point of A.*

Using the concept of a limit point, the *limit of a function* can be defined next.

Definition B.12 (Convergence of function, limit of function) *Let $f : D \to \mathbb{R}$ be a function with domain $D \subseteq \mathbb{R}$. Assume c is a limit point of D. Then $f(x)$ **converges** to $L \in \mathbb{R}$ as x approaches c if and only if for any $\varepsilon > 0$, there exists a $\delta > 0$, such that for $x \in D$*

$$0 < |x - c| < \delta \implies |f(x) - L| < \varepsilon.$$

*If this condition is satisfied, we say that the **limit** of $f(x)$ as $x \to c$ exists and equals L. In this case, we denote the limit as*

$$\lim_{x \to c} f(x) = L.$$

The condition of c being a limit point implies that there are values of $x \in D$ which are arbitrarily close to c. Whether or not $f(x)$ converges to a finite value is also expressed as whether or not the limit exist.

Example B.20 *Consider the function*

$$f(x) = \frac{1}{|1 - x|}, \quad x \in \mathbb{R}\backslash\{1\}.$$

This function is not defined for $x = 1$ but all points in the domain $\mathbb{R}\backslash\{1\} = (-\infty, 1) \cup (1, \infty)$ are limit points. For x close to 1, the value of $|1 - x|$ is close to zero and hence the value of $f(x)$ is very large. In fact, for x close enough to 1, we can make $f(x)$ arbitrarily large. Therefore, the limit $\lim_{x \to 1} f(x)$ does not exist. We could also write

$$\lim_{x \to 1} f(x) = \infty.$$

Sometimes we may need to consider the limit of a function $f(x)$ for the values of x strictly larger than a limit point c. This is denoted the *limit from the right*. Similarly, when x is strictly smaller than c, we denote the corresponding limit the *limit from the left*. This is defined next.

Definition B.13 (Limit from right and left of function) *Let $f : D \to \mathbb{R}$ be a function with domain $D \subseteq \mathbb{R}$. Assume c is a limit point of D. Then $f(x)$ **converges to $L \in \mathbb{R}$ from the right** as x approaches c if and only if for any $\varepsilon > 0$, there exists a $\delta > 0$, such that for $x \in D$*

$$0 < x - c < \delta \;\Rightarrow\; |f(x) - L| < \varepsilon.$$

When a limit from the right of a function exists, it is denoted

$$\lim_{x \to c^+} f(x) = L.$$

*Similarly, $f(x)$ **converges to $L \in \mathbb{R}$ from the left** as x approaches c if and only if for any $\varepsilon > 0$, there exists a $\delta > 0$, such that for $x \in D$*

$$0 < c - x < \delta \;\Rightarrow\; |f(x) - L| < \varepsilon.$$

When a limit from the left of a function exists, it is denoted

$$\lim_{x \to c^-} f(x) = L.$$

For a limit of a function to exist, it must have the same limit both from the right and from the left. This is stated in the next theorem.

Theorem B.1 (Limit, limit from the right, limit from the left) *Let $f : D \to \mathbb{R}$ be a function with domain $D \subseteq \mathbb{R}$, and $c \in D$ a limit point of D. The limit $\lim_{x \to c} f(x)$ exists if, and only if,*

$$\lim_{x \to c^-} f(x) = \lim_{x \to c^+} f(x).$$

In that case, the limit equals $\lim_{x \to c^+} f(x)$ and $\lim_{x \to c^-} f(x)$, i.e.

$$\lim_{x \to c} f(x) = \lim_{x \to c^-} f(x) = \lim_{x \to c^+} f(x).$$

Sometimes it is needed to find the limit of a function which is constructed from other functions. For example, a function may be the sum of two other functions. Then, under certain conditions, the limit of the other functions can be used to find the limit of the function, which is constructed from the these other functions. The next theorem contains results for limits when functions are added, subtracted, multiplied, and divided.

Theorem B.2 (Properties of limit operator) *Assume* $\lim\limits_{x \to c} f(x)$ *and* $\lim\limits_{x \to c} g(x)$ *exist. Then*

$$\lim_{x \to c} (f(x) + g(x)) = \lim_{x \to c} f(x) + \lim_{x \to c} g(x),$$
$$\lim_{x \to c} (f(x) - g(x)) = \lim_{x \to c} f(x) - \lim_{x \to c} g(x),$$
$$\lim_{x \to c} (f(x) \cdot g(x)) = \left(\lim_{x \to c} f(x)\right) \cdot \left(\lim_{x \to c} g(x)\right).$$

If, in addition, $\lim\limits_{x \to c} g(x) \neq 0,$ *then*

$$\lim_{x \to c} \frac{f(x)}{g(x)} = \frac{\lim\limits_{x \to c} f(x)}{\lim\limits_{x \to c} g(x)}.$$

The next theorem relates the limit of a function to the limit of convergent sequences. This relationship is useful when we want to establish a limit, or examine whether a limit exists.

Theorem B.3 (Limit of function and convergent sequences) *Let* $f : D \to \mathbb{R}$ *be a function with domain* $D \subseteq \mathbb{R}$, $L \in \mathbb{R}$, *and* $c \in D$ *a limit point of* D. *Then*

$$\lim_{x \to c} f(x) = L$$

if, and only if,

$$\lim_{i \to \infty} f(x_i) = L$$

for every sequence $\{x_i\}_{i=1}^{\infty}$ *of points in* $D \backslash \{c\}$ *which converges to* c.

Note, the symbol $\backslash$ is the set difference (see Definition A.4). In the second part of the theorem, there are two sequences involved. One is $\{x_i\}_{i=1}^{\infty}$ and the other is a transformation of this sequence, namely, $\{f(x_i)\}_{i=1}^{\infty}$, which is also a sequence of real numbers.

One useful implication of the theorem is as a tool to show that a limit of a function does not exist. If we can find two sequences, x_i and y_i, both with limit c, having the property that the sequences $f(x_i)$ and $f(y_i)$ converge to two different limits, then we know that the limit of the function does not exist. The reason is that the theorem requires *every* sequence $f(x_i)$ to converge to the same L for L to be a limit of the function. It is worth stressing that finding one sequence where $\lim\limits_{i \to \infty} f(x_i) = L$ is not sufficient to conclude that $\lim\limits_{x \to c} f(x) = L$. It does imply, however, that if there is a limit, it must be L.

Example B.21 *Consider the function given by*

$$f(x) = \begin{cases} 0 & \text{if } x \leq 5, \\ 3 & \text{if } x > 5. \end{cases}$$

The functional form of $f()$ *can be written more compactly using the indicator function (see Definition B.21 below),*

$$f(x) = 3 \cdot I(x \leq 5), \quad x \in \mathbb{R}.$$

The domain of this function is all the real numbers, $\mathbb{R}$, so any point $c \in \mathbb{R}$ is a limit point.

Now, using the definition of the limit of a function (Definition B.12), we will show that this function does not have a limit at $x = 5$. One candidate for limit could be $L = 3$. If $L = 3$ is a limit of the function at $x = 5$, then, by Definition B.12, for any $\varepsilon > 0$ there needs to exist a number $\delta > 0$ such that the condition

$$0 < |x - 5| < \delta.$$

implies that

$$|f(x) - L| < \varepsilon.$$

To see that this cannot be the case, suppose that $\varepsilon \in (0,3)$ and that $x < 5$. In that case, no matter the value of δ, we have

$$|f(x) - L| = |0 - 3| = 3 > \varepsilon.$$

This shows that the limit from the left of $f(x)$ at $x = 5$ is not $L = 3$, which rules out the limit of $f(x)$ at $x = 5$ being $L = 3$ (Theorem B.1). The other candidate for a limit at $x = 5$ is $L = 0$. However, in this case, $\varepsilon \in (0,3)$ and $x > 5$ implies that

$$|f(x) - L| = |3 - 0| = 3 > \varepsilon.$$

This shows that the limit from the right of $f(x)$ at $x = 5$ is not $L = 0$, which rules out the limit of $f(x)$ at $x = 5$ being $L = 0$ (Theorem B.1). We conclude that the function does not have a limit at $x = 5$.

An arguably easier way to show that the function does not have a limit at $x = 5$ is using Theorem B.3 and the negative implication that if there are two sequences x_i and y_i leading to two different with limits of $f()$, then no limit exists. First consider a sequence x_i converging to 5 from below (meaning all $x_i < 5$), say,

$$x_i = 5 - 1/i.$$

Then

$$f(x_i) = 0 \text{ for all } i$$

and, thus, the limit of the function for the sequence $x_i = 5 - 1/i$ is 0.

Now, consider a sequence from above (meaning all $y_i > 5$), say,

$$y_i = 5 + 1/i.$$

Then

$$f(x_i) = 3 \text{ for all } i$$

and, thus, the limit of the function for the sequence $x_i = 5 + 1/i$ is 3.

Since there are two sequences x_i and y_i with $\lim_{i \to \infty} x_i = \lim_{i \to \infty} y_i = 5$ leading to two different limits of the function at $x = 5$, then, by Theorem B.3, no limit of the function exists at $x = 5$.

B.2.7   Continuity

Intuitively, we say that a function $f(x)$ is *continuous* if small changes in the input value x, necessarily results in small changes in the function value $f(x)$. The formal definition of continuity is as follows.

Definition B.14 (Continuity of a function) *Let $f : D \to \mathbb{R}$ be a function with domain $D \subseteq \mathbb{R}$. Then $f(x)$ is **continuous at a point** $c \in D$, if and only if for every $\varepsilon > 0$, there exists a $\delta > 0$, such that for $x \in D$,*

$$|x - c| < \delta \;\Rightarrow\; |f(x) - f(c)| < \varepsilon.$$

*If f is continuous at every point of D, then f is said to be **continuous** on D.*

The next theorem provides a relationship between the limit of a function and continuity of the function. Because it links continuity of a function with convergence of sequences, it is sometimes known as the *sequential characterization of continuity.*

Theorem B.4 (Sequential characterization of continuity) *Let $f : D \to \mathbb{R}$ be a function with domain $D \subseteq \mathbb{R}$, and let $c \in D$ be a limit point of D. Then f is continuous at c if and only if for every sequence $\{x_i\}_{i=1}^{\infty}$ in D convergent to c, the sequence $\{f(x_i)\}_{i=1}^{\infty}$ converges to $f(c)$.*

In other words, f is continuous at c if and only if for all sequences $\{x_i\}_{i=1}^{\infty}$ with $\lim_{i \to \infty} x_i = c$, it holds that

$$\lim_{i \to \infty} f(x_i) = f\left(\underbrace{\lim_{i \to \infty} x_i}_{=c}\right) = f(c).$$

In words, if a function is continuous, then one can simply find the limit of $\{f(x_i)\}_{i=1}^{\infty}$ by inserting the limit of x_i into f. The theorem can also be used to show that if there is a convergent sequence $\{x_i\}_{i=1}^{\infty}$ to c and $\lim_{i \to \infty} f(x_i) \neq f(c)$, then f cannot be continuous at c. This is an implication of the next theorem.

Theorem B.5 (Limit of function and continuity) *Let $f : D \to \mathbb{R}$ be a function with domain $D \subseteq \mathbb{R}$, and let $c \subset D$ be a limit point of D. Then f is continuous at c if, and only if*

$$\lim_{x \to c} f(x) = f(c).$$

This theorem does not cover the case where c is an isolated point (i.e. not a limit point). Recall that the limit of a function at c is only defined when c is a limit point. The definition of continuity implies that a function is continuous at a point which is not a limit point.

Example B.22 *Let $f : D \to \mathbb{R}$ be a function with domain $D = [0, 1] \cup \{2\}$. Then, 2 is an isolated point of D and f is therefore continuous at $x = 2$.*

B.2.8 Derivative

Intuitively, the *derivative* of a function $f(x)$ at a point $x = c$ describes the instantaneous rate of change of the function at $x = c$. To formally define this notion, let $f : D \to \mathbb{R}$ be a function and $c \in D$. Define a new function $\phi : D \backslash \{c\} \to \mathbb{R}$, given by

$$\phi(x) = \frac{f(x) - f(c)}{x - c}. \tag{B.1}$$

The function $\phi(x)$ describes the average rate of change of the function $f(x)$ from the point x to the point c.

Example B.23 *Consider a function $f(t)$ with $t \geq 0$ describing the distance a car has traveled in kilometers after t hours. Then $\phi(t)$ denotes the distance traveled from time c to time t, divided by the elapsed time $t - c$. That is, $\phi(t)$ is the average speed, measured in kilometers per hour, at which the car has traveled between the time c and the time t. If, for instance, $c = 1$, $t = 3$, $f(1) = 50$, and $f(2) = 200$, then the car has traveled $f(2) - f(1) = 200 - 50 = 150$ kilometers in $t - c = 3 - 1 = 2$ hours for an average speed of $\phi(2) = 150/2 = 75$ kilometers per hour.*

To assess the *instantaneous* rate of change of f at c, we can investigate the limit of ϕ as x approaches c. This limit, if it exists, is denoted the derivative of f at c. We next provide the formal definition.

Definition B.15 (Derivative, differentiable function) *Let $f : D \to \mathbb{R}$ be a function with domain $D \subseteq \mathbb{R}$, and consider an open interval (a, b), with $a < b$, such that $(a, b) \subseteq D$. Then f is said to be **differentiable at the point** $c \in (a, b)$ if and only if the limit*

$$\lim_{x \to c} \phi(x) = \lim_{x \to c} \left(\frac{f(x) - f(c)}{x - c} \right)$$

*exists. The limit, if it exists, is called the **derivative** of f at c. It is denoted either $f'(c)$ or $\left. \frac{df(x)}{dx} \right|_{x=c}$. If $f()$ is differentiable at all points $c \in D$, then the function f is called* **differentiable**.

Example B.24 *Consider the function*

$$f(x) = x^2, \quad x \in \mathbb{R}.$$

Let $c \in \mathbb{R}$. Then, using that for all $x \in \mathbb{R}$,

$$(x + c)(x - c) = x^2 - c^2,$$

we can write the function (B.1) as

$$\phi(x) = \frac{f(x) - f(c)}{x - c} = \frac{x^2 - c^2}{x - c} = \frac{(x + c)(x - c)}{x - c} = x + c.$$

Therefore, using Theorem B.2,

$$\lim_{x \to c} \phi(x) = \lim_{x \to c} (x + c) = \lim_{x \to c} x + \lim_{x \to c} c = c + c = 2c,$$

showing that $f(x)$ is differentiable in $x = c$ and that $f'(c) = 2c$. Since this holds for all c in the domain of $f()$, the function $f()$ is differentiable.

The process of calculating derivatives is called *differentiation*. If a function $f(x)$ is differentiable, we can write the derivative $f'()$ as a function of x, e.g. $f'(x)$, $\frac{df(x)}{dx}$, $\frac{d}{dx} f(x)$, or, simply, $\frac{df}{dx}$.

Example B.25 (Example B.24, continued) *Consider again the function*

$$f(x) = x^2, \quad x \in \mathbb{R}.$$

We saw in Example B.24 that for all $c \in \mathbb{R}$, we have $f'(c) = 2c$. We may thus view the derivative as a function,

$$f' : \mathbb{R} \to \mathbb{R},$$

given by

$$f'(x) = 2x.$$

In Example B.24, we directly used the definition of the derivative (Definition B.15) to show that the function $f(x) = x^2$ is differentiable with derivative $f'(x) = 2x$. In general, however, it would be quite cumbersome if we should resort to Definition B.15 every time we differentiate a function. Luckily, many of the most commonly used functions have closed-form formulas for their derivatives. Some of the most important ones are given in the table in the next section. The following theorem further contains some properties of derivatives that can make derivations easier.

Theorem B.6 (Rules of differentiation) *Assume $\frac{df(x)}{dx}$ and $\frac{dg(x)}{dx}$ exist. Then*

1. (Linearity) : $\frac{d(\tau_1 f(x)+\tau_2 g(x))}{dx} = \tau_1 \frac{df(x)}{dx} + \tau_2 \frac{dg(x)}{dx}$, *where $\tau_1, \tau_2 \in \mathbb{R}$.*

2. (Product rule) : $\frac{d(f(x)g(x))}{dx} = \left(\frac{df(x)}{dx}\right) g(x) + f(x) \left(\frac{dg(x)}{dx}\right)$.

3. (Chain rule) : $\frac{d(g(f(x)))}{dx} = g'(f(x))f'(x)$.

B.2.9 Inverse of differentiation

Any function F for which its derivative equals f is called an *antiderivative* of f.

Definition B.16 (Antiderivative) *Assume $F(x)$ is differentiable with derivative $f(x)$. Then any function*

$$G(x) = F(x) + d, \quad d \in \mathbb{R},$$

*is an **antiderivative** of the function $f(x)$. It is denoted $G(x) = \int f(x)dx$.*

Example B.26 *Let $f(x) = 2x$. Then one antiderivative of f is $F(x) = x^2$ because, as we saw in Example B.24,*

$$\frac{dx^2}{dx} = 2x.$$

The function $G(x) = x^2 + 3$ is also an antiderivative of $f(x)$ since

$$\frac{d\left(x^2 + 3\right)}{dx} = 2x.$$

Taking the derivative and the antiderivative are inverse operations. That is, if $f(x)$ is a differentiable function, then first differentiating $f()$ and then finding its antiderivative leaves you back at the function $f()$:

$$\int \frac{d}{dx} f(x)dx = f(x) + d.$$

Similarly, if you first take the antiderivative of $f()$ and then the derivative, then you are also back at $f()$:

$$\frac{d}{dx} \int f(x)dx = f(x).$$

Similarly to the case of differentiation, it would be cumbersome if we should resort to the definition of the antiderivative for every new function. Luckily, as was also the case for derivatives, many functions have closed-form formulas for their antiderivatives. Some commonly used functions, their derivatives, and their antiderivatives are listed in the following table.

Function: $f(x)$	Derivative: $\frac{df(x)}{dx}$	Antiderivative: $\int f(x)dx$
$x^a, \quad a \neq -1$	ax^{a-1}	$\frac{1}{a+1}x^{a+1} + d$
x^{-1}	$-x^{-2}$	$\ln(x) + d$
$e^{a \cdot x}, \ a \in \mathbb{R}$	$ae^{a \cdot x}$	$a^{-1}e^{a \cdot x} + d$
$a^x, \quad a > 0$	$\ln(a)a^x$	$\frac{1}{\ln(a)}a^x + d$
$\ln(x)$	x^{-1}	$x\ln(x) - x + d$

B.2.10   Integration

For a positive function, i.e. a function such that $f(x) \geq 0$ for all x in the domain of $f()$, *integration* corresponds to calculating the area between the graph of the function and the first (x) axis. To formally define the integral, we first need the concept of a bounded function.

Definition B.17 (Bounded function) *Let $f : D \to \mathbb{R}$ be a function with domain $D \subseteq \mathbb{R}$. We say that f is **bounded** if and only if there exists a number $M > 0$ such that*

$$|f(x)| \leq M$$

for all $x \in D$.

Now, consider a bounded function defined on the interval $[a, b]$. Let $P_m = \{x_0, x_1, \ldots, x_m\}$ be a *partition* of the interval $[a, b]$, in the sense that

$$a = x_0 < x_1 < \ldots < x_m = b.$$

That is, the partition P_m divides the interval $[a, b]$ into m smaller intervals $[x_0, x_1], \ldots, [x_{m-1}, x_m]$. The *mesh* of the partition, denoted $||P_m||$, is the length of the longest interval in the partition, i.e. $||P_m|| = \max_{i=1,\ldots,m} \Delta x_i$, where $\Delta x_i = x_i - x_{i-1}$. The *Riemann sum*, defined over the partition P_m, is given by

$$S(P_m) = \sum_{i=1}^{m} f(c_i)\Delta x_i, \tag{B.2}$$

where $c_i \in [x_{i-1}, x_i]$. When f is a positive function, the Riemann sum is an approximation of the area under $f(x)$ from $x = a$ to $x = b$, where the approximation is constructed using m rectangles of height $f(c_i)$ and length Δx_i. Intuitively, we expect that this approximation gets better and better the finer we partition the interval $[a, b]$.

Example B.27 *Consider the function*

$$f(x) = 1 - x^2, \quad x \in [0, 1].$$

To approximate the area under $f(x)$ from $x = 0$ to $x = 1$, we may form the equidistant partition $x_0 = 0, x_1 = 2/m, \ldots, x_m = 2$. Let $c_i = (x_i - x_{i-1})/2$ be the mid-point in the i'th interval $[x_{i-1}, x_i]$. Figure B.1 illustrates how the Riemann sum is formed by adding up the areas of m rectangles. The left panel of the figure considers the case of $m = 10$ rectangles, whereas the right panel considers the case of $m = 50$ rectangles. The respective Riemann sums (B.2) are $S(P_{10}) = 0.6675$ and $S(P_{50}) = 0.6667$. The true area under the curve, which we derive using integration in Example B.28 below, is $2/3 = 0.6666\ldots$.

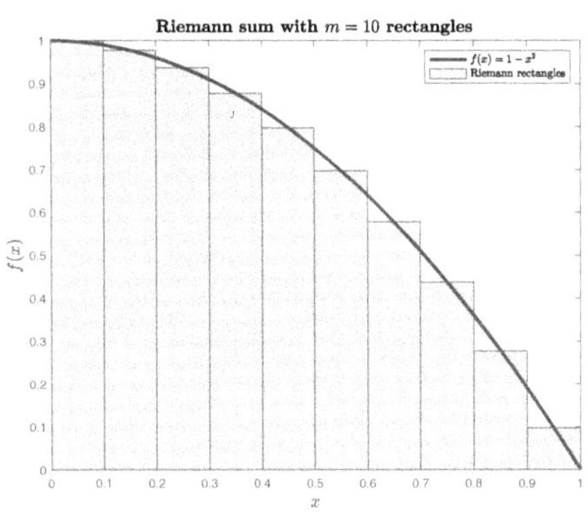
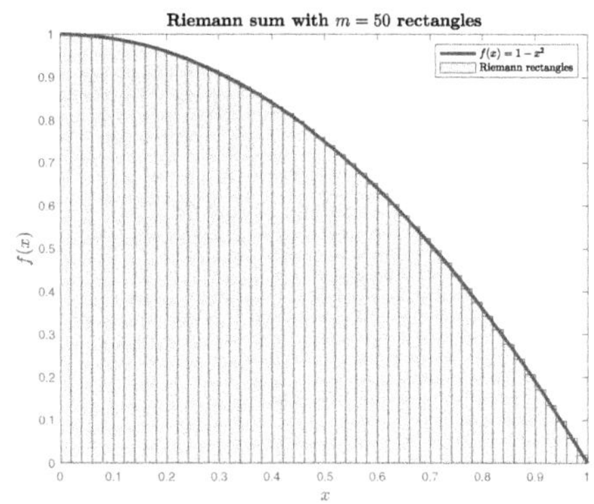

FIGURE B.1

Examples of Riemann sums for the function $f(x) = 1 - x^2$ for $x \in [0, 1]$, see Example B.27. Left: Riemann sum with $m = 10$ rectangles, $S(P_{10}) = 0.6675$. Right: Riemann sum with $m = 50$ rectangles, $S(P_{50}) = 0.6667$.

We can now define the *Riemann integral* as the limit of the Riemann sum as the partitions get finer and finer, i.e. as $m \to \infty$ such that $\|P_m\| \to 0$. Since this is the only type of integral we shall use, we simply refer to it as the *integral*.

Definition B.18 (Riemann integral, integrable function) *Let* $f : [a, b] \to \mathbb{R}$ *be a bounded function. Consider a sequence of partitions* P_m *of* $[a, b]$ *such that* $\lim_{m\to\infty} \|P_m\| = 0$ *and the associated Riemann sums*

$$S(P_m) = \sum_{i=1}^{m} f(c_i)\Delta x_i,$$

where $c_i \in [x_{i-1}, x_i]$. *If the same limit*

$$\lim_{m \to \infty} S(P_m)$$

exists for any sequence of partitions, then we say that f *is* **integrable**. *In this case, we define the* **Riemann integral** $\int_a^b f(x)dx$ *as*

$$\int_a^b f(x)dx = \lim_{m \to \infty} S(P_m).$$

Definition B.18 covers only bounded functions over closed and bounded intervals $[a, b]$. Sometimes, however, we are interested in either integrating functions that are not bounded or functions defined over an interval of infinite length. This leads to the definition of a so-called *improper integral*.

Definition B.19 (Improper Riemann integral) *Consider a function* $f : D \to \mathbb{R}$ *with domain* $D \subseteq \mathbb{R}$.

1. *Assume* f *is integrable on intervals* $[a, b]$ *for any* $a, b \in D$.

   (a) If $D = (-\infty, b]$, *then define* $\int_{-\infty}^b f(x)dx$ *as*

   $$\int_{-\infty}^b f(x)dx = \lim_{a \to -\infty} \int_a^b f(x)dx.$$

(b) If $D = [a, \infty)$, then define $\int_a^\infty f(x)dx$ as

$$\int_a^\infty f(x)dx = \lim_{b \to \infty} \int_a^b f(x)dx.$$

(c) If $D = (-\infty, \infty)$, then define $\int_{-\infty}^\infty f(x)dx$ as

$$\int_{-\infty}^\infty f(x)dx = \lim_{a \to -\infty} \int_a^0 f(x)dx + \lim_{b \to \infty} \int_0^b f(x)dx.$$

2. *Let $D = [a, b] \setminus \{c\}$ and assume that f is integrable on all intervals $[a, d]$ and $[l, b]$ such that $a < d < c$ and $c < l < b$. Then define $\int_a^b f(x)dx$ as*

$$\int_a^b f(x)dx = \lim_{d \to c^-} \int_a^d f(x)dx + \lim_{l \to c^+} \int_l^b f(x)dx.$$

From the definition, we see that improper Riemann integrals are formed by first specifying a (proper) Riemann integral by either avoiding infinite integration limits (1-3 in Definition B.19) or the point where f is undefined (4 in Definition B.19). These proper Riemann integrals are well-defined by Definition B.18. The improper integral is then formed by taking the appropriate limit, when it exists. Some examples of how to do this are provided at the end of this section.

Next we list some properties of integration to aid in integration of a function.

Theorem B.7 (Rules of integration) *Assume the functions are integrable and let $a, b, c \in \mathbb{R}$. Then*

1. $\int_a^b f(x)dx = \int_a^c f(x)dx + \int_c^b f(x)dx.$

2. $\int_a^b \tau \cdot f(x)\ dx = \tau \cdot \int_a^b f(x)dx.$

3. $\int_a^b (f(x) + g(x))\ dx = \int_a^b f(x)dx + \int_a^b g(x)dx.$

4. $\int_a^b f(x)dx = -\int_b^a f(x)dx.$

B.2.11   The Fundamental Theorem of Calculus

Differentiation and integration are closely related. In fact, in a sense they are inverse operations. This relationship is expressed in the so-called *Fundamental Theorem of Calculus.* The theorem has two parts. It is stated next.

Theorem B.8 (Fundamental Theorem of Calculus)
Part I. *Let f be a continuous function on the interval $[a, b]$. Define the function*

$$F(x) = \int_a^x f(y)dy.$$

Then F is continuous on $[a, b]$, differentiable on (a, b), and the derivative of F is f, i.e.

$$F'(x) = f(x)$$

for $x \in (a, b)$.

Part II. *Let $F(x)$ be a differentiable function on $[a, b]$ with derivative $F'(x) = f(x)$. Then the value of $F(b)$ can be expressed as*

$$F(b) = F(a) + \int_a^b f(x)dx \tag{B.3}$$

for any choice of a in the domain of F.

The Fundamental Theorem of Calculus provides a convenient way to evaluate integrals. Instead of using the definition of integrals in the previous subsection, the value of an integral can be evaluated using the antiderivative of f. That is, if f is a function with antiderivative F, then

$$\int_a^b f(x)dx = F(b) - F(a).$$

Sometimes, the notation $[F(x)]_a^b$ or $F(x)|_a^b$ is used in short for $F(b) - F(a)$, i.e. we may write

$$\int_a^b f(x)dx = [F(x)]_a^b = F(x)|_a^b = F(b) - F(a).$$

Example B.28 (Example B.27, continued) *Consider the function*

$$f(x) = 1 - x^2, \quad x \in [0, 1],$$

from Example B.27. Since $\frac{d}{dx}x = 1$ and $\frac{d}{dx}\frac{1}{3}x^3 = x^2$ we conclude using Definition B.16 that an antiderivative of $f(x)$ is

$$F(x) = x - \frac{1}{3}x^3, \quad x \in [0, 1].$$

By Theorem B.8 part II, the integral of $f(x)$ over $[0, 1]$ is

$$\int_0^1 f(x)dx = F(1) - F(0) = 1 - \frac{1}{3} \cdot 1^3 - 0 = \frac{2}{3}.$$

That is, the area between the graph of $f(x)$ and the x-axis over the interval $[0, 1]$ is $\frac{2}{3}$.

There are functions $f(x)$ which do not necessarily have an antiderivative but still can be evaluated as in (B.3). Equivalently, there are functions F which are not necessarily differentiable but still a function f exists such that the expression in (B.3) holds. Those functions F are said to be *absolutely continuous*. This is defined next.

Definition B.20 (Absolutely continuous function) *A function $F()$ is called **absolutely continuous** if and only if there exists a function $f(x)$ such that*

$$F(b) - F(a) = \int_a^b f(x)dx$$

for every choice of a and b in the domain of F, such that $a < b$.

An absolutely continuous function is also continuous whereas a continuous function need not be absolutely continuous. If a continuous function is also differentiable at all but a finite number of points, then it is absolutely continuous. In particular, a differentiable function is absolutely continuous.

We close this section by giving three examples of improper integrals and how these can be calculated. Along the way, examples of proper integrals will also be given. The first two examples illustrate improper integrals arising from infinite integration limits. The last example illustrates how an improper integral may be calculated in case the function being integrated is unbounded.

Example B.29 *Let $\lambda > 0$ and consider the function*

$$f(x) = \lambda e^{-\lambda x}, \quad x \geq 0.$$

We wish to integrate the function $f(x)$ over the domain $D = [0, \infty)$. Note that since $f(x)$ is non-negative, this amounts to finding the area between $f(x)$ and the x-axis over the interval $[0, \infty)$, see the left panel of Figure B.2 for an illustration with $\lambda = 1$.

By the table of antiderivatives given above, an antiderivative of $f(x)$ is

$$F(x) = -e^{-\lambda x}.$$

Let $b > 0$. By Part II of the Fundamental Theorem of Calculus (Theorem B.8), the (proper) Riemann integral of $f(x)$ over $[0, b]$ is

$$\int_0^b f(x)dx = F(b) - F(0) = -e^{-\lambda b} - (-e^0) = 1 - e^{-\lambda b}.$$

The improper Riemann integral is found using 1b in Definition B.19,

$$\int_0^\infty f(x)dx = \lim_{b \to \infty} \int_0^b f(x)dx = \lim_{b \to \infty} \left(1 - e^{-\lambda b}\right) = 1 - 0 = 1,$$

where we used that $\lim_{b \to \infty} e^{-\lambda b} = 0$ when $\lambda > 0$. Since the limit exists, we conclude that the improper integral $\int_0^\infty f(x)dx$ exists and has the value 1.

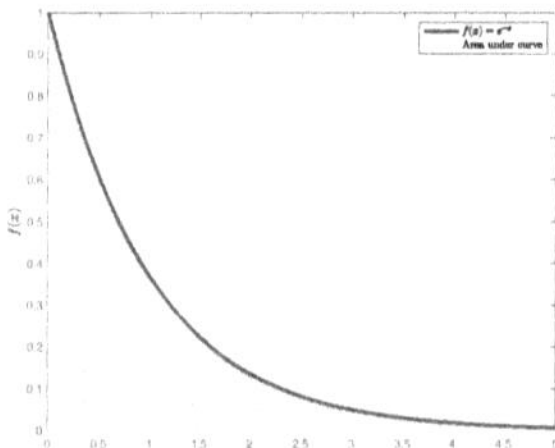 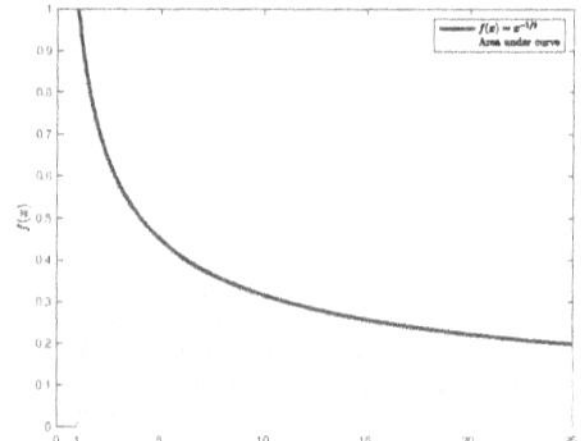 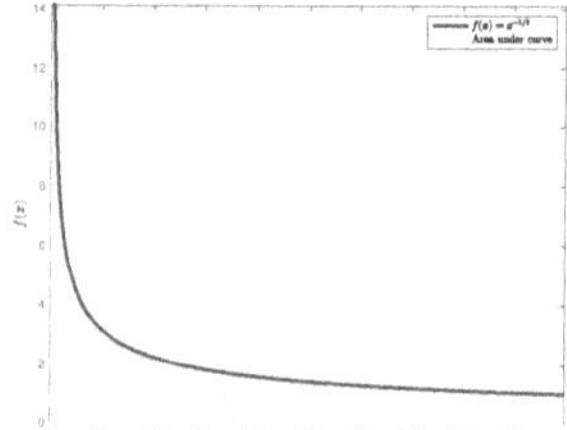

FIGURE B.2

Areas under the positive functions in Examples B.29, B.30, and B.31. Left: Area under $f(x) = \lambda e^{-\lambda x}$ for $x \in [0, 5]$ and $\lambda = 1$ (Example B.29). Middle: Area under $f(x) = x^{-1/2}$ for $x \in [1, 25]$ (Example B.30). Right: Area under $f(x) = x^{-1/2}$ for $x \in (0, 1]$ (Example B.31).

Example B.30 *Consider the function*

$$f(x) = x^{-1/2}, \quad x \geq 1.$$

We wish to integrate the function $f(x)$ over the domain $D = [1, \infty)$. Note that since $f(x)$ is non-negative, this amounts to finding the area between $f(x)$ and the x-axis over the interval $[1, \infty)$, see the middle panel of Figure B.2 for an illustration.

By the table of antiderivatives given above, an antiderivative of $f(x)$ is

$$F(x) = \frac{1}{-1/2 + 1}x^{-1/2+1} = 2x^{1/2} = 2\sqrt{x}.$$

Let $b > 0$. By the Part II of the Fundamental Theorem of Calculus (Theorem B.8), the (proper) Riemann integral of $f(x)$ over $[1, b]$ is

$$\int_1^b f(x)dx = F(b) - F(1) = 2\sqrt{b} - 2\sqrt{1} = 2\left(\sqrt{b} - 1\right).$$

The improper Riemann integral is found using 1b in Definition B.19,

$$\int_1^\infty f(x)dx = \lim_{b \to \infty} \int_1^b f(x)dx = \lim_{b \to \infty} 2\left(\sqrt{b} - 1\right) = 2\left(\infty - 1\right) = \infty,$$

where we used that $\sqrt{b}$ diverges to ∞ as $b \to \infty$. Since the limit does not exist, we conclude that the improper integral $\int_1^\infty f(x)dx$ does not exist either. Informally, the area between $f(x)$ and the x-axis over the interval $[1, \infty)$ is infinitely large.

Example B.31 *Consider the function*

$$f(x) = x^{-1/2}, \quad x \in (0, 1].$$

We wish to integrate the function $f(x)$ over the domain $D = (0, 1]$, which can be written as $D = [0, 1] \setminus \{0\}$ with $f(x)$ being undefined at $x = 0$. Here, the function $f(x)$ is unbounded over D since $\lim_{x \to 0^+} f(x) = \infty$. Note that since $f(x)$ is non-negative, this amounts to finding the area between $f(x)$ and the x-axis over the interval $(0, 1]$, see the right panel of Figure B.2 for an illustration.

By the table of antiderivatives given above, an antiderivative of $f(x)$ is

$$F(x) = \frac{1}{-1/2 + 1}x^{-1/2+1} = 2x^{1/2} = 2\sqrt{x}.$$

Let $d \in (0, 1]$ and note that $F()$ is differentiable on $[d, 1]$. By Part II of the Fundamental Theorem of Calculus (Theorem B.8), the (proper) Riemann integral of $f(x)$ over $[d, 1]$ is

$$\int_d^1 f(x)dx = F(1) - F(d) = 2\sqrt{1} - 2\sqrt{d} = 2\left(1 - \sqrt{d}\right).$$

The improper Riemann integral is found using 2 in Definition B.19,

$$\int_0^1 f(x)dx = \lim_{d \to 0^+} \int_d^1 f(x)dx = \lim_{d \to 0^+} 2\left(1 - \sqrt{d}\right) = 2\left(1 - 0\right) = 2,$$

where we used that $\lim_{d \to 0^+} \sqrt{d} = 0$. Since the limit exists, we conclude that the improper integral $\int_0^1 f(x)dx$ exists and has the value 2.

B.2.12   The indicator function

The indicator function $I()$ is a logical function, which evaluates if a statement is true or false. More precisely, it outputs the number 1 if the statement is true and the number 0 if the statement is false.

Definition B.21 (Indicator function) *Let A be a statement that can be true or false. The function*

$$I(A) = \begin{cases} 1 & \text{if} \quad A \text{ is true}, \\ 0 & \text{if} \quad A \text{ is false}, \end{cases}$$

*is called the **indicator function** of the statement A.*

Example B.32 *Let $x \in \mathbb{R}$ be a number and consider the statement $A = $ "$x = 0$". This statement is true if the number x is zero and false otherwise. Therefore*

$$I(x = 0) = \begin{cases} 1 & \text{if} \quad x = 0, \\ 0 & \text{if} \quad x \neq 0. \end{cases}$$

B.3 Functions of several variables

A function of k variables with domain $D \subseteq \mathbb{R}^k$ and codomain $C \subseteq \mathbb{R}$ takes k values as input and returns a number. We write this as

$$f : D \to C,$$

or, if the codomain is not important for the subsequent analysis,

$$f : D \to \mathbb{R}.$$

Given k inputs $x_1, \ldots, x_k \in D$, the value of the function is denoted $f(x_1, \ldots, x_k)$, or $f(x)$, where $x = (x_1, \ldots, x_k)$.

Example B.33 *Let*
$$f(x_1, x_2, x_3) = 5x_1 - 6x_2 + 4x_3.$$

Then
$$f(1, 2, 3) = 5 \cdot 1 - 6 \cdot 2 + 4 \cdot 3 = 5.$$

B.3.1 Partial differentiation

Suppose we are interested in finding the rate of change of f with respect to one of these variables, say x_j, at a point $c = (c_1, \ldots, c_k) \in D$. To this end, define a new function $\psi : [a, b] \to \mathbb{R}$ given by

$$\psi(x_j) = f(c_1, \ldots, c_{j-1}, x_j, c_{j+1}, \ldots, c_k), \tag{B.4}$$

where $[a, b]$ is an interval containing c_j. This is a function of one variable. The rate of change of $\psi(x_j)$ with respect to x_j can be found using the definition of a derivative of a function of one variable (Definition B.15).

Example B.34 *Let*
$$f(x_1, x_2, x_3) = 5x_1 - 6x_2^2 + 4x_3^3.$$

Suppose the interest is in the rate of change of $f()$ with respect to x_2 at a point $(c_1, c_2, c_3) = (3, 2, 1)$. Then

$$\psi(x_2) = f(c_1, x_2, c_3) = 5 \cdot 3 - 6 \cdot x_2^2 + 4 \cdot 1^3 = 19 - 6 \cdot x_2^2.$$

The derivative of $\psi(x_2)$ with respect to x_2 is

$$\frac{d\psi(x_2)}{dx_2} = -6x_2.$$

Evaluate this at $c_2 = 2$ to get the rate of change of $\psi(x_2)$ in the point $x_2 = c_2 = 2$,

$$\left. \frac{d\psi(x_2)}{dx_2} \right|_{x_2 = c_2} = -12.$$

The derivative of $\psi(x_j)$, evaluated at the point c_j, is called the *partial derivative* of f with respect to x_j at the point c.

Definition B.22 (Partial derivative) *Let $D \subseteq \mathbb{R}^k$ and consider the function $f : D \to \mathbb{R}$ of k variables. Let $\psi(x_j)$ be defined according to (B.4). Assume the domain of $\psi(x_j)$ contains an open interval (a, b). Then f is said to be **partially differentiable** with respect to x_j at $c \in D$ with $c_j \in (a, b)$, if and only if*

$$\lim_{x_j \to c_j} \left(\frac{\psi(x_j) - \psi(c_j)}{x_j - c_j} \right)$$

*exists. The limit, if it exists, is called the **partial derivative** of f with respect to x_j at c. It is denoted $f_j'(c)$ or $\left. \frac{\partial f(x)}{\partial x_j} \right|_{x=c}$.*

We used the $\psi()$ function to stress that partial derivative means choosing one of the x_j's for which the partial derivative is to be calculated, and then treating the function $f(x_1, \ldots, x_k)$ as if it only is a function of x_j, namely, the function $\psi(x_j)$. If we do not want to involve the $\psi()$ function, then we could also have defined the partial derivative of f with respect to x_j directly as

$$\left. \frac{\partial f(x_1, \ldots, x_k)}{\partial x_j} \right|_{x=c} = \lim_{x_j \to c_j} \left(\frac{f(c_1, \ldots, c_{j-1}, x_j, c_{j+1}, \ldots, c_k) - f(c_1, \ldots, c_k)}{x_j - c_j} \right).$$

We may want to evaluate the partial derivative of f with respect to x_j at other points than c. Then we often simply write the partial derivative with $x_1, \ldots, x_k$ in place of $c_1, \ldots, c_k$.

Example B.35 *Let*

$$f(x_1, x_2) = 3 \cdot x_1^2 - 5 \cdot x_2 + 4 \cdot x_1 \cdot x_2.$$

Then the partial derivative of f with respect to x_1 at $c = (c_1, c_2)$ is

$$\left. \frac{\partial f(x_1, x_2)}{\partial x_1} \right|_{x=c} = \left. \frac{d \left(3 \cdot x_1^2 - 5 \cdot c_2 + 4 \cdot x_1 \cdot c_2 \right)}{dx_1} \right|_{x_1 = c_1} = 6 \cdot c_1 + 4 \cdot c_2.$$

If we want the partial derivative of f with respect to x_1 at any point, then replace c_1 and c_2 with x_1 and x_2

$$\frac{\partial f(x_1, x_2)}{\partial x_1} = 6 \cdot x_1 + 4 \cdot x_2.$$

In fact, we do not need to substitute c for x if we remember to treat any variable $x_1, \ldots, x_{j-1}, x_{j+1}, \ldots, x_k$ as constant when we take the partial derivative of f with respect to x_j. For example, the partial derivative of $f(x_1, x_2)$ with respect to x_2, as a function of (x_1, x_2), may be found by simply differentiating $f(x_1, x_2)$ with respect to x_2, while treating x_1 as a constant,

$$\frac{\partial f(x_1, x_2)}{\partial x_2} = -5 + 4 \cdot x_1.$$

The discussion above shows that the partial derivative of a multivariate function of k variables $f : \mathbb{R}^k \to \mathbb{R}$ with respect to one of the variables, say, x_j is again a multivariate function of k variables,

$$\frac{\partial f}{\partial x_j} : \mathbb{R}^k \to \mathbb{R},$$

given by

$$\frac{\partial f}{\partial x_j}(x_1, \ldots, x_n).$$

We may also be interested in the rate of change of the partial derivatives. That is, when it exists, we could consider the partial derivative of a partial derivative, the so-called *second-order partial derivative*. If the partial derivative of a function is itself partially differentiable, we say that the function is *twice partially differentiable*. The second-order partial derivative with respect to x_k and x_m is denoted

$$\frac{\partial^2 f(x_1, \dots, x_k)}{\partial x_j \partial x_m}$$

and it is calculated by first taking the partial derivative with respect to x_j and then take the partial derivative of that function with respect to x_m,

$$\frac{\partial^2 f(x_1, \dots, x_k)}{\partial x_j \partial x_m} = \frac{\partial \left(\frac{\partial f(x_1, \dots, x_k)}{\partial x_j} \right)}{\partial x_m}.$$

Formally, the second-order partial derivatives will depend on whether one first differentiates with respect to x_j or with respect to x_m. Under quite weak conditions, however, it turns out that the order of differentiation does not matter. This result is sometimes known as *Young's theorem*.

Theorem B.9 (Young's theorem) *Let $D \subseteq \mathbb{R}^k$ and consider the function $f : D \to \mathbb{R}$ of k variables. Assume that $\frac{\partial^2 f(x_1, \dots, x_k)}{\partial x_j \partial x_m}$ and $\frac{\partial^2 f(x_1, \dots, x_k)}{\partial x_m \partial x_j}$ exist and are continuous functions of $(x_1, \dots, x_k)$. Then*

$$\frac{\partial^2 f(x_1, \dots, x_k)}{\partial x_j \partial x_m} = \frac{\partial^2 f(x_1, \dots, x_k)}{\partial x_m \partial x_j}$$

for all $j, m = 1, \dots, k$.

Example B.36 *In Example B.35, where*

$$f(x_1, x_2) = 3 \cdot x_1^2 - 5 \cdot x_2 + 4 \cdot x_1 \cdot x_2$$

we have the two first-order partial derivatives

$$\frac{\partial f}{\partial x_1} = 6 \cdot x_1 + 4 \cdot x_2$$

and

$$\frac{\partial f}{\partial x_2} = -5 + 4 \cdot x_1.$$

The second-order partial derivative with respect to x_1 twice is

$$\frac{\partial^2 f}{\partial x_1^2} = \frac{\partial f}{\partial x_1}(6 \cdot x_1 + 4 \cdot x_2) = 6.$$

The second-order partial derivative with respect to x_2 twice is

$$\frac{\partial^2 f}{\partial x_2^2} = \frac{\partial f}{\partial x_2}(-5 + 4 \cdot x_1) = 0.$$

The second-order partial derivative with respect to x_1 and x_2 is

$$\frac{\partial^2 f}{\partial x_1 \partial x_2} = \frac{\partial}{\partial x_2} \frac{\partial f}{\partial x_1} = \frac{\partial}{\partial x_2}(6 \cdot x_1 + 4 \cdot x_2) = 4.$$

The second-order partial derivative with respect to x_2 and x_1 is

$$\frac{\partial^2 f}{\partial x_2 \partial x_1} = \frac{\partial}{\partial x_1} \frac{\partial f}{\partial x_2} = \frac{\partial}{\partial x_1}(-5 + 4 \cdot x_1) = 4.$$

That $\frac{\partial^2 f}{\partial x_1 \partial x_2} = \frac{\partial^2 f}{\partial x_2 \partial x_1}$ was to be expected, since the second-order partial derivatives are constant (and hence continuous) and therefore Young's theorem implies that the order of differentiation does not matter, i.e. $\frac{\partial^2 f}{\partial x_1 \partial x_2} = \frac{\partial^2 f}{\partial x_2 \partial x_1}$.

B.3.2 Integration over several variables

In this section, we briefly review how to integrate functions of several variables. We will illustrate the concepts using a function $f(x_1, x_2)$ of two variables. The extension to multivariate functions $f(x_1, \ldots, x_k)$ of k variables then follows.

Consider a function of two variables $f(x_1, x_2)$ and suppose we want to integrate this function for x_1 in the interval (a_1, b_1) and x_2 in the interval (a_2, b_2). This integral is denoted as

$$\int \int_{[a_1,b_1] \times [a_2,b_2]} f(x_1, x_2) d(x_1, x_2).$$

Due to a result known as *Fubini's theorem*, then, under very weak conditions, integration over several variables can be done by integrating one variable at a time and the order of integration can be exchanged, i.e. it does not matter if we first integrate with respect to x_1 and then x_2 or vice versa. Formally,

$$\int \int_{[a_1,b_1] \times [a_2,b_2]} f(x_1, x_2) d(x_1, x_2) = \int_{a_1}^{b_1} \int_{a_2}^{b_2} f(x_1, x_2) dx_2 dx_1 = \int_{a_2}^{b_2} \int_{a_1}^{b_1} f(x_1, x_2) dx_1 dx_2.$$

We can therefore use the techniques from integration of a function of one variable by dividing several integrals into a sequence of separate integrals. This is done as follows

$$\int_{a_1}^{b_1} \underbrace{\left[\int_{a_2}^{b_2} f(x_1, x_2) dx_2 \right]}_{=g(x_1)} dx_1 = \int_{a_1}^{b_1} g(x_1) dx_1,$$

where

$$g(x_1) = \int_{a_2}^{b_2} f(x_1, x_2) dx_2.$$

First solve to get $g(x_1)$ by integrating $f(x_1, x_2)$ with respect to x_2, treating x_1 as if it is a constant. Sometimes we say that the function $g(x_1)$ only depends on x_1 because x_2 has been "integrated out". After $g(x_1)$ has been found, then integrate this function with respect to x_1.

Example B.37 *Let*

$$f(x_1, x_2) = x_1 x_2.$$

Then

$$\int_{a_1}^{b_1} \int_{a_2}^{b_2} x_1 x_2 dx_2 dx_1 = \int_{a_1}^{b_1} g(x_1) dx_1,$$

where

$$g(x_1) = \int_{a_2}^{b_2} x_1 x_2 dx_2.$$

The solution is

$$g(x_1) = x_1 \int_{a_2}^{b_2} x_2 dx_2 = x_1 \left[\frac{1}{2}x_2^2\right]_{a_2}^{b_2} = x_1 \left(\frac{1}{2}b_2^2 - \frac{1}{2}a_2^2\right).$$

Then solve

$$
\begin{aligned}
\int_{a_1}^{b_1} g(x_1) dx_1 &= \int_{a_1}^{b_1} x_1 \left(\frac{1}{2}b_2^2 - \frac{1}{2}a_2^2\right) dx_1 \\
&= \left(\frac{1}{2}b_2^2 - \frac{1}{2}a_2^2\right) \int_{a_1}^{b_1} x_1 dx_1 \\
&= \left(\frac{1}{2}b_2^2 - \frac{1}{2}a_2^2\right) \left[\frac{1}{2}x_1^2\right]_{a_1}^{b_1} \\
&= \left(\frac{1}{2}b_2^2 - \frac{1}{2}a_2^2\right) \left(\frac{1}{2}b_1^2 - \frac{1}{2}a_1^2\right).
\end{aligned}
$$

Finally, conclude that

$$\int_{a_1}^{b_1} \int_{a_2}^{b_2} x_1 x_2 dx_2 dx_1 = \left(\frac{1}{2}b_2^2 - \frac{1}{2}a_2^2\right) \left(\frac{1}{2}b_1^2 - \frac{1}{2}a_1^2\right).$$

As mentioned above, the order of integration does not matter, so we could just as well have started by integrating with respect to x_1, which would yield the same result:

$$\int_{a_2}^{b_2} \int_{a_1}^{b_1} x_1 x_2 dx_1 dx_2 = \left(\frac{1}{2}b_2^2 - \frac{1}{2}a_2^2\right) \left(\frac{1}{2}b_1^2 - \frac{1}{2}a_1^2\right).$$

B.3.3 Differentiation of an integral

Consider the function

$$g(y) = \int_{a(y)}^{b(y)} f(x,y) dx,$$

where the integration boundaries a and b are allowed to be a function of y. The function $g()$ is a function of y only because x is "integrated out".

Under quite weak conditions, the so-called *Leibniz integral rule* states the form of the derivative of the function $g(y)$ with respect to y, namely

$$\frac{dg(y)}{dy} = f(b(y), y)\frac{db(y)}{dy} - f(a(y), y)\frac{da(y)}{dy} + \int_{a(y)}^{b(y)} \frac{\partial f(x,y)}{\partial y} dx.$$

Note that in the special case where the boundaries $a(y)$ and $b(y)$ are not functions of y, then

$$\frac{d}{dy} \int_a^b f(x,y) dx = \int_a^b \frac{\partial f(x,y)}{\partial y} dx.$$

Example B.38 *Let*

$$g(y) = \int_0^{10} x \cdot y \, dx.$$

Then

$$\frac{dg(y)}{dy} = \int_0^{10} \frac{\partial (x \cdot y)}{\partial y} dx = \int_0^{10} x dx = \left[0.5 \cdot x^2\right]_0^{10} = 50.$$

Alternatively, we could have solved the integral first

$$g(y) = \int_0^{10} x \cdot y \ dx = y \cdot \left[0.5 \cdot x^2\right]_0^{10} = y \cdot 50,$$

and then taken the derivative of $g(y)$ after the integral has been solved

$$\frac{dg(y)}{dy} = \frac{d(50 \cdot y)}{dy} = 50.$$

B.4 Proofs

For brevity, we have chosen not to include proofs of the results given in this appendix. The interested reader may find the proofs in intermediate textbooks on real analysis.

C

Appendix C: Simulation from a distribution

When analyzing a model or interpreting estimation results, we may need to evaluate a distribution. If the distribution is intractable to evaluate, for instance, if there is no closed form solution, a commonly used approach is to simulate a sample from the distribution using numerical methods, and use this sample to approximate the distribution of relevant aspects of it. For example, with Bayesian methods (Chapter 19), the posterior distribution often has a form which is not easy to analyze directly. In this appendix, we discuss some of the methods available for simulation from a distribution.

The usefulness of simulated draws from a distribution are discussed under Monte Carlo simulation (Chapter 11). For example, with simulated draws we can calculate the mean, variance, and median of a distribution. Consider a random variable W with distribution given by the CDF F_W, and let $W_1^{\#}, \ldots, W_M^{\#}$ be M draws from this distribution, where we can decide the value of M. Then the mean of W may be approximated by

$$E(W) \approx \frac{1}{M} \sum_{i=1}^{M} W_i^{\#}. \tag{C.1}$$

More generally, the mean of any function $g()$ of W may be approximated by

$$E(g(W)) \approx \frac{1}{M} \sum_{i=1}^{M} g(W_i^{\#}).$$

The median can be approximated by the middle value of the ordered random variables $W_{(1)}^{\#}, \ldots, W_{(M)}^{\#}$. Other quantiles may be approximated analogously. By a Law of Large Numbers (e.g. Theorem 11.4), the approximation can be made arbitrarily good by choosing the number of draws M sufficiently large.

In the remainder of this appendix, we discuss various methods to simulate the draws $W_1^{\#}, \ldots, W_M^{\#}$ from a distribution F_W.

C.1 Pseudo-random numbers

In this section, we consider how to draw "random" numbers based on a so-called *pseudo-random number generator*. This will be useful for all the methods to come because they are all based on using random numbers. The most direct use of random numbers to make a draw $W^{\#}$ from a generic distribution with CDF $F_W()$ is to use Theorem 11.7, presented in Chapter 11, which implies that

$$U \sim U(0,1) \Rightarrow W^{\#} = F_W^{-1}(U) \sim F_W, \tag{C.2}$$

DOI: 10.1201/9781003591191-C

where $F_W^{-1}()$ is the inverse of the CDF F_W. Thus, if we have a method for drawing uniform random numbers between 0 and 1, then we can make draws from the distribution of W, so long as we can calculate the inverse CDF $F_W^{-1}()$.

A pseudo-random number generator can be implemented on a computer and most statistical software packages offer preprogrammed number generators. We focus on drawing pseudo-random numbers from a uniform $U(0,1)$ distribution. Since a computer fundamentally operates completely deterministically, we search for a method that can generate numbers in the interval $[0,1]$ that looks like they are drawn from a $U(0,1)$ distribution, in the sense that the generated numbers are indistinguishable from truly random $U(0,1)$ numbers, according to certain statistical criteria.

To illustrate the basic working of a pseudorandom number generator, we review the so-called *Lehmer random number generator*. Let z_0 be some positive integer. A new number, denoted z_1, is generated from z_0 using the formula

$$z_1 = ((a \cdot z_0) \bmod m),$$

where a and m are suitable integers, and the "mod" operator returns the remainder, defined as the difference between a number and the nearest integer no larger than that number, which is divisible by m. For instance, $8 \bmod 3 = 2$ since 6 is the nearest integer no larger than 8, which is divisible by 3, and $8 - 6 = 2$. The range of the mod operator is always an integer belonging to the set $\{0, \ldots, m-1\}$.

To obtain the next number z_2 in the sequence, set

$$z_2 = ((a \cdot z_1) \bmod m).$$

Thus, the numbers $z_3, z_4, \ldots$ can be generated by the iterative formula

$$z_{i+1} = ((a \cdot z_i) \bmod m).$$

The numbers z_i can be converted into numbers u_i in the interval $[0,1]$ by setting

$$u_i = \frac{z_i}{m-1}.$$

To implement this pseudorandom number generator, it is necessary to choose values of z_0, a, and m. The number z_0 is called the *seed* of the pseudorandom number generator. Good choices of m and a have been shown to be $m = 2^{31} - 1$ and $a = 48271$. With these choices, the number generator produces $2^{31} - 2$ different numbers before it repeats itself. This is called the *period* of a number generator.

C.2   Rejection sampling

In this section, we consider making draws of the random variable W, when the density function f_W of W is easy to evaluate whereas the CDF F_W is not easy to evaluate and, thus, neither is its inverse F_W^{-1}. That is, it is intractable to use (C.2) to draw from the distribution of W.

The basis of rejection sampling is to make use of an auxiliary distribution from which it is easy to draw observations. Let $g()$ be the PDF of the auxiliary distribution. The choice of auxiliary distribution must be such that there exists a constant $m > 0$ fulfilling

$$f_W(w) \le m \cdot g(w), \quad \text{for all } w. \tag{C.3}$$

That is, $g()$ is a PDF which, after multiplication by the constant $m > 0$, bounds, or envelops, the PDF $f_W()$. Next is an example of how to bound a PDF.

Example C.1 *Let W be distributed according to the PDF*

$$f_W(w) = \begin{cases} 0.5 - 0.125 \cdot w & if \quad 0 \leq w \leq 4, \\ 0 & otherwise. \end{cases}$$

To bound this density, consider the PDF of the $U(0,4)$ distribution. Denote this PDF by $g(w)$. It is given by

$$g(w) = \begin{cases} 0.25 & if \quad 0 \leq w \leq 4, \\ 0 & otherwise. \end{cases}$$

If we choose $m = 2$ or larger, then condition (C.3) is satisfied since

$$f_W(w) \leq 2 \cdot g(w) = 2 \cdot 0.25 = 0.5, \quad for \ all \ 0 \leq w \leq 4,$$

and $f_W(w)$ can at most be 0.5, namely, for $w = 0$. For $w < 0$ and $w > 4$, both $f_W(w)$ and $g(w)$ equal 0 and, thus, condition (C.3) is satisfied.

To generate a draw from $f_W()$, the method of *rejection sampling* proposes to make a draw from a distribution with PDF $g()$ and then accepting or rejecting this draw according to some pre-determined probability that depends on $f_W()$, $g()$, and m. For this reason, the auxiliary PDF $g()$ is called a *proposal density*. If the proposed sample is accepted, it may be considered to be drawn from $f_W()$; if it is rejected, a new draw from $g()$ is generated and the process repeated until a draw is accepted. Formally, the rejection sampling algorithm is as follows.

Algorithm C.2.1 (Rejection sampling)

1. *Randomly make a draw Q from the distribution $g(w)$ and define the* acceptance *probability*

$$p_{accept}(Q) = \frac{f_W(Q)}{m \cdot g(Q)}. \tag{C.4}$$

2. *Randomly make a draw from $U \sim U(0,1)$.*

3. *Accept Q as a draw of W if*
$$U \leq p_{accept}(Q),$$

   where $p_{accept}(Q)$ is given in (C.4). Otherwise reject Q as a draw of W.

4. *Repeat 1. to 3. until a draw is accepted.*

5. *For $m = 1, \ldots, M$, repeat 1. to 4. yielding $W_1^{\#}, \ldots, W_m^{\#}$ as the accepted Qs.*

The probability of a draw from $g(w)$ being accepted is $1/m$. This can be shown as follows. In step 3, for a given draw Q from $g(w)$, the probability that U is less than that draw is equal to the acceptance probability (C.4), i.e.

$$\Pr\left(U \leq p_{accept}(Q) \mid Q\right) = p_{accept}(Q) = \frac{f_W(Q)}{m \cdot g(Q)}.$$

Unconditionally, the probability of any one draw being accepted is

$$
\begin{aligned}
\Pr\left(U \leq \frac{f_W(Q)}{m \cdot g(Q)}\right) &= \int \Pr\left(U \leq \frac{f_W(Q)}{m \cdot g(Q)} \,\middle|\, Q = q\right) \cdot g(q)\, dq \\
&= \int \frac{f_W(q)}{m \cdot g(q)} \cdot g(q)\, dq = \int \frac{f_W(q)}{m}\, dq \\
&= \frac{1}{m},
\end{aligned}
$$

where we used that $\int f_W(q)dq = 1$ because $f_W()$ is a PDF. This shows that the lower we can choose m, the more efficient the algorithm is. Indeed, if $m \approx 1$ almost all proposals Q will be accepted; conversely, if $m \gg 1$ then almost all proposals will be rejected. In the latter case, it may be very time-consuming to generate a large number M of draws from $f_W()$ using rejection sampling, because a large proportion of proposals are rejected.

Example C.2 (Example C.1, continued) *We may use proposals from the $U(0,4)$ distribution along with $m = 2$ to sample from $f_W()$ using rejection sampling. The acceptance probability for each proposal is $\frac{1}{m} = \frac{1}{2} = 50\%$. Hence, on average it is necessary to make twice as many draws from $g(w)$ to obtain the number of draws desired from f_W. You are asked to use rejection sampling for simulating from $f_W()$ in Problem C.7.1 at the end of this chapter.*

Next we show why rejection sampling works, that is, why the Qs accepted in the rejection sampling algorithm is distributed according to f_W. Denote by $W_A^{\#}$ an accepted Q. Then

$$
\begin{aligned}
\Pr\left(W_A^{\#} \leq c\right) &= \Pr(Q \leq c \mid Q \text{ accepted }) \\
&= \Pr\left(Q \leq c \,\middle|\, U \leq \frac{f_W(Q)}{m \cdot g(Q)}\right) \\
&= \frac{\Pr\left(Q \leq c,\, U \leq \frac{f_W(Q)}{m \cdot g(Q)}\right)}{\Pr\left(U \leq \frac{f_W(Q)}{m \cdot g(Q)}\right)}.
\end{aligned}
$$

The denominator was calculated above to yield $\frac{1}{m}$. By rewriting the numerator using a conditional probability as follows, we get

$$
\begin{aligned}
\Pr\left(W_A^{\#} \leq c\right) &= \frac{\Pr\left(Q \leq c,\, U \leq \frac{f_W(Q)}{m \cdot g(Q)}\right)}{\Pr\left(U \leq \frac{f_W(Q)}{m \cdot g(Q)}\right)} \\
&= \frac{\int_{-\infty}^{\infty} \Pr\left(Q \leq c,\, U \leq \frac{f_W(Q)}{m \cdot g(Q)} \,\middle|\, Q = q\right) g(q)\, dq}{\frac{1}{m}} \\
&= m \int_{-\infty}^{c} \Pr\left(U \leq \frac{f_W(Q)}{m \cdot g(Q)} \,\middle|\, Q = q\right) g(q)\, dq \\
&= m \int_{-\infty}^{c} \frac{f_W(q)}{m \cdot g(q)} \cdot g(q)\, dq \\
&= \int_{-\infty}^{c} f_W(q)\, dq \\
&= F_W(c),
\end{aligned}
$$

the CDF of W. Hence, the distribution of the accepted draws $W_A^{\#}$ constructed by the rejection sampling method has the same distribution as W.

As mentioned above, the rejection sampling may be very inefficient if a high value of m is needed. It may also be inefficient if we want to draw from a K-dimensional joint distribution. The algorithms introduced in the next sections are designed to remedy these inefficiencies.

C.3 The Markov Chain Monte Carlo method

The *Markov Chain Monte Carlo* (MCMC) method is designed to make draws from a distribution $f_W(w)$ by means of a Markov chain, where drawing from the transition probability of that Markov chain behaves as if drawing from $f_W(w)$. This is useful because in many cases it is intractable to draw from $f_W(w)$ but tractable to draw from the transition probability of the Markov chain. In this section, we only consider continuous random variables.

C.3.1 Markov chains

To present the MCMC method, we provide a brief discussion of so-called homogenous Markov chains. Let W_m, $m = 1, \ldots, M$ be a sequence of random variables. Such a sequence is said to be a *first-order Markov chain* if the distribution of W_m conditional on $W_{m-1}, \ldots, W_1$ only depends on W_{m-1}. That is,

$$f_{W_m|W_{m-1},\ldots,W_1}(w_m|w_{m-1}, \ldots, w_1) = f_{W_m|W_{m-1}}(w_m|w_{m-1}).$$

The Markov chain is said to be *homogeneous* if the transition probability $f_{W_m|W_{m-1}}(w|x)$ does not depend on m:

$$f_{W_m|W_{m-1}}(w|x) = p(w|x), \quad \text{for all } w \text{ and } x$$

for some conditional density $p(w|x)$, called the *transition density* or *transition probability* of the Markov chain. In the following, we assume a homogeneous Markov chain with transition probability $p(w|x)$. Then the marginal distribution of W_m is

$$f_{W_m}(w) = \int p(w|x) \cdot f_{W_{m-1}}(x) \, dx$$

A Markov chain is said to be *ergodic* if the marginal distribution $f_{W_m}(w)$ converges to a PDF $g(w)$, i.e. if

$$f_m(w) \to g(w), \quad \text{for } m \to \infty \text{ and for all } w.$$

In this case, the PDF $g(w)$ is said to be a *stationary distribution* of the Markov chain. It is also a solution to the integral equation

$$g(w) = \int p(w|x) \cdot g(x) \, dx.$$

It can be useful to represent a Markov chain as an iterative scheme of random variables. In general, a homogeneous first-order Markov chain can be written as

$$W_m = h(W_{m-1}, \varepsilon_m),$$

where h is an appropriately chosen function and ε_m is a sequence of iid random variables. The next example illustrates how a Markov Chain can be written in this iterative scheme, and how that can be used to find the stationary distribution.

Example C.3 *Consider the random variable W_m, defined by*

$$W_m = \beta_0 + \beta_1 W_{m-1} + \varepsilon_m,$$

where ε_m, $m = 1, \ldots, M$ is a sequence of independent and identically distributed random variables with the following normal distribution

$$\varepsilon_m \sim N\left(\mu_\varepsilon, \sigma_\varepsilon^2\right).$$

This is a first-order Markov chain with transition probability $p(w_m|w_{m-1})$, given by

$$W_m|W_{m-1} = w_{m-1} \sim N\left(\beta_0 + \beta_1 w_{m-1}, \sigma_\varepsilon^2\right).$$

To find the stationary distribution, write the iterative scheme as a function of the ε_m's and W_0 only

$$
\begin{aligned}
W_m &= \beta_0 + \beta_1 W_{m-1} + \varepsilon_m \\
&= \beta_0 + \beta_1\left(\beta_0 + \beta_1 W_{m-2} + \varepsilon_{m-1}\right) + \varepsilon_m \\
&= \beta_0(1 + \beta_1) + \beta_1^2 W_{m-2} + \varepsilon_m + \beta_1\varepsilon_{m-1} \\
&\quad\vdots \\
&= \beta_0(1 + \beta_1 + \ldots + \beta_1^{m-1}) + \beta_1^m W_0 + \varepsilon_m + \beta_1\varepsilon_{m-1} + \ldots + \beta_1^{m-1}\varepsilon_1.
\end{aligned}
$$

The distribution of W_m conditional on W_0 is

$$W_m|W_0 = w_0 \sim N\left(\beta_0(1 + \beta_1 + \ldots + \beta_1^{m-1}) + \beta_1^m w_0, (1 + \beta_1 + \ldots + \beta_1^{m-1})^2\sigma_\varepsilon^2\right).$$

For this to converge to a stationary distribution, assume $0 \leq \beta_1 < 1$. Then for $m \to \infty$,

$$W_m|W_0 = w_0 \xrightarrow{d} N\left(\frac{1}{1 - \beta_1}\beta_0, \left(\frac{1}{1 - \beta_1}\right)^2\sigma_\varepsilon^2\right).$$

In Example C.3, the stationary distribution does not depend on the initial value of the iterative scheme, w_0. This is no coincidence. In general, it is true that the stationary distribution obtained from iteration of an ergodic Markov chain, does not depend on the initial value of the Markov chain.

C.3.2 Making draws from a Markov chain

In MCMC methods, we circumvent the above discussion of an ergodic homogeneous Markov chain. The starting point is a distribution of W with PDF $f_W(w)$ from which we want to make a draw. We then construct a Markov chain, which has f_W as its stationary distribution. In particular, we seek a transition probability $p(w|x)$ that leads to the stationary distribution $f_W(w)$. The idea of the MCMC method is that making a draw from f_W can equivalently be obtained by making a draw from the transition probability $p(w|x)$. We defer the issue on how to choose a transition probability $p(w|x)$ to the following sections, where we consider various methods to construct the transition probability. In this section, we focus on the equivalence of drawing from the transition probability and drawing from f_W.

Consider a first-order homogeneous Markov chain initialized by the PDF $g_0(w)$. Then the following PDFs of the elements $W_1, \ldots, W_m$ in the Markov chain are

$$g_1(w) = \int p(w|x) \cdot g_0(x) \, dx,$$

$$g_2(w) = \int p(w|x) \cdot g_1(x) \, dx,$$

$$g_3(w) = \int p(w|x) \cdot g_2(x) \, dx,$$

$$\vdots$$

$$g_m(w) = \int p(w|x) \cdot g_{m-1}(x) \, dx.$$

Assuming the transition probability $p(w|x)$ is chosen to be a transition probability of an ergodic homogeneous Markov chain with stationary distribution f_W, then

$$g_m(w) \to f_W(w) \quad \text{for } m \to \infty \text{ and for all } w.$$

The stationary distribution can be shown to be independent on the initial PDF $g_0(w)$ as long as the support of the initial PDF is a subset of the support of f_W.

The usefulness of the above is that we may make draws from f_W by making iterated draws from the transition probability according to the iterative scheme

$$\text{draw } W_0^{\#} \text{ from } g_0(w),$$
$$\text{draw } W_1^{\#} \text{ from } p(w|W_0^{\#}),$$
$$\text{draw } W_2^{\#} \text{ from } p(w|W_1^{\#}),$$
$$\vdots$$
$$\text{draw } W_m^{\#} \text{ from } p(w|W_{m-1}^{\#}).$$

From some m onwards, $W_m^{\#}$ behaves as if it is drawn from f_W. This leads to the MCMC algorithm

Algorithm C.3.1 (Markov Chain Monte Carlo)

1. *Choose $M_{burn} \geq 0$ and set $M^* = M + M_{burn}$, where M is the desired number of observations to be generated.*

2. *Choose a starting value denoted $W_0^{\#}$.*

3. *For $i = 1, \ldots, M^*$: Draw $W_i^{\#}$ from the transition probability $p(w \mid W_{i-1}^{\#})$.*

4. *Discard the initial M_{burn} observations $i = 1, \ldots, M_{burn}$, leaving the M random draws $W_{(M_{burn}+1)}^{\#}, \ldots, W_{M^*}^{\#}$ as draws from the stationary distribution f_W of the Markov chain.*

The reason for the fourth step, where the so-called *burn-in sample* $W_0^{\#}, \ldots, W_{M_{burn}}^{\#}$ is discarded, is to lessen the impact of the initial starting value $W_0^{\#}$ and make sure sufficient iterations have been made such that the draws are approximately from f_W. From $M_{burn}+1$ onwards, we can use all the draws as draws from f_W.

There can be many transition probabilities that lead to the same stationary distribution. This is illustrated in the next example.

Example C.4 *Suppose we want to make a draw from a normal distribution using the MCMC method. Let*

$$W \sim N(\mu, \sigma^2).$$

We can use the result in Example C.3 to construct a transition probability. In Example C.3, it was found that the transition probability $p(w_m|w_{m-1})$, given by

$$W_m|W_{m-1} = w_{m-1} \sim N\left(\beta_0 + \beta_1 w_{m-1}, \sigma_\varepsilon^2\right), \tag{C.5}$$

leads to the stationary distribution

$$W_\infty \sim N\left(\frac{1}{1-\beta_1}\beta_0, \left(\frac{1}{1-\beta_1}\right)^2 \sigma_\varepsilon^2\right),$$

assuming that $0 \le \beta_1 < 1$. Therefore, if we choose β_0, β_1 and σ_ε^2, such that

$$\mu = \frac{1}{1-\beta_1}\beta_0$$

and

$$\sigma^2 = \left(\frac{1}{1-\beta_1}\right)^2 \sigma_\varepsilon^2,$$

then we can use the transition probability $p(w_m|w_{m-1})$ (C.5) in Algorithm C.3.1 to draw samples $N(\mu, \sigma^2)$ using the MCMC method. There are three unknowns β_0, β_1, and σ_ε^2 and two knowns μ and σ^2, so there are infinitely many solutions that satisfy the two equations. For any choice of β_1 satisfying $0 \le \beta_1 < 1$, set

$$\beta_0 = (1-\beta_1)\mu$$

and

$$\sigma_\varepsilon^2 = (1-\beta_1)^2 \sigma^2.$$

For example, if $W \sim N(10, 8)$, then for $\beta_1 = 0.5$, we have

$$\beta_0 = (1-0.5)10 = 5,$$

and

$$\sigma_\varepsilon^2 = (1-0.5)^2 8 = 2,$$

the transition probability

$$W_m|W_{m-1} = w_{m-1} \sim N\left(5 + 0.5w_{m-1}, 2\right)$$

defines a Markov chain with $W \sim N(10, 8)$ as stationary distribution. Thus, after a burn-in sample, we can generate draws from $W \sim N(10, 8)$ by making draws from the transition probability. If we choose another value of β_1, and adjust β_0 and σ_ε^2 accordingly, we will get a different transition probability but the same stationary distribution.

Notice, this example is included to show that many transition probabilities can lead to the same stationary distribution. This example does not, however, illustrate an advantage of making draws from the transition probability since it is at least as tractable to make draws directly from $W \sim N(10, 8)$.

There is one big difference between using the MCMC method to generate draws from a random variable W, compared with simulation methods based on independent draws, such as the inverse CDF method (C.2) and rejection sampling. In the latter cases, the simulated draws $W_1^\#, \ldots, W_M^\#$ are iid by construction. They are, in particular, statistically independent. In contrast, in the MCMC case, the draws will, in general, be correlated. The reason is the form of the MCMC algorithm where $W_{m-1}^\#$ is used to generate $W_m^\#$, thus generating a Markov chain. This can have implications for the value of the number of samples M, necessary to accurately use simulation-based methods to approximate aspects of a distribution, e.g. the mean as in (C.1). In general, it takes more MCMC samples to accurately estimate aspects of a distribution, compared to samples generated using methods resulting in draws that are statistically independent. In Problem C.7.2 at the end of this chapter, you are asked to explore this in the context of Example C.4

In many contexts, we need to draw from a joint distribution $f_{W_1,\ldots,W_K}(w_1,\ldots,w_K)$ of K random variables $W = (W_1,\ldots,W_K)$. To use the MCMC method in such a context, a joint transition probability

$$p(w|x) = p\left(w_1,\ldots,w_K \mid x_1,\ldots,x_K\right)$$

must be specified. Depending on how the joint transition probability is specified, we may either draw from $W_1,\ldots,W_K$ simultaneously, e.g. as with the Metropolis algorithm (Section C.5), or we may draw them one at a time, e.g. as with the Gibbs sampler algorithm (Section C.4). These algorithms are also examples of generic ways to construct transition probabilities compared to e.g. Example C.4, where the transition probability is constructed using the derivation showing that a normal transition probability leads to a normal distribution.

In the following sections, we consider various methods to construct the transition probability $p(w|x)$ such that the MCMC method can be used to make draws from f_W.

C.4   The Gibbs sampler

The *Gibbs sampler* is an MCMC method where the transition probability has a particular functional form. It is applicable to make draws from a joint PDF $f_{W_1,\ldots,W_K}(w_1,\ldots,w_K)$ of K random variables $W = (W_1,\ldots,W_K)$.

The K-dimensional transition probability for the Gibbs sampler is defined as follows. Let W_{-k} denote all the random variables $W_1,\ldots,W_K$ except W_k. Hence, W_{-k} consists of the $K-1$ random variables

$$W_{-k} = (W_1,\ldots,W_{k-1},W_{k+1},\ldots,W_K).$$

Let $f_{W_k|W_{-k}}$ be the PDF of W_k conditional on W_{-k}. There are K of such conditional probabilities, namely,

$$f_{W_1|W_{-1}}\left(w_1|w_{-1}\right),$$
$$f_{W_2|W_{-2}}\left(w_2|w_{-2}\right),$$
$$\vdots$$
$$f_{W_K|W_{-K}}\left(w_K|w_{-K}\right).$$

Then we define the transition probability for the Gibbs sampler as

$$
\begin{aligned}
p\left(w_1,\ldots,w_m \mid x_1,\ldots,x_K\right) \;=\; & f_{W_1|W_{-1}}\left(w_1|x_2\ldots,x_K\right) \\
& \cdot f_{W_2|W_{-2}}\left(w_2|w_1,x_3,\ldots,x_K\right) \\
& \cdot f_{W_3|W_{-3}}\left(w_3|w_1,w_2,x_4,\ldots,x_K\right) \\
& \cdots \\
& \cdot f_{W_K|W_{-K}}\left(w_K|w_1,\ldots,w_{K-1}\right).
\end{aligned}
\tag{C.6}
$$

It can be verified (Problem C.7.3) that $p\left(w_1,\ldots,w_m \mid x_1,\ldots,x_K\right)$ defined in this way is a PDF. Under certain assumptions, it can be shown that a Markov chain with transition probability given by (C.6) has $f_{W_1,\ldots,W_K}\left(w_1,\ldots,w_K\right)$ as stationary distribution. We therefore can proceed to implement drawing from this Markov chain by the MCMC method.

The Gibbs sampler makes draws from the transition probability (C.6) by drawing from each conditional probability $f_{W_k|W_{-k}}$ starting for $k=1$ and ending at $k=K$. In addition, the scheme is such that the draw in round m made for W_1, i.e. $W_{1,m}^{\#}$, is used in the draw of W_2, i.e. $W_{2,m}^{\#}$, instead of the draw of W_1 from round $(m-1)$, i.e. $W_{1,m-1}^{\#}$. Hence, the draws in round m are

$$
W_{1,m}^{\#} \text{ is drawn from } f_{W_1|W_{-1}}\left(w_1 \mid W_{2,m-1}^{\#},\ldots,W_{K,m-1}^{\#}\right),
$$

$$
W_{2,m}^{\#} \text{ is drawn from } f_{W_2|W_{-2}}\left(w_2 \mid W_{1,m}^{\#},W_{3,m-1}^{\#},\ldots,W_{K,m-1}^{\#}\right),
$$

$$
W_{3,m}^{\#} \text{ is drawn from } f_{W_3|W_{-3}}\left(w_3 \mid W_{1,m}^{\#},W_{2,m}^{\#},W_{4,m-1}^{\#},\ldots,W_{K,m-1}^{\#}\right),
$$

$$
\vdots
$$

$$
W_{K,m}^{\#} \text{ is drawn from } f_{W_K|W_{-K}}\left(w_K \mid W_{1,m}^{\#},\ldots,W_{K-1,m}^{\#}\right)
$$

The result is the m'th draw $W_m^{\#} = \left(W_{1,m}^{\#},\ldots,W_{K,m}^{\#}\right)$, based on the $(m-1)$'th draw $W_{m-1}^{\#} = \left(W_{1,m-1}^{\#},\ldots,W_{K,m-1}^{\#}\right)$. Note that, due to the Gibbs sampling scheme, the m'th draw does not depend on $W_{1,m-1}^{\#}$.

For the Gibbs sampler, it is only necessary to draw observations from the one-dimensional distributions $f_{W_k|W_{-k}}$, $k=1,\ldots,K$. For the Gibbs sampler to be useful in practice, it must be tractable to make draws from all the conditional probabilities $f_{W_k|W_{-k}}$.

Example C.5 (Sampling from a bivariate beta distribution using the Gibbs sampler) *Suppose W_1 conditional on W_2, and W_2 conditional on W_1, follow the beta distributions*

$$
\begin{aligned}
W_1 \mid W_2 \;&\sim\; Beta(\alpha,1+W_2), \\
W_2 \mid W_1 \;&\sim\; Beta(\beta,1+W_1),
\end{aligned}
$$

where $\alpha,\beta>0$ are parameters. This situation could, for instance, be encountered if there are two unknown parameters, W_1 and W_2, which are correlated and both restricted to lie in the unit interval $[0,1]$. For instance, W_1 and W_2 may be parameters in Bernoulli distributions in a Bayesian analysis.

Since the conditional distributions $f_{W_k|W_{-k}}\left(w_k \mid w_{-k}\right)$ are beta distributions, it is straight forward to use the Gibbs sampler to produce samples from $W=(W_1,W_2)$. You are asked to do this in Problem C.7.4 at the end of this chapter.

C.5   The Metropolis algorithm

The *Metropolis algorithm* constructs the transition probability in the Markov chain by sampling from a proposal density $g(x|y)$ from which it is tractable to draw, and then implementing a subsequent rejection sampling step.

The Metropolis algorithm works as follows. Suppose making a draw from the proposal distribution $g(w \mid x)$ is tractable and that it is symmetric in the sense that $g(w|x) = g(x|w)$ for all w, x. Given a draw $W_{m-1}^{\#}$ from W, sample Q from the proposal distribution $g(w \mid W_{m-1}^{\#})$. Define the acceptance probability

$$p_{accept}\left(Q, W_{m-1}^{\#}\right) = \min\left(1, \frac{f_W(Q)}{f_W\left(W_{m-1}^{\#}\right)}\right). \tag{C.7}$$

Then, with probability $p_{accept}\left(Q, W_{m-1}^{\#}\right)$, we accept Q as our new sample $W_m^{\#}$. If the sample is not accepted, $W_{m-1}^{\#}$ is re-used as the new sample $W_m^{\#}$. To summarize, for $i = 1, 2, \ldots, M^*$, we set

$$W_m^{\#} = \begin{cases} Q, & \text{if } U \le p_{accept}\left(Q, W_{m-1}^{\#}\right), \\ W_{m-1}^{\#}, & \text{otherwise,} \end{cases}$$

where Q is sampled from $g(w \mid W_{m-1}^{\#})$, $p_{accept}\left(Q, W_{m-1}^{\#}\right)$ is given in (C.7), and $U \sim U(0, 1)$.

Note that the acceptance probability (C.7) relies on the target distribution $f_W()$ only through the ratio $\frac{f_W(Q)}{f_W\left(W_{m-1}^{\#}\right)}$. This makes the algorithm particularly convenient for sampling from posterior distributions in Bayesian contexts, since the normalizing constant in a posterior distribution, i.e. $f_{\tilde{Y}}(\tilde{y})$ in (19.1), is often difficult to calculate, see Example 19.9 in Chapter 19 for an illustration. The ratio of $f_W()$ in (C.7) means that such constants cancel and are thus not necessary for simulating from the posterior distribution using the Metropolis algorithm.

The Metropolis algorithm is applicable for making draws from a joint distribution, i.e. in the case where $W = (W_1, \ldots, W_K)$. It makes the draws simultaneously in contrast to the Gibbs Sampler, in which there is a sequential scheme for drawing from K random variables in each round m.

A downside of the Metropolis algorithm is that the proposal distribution needs to be symmetric in the sense that $g(w|x) = g(x|w)$ for all w, x. This is, for instance, the case for the normal distribution, but it will not be satisfied for all proposal distributions. This requirement is relaxed for the algorithm discussed in the next section.

C.6   The Metropolis-Hastings algorithm

The *Metropolis-Hastings algorithm* is similar in structure to the Metropolis algorithm, except that the proposal density $g(x|y)$ is not required to be symmetric. Therefore, for theoretical reasons, the acceptance probability needs to be adjusted accordingly. Define

$$p_{accept}\left(Q, W_{m-1}^{\#}\right) = \min\left(1, \frac{f_W(Q)}{f_W\left(W_{m-1}^{\#}\right)} \cdot \frac{g\left(Q|W_{m-1}^{\#}\right)}{g\left(W_{m-1}^{\#}|Q\right)}\right). \tag{C.8}$$

The Metropolis-Hastings algorithm is similar to the Metropolis algorithm, but using the acceptance probability (C.8) in place of (C.7). It is easily seen from (C.8) that if the proposal density is symmetric, then the Metropolis-Hastings acceptance probability (C.8) reduces to the Metropolis acceptance probability (C.7). That is, in the case of a symmetric proposal distribution $g(x|y)$, the Metropolis-Hastings algorithm is equivalent to the Metropolis algorithm.

To summarize the Metropolis-Hastings algorithm, for $i = 1, 2, \ldots, M^*$, we set

$$W_m^{\#} = \begin{cases} Q & \text{if } U \leq p_{accept}\left(Q, W_{m-1}^{\#}\right), \\ W_{m-1}^{\#} & \text{otherwise,} \end{cases}$$

where Q is sampled from $g(w \mid W_{m-1}^{\#})$, $p_{accept}\left(Q, W_{m-1}^{\#}\right)$ is given in (C.8), and $U \sim U(0, 1)$. Similarly to the Metropolis algorithm, the acceptance probability (C.8) relies on only the ratios of $f_W()$ and $g()$, meaning that it is not necessary to know the normalizing constants.

Example C.6 (Metropolis-Hastings sampling from the posterior distribution of Example 19.9) *In Example 19.9 of Chapter 19, a Laplace prior and normal population led to a posterior distribution given by*

$$f_{\theta|\widetilde{Y}}(\theta \mid \widetilde{y}) = \frac{\exp\left(-\frac{1}{2\sigma_0^2}\sum_{i=1}^{n}(\widetilde{y}_i - \theta)^2\right)\exp(-|\theta - \mu|/\beta)}{\int \exp\left(-\frac{1}{2\sigma_0^2}\sum_{i=1}^{n}(\widetilde{y}_i - \theta)^2\right)\exp(-|\theta - \mu|/\beta)d\theta}. \tag{C.9}$$

We wish to sample from this distribution using the Metropolis-Hastings algorithm. Note that the denominator in (C.9) does not depend on θ, i.e. it is a normalizing constant that is not needed for the Metropolis-Hastings acceptance probability (C.8). As a function of θ, the posterior is proportional to the numerator in (C.9), which is often written as

$$f_{\theta|\widetilde{Y}}(\theta \mid \widetilde{y}) \propto \exp\left(-\frac{1}{2\sigma_0^2}\sum_{i=1}^{n}(\widetilde{y}_i - \theta)^2\right)\exp(-|\theta - \mu|/\beta).$$

As hyperparameters for our prior distribution, we choose $\mu = 100$ and $\beta = 10$. As our proposal distribution $g(x|y)$, we use a Gaussian distribution, i.e.

$$Q \mid W_{m-1}^{\#} \sim N(W_{m-1}^{\#}, \tau^2),$$

where we set $\tau^2 = 100$. This value of τ^2 has been tuned so as to reach an average acceptance probability of around 40%, which has been found to work well for the Metropolis-Hastings algorithm. Note that because we use a symmetric proposal distribution, the Metropolis-Hastings algorithm reduces to the Metropolis algorithm, i.e. the acceptance probability (C.7) may be used.

Suppose that $\sigma_0^2 = 100$ and that we observe the simple random sample

$$(\widetilde{y}_1, \ldots, \widetilde{y}_{10}) = (116.62, 130.67, 131.76, 125.37, 112.04, 116.77, 109.03, 131.17, 123.77, 130.30),$$

which is such that $\overline{y} = 122.75$. Then, all elements are in place for using the Metropolis-Hastings algorithm to simulate from the posterior (C.9). You are asked to do this in Problem

C.7.6 at the end of this chapter. To illustrate how the output of such an exercise may look, Figure C.1 contains Metropolis-Hastings output for $M = 10000$ samples, where $M_{burn} = 5000$ samples were used as burn in, i.e. $M^* = M + M_{burn} = 15000$ total samples were drawn using the Metropolis-Hastings algorithm. The initial value was set to $W_0 = 100$. The top left panel of Figure C.1 shows the main output, namely the Laplace prior and the simulated posterior, plotted using a Gaussian kernel density estimator. The top right panel shows an autocorrelation plot, namely the autocorrelation between the m'th sample $W_m^\#$ and the sample l lags away, $W_{m-l}^\#$, for $l = 0, 1, \ldots, 20$. As the panel shows, there is autocorrelation between samples. This is to be expected because, by design, an MCMC algorithm produces dependence between elements. However, it should preferably be the case that the autocorrelation quickly diminishes, so that the MCMC sample is sufficiently diverse. This is the case here, since after roughly 10 lags, no autocorrelation is present in the Markov chain. Lastly, the bottom panel simply shows the Monte Carlo samples $W_m^\#$ for $m = 1, \ldots, M$. This plot can be useful for diagnosing whether the chain is "mixing" sufficiently well, i.e. exploring the relevant parts of the distribution being simulated. If the chain had a tendency of being "stuck", going through long periods where $W_m^\# = W_{m-1}^\#$, this could be diagnosed in a plot like this.

The Monte Carlo estimate of the posterior mean is the sample mean of the Monte Carlo samples $W_1^\#, \ldots, W_M^\#$. In this case, this is 121.84. The average acceptance probability is 36%.

C.7   Exercises

Problem C.7.1 *Consider the setup in Examples C.1 and C.2.*

1. *Derive the following aspects of the distribution of W analytically: $E(W), E(W^2), Var(W)$, the median of W, and the mode of W.*

2. *Use rejection sampling (Section C.2) to simulate $M = 10,000$ iid random variables $W_1^\#, \ldots, W_M^\#$ with the same distribution as W. (Hint: You may, e.g. use the uniform distribution as proposal density, as suggested in Example C.2. Most statistical software packages have an option to simulate uniform random variables.)*

3. *Calculate the average number of proposed samples until acceptance in your rejection sampling implementation. Compare with the theoretical acceptance probability of $\frac{1}{m}$.*

4. *Use the Monte Carlo sample $W_1^\#, \ldots, W_M^\#$ and a kernel density estimator to estimate the PDF $f(w)$ of W. Compare with the true PDF of W.*

5. *Use the Monte Carlo sample $W_1^\#, \ldots, W_M^\#$ to approximate the following aspects of the distribution of W: $E(W), E(W^2), Var(W)$, the median of W, and the mode of W. Compare with your analytical results.*

Problem C.7.2 *Consider the setup in Example C.4. In this problem, you are asked to compare simulation from the normal distribution $N(\mu, \sigma^2)$ using the MCMC method with simulations from the same distribution using direct simulation resulting in iid samples. In particular, we consider Monte Carlo approximation of the mean μ using the two simulation*

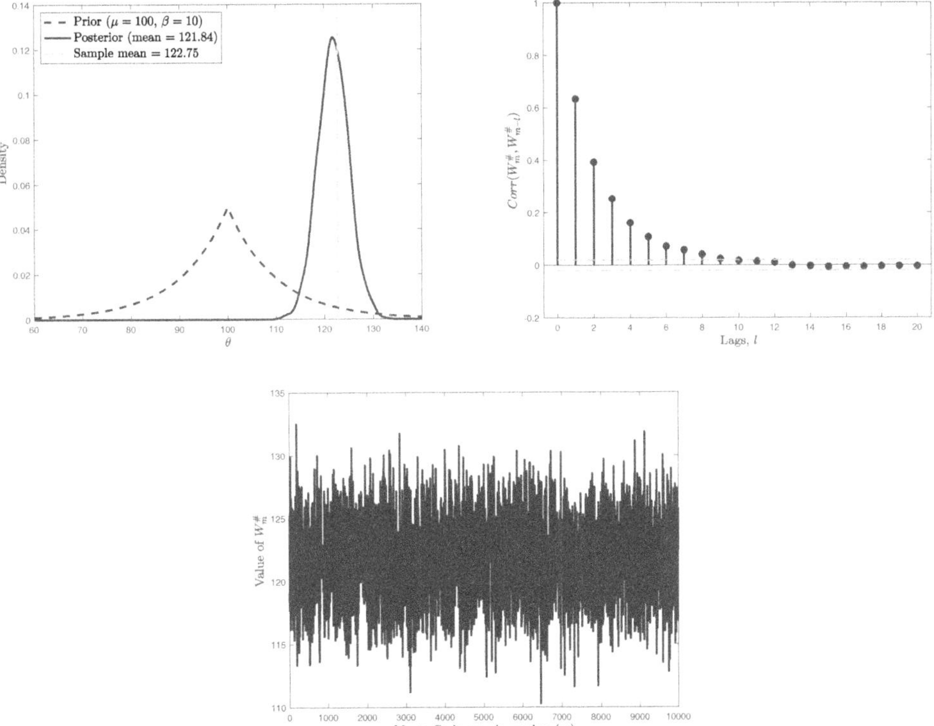

FIGURE C.1

Output from Metropolis-Hastings algorithm, Example C.6. Top left: Laplace prior (dashed) and posterior (solid) estimated by Metropolis-Hastings simulation and plotted using a Gaussian kernel density estimator with bandwidth $h = 0.75$. The vertical line is the sample mean $\overline{y} = 122.75$. Top right: Monte Carlo estimates of the correlations $Corr(W_m^{\#}, W_{m-l}^{\#})$ for $l = 0, \ldots, 20$. Horizontal dashed lines denote a 95%-confidence band, located at $\pm 1.96/\sqrt{M}$. Bottom: Simulated Metropolis-Hastings samples $W_m^{\#}$, $m = 1, \ldots, 10000$.

methods in (C.1), *i.e.*

$$\widehat{\mu}_{direct}^{\#} = \frac{1}{M_{direct}} \sum_{i=1}^{M_{direct}} W_i^{\#,direct},$$

$$\widehat{\mu}_{mcmc}^{\#} = \frac{1}{M_{mcmc}} \sum_{i=1}^{M_{mcmc}} W_i^{\#,mcmc},$$

where $W_i^{\#,direct}$ are draws from $N(\mu, \sigma^2)$ using direct (iid) simulation, $W_i^{\#,mcmc}$ are draws from $N(\mu, \sigma^2)$ using the MCMC algorithm, and $M_{direct}, M_{mcmc} \geq 1$ are the number of samples generated by the respective methods. When considering the MCMC algorithm, you may assume that the draws are from W_{∞}, i.e. M_{burn} has been set to a very large number.

1. *Calculate the expected value of $\widehat{\mu}_{direct}^{\#}$ and $\widehat{\mu}_{mcmc}^{\#}$, i.e. $E(\widehat{\mu}_{direct}^{\#})$ and $E(\widehat{\mu}_{mcmc}^{\#})$. Are the simulation-based estimators of μ unbiased?*

2. Calculate the variances of $\widehat{\mu}^{\#}_{direct}$ and $\widehat{\mu}^{\#}_{mcmc}$, i.e. $Var(\widehat{\mu}^{\#}_{direct})$ and $Var(\widehat{\mu}^{\#}_{mcmc})$. *(Hint: Recall that the MCMC draws $W_i^{\#,mcmc}$ are correlated.)* For $M_{direct} = M_{mcmc}$, which simulation-based estimator of μ is most accurate?

3. For a given value $M_{mcmc} \geq 1$, what is the value of M_{direct} such that $Var(\widehat{\mu}^{\#}_{direct}) = Var(\widehat{\mu}^{\#}_{mcmc})$?

Note: In 3, the value M_{direct}, i.e. the number of independent samples necessary to produce an estimator which is as accurate as the MCMC estimator, is sometimes called the effective sample size *of the MCMC approach. Thus, M_{mcmc} is the actual MCMC sample size and the value of M_{direct} such that $Var(\widehat{\mu}^{\#}_{direct}) = Var(\widehat{\mu}^{\#}_{mcmc})$, is the effective MCMC sample size.*

Problem C.7.3 *Prove that the transition probability given by (C.6) is a PDF.*

Problem C.7.4 *Let $W = (W_1, W_2)$ and consider the setup in Example C.5, i.e.*

$$
\begin{aligned}
W_1 &\sim Beta(\alpha, 1 + W_2), \\
W_2 &\sim Beta(\beta, 1 + W_1),
\end{aligned}
$$

where $\alpha = 2, \beta = 1$.

1. *Use the Gibbs sampler to generate $M = 10,000$ samples from W. (Hint: Most statistical software packages have a procedure for sampling from the beta distribution.)*

2. *Plot the M samples in a scatter plot with the value of W_1 on the x-axis and the value of W_2 on the y-axis. Does W_1 appear to be correlated with W_2?*

3. *Use a kernel density estimator to estimate the marginal PDFs of both W_1 and W_2. Plot the estimated PDF in a graph.*

4. *Use the M samples to produce Monte Carlo estimates of the following population aspects: $E(W_1)$, $E(W_2)$, $E(W_1 \cdot W_2)$, $Var(W_1)$, $Var(W_2)$, $Corr(W_1, W_2)$, $\Pr(W_1 > 0.5)$, $\Pr(W_2 > 0.5)$, $\Pr(W_1 \cdot W_2 > 0.5)$, and $\Pr(W_1 > W_2)$.*

5. *Use also the Monte Carlo sample to estimate the correlation between the successive Gibbs samples, i.e. $Corr(W^{\#}_{1,m-1}, W^{\#}_{1,m})$, $Corr(W^{\#}_{1,m-1}, W^{\#}_{2,m})$, and $Corr(W^{\#}_{2,m-1}, W^{\#}_{1,m})$.*

Problem C.7.5 *Let $W = (W_1, W_2) \sim N(\mu_1, \mu_2, \sigma_1^2, \sigma_2^2, \rho)$ have a bivariate normal distribution with $\mu_1 = 1$, $\mu_2 = 0$, $\sigma_1^2 = \sigma_2^2 = 1$, and $\rho = -0.75$.*

1. *Use the Gibbs sampler to generate $M = 10,000$ samples from W. (Hint: Derive the conditional distributions of W_i given W_{-i}, $i = 1, 2$.)*

2. *Plot the M samples in a scatter plot with the value of W_1 on the x-axis and the value of W_2 on the y-axis. Does W_1 appear to be correlated with W_2?*

3. *Use a kernel density estimator to estimate the marginal PDFs of both W_1 and W_2. Plot the estimated PDF in a graph.*

4. *Use the M samples to produce Monte Carlo estimates of the following population aspects: $E(W_1)$, $E(W_2)$, $E(W_1 \cdot W_2)$, $Var(W_1)$, $Var(W_2)$, $Corr(W_1, W_2)$, and $\Pr(W_1 > W_2)$.*

5. *Use also the Monte Carlo sample to estimate the correlation between the successive Gibbs samples, i.e. $Corr(W^{\#}_{1,m-1}, W^{\#}_{1,m})$, $Corr(W^{\#}_{1,m-1}, W^{\#}_{2,m})$, and $Corr(W^{\#}_{2,m-1}, W^{\#}_{1,m})$.*

Problem C.7.6 *Consider the setup of Example C.6.*

1. *Re-do the Metropolis-Hastings simulation analysis of Example C.6.*

2. *Use the MCMC sample to construct a 95%-posterior density interval.*

3. *Try varying the variance parameter τ^2 used in the proposal density and examine what the effects are on the average acceptance probability and on the posterior distribution.*

4. *Try varying the prior distribution by varying μ and β and assess the impact on the subsequent analysis.*

Index

Addition rule of probabilities, 8
Analogy principle, 223

Bayes decision rule, 406
Bayes factor, 408
Bayes' Theorem, 71
Bernoulli distribution
    Probability mass function, 138
Bernoulli population, 138
Beta distribution
    Density function, 146
    Distribution function, 147
Beta function, 146
Bimodal distribution, 89
Binomial coefficient, 139
Binomial distribution
    Probability mass function, 139
Bootstrap
    Bootstrap population, 238
    Bootstrap sample, 238
Borel Sigma-algebra, 424

Central Limit Theorem, 245
Central moment, 83
Chain rule for probabilities, 68
Chi square distribution, 151
    Non-central, 152
Conditional expectation, 103
Conditional expectation function, 103, 106
Conditional probability
    Events, 60
Conditional quantile, 112
Conditional variance, 109
Confidence Interval
    Asymptotic, 266, 267
    Asymptotically valid, 266
    Bootstrap, 264, 265
    Definition, 256
    Population mean, 260, 262
Continuity
    Absolutely continuous, 445
Continuous mapping theorem, 248
Convergence in distribution, 243

Correlation coefficient, 100
Covariance, 97
Critical region, 306
Critical value, 306
Cumulative distribution function, 18
    Conditional, 65
    Joint, 41

Decision rule
    Admissible, 289
    Definition, 277
    Size, 281
Delta method, 249
Density function, 26
    Conditional, 64
    Joint, 49, 55
    Marginal, 51
Distribution
    Laplace, 394

Empirical distribution function, 220
Empirical probability mass function, 221
Erlang distribution, 142
Estimand, 190
Estimate, 190
Estimator, 190
    Analog, 223
    Bias, 194
    Consistent, 197
    Efficient, 201
    Interval, 256
    k central moment, 227
    k moment, 227
    Mean absolute error, 196
    Mean squared error, mse, 194
    Point, 256
Event, 6
Event algebra, 7
Expected posterior loss, 406
Expected value, 76
Exponential distribution
    Density function, 141

F-distribution, 154

Factorial, 139
Fisher information, 371
Fubini's Theorem, 451
Function
    Antiderivative, 441
    Codomain, 430
    Continuity, 439
    Derivative, 440
    Domain, 430
    Improper integral, 443
    Infimum, 431
    Integral, 443
    Inverse image, 432
    Limit, 435
    Limit from left, 436
    Limit from right, 436
    Piecewise, 10
    Range, 430
    Supremum, 431
Fundamental theorem of calculus, 444
Fundamental theorem of statistics, 221

Gamma function, 146
Gaussian distribution, 148

Hypothesis
    Composite, 275
    Simple, 275

iid, 177
Indicator function, 447
Information inequality, 357
Interquartile range, 88
Inverse CDF, 85
Inverse function, 433
Inverse image
    point, 12

Kullback-Leibler divergence, 359
Kurtosis, 83

Law of Iterated Expectations, 106
Law of Large Numbers, 199
Leibniz integral rule, 452
Likelihood function, 339, 345
    Realized, 339, 346
Likelihood ratio, 326
Likelihood ratio test, 327
Limit point, 435
Limiting distribution, 244
Log-Likelihood function, 341
Log-likelihood function, 346

Log-normal distribution, 218
Loss
    Expected posterior, 406
Loss function
    0-1 loss, 196
    Estimator, 193

Markov Chain Monte Carlo, 458
Markov property, 352
Maximum a posteriori estimator, 404
Maximum likelihood estimator, 341, 347
MCMC, 458
Mean, 76
Mean independence, 131
    Conditional, 132
Mean value theorem, 379
Median, 85
Metropolis algorithm, 464
Metropolis-Hastings algorithm, 464
Modal value, 88
Mode, 88
Moment, 83
Monte Carlo
    Distribution, 204
    Simulation, 203

Normal distribution, 148
    Bivariate, 161
    Standard Bivariate Normal
        Distribution, 166
    Standard normal distribution, 149
    Sum of iid random variables, 150
Nuisance parameter, 276

Order statistics, 225

p-quantile estimator, 225
p-value, 319, 320
Parameter space, 197
Partial derivative, 449
Poisson distribution, 143
Population, 2
Posterior distribution, 390
Power function, 278
Prior
    Jeffreys , 401
Prior distribution, 389
Probability density function
    Conditional, 66, 67
Probability function, 22
    Conditional, 62
    Joint, 42, 48

Probability mass function, 22
    Conditional, 66, 67
Probability measure, 8
Probability model, 9
Probability space, 9

Quantile, 85

Random variable, 10
    Continuous, 25
    Discrete, 21
    Mixed type, 32
Range of random variable, 88
Risk
    Decision rule, 286
Risk function
    Estimator, 194

Sample
    iid, 177
    Random, 175
    Realized, 175
    Simple random sample, 177
Sample average, 191
Sample distribution, 175
Sample space, 5
Sample statistics, 189
Sample variance, 229
Sampling mechanism, 2
Sampling with replacement, 178
Sampling without replacement, 178
Score function, 342, 347
Sequence of numbers, 434
    Limit, 434
Set, 415
    Complement, 418
    Difference, 418
    Disjoint, 420
    Infimum, 430
    Intersection, 418, 420, 421
    Subset, 416
    Supremum, 429
    Union, 418, 420, 421
Sigma-algebra, 7
Significance level, 296
Skewness, 83
Slutsky's Theorem, 248
Standard deviation, 81
Standard error
    Definition, 251
    Mean, 251

Statistical experiment, 2
Statistical independence
    CDF, 123
    Conditional, 128
        Continuous random variables, 128
        Discrete random variables, 128
    Continuous random variables, 123
    Discrete random variables, 123, 124
    Events, 122
    Mutual, 126
        Continuous random variables, 126
        Discrete random variables, 126
    pairwise, 126
Support of distribution, 88

t-distribution, 153
    Non-central, 153
t-test, 305
    Composite, 314
t-test statistics, 304
Test rule
    Consistent, 282
    Minimax, 291
Type I error, 280
Type II error, 280

Uniform distribution
    Density function, 145
    Distribution function, 145
    Estimation, 344
Unimodal distribution, 89

Variance, 81

Young's Theorem, 450

For Product Safety Concerns and Information please contact our EU
representative GPSR@taylorandfrancis.com
Taylor & Francis Verlag GmbH, Kaufingerstraße 24, 80331 München, Germany

www.ingramcontent.com/pod-product-compliance
Ingram Content Group UK Ltd.
Pitfield, Milton Keynes, MK11 3LW, UK
UKHW052359060726
473021UK00005B/134